# Advancements in Biogeography

# Advancements in Biogeography

Edited by **Neil Griffin**

R CALLISTO REFERENCE

New York

Published by Callisto Reference,
106 Park Avenue, Suite 200,
New York, NY 10016, USA
www.callistoreference.com

**Advancements in Biogeography**
Edited by Neil Griffin

International Standard Book Number: 978-1-63239-033-2 (Hardback)

Printed in the United States of America.

# Contents

**Permissions**

**List of Contributors**

# Preface

The purpose of the book is to provide a glimpse into the dynamics and to present opinions and studies of some of the scientists engaged in the development of new ideas in the field from very different standpoints. This book will prove useful to students and researchers owing to its high content quality.

Biogeography involves dealing with the study of varied life forms present on the face of the Earth. This book is a compilation of research works of more than 30 scientific experts on biogeography from across the globe. It intends to elaborate on spatial and temporal variation of biological assemblages in relation to landscape intricacy and environmental changes. This book adopts four major themes: bio-geographic theory and tests of ideas and concepts; the territorial biogeography of individual taxa, biogeography of complex landscapes, and the deep-time evolutionary biogeography of macro-taxa. In addition, the book also presents new information about unusual landscapes, the natural history of a wide array of lesser known plant and animal species, and global conservation issues. It is well illustrated with various maps, graphics, and photographs, and provides significant information on various topics. It will work as a useful tool for experts and general public interested in global biogeography, taxonomy, development and conservation.

At the end, I would like to appreciate all the efforts made by the authors in completing their chapters professionally. I express my deepest gratitude to all of them for contributing to this book by sharing their valuable works. A special thanks to my family and friends for their constant support in this journey.

**Editor**

# Part 1

## Biogeographic Theory:
## Testing Concepts and Processes

# 1

# Biogeographic Hierarchical
# Levels and Parasite Speciation

Hugo H. Mejía-Madrid
*Universidad Nacional Autónoma de México/Facultad de Ciencias*
*México*

## 1. Introduction

The historical biogeography of freshwater fish helminth parasites is linked to speciation. It is well recognized and documented that bursts of speciation are often limited to certain periods of geological time (Eldredge & Gould 1972; Gould & Eldredge; 1977, Gould, 2002; Hoberg & Brooks, 2008) and is highly dependent on the space in which it occurs (Brooks & McLennan, 2002; Hoberg & Brooks, 2008; Lieberman, 2003; Vrba, 2005). This holds for free-living organisms and parasites as well (Brooks & McLennan, 1993b; Hoberg & Brooks, 2008). Empirical evidence supports the contention that freshwater fish helminth parasites have mainly speciated allopatrically through two processes, vicariance and dispersal via host switching when the host has moved into novel areas, i.e. dispersal across pre-existing barriers (physical change *sensu* Vrba, 2005) (Brooks and McLennan, 1991, 1993b, 2002; Choudhury & Dick, 2001; Poulin, 1998). This suggests that the evolutionary biology of helminth parasites should have a strong biogeographical component, one that acts above the species level and to a lesser extent may have been driven by coevolutionary phenomena (Brooks and McLennan, 2002; Pérez-Ponce de León & Choudhury, 2005; Hoberg & Brooks, 2008). It is currently recognized that two processes, linked cyclically in time and space, have produced these patterns in parasite historical biogeography: taxon pulses (TP; Erwin, 1981; Hoberg & Brooks, 2008; 2010) and ecological fitting (EF; Janzen, 1985; Hoberg & Brooks, 2008, 2010).

TP have a strong biogeographical component. TP coupled with EF can explain several phenomena linked to parasite diversity in space and time, parasite richness across wide-ranging geographical areas, and the geography of diseases (emerging infectious diseases, EID; Brooks & Ferrao, 2005; Hoberg & Brooks, 2008). The coupling of TP and EF has scarcely been explored outside the context of recent events in parasite epidemiology (Hoberg & Brooks, 2008, 2010). The pattern that can identify the occurrence of TP in deep phylogenies has been little explored before the Cenozoic period, except for tethrabothriidean cestodes (Hoberg & Brooks, 2008). Outbursts of speciation are probably linked to both micro- and macro- spatial and evolutionary scales; on a macroevolutionary scale TP probably are linked to punctuated events of speciation, with a predominance of peripheral isolates speciation or postdispersal speciation after an expansion phase and not especially to *in situ* speciation due to isolation and environmental heterogeneity (Hoberg & Brooks, 2008, 2010; Vrba, 2005).

EF combines both biogeographical and ecological components (Hoberg & Brooks, 2008). This purportedly common phenomenon suggests that parasite evolution might be linked to resource tracking more than to coevolutionary phenomena (Agosta & Klemens, 2008; Agosta et al., 2010; Brooks & McLennan; Brooks et al., 2006; Hoberg & Brooks, 2008). This has not only short-term implications in parasite evolution but also long-term implications that provide insight into deep phylogenies and therefore historical biogeographical analyses. A historical biogeographical pattern can reveal instances of EF in a relatively straightforward fashion; namely, a parasite with a wide-ranging distribution, but limited to few host taxa, when the host taxon is much more diverse than the parasites inhabiting it (Brooks & McLennan, 2002). The process that generates this pattern involves parasite exploitation of newly available resources without having to evolve novel capabilities for host utilization (Hoberg & Brooks, 2008).

Freshwater fish parasitic helminths have been used as examples of host-parasite interactions for nearly three decades at the micro and macroevolutionary level, and at micro and macrobiogeographical scales (Brooks & Mc Lennan, 1993, 2002; Choudhury & Dick, 1996, 1996, 20001; Mejía-Madrid et al., 2007a,b; Pérez-Ponce de León & Choudhury, 2005; Pérez-Ponce de León et al., 2007; Rosas-Valdez et al., 2008; Choudhury, 2009). Nevertheless, the influence of TP and EF has not been addressed directly to explain pattern and process in deep phylogenies of fish parasites, in contrast to well-explored hypotheses that deal with primates (Brooks & Glen, 1982; Brooks & McLennan, 2003; Folinsbee & Brooks, 2007), Beringian mammal parasites (Hoberg & Brooks, 2008), and Palearctic parasites (Nieberding, 2004, 2005). Historical biogeography of freshwater fish helminth parasites would benefit much from such theoretical approaches.

The first aim of this chapter is to extend the phylogenetic and historical biogeographical analysis of *Rhabdochona* Railliet, 1916 species to include a more detailed account of the recent theoretical developments of TP and EF relative to freshwater fish helminths. It is entertained herein that the inclusion of such developments will help clarify to a certain extent how the deep phylogeny of a monophyletic clade of freshwater nematode parasites is related to phenomena that have not been previously considered, but are closely related to their historical biogeography. The second aim is to interpret these results across a wide spectrum of natural history data within a phylogenetic and historical biogeographical framework, including: speciation, comparison of modern distributions of hosts and parasites with fossil distribution of marine and freshwater fishes, their diversification intervals, sequential heterochrony, the spatial scale at which the phylogeny takes place, and phylogeography. Whereas it is clear that the present chapter focuses on the interpretation of hierarchical patterns in historical biogeography (Sanmartin et al., 2001), the uncertainties associated with the patterns presented here cannot be assessed at this stage of discovery.

## 1.1 Definition of biological terms employed

The historical biogeographical analysis employed here is based mainly on the "discovery based" protocol of van Veller & Brooks (2001), Halas et al. (2005), Hoberg, (2001), Hoberg & Brooks (2008), and Lieberman (2003). This approach is preferred in the present case because it includes all empirical information available to explain patterns of deep historical biogeography and includes no *a priori* assumptions of geological evolution or host evolution (Hoberg & Brooks, 2010). Such approach has been called phylogenetic biogeography

(Hennig, 1966; Brundin, 1981; van Veller & Brooks, 2001). Terms related to TP and EF follow Hoberg & Brooks (2008).

The historical species concept here used is the PSC1 (Phylogenetic Species Concept 1, Cracraft, 1989; Brooks & McLennan, 2002; Coyne and Orr, 2004) because where there is ecological fitting and long standing stasis, probable ancestors coexist with descendants for a considerable amount of time, during which time the process of host switching to novel resources and subsequent speciation takes place (Brooks & McLennan, 2002). Despite the fact that phylogenetic systematics has a strong gradualistic basis (Wagner & Erwin, 1995; Hennig, 1966; Wiley & Lieberman, 2011; but see Eldredge & Cracraft for a different point of view) no *a priori* considerations on the scale of evolution are entertained here, e.g., phyletic gradualism or punctuated equilibrium. Nevertheless, PSC1 is simply interpreted as pattern and the processes considered herein imply speciation promoted by physical change (Vrba, 2005).

Net diversification interval (NDI) was calculated for different parasitic nematode taxa after Stanley (1975, 1998; Coyne & Orr, 2004). These calculations employ data on numbers of extant helminth parasite species, especially those within monophyletic clades where rate of description of new species has achieved or is near stabilization. The date of calibration is taken from the most ancient fossil host related to present day clades. As siluriforms are considered the original hosts of *Rhabdochona* spp. (Moravec, 2010), the calibration point is taken to be 140 mya (Ferraris, 2007; Lundberg et al., 2007).

Heterochrony - changes in the relative time of appearance and rate of development for characters already present in ancestors (Gould, 1977) - is understood here as sequential heterochrony which conceptually incorporates timing of metamorphosis from one growth stage to another (McNamara & McKinney, 2005). Sequential heterochrony can account for the origin of certain characters of *Rhabdochona* spp. in relation to host switching and dispersal of hosts.

Finally, the phylogeography of one species of American *Rhabdochona* will be addressed in order to demonstrate that these particular nematode parasites have low speciation rates when compared to the number of species of hosts they inhabit.

## 2. The historical biogeography of *Rhabdochona* species: An overview

*Rhabdochona* species are a world-wide group of spirurid nematodes that inhabit all continents as intestinal parasites of freshwater fishes (Moravec et al., 2011). Recent molecular studies have removed them from Thelazioidea (Černotíková, et al., 2011; Nadler et al., 2007). Their outstanding morphological characters include a wide prostom, a character shared with other nematodes, several longitudinal cuticular ridges internal to the prostom that anteriorly (prorhabdion) form teeth, sessile caudal male papillae arranged in paired ventrolateral rows, eggs with different ornamental covers, and peculiarly-shaped male spicules. Some of the aforementioned characters are shared with other putatively phylogenetically related groups (Černotíková, et al., 2011; Mejía-Madrid et al., 2007a; Nadler et al., 2007), i.e., polar filaments on egg surface (*Cystidicola* spp., some species of *Spinitectus*), a wide prostom (*Megachona chamelensis* Mejía-Madrid & Pérez-Ponce de León, 2007), and caudal papillae (Physalopteridae, Cystidicolidae, and Spinitectidae). Despite the generality of most of the characters used for classifying *Rhabdochona* spp., spicular morphology remains peculiar and possesses variation almost unique to this group of nematodes (*Spinitectus* spp. shows a similar variation). The form of this character is species specific to *Rhabdochona* spp. (Mejía-Madrid et al., 2007a; Moravec, 2010; Rasheed, 1965). Indeed, the first phylogenetic systematic analysis of this group recovered spicule form as a consistent character (Mejía-Madrid et al., 2007a). The intraspecific variability of the aforementioned character is quite limited, as a study of different spicules of North American species indicates (Mejía-Madrid, unpublished data).

*Rhabdochona* spp. belong to the family Rhabdochonidae. Among the 10 genera of Rhabdochonidae, 8 contain species that parasitize chondrichthyans and teleosts, and from these only 3 (*Beaninema* Caspeta-Mandujano et al., 2001; *Prosungulonema* Roytman, 1963, and *Rhabdochona*) include species that parasitize freshwater fishes (Mejía-Madrid & Pérez-Ponce de León, 2007). Nevertheless, *Rhabdochona* is the most diverse genus of this family, with 92 valid species (Moravec et al., 2011).

A phylogenetic and historical biogeographical analysis of *Rhabdochona* species is now due mainly because the systematic research on the whole genus is reaching a stage of maturity that is reflected in the stabilization of species rate discovery (Mejía-Madrid, unpublished data), and because of the quality of new descriptions and redescriptions (Sánchez-Álvarez et al., 1998; Caspeta-Mandujano & Moravec, 2000, 2001; Mejía-Madrid & Pérez-Ponce de León, 2003; Mejía-Madrid et al., 2007a; Moravec & Muzzall, 2007; Moravec et al., 2011). Such detailed morphological descriptions are essential for a clear distinction between species. Additionally, the molecular database of *Rhabdochona* spp. from Asia, Europe, and America is increasing (Černotíková et al., 2011; Mejía-Madrid & Nadler unpublished data; Wijowá et al., 2007). In the present analysis 37 out of 92 (40%) valid species have been included, mainly because this set of species is fairly well described for their main discriminant character, the male left spicule, as well as for other key characters (Mejía-Madrid et al., 2007a; Moravec, 2010; Moravec et al., 2011). The American species are completely represented in the present analyses, but I include representative species distributed worldwide, with the exception of *R. papuanensis* Moravec, Ríha & Kuchta, 2008.

The historical biogeographic analysis presented here is based on the updated matrix used for generating the phylogenetic framework of *Rhabdochona* spp. presented in Mejía-Madrid et al. (2007a) with additional character coding derived from recently redescribed species from the Americas and Asia (Moravec & Muzzall, 2007; Moravec, 2010; Moravec, et al., 2011; Figure 1). The results presented herein represent a new phylogenetic framework for *Rhabdochona* spp., with two outstanding characteristics: the phylogeny is fairly well resolved and the degree of resolution is higher than that previously recovered.

Historical biogeographical analysis of *Rhabdochona* spp. reveals an ancient origin for the group that probably predates current continental configurations (Mejía-Madrid et al., 2007a; Moravec, 2010; Figures 2-4). Extant species distributions reflect past distributions, nevertheless these are the product not only of vicariance but also of past dispersal in a limited geographical range: however, these are difficult to distinguish from phylogenies alone (Brooks & McLennan, 2002 and references therein; Brooks & Ferrao, 2005; Wagner & Erwin, 1995). A reticulated historical biogeographical pattern is apparent when the phylogeny of *Rhabdochona* is interpreted graphically. This pattern reveals that if a) *Rhabdochona* species tend to remain relatively near their area of origin (Roy et al., 2009), closely related species in the phylogeny should inhabit neighbouring areas. This can be interpreted as vicariance and therefore b) where species that do not conform to this pattern (e.g., exhibiting disjunct distributions, but with closely related species widely separated geographically), this can be interpreted as the product of past or relatively recent dispersal.

Furthermore, in the present case, *Rhabdochona* spp. might be interpreted as 'ancient' ecological relics (Hoberg & Brooks, 2008) in that their original hosts underwent widespread extinction and survived into deep (not shallow) time through colonization and secondary radiation into a novel, but ecologically equivalent, host group, namely, the cyprinids, where they have succeeded ever since host switching from the silurids.

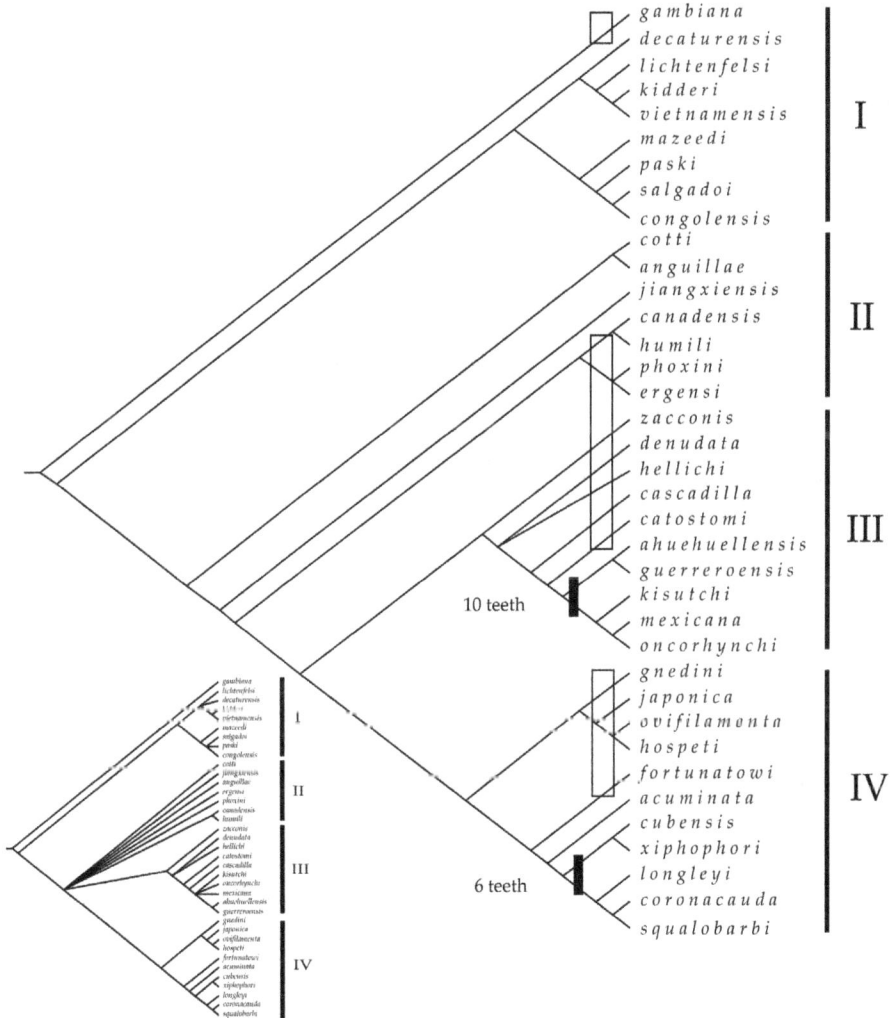

Fig. 1. Phylogeny of *Rhabdochona* species. Majority rule tree of 538 trees. Inset: strict consensus tree. Rectangles represent cypriniform hosts. Black hashmarks represent characters shared by members of clade.

The basal clade of *Rhabdochona* (Figure 1, clade I) is represented by the African *R. gambiana*, a species that parasitizes basal cyprinids from rivers in West Africa, and from the African Rift Lakes (Moravec, 1972). It is followed by an assemblage of African species that parasitize siluriform and characid fishes: *R. paski*, *R. congolensis* in several basins in South Africa and the African Rift Lakes, plus the Indian species, *R. mazeedi* specific to at least 3 families of siluroids, (Moravec, 1972, 1975, Moravec et al., 2010; Puylaert, 1973). Within this clade a species of *Rhabochona* from *Profundulus labialis* Günther distributed in southern Mexico is consistently found, namely *R. salgadoi*. An emerging pattern of phylogenetic and historical biogeographical relationships between species of Africa and Southern Mexico is becoming apparent, including a diverse array of coexisting floristic and faunal assemblages. Those assemblages appear to have originated in other parts of the world, and exhibit extralimital affinities and origins (Lundberg et al., 2007). For example, the freshwater siluriform fish, *Lacantunia enigmatica* Rodiles-Hernández, Hendrickson, Lundberg & Humphries is endemic to Southern Mexico but is morphologically and molecularly related to the claroteids of Africa but to no other siluriform fish in North America (Lundberg et al., 2007). This distribution pattern has been recovered in phylogenies of Upper Jurassic, Cretaceous and Lower Palaeocene freshwater and marine fossil teleosts (Forey, 2010) and it seems not to be uncommon among other fossil faunas. These areas were probably not contiguous during the geological periods mentioned, but likely were passageways when environmental conditions favoured dispersal (Lundberg et al., 2007).

The sister subclade of the "African" clade just described is that of a group of species that indicate a relationship between North America and eastern Asia. To what extent this clade reflects former exchanges between the fish faunas of the Far East and eastern North America remains a matter of debate (Sanmartin et al., 2001). Modern disjunct fish distributions probably date back to the Late Mesozoic and might have resulted from a cross-Atlantic division (Sanmartín et al., 2001). In the aforementioned sister subclade three North American parasite species, *R. lichtenfelsi*, *R. decaturensis* and *R. kidderi* (the latter parasitic in cichlids and pimelodids of cenotes and caves, respectively, refugia for relict fish species) are related to an eastern Asian species (*R. vietnamensis*). All of these parasite species occur among diverse fish families where cyprinids are completely absent. The distribution of this clade is similar to the modern relict distribution of horseshoe crabs and other marine fishes. To what extent *Rhabdochona* spp. were dispersed by marine seaways will be discussed below.

Clade II is still unresolved in the strict consensus tree (Figure 1). This clade comprises a group of 5 species in a polytomy and another with two sister species (*R. canadensis* and *R. humili*) that probably were dispersed across a Northern proto-Atlantic.

Clades III and IV are fully resolved and comprise *Rhabdochona* spp. that predominantly infect cyprinids, and to a lesser extent, catostomids, silurids, poeciliids, and characids (Figures 1 and 4). Clade III comprises species that parasitize some Cypriniformes in central and eastern Asia plus Eastern Europe and predominantly salmonids from eastern Asia, Western North America and goodeids down into Central Mexico. One or several episodes of dispersal through Beringia may have left tracks in the distribution of *Rhabdochona* spp. Host switching from cyprinids to salmonids during invasion could have been the mechanism that triggered speciation within this clade for there seems to be a consistent evolutionary novelty among these species, namely, 10 teeth in prostom.

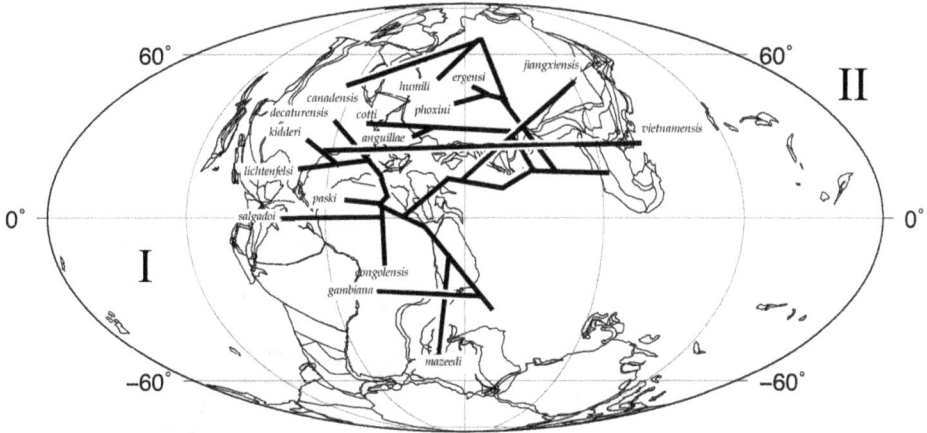

Fig. 2. Basal clades I and II of the phylogeny of *Rhabdochona* species superimposed on a world paleomap of the Lower Cretaceous c. 140 mya. Species names are approximately located in their area of distribution for clarity. Only the outline of tectonic plates is shown. Map and localities generated by http://www.odsn.de/odsn/services/paleomap/paleomap.html

Clade IV is represented by species that are distributed from central Asia to eastern Europe (*R. fortunatowi*, *R. gnedini*) and North America (*R. ovifilamenta*), South America, Mexico (*R. acuminata*), Cuba (*R. cubenesis*), and the southern United States (*R. longleyi*, inhabitant of cave-dwelling blind catfishes). This series of dispersal events probably was limited to the western and northern Tethys, because there are only two or three species of *Rhabdochona* in South America, where the diversity of freshwater fishes is the largest in the world mainly because of the extension and complexity of the geological history of this sub-continent (Brooks, 1992; Lundberg et al., 1998). Additionally, the distribution of Clade IV seems to follow the tracks of *Spinitectus* spp. (Choudhury & Dick, 2001) and proteocephalid distributions in the modern Holarctic and Palaeartic biogeographic zones (Brooks, 1978; de Chambrier, 2004; Rego et al., 1998). The latter northern helminth faunas are well represented in South America, whereas *Rhabdochona* spp. are almost entirely absent. It is probable that the presence of cyprinids by the end of the Mesozoic Era set the stage for dispersal of *Rhabdochona* spp. in the Tertiary.

The main fish families of each species of *Rhabdochona* analyzed (Figure 4) reveal an apparent history of host-switching, but with an underlying structure: all groups from clade II onwards comprise basal clades where nematodes predominantly parasitize cyprinids (Figure 1), except Clade III that involves a clear salmonid host component and a subclade in IV that includes *cubensis*, *xiphophori*, *longleyi*, and *squalobarbi* and in which teeth number (6) predominates in parasites that inhabit non-cyprinid hosts.

A general comparison was undertaken between the parasite host area phylogeny of *Rhabdochona* spp. (Figure 4) and phylogenies of proteocephalids (de Chambrier et al., 2004; Rego et al., 1998), *Spinitectus* spp. (Choudhury & Dick, 2001), fossil teleosts (both freshwater and marine; Forey, 2010), and Cyprinidontiformes (Parenti, 1981). Discovery based-methods were employed to generate general and particular area hypothesis (BPA and PACT 2.0 software; Wojcicki & Brooks, 2005). Several attempts were made to compare different trees, but as areas recorded for helminth parasites are so distinct (despite the fact that area names were adjusted in each case), only a single area cladogram could be recovered from the PACT analysis between *Rhabdochona* and Cyprinodontiformes. On the other hand, a single cladogram for fossil fishes (Forey, 2010) emerged and was therefore compared in broad terms with the area cladogram of *Rhabdochona* spp.

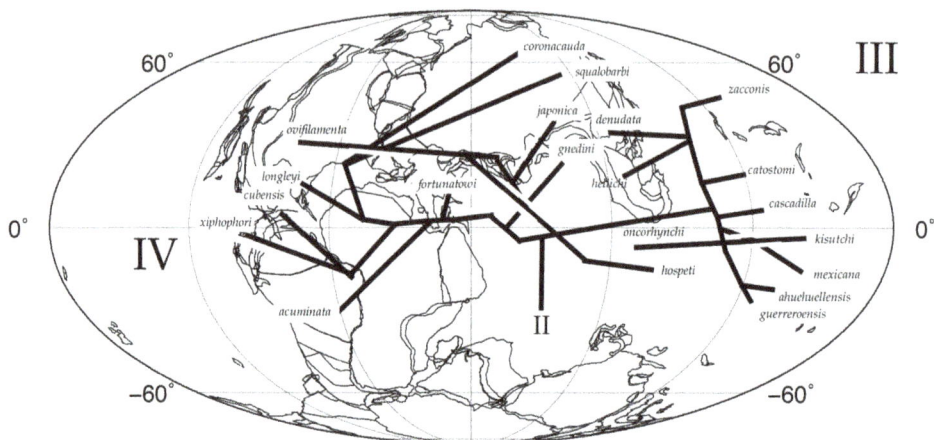

Fig. 3. Clades III and IV of the phylogeny of *Rhabdochona* species superimposed on a world paleomap of the Lower Cretaceous c. 140 mya. Species names are approximately located in their area of distribution for clarity. Only the outline of tectonic plates is shown. Map and localities generated by http://www.odsn.de/odsn/services/paleomap/paleomap.html

The PACT tree of Cyprinodontiformes and *Rhabdochona* spp. recovers a series of postdispersal speciation origins for species within clades I and II (Figure 5). This pattern can be inferred as well as when the areas are inspected on the 140 mya map (Figures 2 and 3). Most of the areas are not neighbouring but certainly are coastal marine tracks of southern, western and northern Tethys. It must be noted that there is a time gap between the origin of Old World aplocheiloids and New World aplocheiloids. In between, clades I and II of *Rhabdochona* spp. diversified before Africa and South America separated. So probably later on, geologically speaking, *Rhabdochona* spp. dispersed into western Tethys, where most probably *R. salgadoi* originated.

When compared to the fossil fish PACT tree (Figure 6), *Rhabdochona* spp. can only be compared in very broad terms. The reticulated pattern of areas of fossil fishes is in itself complex. Two episodes of widespread dispersal and probably postdispersal speciation can be inferred: the aforementioned clade I and part of clade IV. While nematode species may conform to a Sea of Tethys distribution, the phylogeny on its own cannot account by itself for a general explanation of a taxon pulse or pulses that took place probably 140 mya. Despite the fact that cyprinodontiforms are freshwater fishes (although some inhabit brackish waters) and therefore have been interpreted as having evolved by vicariance (Parenti, 1981 only invoked vicariance for the present distribution of cyprinodontiforms) the PACT tree could recover postdispersal speciation in the nematode parasites. This indicates that *Rhabdochona* spp. within clade I probably speciated in eastern Africa freshwater drainages of Gondwana and some species reached western Africa later by dispersal when the Congo basin flowed west (Stankiewicz & de Wit, 2006). A reticulated biogeographic

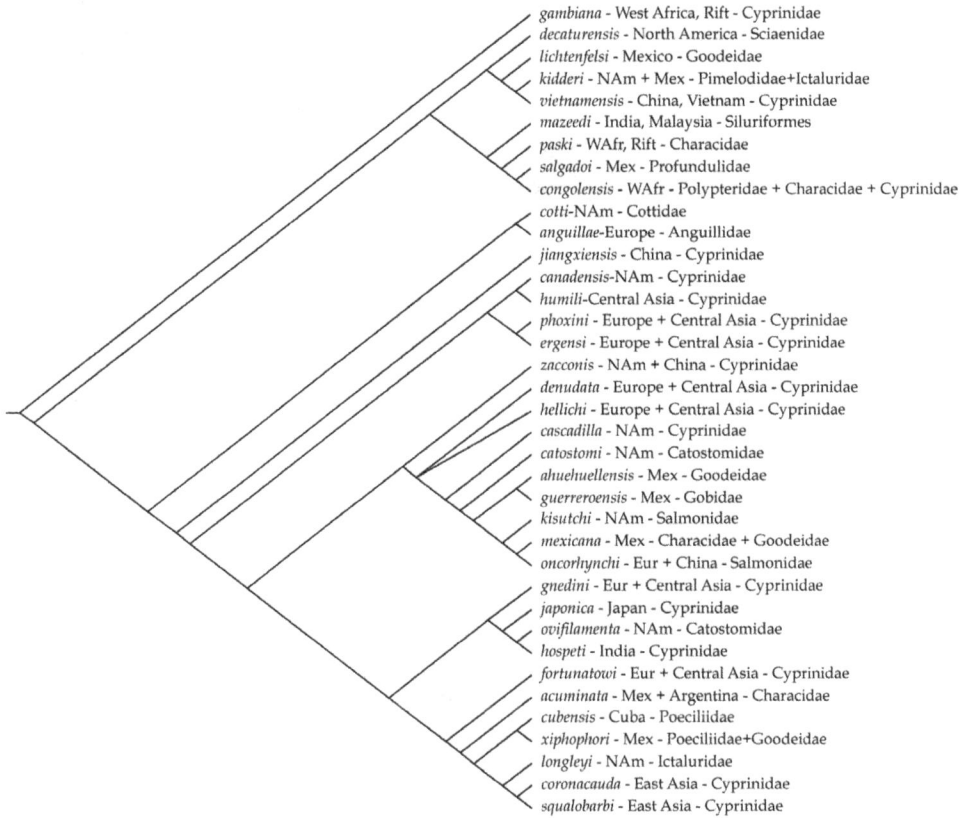

gambiana - West Africa, Rift - Cyprinidae
decaturensis - North America - Sciaenidae
lichtenfelsi - Mexico - Goodeidae
kidderi - NAm + Mex - Pimelodidae+Ictaluridae
vietnamensis - China, Vietnam - Cyprinidae
mazeedi - India, Malaysia - Siluriformes
paski - WAfr, Rift - Characidae
salgadoi - Mex - Profundulidae
congolensis - WAfr - Polypteridae + Characidae + Cyprinidae
cotti-NAm - Cottidae
anguillae-Europe - Anguillidae
jiangxiensis - China - Cyprinidae
canadensis-NAm - Cyprinidae
humili-Central Asia - Cyprinidae
phoxini - Europe + Central Asia - Cyprinidae
ergensi - Europe + Central Asia - Cyprinidae
zacconis - NAm + China - Cyprinidae
denudata - Europe + Central Asia - Cyprinidae
hellichi - Europe + Central Asia - Cyprinidae
cascadilla - NAm - Cyprinidae
catostomi - NAm - Catostomidae
ahuehuellensis - Mex - Goodeidae
guerreroensis - Mex - Gobidae
kisutchi - NAm - Salmonidae
mexicana - Mex - Characidae + Goodeidae
oncorhynchi - Eur + China - Salmonidae
gnedini - Eur + Central Asia - Cyprinidae
japonica - Japan - Cyprinidae
ovifilamenta - NAm - Catostomidae
hospeti - India - Cyprinidae
fortunatowi - Eur + Central Asia - Cyprinidae
acuminata - Mex + Argentina - Characidae
cubensis - Cuba - Poeciliidae
xiphophori - Mex - Poeciliidae+Goodeidae
longleyi - NAm - Ictaluridae
coronacauda - East Asia - Cyprinidae
squalobarbi - East Asia - Cyprinidae

Fig. 4. Parasite-host area cladogram of *Rhabdochona* species.

pattern of fossil fishes distribution might indicate that freshwater and marine dispersion throughout the western Tethys took hold once Laurasia and Gondwana drifted apart (Forey, 2010). *Rhabdochona* spp. seems to have been influenced by this wide-ranging dispersal routes that included South America, Cuba, North America, and all the way to eastern Asia (i.e., the northern Tethys sea and freshwater drainages that flowed into it). Therefore, it seems that *Rhabdochona* was originally a freshwater parasite that has had a deep history of marine dispersion.

Two hypotheses emerge from the historical biogeography pattern seen in *Rhabdochona* spp. First, the original hosts of *Rhabdochona* species were not cyprinids, but some other group of teleosts, probably silurids as mentioned above. The gap seen in host trends between clades I and part of II can be explained by the fact that the other hosts that might have been parasitized by these nematodes probably became extinct by the end of the Cretaceous, along many other groups of vertebrates and plants.

Second, cyprinids, already present in the Cretaceous were already parasitized with *Rhabdochona* through host switching from siluriforms. As cyprinids began to diversify in the early Tertiary, *Rhabdochona* remained in the areas they once inhabited, probably infecting a

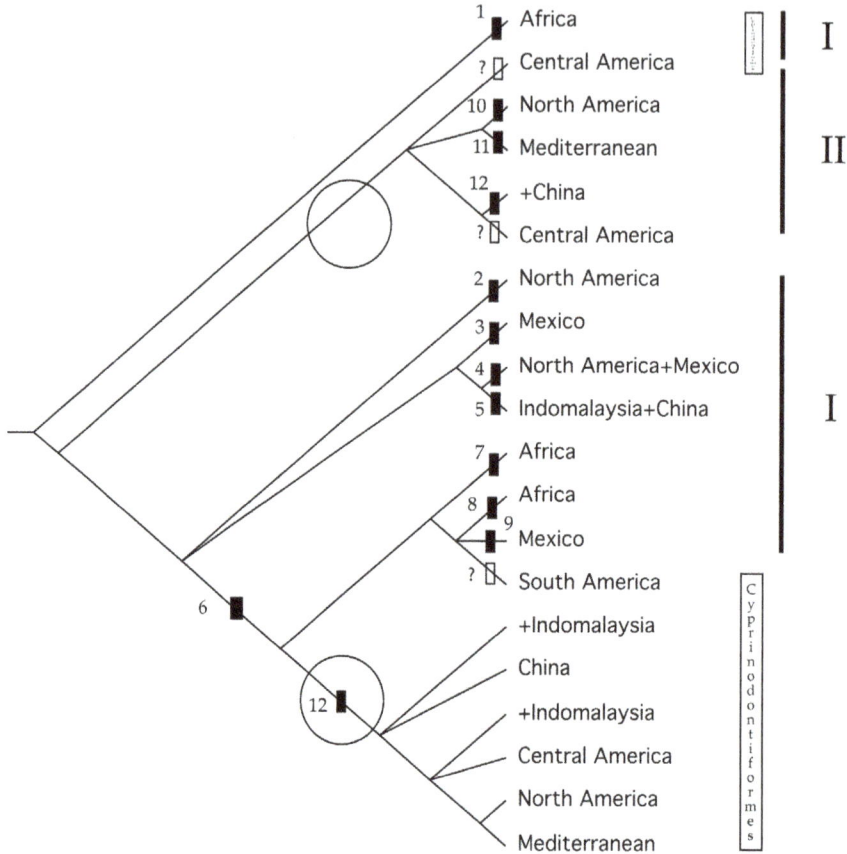

Fig. 5. PACT generated tree from *Rhabdochona* spp. and Cyprinodontiformes area cladograms (data of fishes from Parenti, 1981). Hashmark numbers indicate *Rhabdochona* species: 1=*gambiana*; 2=*decaturensis*, 3 *lichtenfelsi*, 4=*kidderi*; 5=*vietnamensis*; 6=*mazeedi*; 7=*paski*; 8=*salgadoi*; 9=*congolensis*; 10=*cotti*; 11=*anguillae*; 12=*jiangxiensis*. Circles indicate redundant area clades. ?=no response to dispersal/vicariance.

wider range of hosts, but remained in the cyprinids that finally dispersed them to other drainages (ecological relicts). Nevertheless, there seems to be no empirical evidence of cyprinids dispersing through marine waters in the present or past (Cavender et al., 1998; Chen et al., 2008; Hurley et al., 2007; Peng et al., 2006). So, other hosts, probably diadromous fishes, helped disperse *Rhabdochona* spp. in what seems an ancient TP that took place when the major landmasses of Laurasia and Gondwana separated. In this case *Rhabdochona* spp. might have only been sea dispersed but remained in freshwater hosts, as in the case of the diadromous salmonid species, in which marine hosts serve only as dispersal agents, but the parasites only develop in freshwater in appropriate intermediate hosts, i.e., ephemeropterans.

During the Tertiary, *Rhabdochona* could have been dispersed by diadromous salmonids from eastern Asia to western North America during the formation of one of the Beringian land bridges (Hoberg & Brooks, 2008). Similarly, ancestral acipenserids were diadromous and dispersion of their helminths to North America from Eurasia explains their present distribution (Choudhury & Dick, 2001). In Mexico, although salmonids are mainly restricted to Northwestern Mexico (Miller & Smith, 2005), fossils of these fishes dated from the Late Pleistocene have been found futher south and near modern Lake Chapala (Cavender & Miller, 1982; Miller & Smith, 1986; Miller, 2005) and most probably left parasites that evolved in Mesa Central of Mexico and speciated through host switching into species like *R. mexicana*, *R. ahuehuellensis*, and *R. guerreroensis*.

### 2.1 Taxon pulses as inferred from the phylogeny of *Rhabdochona* species

Taxon pulse patterns seem to conform to long-term biogeographical phenomena (Erwin, 1988; Halas et al., 2005; Marshall & Liebherr, 2000; Brooks & Ferrao, 2005). Clade by clade, *Rhabdochona* species reflect various episodes of TP in the Mesozoic and Cenozoic eras (Figures 2-6). *Rhabdochona* species seem to have had a long history of expansion when it originated in southern Tethys coasts or rivers (Figure 3). It can be inferred that much speciation could have occurred around 140 mya, but many of its representatives have become extinct. The historical biogeographical relationships between African, Asian, Mesoamerican, North American, and European species in these basal clades might indicate that species of *Rhabdochona* are a relict group of nematode parasites of extinct fishes (ecological relicts *sensu* Hoberg & Brooks, 2008). The alternative view is that few species originated ever since, and *Rhabdochona* as a genus has never been really species rich, if compared with closely related groups, like ascarids (Ascaridida) or filarids (Chabaud, 1975; Nadler et al., 2007).

*Rhabdochona* spp. represents a case where TP probably is the pattern that best explains early postdispersal speciation events during the Cretaceous. If this nematode group evolved as parasites of long extinct teleost fishes, probably ancient silurids, and developed broad ranges (i.e., *R. denudata*) with cyprinid dispersal by EF, it can be inferred that they evolved by host-switching (Mejía-Madrid et al., 2007a). There might be no other more parsimonious explanation as to why *Rhabdochona* spp. represent an ancient nematode stock predating their modern hosts.

Clade III shows a most recent history of TP, i.e., dispersal across Beringia that probably dates back to the last glaciation. Salmonid fishes inhabited naturally Central Mexico most probably during the Pliocene, just before the recorded draught that originated the modern Sonoran Desert (Miller & Smith, 1986). Remnants of these former freshwater fauna include *Oncorhynchus chrysoleucas* from Northwestern Mexico. Fossils from salmonids have been found in former Lake Chapala in western Mexico. So, most probably there is strong evidence that *Rhabdochona* spp. that parasitize freshwater fishes of Central Mexico related to species of salmonids that range from eastern Asia to North America, are probably a consequence of dispersal along a broad front during an episode of expansion from eastern Asia to North America (Mejía-Madrid et al., 2007a). The similarity in their morphology, especially the presence of a closely similar left spicule form in males, makes them a good candidate for a pattern of a recent episode of TP just before habitat fragmentation due to the desertification of southern North America that left as evidence of a once humid and freshwater landscape nematodes that parasitize modern freshwater fish faunas. This pattern represents a contraction of once widespread taxa, such as goodeid fishes (Domínguez &

Doadrio, 2004; Webb et al., 2003) and has helped to determine an historical link in the Nearctic connection (Pérez-Ponce de León & Choudhury, 2002). Such connection is a by-product of previous periods of biotic expansions and contractions with additional geographical heterogeneity (Halas et al., 2005).

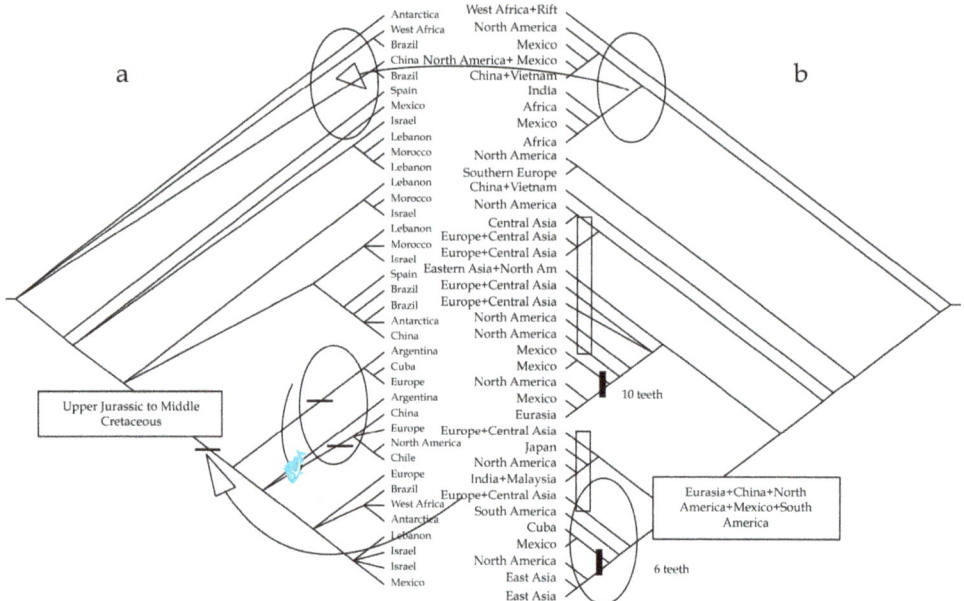

Fig. 6. Comparison of area cladograms. a) PACT tree of 3 fossil fish phylogenies (data after Forey, 2010); b) Area cladogram of *Rhabdochona* spp. Broad similarities between African clades and Western and Central Tethys are shown. Refer to text.

## 2.2 Ecological fitting and widespread host switching of *Rhabdochona* species

EF can explain the low diversity of this particular parasite nematode genus as compared to the diversity of their hosts. This could be reinforced by the fact that EF promotes and maintains evolutionary stasis alternatively to speciation in certain biological interactions (Agosta & Klemens, 2008; Gould, 2002). Nowhere in evolutionary phenomena can EF be more pervasive an argument of parasite evolution than at the macroevolutionary level. Extinction must be taken with caution for explaining host occupation of *Rhabdochona* spp. Nevertheless, many parasites species probably became extinct at the end of the K/T boundary along with their hosts, although some survived thanks to host switching (Dunn et. al., 2009).

Host switching may provide the most parsimonius explanation of the widespread lack of cophylogenetic patterns recovered from freshwater and marine fish parasite analyses. Host switching can be completed with or without speciation of the parasite associate. The pattern seen in *Rhabdochona* is far from reflecting any cospeciation pattern. There are few species of *Rhabdochona*, most infecting basal clades of cyprinids (Moravec, 2010), which reinforces the explanations entertained above. As far as can be said, *Rhabdochona* spp. infected cyprinid fishes or their ancestors by EF once their original hosts become extinct at different periods during the Cretaceous. This is exemplified by *R. anguilla*, which is a parasite specific to European freshwater eels. Freshwater eels seem to have originated in the Far East (Bastrop et al., 2000; Tsukamoto et al., 2002) and then dispersed through the Indian Ocean to the rest of Asia and into Europe. *Rhabdochona anguilla* is not present in Asian eels. The aforementioned then was a former parasite of another teleost, because it appears to be the sister species of *R. cotti*, a parasite of eastern North American Cottidae, a fish family with mostly marine representatives.

## 3. Speciation and stasis of *Rhabdochona* species

Parasites leave few fossils and lack a substantial fossil record when compared to free-living, hard skeleton species. So, diversification or speciation times of parasites based on modern species numbers are outstanding. A molecular clock approach has rarely been used in the case of nematodes, either free-living or parasitic as dates of divergence vary from taxon to taxon (Kiontke & Fitch, 2005). In order to explain how species diversity affects the historical biogeography of *Rhabdochona* spp., speciation intervals (as indicated above) were calculated to address: a) speciation periods and therefore, stasis times of *Rhabdochona* spp. and b) the depauperate condition of *Rhabdochona* compared to the number of potential extant hosts into which they could have diversified. A calculation of this type is more reliable when a fossil record exists, but such a record does not exist for nematodes (Poinar, 1984). I propose that a calculation of NDI (see Coyne & Orr, 2004) can indicate a first approximation of the rates and intervals of diversification of certain parasite clades. NDI values were calculated for monophyletic parasite clades. Some taxa of Platyhelminthes are used to compare the NDI of different groups of parasites not related to nematodes (Table 1).

The equation employed is:

$$t/\ln Nt = NDI \tag{1}$$

where t is the age of the parasite taxon (calibrated from its putative original hosts), Nt, the number of extant species (Coyne & Orr, 2004; Gould, 2002; Stanley, 1975, 1998). When compared to other invertebrate species where a fossil record exists, metazoan parasites score near or higher than their free-living relatives in their diversification intervals (Coyne and Orr, 2004; Stanley, 1998). This would mean that nematode and plathyhelminth parasites have longer diversification intervals than their hosts, in general. NDI calculations are dependant on the number of species described or estimated and cannot incorporate extinct species.

In terms of historical biogeography, ecological fitting can be explained in terms of long diversity intervals, e.g., long periods intervening between one diversification event from another, where speciation is probably low and is replaced by resource tracking. This may mean that *Rhabdochona* spp., have been around since the Cretaceous parasitizing different groups of fishes mainly by EF. Such data, when compared to that of their hosts, can only indicate that long diversification intervals are a common feature in metazoan parasites, and may be interpreted as longer stasis periods as compared to that of their fish or other vertebrate hosts.

| Monophyletic clade[1] Helminth parasites | Number of extant species | t (ranges in millions of years) [3] | NDI | Host calibration | References on host origin/parasite clade |
|---|---|---|---|---|---|
| Monogenea[2] | 4998 | 466 | 54.7 | Origin of Agnatha | our estimates |
|  |  | 460.5 | 54.0 |  |  |
| Digenea | 11846 | 416 | 44.3 | Origin of Placodermi | Brooks & McLennan, 1993 |
|  |  | 411 | 43.8 |  |  |
| Eucestoda | 3851 | 350 | 42.4 | Origin of Vertebrata | Hoberg et al., 1999 |
|  |  | 420 | 50.8 |  |  |
| Ascaridoidea | 818 | 306.5 | 45.7 | Oldest reptile fossils | our estimates |
|  |  | 311.7 | 46.4 |  |  |
| Oxyuroidea | 725 | 416 | 63.1 | First terrestrial arthropods |  |
| Dracunculoidea (Moravec, 2006) | 160 | 306.5 | 60.3 | Oldest reptile fossils |  |
|  |  | 311.7 | 61.4 |  |  |
| Camallanidae | 348 | 306.5 | 52.3 | Oldest reptile fossils |  |
|  |  | 311.7 | 53.2 |  |  |
| Physalopteroidea | 291 | 374.5 | 66.0 | First tetrapods |  |
| Trichostrongylidae | 130 | 374.5 | 76.9 |  |  |
| *Rhabdochona* (Moravec, 2011) | 92 | 140 | 30.9 | Silurids |  |
| Fish hosts [4] |  |  |  |  |  |
| Cyprinidae | 23000 | 48.6 | 4.8 |  | our estimates/Stanley 1998 |
|  |  | 55.8 | 5.5/5.6 |  |  |
| Siluridae | 104 | 11.6 | 2.49 |  | our estimates |
|  |  | 5.3 | 1.1 |  |  |
| Siluriformes | 3000 | 99.6 | 12.4 |  |  |
|  |  | 65.5 | 8.1 |  |  |
| Ictaluridae | - | - | 13.2 |  | Stanley 1998 |
| *Gambusia* | - | - | 1.6 |  | Stanley 1998 |
| Salmonidae[4] | 225 | 48.6 | 8.9 |  | our estimates |
|  |  | 40.8 | 7.4 |  |  |

Table 1. Net diversification rates (NDI) of some metazoan parasite clades (helminths) and some common hosts of *Rhabdochona*. [1]Clades as they appear in phylogenies regardless of hierarchical nomenclature. [2]Parasite species counts according to http://insects.tamu.edu/research/collection/hallan/0SYNOPT1.htm unless otherwise stated. [3]After *The Fossil Record 2* (Benton, 1993) except for Siluridae.
[4] http://www.fishbase.org. See text for explanation.

From the foregoing data, it may be inferred that parasites do not have time lags in their speciation as compared to their hosts (Brooks, 1981; Brooks & McLennan, 1993; Manter, 1963; Choudhury et al., 2002). As invertebrates, metazoan parasite species have longer stasis periods, as seen through NDI, than their vertebrate hosts. EF appears to have been the structuring force behind parasite colonization/dispersal. This helps explain why not every freshwater fish has a *Rhabdochona* species, and why there are only 8000 species of Digenea, while there are more than 40, 000 species of potential vertebrate hosts to be infected (Hugot et al., 2001). A similar observation was inferred from gyrodactylid phylogenies (Simková et al., 2006; Zietara, & Lumme, 2002).

Additionally, nematode parasites may not speciate along with their vertebrate hosts because they have longer stasis periods than do their hosts. Nematode parasite survival strategy involves moving among host species via EF without speciating. Parasite speciation might take place when parasites switch host families. This might be the real reason why Fahrenholz's and Manter's rules (co-speciation of parasites and hosts and parasite evolution lags behind host evolution, respectively) are not supported among these parasites and why research programs based on cospeciation are unlikely to succeed.

A consequence of EF is the coexistence of ancestral and descendant species in space and in time (Brooks & McLennan, 2002). The PSC1 concept includes survival or coexistence of ancestors along with their descendant species. It seems that this is not uncommon in the paleontological record, although it has rarely been acknowledged among extant species (Wagner & Erwin, 1995). In the case of relatively recent episodes of putative speciation events in *Rhabdochona* spp. one example involves a pair of sister species in clade III: *R. ahuehuellensis* and *R. guerreroensis*. The former species posseses no autapomorphies and its characters are shared with its putative ancestor (node), i.e., *R. ahuehuellensis* seems to be ancestral to *R. guerreroensis*.

EF also supports the hypothesis that parasite faunas in freshwater fishes are largely circumscribed to higher levels of monophyletic host taxa (Pérez-Ponce de León and Choudhury, 2005). Such a pattern has previously been noted in the monogeneans (Zietara & Lumme, 2002). This indicates that there is little coevolutionary pattern at the species level, and speciation takes place at the parasite-species/host-family level, coupled with long-distance postdispersal speciation. Thus, a higher level of host and geographic expansion can be discerned in *Rhabdochona* and probably other freshwater fish helminth parasites that promotes speciation at the family (clade) level of hosts, when hosts move into new habitats at macrobiogeographical scales, but not at lower levels.

## 4. The phylogeography of *Rhabdochona lichtenfelsi*

The phylogeography of the American species, *R. lichtenfelsi* (Mejía-Madrid et al., 2007b) revealed low divergence between the distinct subclades identified in its range, despite the fact that this species has colonized at least 15 different goodeine host species in 10 genera during the past million years. It even inhabits geographically distant freshwater drainages in Central Mexico. The haplotypes of this nematode species were explored in 3 different genera of goodeines and could not be differentiated among them (Mejía- Madrid et al., 2007b). Recent colonization and expansion of their hosts enabled *R. lichtenfelsi* to invade new drainages north of its ancient distribution, revealing dispersal/colonization within the last million years. Despite this dispersal phenomenon, *R. lichtenfelsi* has not responded to vicariance, in contrast to their goodeine hosts, which have radiated to nearly 40 genera in the past 16 my (Doadrio & Domínguez, 2004; Webb et al., 2004).

## 5. Geomorphological scaling effects in the phylogeny of *Rhabdochona*

Aquatic habitat areas involved in broad historical biogeographical interpretations change at very different scales. These spatial scales actually represent disparate orders of magnitude. Historical biogeography comprises two levels of spatial variation: a macroscale involving areas that range from $10^1$ to $10^7$ km$^2$, and a microscale involving magnitudes from $10^{-8}$ to $10^1$ km$^2$. The degree of magnitude of differences between areas is related to the hierarchical level of evolution addressed in historical biogeographical analyses. Vicariance is a phenomenon that involves the physical separation of landmasses at continental spatial scales, but not necessarily in the same geological time scales (Cavin et al., 2008; Forey, 2010), or that take place simultaneously at scales above 10 km$^2$. These include minor to major tectonic units ranging from continent and ocean basin building or to origination of fault blocks, volcanoes, troughs, sedimentary sub-basins, or individual mountain ranges. By microscale I refer to scenarios that some proponents of the BSC had in mind, involving large-scale erosional or depositional units, like the development of river deltas, major valleys, or piedmonts within major landmasses. The test for the role of different historical biogeographical scenarios includes the duration of these various vicariance-dispersal phenomena within the range of geomorphological microscale or macroscale events.

Metazoan parasites undergo various forms of speciation according to the geomorphological scale in which the taxa radiate. But which spatial scales have been involved in the events leading to the crown group of *Rhabdochona*? *Rhabdochona* spp. might speciate only when hosts diverge or disperse after invading new habitats, i.e., when long-distance host dispersal occurs. Therefore, active speciation by peripheral isolates or postdispersal speciation might have structured the present phylogenetic relationshisps within this genus. The basal clades of *Rhabdochona* spp. (I and II) might conform to postdispersal speciation along considerable distances. Yet, as this dispersal was taking place, landmasses were actively separating by the Upper Cretaceous that probably enhanced speciation probability during the early phases of the evolution of *Rhabdochona* spp. This seemingly occurred when marine and freshwater teleosts began to diversify (Forey, 2010). So, *Rhabdochona* historical biogeography has tracked the breakup of Pangea as well as the radiation of their original host families.

The origination of some clades of *Rhabdochona*, namely the crown clades III and IV (Figures 1 and 4), is related to macrobiogeographical spatial scale events and wide-ranging changes in host habitat, due to dispersal and probably host-switching, along with further host range extension, as well as heterochronic macroevolutionary events that involved a change in the timing of maturation (paedomorphosis), as described below.

## 6. Heterochrony and evolutionary novelties in *Rhabdochona* species

Clade IV of the phylogeny (Figure 1) includes another diverse array of species of *Rhabdochona* that parasitize mainly cyprinids and to a lesser extent characids, poeciliids, and catostomids. Curiously enough, a crown group that comprises Mexican, Cuban, and South

American species that possess 6 teeth in their prostom, appear to be closely related to a sister pair comprised of R. *coronacauda* and R. *squalobarbi* (which possess lateral body alae, a character regarded as plesiomorphic by Moravec, 2010) from Europe and the Far East. The basal species of this group is represented by R. *fortunatowi*, a species whose males possess a left spicule that is very characteristic of a group that comprises the Afghan species, R. *tigridis* (not included in this analysis), R. *acuminata* from South America, and R. *xiphophori* from Mexico.

The other clade that suggests speciation events involving change of hosts and location is that of the clade involving the salmonid species of *Rhabdochona*. This clade has undergone a change in tooth number from 14 to 10, and host switching from cyprinids to salmonids.

*Rhabdochona* phylogeny thus reveals an interesting pattern of speciation by heterochrony. The life cycle of *Rhabdochona* involves a 4th larval stage characterized by the presence of 6 teeth in the stoma and simple deirids (Moravec, 2010 and references therein). When this 4th stage moults into an adult, they exhibit a duplication of lateral teeth seen in many "14 tooth species" and the dorsal-ventral teeth multiply as well. The phylogeny of *Rhabdochona* recovers an entire clade that possesses the these larval characteristics. These apomorphies probably owe their appearance to paedomorphosis. This heterochronic event is seemingly coupled to the acquisition of hosts in several fish families different from cyprinids. Furthermore, this pattern is in accordance with a change in the environmental or geomorphologic conditions that probably accompanied these speciation events, related to the aforementioned TP event. Additionally, the common presence of homoplasic characters in *Rhabdochona* phylogeny suggests reversals originating through heterochronic events, which enabled these nematode parasites to disperse/colonize to new host families. Therefore, heterochrony (Brooks & McLennan, 1993a; Gould, 1977) is probably one of the mechanisms underlying speciation of *Rhabdochona* spp., coupled with colonization of new host families after these hosts dispersed distances greater than a radius of $10^2$ km$^2$ or more.

The ontogenetic age structure of metazoan parasites has scarcely been studied in macroevolutionary terms (Brooks & McLennan, 1993a). Developmental constraints might be one of the causes of conservatism in living fossils (Avise et al., 1994). *Rhabdochona* spp. cannot qualify as living fossils, although some authors have stated that absence of a record is not evidence that a form, a lineage, or a molecule is not or has not been persistent (Liow, 2006). However, these taxa may qualify as an extremely conservative group (Hoberg & Brooks, 2008) in which few morphological changes have allowed species within this nematode genus to disperse through its hosts to multiple continents after the breakup of Pangea, and later due to TP. It is probable that some species of *Rhabdochona* have remained in stasis for nearly 31 my.

## 7. Conclusions

A combination of events promoted by TP, host switching and postdispersal speciation has resulted in the modern distribution of *Rhabdochona* spp. Ecological fitting helps explain why in past expansion phases of these parasites into new hosts no apparent speciation has occurred. *Rhabdochona* represents a group of species that has remained relatively unchanged as it has invaded new fish hosts, as inferred from morphology, life cycle, and host habitat preference analyses.

EF has played a major role in the evolution of freshwater fish helminth host parasite systems by maintaining parasite species stasis while they track plesiomorphic resources (Agosta et al., 2010; Brooks & McLennan, 2002; Brooks & Ferrao, 2005; Brooks et al., 2006). This might explain why parasite adaptive radiation is not as extensive as has been previously believed (Brooks & McLennan, 1993). Nematode parasite radiations probably have been driven more by speciation via TP and punctuated phenomena than by phyletic gradualism. Vertebrate nematode parasites likely have longer periods of stasis than do their hosts. Therefore, the effect of stasis via EF results in an apparent "lag" pattern in parasite evolution, as compared to evolution of their hosts. Lastly, heterochrony probably is a primary mechanism in nematode parasite postdispersal speciation, when coupled with long distance dispersal of their hosts.

## 8. Acknowledgment

The writing of this chapter was supported by a CONACYT-SNI 43282 and Programa de Estímulos a la Investigación, Desarrollo Tecnológico e Innovación del CONACYT grants.

## 9. References

Agosta, S. J. & Klemens, J. A. (2008). Ecological fitting by phenotypically flexible genotypes: implications for species associations, community assembly and evolution. *Ecology Letters*, Vol.11, No.11, (November 2008), pp. 1123–1134, ISSN 1461-0248

Agosta, S. J., Janz, N. & Brooks, D. R. (2010). How specialists can be generalists: resolving the "parasite paradox" and implications for emerging infectious disease. *Zoologia*, Vol.27 No.2, (April 2010), pp. 151–162, ISSN 0101-8175, 1806-969X

Avise, J. C., Nelson, W. S. & Sugita, H. (1994). A speciational history of "living fossils": molecular evolutionary patterns in horseshoe crabs. *Evolution*, Vol.48, No.6 (April 1994), pp. 1986-2001, ISSN 1558-5646

Bastrop R., Strehlow, B., Jurss, K., Sturmbauer, C. (2000). A new molecular phylogenetic hypothesis for the evolution of freshwater eels. *Molecular Phylogenetics and Evolution* Vol.14, No.2, (February 2000), pp. 250-258, ISSN 1055-7903

Benton, M.J. (ed.) (1993). *The Fossil Record ?* Chapman & Hall, London, ISBN: 0-412-39380-8

Brooks, D. R. (1978). Evolutionary history of the cestode order Proteocephalidea. *Systematic Zoology*, Vol.27, No.3, (June, 1978), pp. 312-323, ISSN 0039-7989

Brooks, D. R. (1981). Hennig's parasitological method: a proposed solution. *Systematic Zoology*, Vol.30, No.3 (September 1981), pp. 229-249, ISSN 0039-7989

Brooks, D.R. & Glen, D.R., (1982). Pinworms and primates: a case study in coevolution. *Proceedings of the Helminthological Society of Washington*, Vol.49, No.1 (January 1982), pp. 76-85, ISBN/ISSN: 0018-0130

Brooks, D. R. (1992). Origins, diversification, and historical structure of the helminth fauna inhabiting neotropical freshwater stingrays (Potamotrygonidae). *Journal of Parasitology*, Vol.78, No.4, (August 1992), pp. 588-595, ISSN: 1937-2345 (online), ISSN 0022-3395 (print)

Brooks D. R. & McLennan, D. A. (1993a). Comparative study of adaptive radiations with an example using parasitic flatworms (Paltyhelminthes: Cercomeria). *The American Naturalist*, Vol.142, No.5, (November 1993), pp. 755-778, ISSN 00030147

Brooks D. R. & McLennan, D. A. (1993b). *Parascript. Parasites and the language of evolution*, Smithsonian Institution Press, ISBN 1-56098-215-2, Washington and London

Brooks, D. R., O'Grady, T. O. & Glen, D. R. (1985). Phylogenetic analysis of the Digenea (Platyhelminthes: Cercomeria) with comments on their adaptive radiation. *Canadian Journal of Zoology* 63: 411-443, ISSN 0008-4301

Brooks, D.R. & McLennan, D.A. (2002). *The Nature of Diversity: An Evolutionary Voyage of Discovery*, University of Chicago Press, ISBN 0-226-075907-3, Chicago, U.S.A.

Brooks, D.R. & McLennan, D.A.. (2003). Extending phylogenetic studies of coevolution: secondary Brooks parsimony analysis, parasites, and the great apes. *Cladistics*, Vol.19, No.2 (April 2002), pp. 104-119, ISSN 1096-0031 online

Brooks, D. R. & Ferrao. A. L. (2005). The historical biogeography of co-evolution: emerging infectious diseases are evolutionary accidents waiting to happen. *Journal of Biogeography*, Vol.32, No.8, (August 2005), pp. 1291–1299, ISSN 0305-0270 (print), ISSN: 1365-2699 (online).

Brooks, D. R., León-Règagnon, V., McLennan, D. A. & Zelmer, D. (2006). Ecological Fitting as a Determinant of the Community Structure of Platyhelminth Parasites of Anurans. *Ecology*, Vol.87, Vol. 7 Supplement, (July 2006), pp. S76-S85, ISSN 0012-9658

Brundin, L. 1981. Croizat's biogeography versus phylogenetic biogeography, In: *Vicariance biogeography: A critique*, G. Nelson & D. E. Rosen, (Eds.), 94-138, Columbia University Press, ISBN 0231048084, New York.

Caspeta-Mandujano J.M., Moravec F., & Salgado-Maldonado G. (2000). *Rhabdochona mexicana* sp. n. (Nematoda: Rhabdochonidae) from the intestine of characid fishes in Mexico. *Folia Parasitologica* (Praha), Vol.47, No.3, (February 2000), pp. 211-215, ISSN 0015-5683 (print), ISSN 1803-6465 (online).

Caspeta-Mandujano J.M., Moravec F., & Salgado-Maldonado G. (2001). Two new species of rhabdochonids (Nematoda: Rhabdochonidae) from freshwater fishes in Mexico, with a description of a new genus. *Journal of Parasitology*, Vol. 87, No.1, (January 2001), pp. 139–143 ISSN: 1937-2345, ISSN: 0022-3395

Cavender, T. M. 1998. Development of the North American Tertiary freshwater fish fauna with a look at parallel trends found in the European record. *Italian Journal of Zoology*, Vol.65, Supplement 1, (January 2009), pp. 149-161, ISSN 1125-0003

Cavender, T. M. & Miller, R. R. 1982. *Salmo australis*, a new species of fossil salmonid from Southwestern Mexico. *Contributions from the Museum of Paleontology, The University of Michigan*, Vol.26, No.1, (December 1982), pp. 1-17, ISSN 0097-3556

Cavin, L., Longbottom, A. & Richter, M. (eds.) (2008). *Fishes and the Break-up of Pangea*. Geological Society, London, Special Publications, 295, ISBN 978-1-86239-248-9.

Černotíková, E., A. Horák, & F. Moravec. (2011). Phylogenetic relationships of some spirurine nematodes (Nematoda: Chromadorea: Rhabditida: Spirurina) parasitic in fishes inferred from SSU rRNA gene sequences. *Folia Parasitologica*, Vol.58, No.2 (2011), pp. 135–148, ISSN 0015-5683

Chabaud A.G. 1975. Camallanoidea, Dracunculoidea, Gnathostomatoidea, Physalopteroidea, Rictularioidea and Thelazioidea. Keys to genera of the order Spirurida, Part 1. In: *CIH keys to the nematode parasites of vertebrates 3*. (C.R. Anderson, A.G. Chabaud and S. Willmott Eds.), 1-27, Commonwealth Agricultural Bureaux, Farnham Royal, ISSN 0305-2729, Bucks, United Kingdom

Chen, W-J., Miya, M., Saitoh, K., & Mayden, R. L. (2008). Phylogenetic utility of two existing and four novel nuclear gene loci in recontructing Tree of Life of ray-finned fishes: The order Cypriniformes (Ostariophysi) as a case study. *Gene*, Vol.423, No. 2, (July 2008), pp. 125-134, ISSN 0378-1119

Choudhury, A. (2009). A new species of deropristiid (Trematoda: Deropristiidae) from the lake sturgeon, *Acipenser fulvescens*, in Wisconsin, and its biogeographical implications. *Journal of Parasitology*, Vol.95, No.5, (September 2009), pp. 1159-1164, ISSN 1937-2345, ISSN 0022-3395

Choudhury, A. & Dick, T.A. (1998). Systematics of the Deropristiidae Cable and Hunninen, 1942 (Trematoda) and biogeographical associations with sturgeons (Osteichthyes: Acipenseridae). *Systematic Parasitology*, Vol.41, No.1, (September 1998), pp. 21-39, ISSN 0165-5752

Choudhury, A. & Dick, T.A. (1996). Observations on the systematics and biogeography of the genus *Truttaedacnitis* (Nematoda: Cucullanidae). *Journal of Parasitology*, Vol.82, No.6, (December 1996), pp. 965-976, ISSN 1937-2345

Choudhury, A. & Dick, T.A. (2000). Richness and diverstiy of helminth communities in tropical freshwater fishes: empirical evidence. *Journal of Biogeography* Vol.27, No.4, (July 2000), pp. 935-956, ISSN 0305-0270

Choudhury, A. & Dick, T.A. (2001). Sturgeons and their parasites: Patterns and processes in historical biogeography. *Journal of Biogeography*, Vol.28, No.11-12, (November 2001), pp. 1411 -1439, ISSN 0305- 0270

Choudhury, A., Moore, B. R. & Marques, F. L. (2002). Vernon Kellogg, host switching and cospeciation in parasites: rescuing straggled ideas. *Journal of Parasitology*, Vol.88, No.5, (October 2002), pp. 1045-1048, ISSN 1937-2345

Coyne, J. A. & Orr, H. A. (2004). *Speciation*. Sinauer Associates, Inc. Publishers, ISBN 978-0-87893-089-0, Sunderland, Massachusetts, U.S.A. .

Cutter, A. D., Wasmuth, J. D. & Washington, N. L. (2008). Patterns of Molecular Evolution in Caenorhabditis Preclude Ancient Origins of Selfing. *Genetics*, Vol.178, No.4 (April 2008), pp. 2093–2104, ISSN: 0016-6731

Cracraft, J. (1989). Speciation and its ontology: the empirical consequences of alternative species conepts for understanding patterns and processes of differentiation. In: *Speciation and Its Conooqucnceo*, D. Otte, & J. A. Endler, (Eds.), 28-59, Sinauer Associates Inc, ISBN: 0-87893-658-0, Sunderland, Massachusetts, U.S.A.

de Chambrier, A., Zehnder, M., Vaucher, C. & Mariaux, J. (2004). The evolution of the Proteocephalidea (Platyhelminthes, Eucestoda) based on an enlarged molecular phylogeny, with comments on their uterine development. *Systematic Parasitology*, Vol.57, No.3 (March 2004), pp. 159-171, ISSN 0165-5752

Doadrio, I. & Domínguez, O. (2004). Phylogenetic relationships within the fish family Goodeidae based on cytochrome b sequence data. *Molecular Phylogenetics and Evolution*, Vol.31, No.2, (May 2004), pp. 416-430, ISSN 1055-7903

Dunn, R. R., Harris, C. H., Colwell, R. K., Koh, L. P. & Sodhi, N .S. (2009). The sixth mass coextinction: are most endangered species parasites and mutualists? *Proceedings of the Royal Society B*, Vol.276, No.1670, (September 2009), pp. 3037-3045, ISSN 1471-2954

Eldredge, N. & Gould, S. J. (1972). Punctuated equilibria: an alternative to phyletic gradualism. In: *Models in Paleobiology*, T. J. M. Schopf, (Ed.), 82-115, Freeman,

Cooper and Company, ISBN-10 0877353255, ISBN-13 978-0877353256, San Francisco, U.S.A.

Eldredge, N. & J. Cracraft. (1980). *Phylogenetic Patterns and the Evolutionary Process. Method and Theory in Comparative Biology*, Columbia University Press, ISBN 0-231-03802-X, New York, U.S.A.

Erwin, T. L. (1981). Taxon pulses, vicariance, and dispersal: an evolutionary synthesis illustrated by carabid beetles. In: *Vicariance biogeography: a critique*, G. Nelson, & D.E. Rosen, (Eds.), 159–196, Columbia University Press, ISBN: 0-231-04808-4, New York, U.S.A.

Ferraris, C. J. (2007). Checklist of catfishes, recent and fossil (Osteichthyes: Siluriformes, and catalogue of siluriform primary types. *Zootaxa* Vol.1419, (March, 2007), pp. 1-628, ISSN 1175-5334

Folinsbee, K. E. & Brooks, D. R. (2007). Miocene hominoid biogeography: pulses of dispersal and differentiation. *Journal of Biogeography*, Vol.34, No.3, (March 2007), pp. 383-397, ISSN 0305-0270

Forey, P. L. (2010). Tethys and Teleosts. In: *Beyond Cladistics. The Branching of a Paradigm*. D. Williams, & S. Knapp, (Eds.), 243-266, University of California Press, ISBN 978-0-520-26772-5, Berkeley, Los Angeles, London

Gould, S. J. & Eldredge, N. (1977). Punctuated equilibria: the tempo and mode of evolution econsidered. *Paleobiology*, Vol.3, No.2, (Spring 1977), pp. 115-151, ISSN 0094-8373

Gould, S. J. (1977). *Ontogeny and Phylogeny*, The Belknap Press of Harvard University Press, ISBN 0-674-63941-3, Harvard, Cambridge, Massachusetts and London, England

Gould, S. J. (2002). *The structure of evolutionary theory*, The Belknap Press of Harvard University Press, ISBN 0-674-00613-5, Cambridge, Massachusetts and London, England

Halas, D., Zamparo, D., & Brooks, D. R. (2005). A historical biogeographical protocol for studying biotic diversification by taxon pulses. *Journal of Biogeography*, Vol.32, No.2, (February 2005), pp. 249-260, ISSN 0305-0270

Hennig, W. (1966). Phylogenetic systematics, University of Illinois Press, ISBN 0-252-00745-X, Urbana, Illinois, U.S.A.

Hoberg, E. P., Gardner, S. L., & Campbell, R. A. (1999). Systematics of the Eucestoda: Advances toward a new phylogenetic paradigm, and observations on the early diversification of tapeworms and vertebrates. *Systematic Parasitology*, Vol.41, No.1, (January 1999), pp. 1-12, ISSN 0165-5752

Hoberg, E. P. (2005). Coevolution and biogeography among nematodirinae (Nematoda: Trichostrongylina) lagomorpha and artiodactyla (Mammalia): exploring determinants of history and structure for the Northern Fauna across the Holarctic. *Journal of Parasitology*, Vol.91, No.2, (April 2005), pp. 358-369, ISSN 1937-2345

Hoberg, E. P. & Brooks, D. R. (2008). A macroevolutionary mosaic: episodic host-switching, geographical colonization and diversification in complex host-parasite systems. *Journal of Biogeography*, Vol.35, No.9, (September, 2008), pp. 1533-1550. ISSN 0305-0270

Hoberg, E. P. & D. R. Brooks. (2010). Chapter 1: Beyond vicariance: integrating taxon pulses, ecological fitting, and oscillation in evolution and historical biogeography, In: *The Biogeography of Host-Parasite Interactions*, S. Morand and B. R. Krasnov, (Eds.), 7-20, Oxford University Press, ISBN 978-0-19-956135-3, Oxford, United Kingdom

Hugot, J-P, Baujard, P., & Morand, S. (2001). Biodiversity in helminths and nematodes as field of study: an overview. *Nematology*, Vol.3, No.3 (September 2001), pp. 199-208, ISSN 1388-5545

Hurley, I. A., Lockridge Mueller,R., Dunn, K. A., Schmidt, E. J., Friedman, M., Ho, R. K. , Prince, V. E. , Yang, Z., Thomas, M. G. & Coates, M. I. (2007).A new time-scale for ray-finned fish evolution. *Proceedings of the Royal Society* B, Vol.274, No.1609, (February), pp. 489-498, ISSN 1471-2954

Jablonski, D. (2008). Species Selection: Theory and Data. *Annual Review of Ecology and Systematics*, Vol.39, No.2008, (December 2008), pp. 501–524, ISSN 0066-4162

Janzen, D. H. (1985). On Ecological Fitting. *Oikos*, Vol.45, No.3, (1985), pp. 308-310, ISSN 0030-1299

Kiontke, K. & Fitch, D.H.A. (August 11, 2005). The Phylogenetic relationships of Caenorhabditis and other rhabditids, In: WormBook, 08.11.2005, Available from The C. elegans Research Community, WormBook, doi/10.1895/wormbook.1.11.1

Lieberman, B. S. (2003). Paleobiogeography: the relevance of fossils to biogeography. *Annual Review of Ecology, Evolution and Systematics*, Vol.34, No.1, (November 2003), pp. 51-69, ISSN 1543-592X

Liow, L. H. . (2006). Oddities, wonders, and other tall tales of "living fossils", Unpublished PhD Thesis, The Faculty of the Division of the Biological Sciences and The Pritzker School, University of Chicago, Illinois, U.S.A.

Lundberg, J. G., Marshall, L.G., Guerrero, J., Horton, B., Malabarba, M. C. S. L., & Wessenlingh, F. (1998). The Stage for Neotropical Fish Diversification: A History of Tropical South American Rivers, In: *Phylogeny and Classification of Neotropical Fishes*, L.R. Malabarga, , R. E. Reis, R. P. Vari, Z. M. Lucena, & C. A. S. Lucena, (Eds), 13-48, Edipucrs, ISBN 8574300357, Porto Alegre, Brazil

Lundberg, J. G., Sullivan, J. P., Rodiles-Hernández, R. & Hendrickson, D. A. (2007). Discovery of African roots for the Mesoamerican Chiapas catfish, Lacantunia enigmatica, requires an ancient intercontinental passage. *Proceedings of the Academy of Natural Sciences of Philadelphia*, Vol.153, No.1, (September 2007), pp. 39-53, ISSN 0097-3157

Manter, H.W. (1963). The Zoogeographical Affinities of Trematodes of South American Freshwater Fishes. *Systematic Zoology*, Vol.12, No.2, (June 1963), pp. 45-70, ISSN 0039-7989

Marshall, C.J. & Liebherr, J. K. (2000). Cladistic biogeography of the Mexican Transition zone. *Journal of Biogeography*, Vol.27, No.1, (January 2000), pp. 203-216, ISSN 0305-0270

McNamara, K.J., & McKinney, M.L. (2005). Heterochrony, disparity and macroevolution, In: *Macroevolution: Diversity, Disparity, Contingency. Essays in Honor of Stephen Jay Gould*, E. S. Vrba, & N. Eldredge, 17-26, The Paleontological Society, ISBN 1-891276-49-2, ISSN 0094-8373, Lawrence, Kansas, U.S.A.

Mejía-Madrid H., & Pérez-Ponce de León, G. (2003). *Rhabdochona ahuehuellensis* n. sp. (Nematoda: Rhabdochonidae) from the Balsas goodeid, *Ilyodon whitei* (Osteichthyes: Gooedeidae [sic]), in Mexico. *Journal of Parasitology*, Vol.89, No.2, (March 2003), pp. 356–361, ISSN 1937-2345

Mejía-Madrid, H. H., Choudhury, A. & Pérez-Ponce de León, G. (2007a). Phylogeny and biogeography of *Rhabdochona* Railliet, 1916 (Nematoda: Rhabdochonidae) species from the Americas. *Systematic Parasitology*, Vol.67, No.1, (January 2007), pp. 1-18, ISSN 0165-5752

Mejía-Madrid, H. H., Vázquez-Domínguez, E. & Pérez-Ponce de León, G. (2007b). Phylogeography and freshwater basins in central Mexico: recent history as revealed by the fish parasite *Rhabdochona lichtenfelsi* (Nematoda). *Journal of Biogeography*, Vol.34, No.5, (May 2007), pp. 787-801, ISSN 0305-0270

Mejía-Madrid H., & Pérez-Ponce de León G. (2007). A new rhabdochonid from the blue striper (Nematoda: Rhabdochonidae) off the coast of Mexico. *Journal of Parasitology*, Vol.89, No.1, (February 2007), pp. 356–361, ISSN 1937-2345

Miller, R. R. & Smith, M L. (1986). Origin and Geography of the Fishes of Central Mexico, In: *The Zoogeography of North American Freshwater Fishes*, C. H. Hocutt & E. O. Wiley (Eds.), 487-518, John Wiley & Sons, ISBN 0-471-86419-6, New York, Chichester, Brisbane, Toronto, Singapore

Miller, R. R. (2005). *Freshwater Fishes of México*, The University of Chicago Press, ISBN 0-226-52604-6, Chicago and London

Moravec, F. (1972). A revision of the African species of the nematode genus Rhabdochona Railliet, 1916. Vestník Ceskoslovenské Spolecnosti Zoologické Vol.36, pp. 196-208, ISBN/ISSN 0042-4595

Moravec F. (1975). *Reconstruction of the nematode genus* Rhabdochona *Railliet, 1916 with a review of the species parasitic in fishes of Europe and Asia*, Studie ČSAV No.8. Academia, Prague

Moravec F. (2006). *Dracunculoid and anguillicoid nematodes parasitic in vertebrates*. Academia, ISBN 80-200-1431-4, Prague

Moravec, F., & Muzzall, P. (2007). Redescription of *Rhabdochona cotti* (Nematoda, Rhabdochonidae) from *Cottus caeruleomentum* (Teleostei, Cottidae) in Maryland, USA, with remarks on the taxonomy of North American *Rhabdochona* spp. *Acta Parasitologica*, Vol.52, No.1, (January 2007), pp. 51–57, ISSN 1230-2821

Moravec, F. (2010). Some aspects of the taxonomy, biology, possible evolution and biogeography of nematodes of the spirurine genus *Rhabdochona* Railliet, 1916 (Rhabdochonidae, Thelazioidea). *Acta Parasitologica*,Vol.55, No.2, (February 2010), pp. 144–160; ISSN 1230-2821

Moravec, F., Levron, C. & de Buron, I. (2011). Morphology and taxonomic status of two little-known nematode species parasitizing North American Fishes. *Journal of Parasitology*, Vol.97, No.2, (April 2011), pp. 297-304, ISSN 1937-2345

Nadler, S.A., Carreno, R.A., Mejía-Madrid, H., Ullberg, J., Pagan, C., Houston, R., & Hugot, J.-P. (2007). Molecular phylogeny of clade III nematodes reveals multiple origins of tissue parasitism. *Parasitology*, Vol.134, No.10, (September 2007), pp. 1421-1442, ISSN 0031-1820

Nieberding, C., Morand, S., Libois, R. & Michaux, J. R. (2004). A parasite reveals cryptic phylogeographic history of its host. *Proceedings Royal Society London* B, Vol.271, No.1557, (December 2004), pp. 2559-2568, ISSN 1471-2954

Nieberding, C., Morand, S., Douaduy, C.J., Libois, R. & Michaux, J. R. (2005). Phylogeography of a Nematode (*Heligmosomoides polygyrus*) in the Western Palearctic region: Persistence of Northern cryptic populations during ice ages? *Molecular Ecology*, Vol.14, No.3, (March 2005), pp. 765-779, ISSN 0962-1083

Parenti, L. R. (1981). A phylogenetic and biogeographic analysis of Cyprinodontiform Fishes (Teleostei, Atherinomorpha). *Bulletin of the American Museum of Natural History*, Vol.168, No. 4, (n.m.), pp. 1-99, ISSN 0003-0090

Pérez-Ponce de León, G. & Choudhury, A. (2002). Adult endohelminth parasites of ictalurid fishes (Osteichthyes: Ictaluridae) in Mexico: Empirical evidence for biogeographical patterns. *Comparative Parasitology*, Vol.69, No.1, (January 2002), pp. 10 -19, ISSN 1525-2647

Pérez-Ponce de León, G., & Choudhury, A. (2005). Biogeography of helminth parasites of freshwater fishes in Mexico: the search for patterns and processes. *Journal of Biogeography*, Vol.32, No.4, (April 2005), pp. 645-659, ISSN 0305-0270

Pérez-Ponce de León, G., Choudhury, A., Rosas-Valdez, R. & Mejía-Madrid, H. H. (2007). The systematic position of *Wallinia* spp. and *Margotrema* spp. (Digenea), parasites of Middle-American and Neotropical freshwater fishes, based on the 28S ribosomal RNA gene. *Systematic Parasitology*, Vol.68, No.1, (September 2007), pp. 49-55, ISSN 0165-5752

Peng, Z, S. He, J. Wang, Wang, W. & Diogo, R. (2006). Mitochondrial molecular clocks and the origin of the major Otocephalan clades (Pisces: Teleostei): A new insight. *Gene*, Vol.370, No.29, (March 2006), pp. 113-124, ISSN 0378-1119

Poinar Jr., G. O. (1984). Fossil evidence of nematode parasitism. *Revue de Nématologie*, Vol.72, No.2, (n.m.), pp. 201-203, ISSN 0183-9187.

Poulin, R. (1998). Large-scale patterns of host use by parasites of freshwater fishes. *Ecology Letters*, Vol.1, No.2, (September 1998), pp. 118-128, ISSN 1461-023X

Poulin, R. & Mouillot, D. (2003). Host introductions and the geography of parasite taxonomic diversity. *Journal of Biogeography*, Vol.30, No.6, (June 2003), pp. 837–845, ISSN 0305-0270

Poulin, R. & Morand, S. (2004). *Parasite Biodiversity*. Smithsonian Books, Smithsonian Institution, ISBN 1-58834-170-4,Washington, USA.

Puylaert, F.A. (1973). Rhabdochonidae parasites de poissons africains d'eau douce et ce groupe. *Revue de Zoologie et de Botanique Africaine*, Vol.87, No.4, (December 1973), pp. 647–665, ISSN 0035-1814

Rasheed, S. (1965). A preliminary review of the genus *Rhabdochona* Railliet, 1916 with description of a new and related genus. *Acta Parasitologica Polonica*, Vol.13, No.42, (December 1965), pp. 407–424, ISSN 0065-1478

Rego, A. A., de Chambrier, A., Hanzelová, V., Hoberg, E., Scholz, T, Weekes, P. & Zehnder, M. (1998). Preliminary phylogenetic analysis of subfamilies of the Proteocephalidea (Eucestoda). *Systematic Parasitology*, Vol.40, No.1 (May 1998), pp. 1-19, ISSN 0165-5752

Rosas-Valdez, R., Domínguez-Domínguez, O., Choudhury, A. & Pérez-Ponce de León, G. (2008). Helminth Parasites of the Balsas Catfish *Ictalurus balsanus* (Siluriformes: Ictaluridae) in Several Localities of the Balsas River Drainage, Mexico: Species Composition and Biogeographical Affinities. *Comparative Parasitology*, Vol.74, No.2, (July 2007), pp. 204–210, ISSN: 1525-2647

Roy, K., Hunt, G., Jablonski, D., Krug, A. Z., & Valentine, J. W. (2009). A macroevolutionary perspective on species range limits. *Proceedings of the Royal Society* B, Vol.276, No.1661, (February 2009), pp. 1485–1493, ISSN 1471-2954

Sánchez-Alvarez A., García-Prieto L., & Pérez-Ponce de León G. (1998). A new species of *Rhabdochona* Railliet, 1916 (Nematoda: Rhabdochonidae) from endemic goodeids (Cyprinodontiformes) from two Mexican lakes. *Journal of Parasitology*, Vol.84, No.4, (August 1998), pp. 840–845, ISSN 1937-2345

Sarmartín, I. , Enghoff, H., & F. Ronquist. (2001). Patterns of animal dispersal, vicariance and diversification in the Holarctic. *Biological Journal of the Linnean Society*, Vol.73, No.4, (August 2001), pp. 345–390, ISSN 1095-8312

Simková, A, Verneau, O., Gelnar, M. & Morand, S. (2006). Specificity and specialization of congeneric monogeneans parasitizing cyprinid fish. *Evolution*, Vol.60, No.5, (May 2006), pp. 1023-1037, ISSN 1558-5646

Stankiewicz, J. & de Wit, M.J. (2006). A proposed drainage evolution model for Central Africa - did the Congo flow east? *Journal of African Earth Sciences*, Vol.44, No.1, (January 2006), pp. 75-84, ISSN 1464-343X

Stanley, S. M. (1975). A theory of evolution above the species level. *Proceedings of the National Academy of Sciences* Vol.72, No.2, (February 1975), pp. 646-650, ISSN 1091-6490

Stanley, S. M. (1989). Chapter 8. Fossils, Macroevolution, and Theoretical Ecology, In: *Perspectives in Ecological Theory*, J. Roughgarden, R. M. May, & S. A. Levin, (Eds.)., 125-134, Princeton University Press, ISBN 0- 691-08508-0, Princeton, New Jersey.

Stanley, S. M. (1998). *Macroevolution. Pattern and Process*, The Johns Hopkins University Press, ISBN 0-8018-5735-X, Baltimore and London.

Tsukamoto, K., Aoyama, J. & Miller, M. J. (2002). Migration, speciation, and the evolution of diadromy in anguillid eels. *Canadian Journal of Fisheries and Aquatic Science*, Vol.59, No.12, (December 2002), pp. 1989- 1998, ISSN 0706-652X

van Veller, M.G.P. & Brooks, D.R. (2001) When simplicity is not parsimonious: a priori and a posteriori approaches in historical biogeography. *Journal of Biogeography*, Vol.28, No.1, (January 2001), pp. 1–12, ISSN 0305-0270

Vrba, E. S. (2005). Mass turnover and heterochrony events in response to physical change., In: *Macroevolution: Diversity, Disparity, Contingency. Essays in Honor of Stephen Jay Gould*, E. S. Vrba, & N. Eldredge, 157-174, The Paleontological Society, ISBN 1-891276-49-2 , ISSN 0094-8373, Lawrence, Kansas, U.S.A.

Wagner, P.J. & Erwin, D.H. (1995). Phylogenetic patterns as tests of speciation hypotheses, In: *New approaches for studying speciation in the fossil record*, D.H. Erwin & R.L. Anstey, (Eds.), 87–122, Columbia University Press, ISBN 0-231-08248-7, New York

Webb, S. A., Graves, J. A., Macias-Garcia, C., Magurran, A. E., Foighil, D. Ó & Ritchie, M. G. (2004). Molecular phylogeny of the livebearing Goodeidae (Cyprinodontiformes). *Molecular Phylogenetics and Evolution*, Vol.30, No.3, (March 2004), pp. 527-544, ISSN 1055-7903

Wiley, E. O. & Lieberman, B. S. (2011). *Phylogenetic Systematics. The Theory and Practice of Phylogenetic Systematics*. Second Edition. Wiley-Blackwell. 406 p. ISBN 978-0-470-90596-8.

Wijová M., Moravec F., Horák A., & Lukeš J. (2006). Evolutionary relationships of Spirurina (Nematoda: Chromadorea: Rhabditida) with special emphasis on dracunculoid nematodes inferred from SSU rRNA gene sequences. *International Journal for Parasitology*, Vol.36, No.9, (August 2006), pp. 1067–1075, ISSN 0020-7519

Wojcicki M. & Brooks. D. R. (2005). PACT: an efficient and powerful algorithm for generating area cladograms. *Journal of Biogeography*, Vol32, No.5, (May 2005), pp. 755–774, ISSN 0305-0270

Zietara, M. S. & Lumme, J. (2002). Speciation by host switch and adaptive radiation in a fish parasite genus *Gyrodactylus* (Monogenea, Girodactylidae). *Evolution*, Vol.56, No.12, (December 2002), pp. 2445-2458, ISSN 1558-5646

# Influences of Island Characteristics on Plant Community Structure of Farasan Archipelago, Saudi Arabia: Island Biogeography and Nested Pattern

Khalid Al Mutairi[1], Mashhor Mansor[1], Magdy El-Bana[2,3,*],
Saud L. Al-Rowaily[2] and Asyraf Mansor[1]

[1]*School of Biological Sciences, Universiti Sains Malaysia, Penang,*
[2]*Department of Plant Production, College of Agricultural & Food Sciences,*
*King Saud University, Riyadh,*
[3]*Department of Biological Sciences, Faculty of Education at El-Arish,*
*Suez Canal University, El-Arish,*
[1]*Malaysia*
[2]*Saudi Arabia*
[3]*Egypt*

## 1. Introduction

Biogeographers have long been fascinated by the factors influencing numbers of species on islands. The increase in species number with area is one of the oldest known ecological patterns, first documented by Watson and deCandolle in the mid-nineteenth century (Rosenzweig, 1995). Island biogeographers lacked a cohesive body of theory, until island biogeography equilibrium theory (MacArthur & Wilson, 1967) attempted to explain variation in species richness between islands of different area and isolation. The theory predicts that species richness decreases with decreasing island area and increasing isolation as these two variables influence immigration and extinction (Rosenzweig, 1995). Numerous studies have examined and argued the stability of these relationships on different island groups and for different taxonomic categories.

However, the equilibrium theory should be expanded to include other aspects of insularity other than area and isolation in order to fully understand the mechanisms of island biogeography (Whittaker, 2000). In addition to area, distance, and elevation, numerous other variables have been examined as potential predictors of insular species richness, such as habitat diversity (Rafe et al., 1985; Kohn & Walsh, 1994), rainfall (Heatwole, 1991), soil type (Johnson & Simberloff, 1974), energy (Wright, 1983) and disturbance (El-Bana, 2009).

Although classical island biogeographical theory has been questioned (Gilbert, 1980; Whittaker, 2000) and a call for a new paradigm of island biogeography has been issued

---

* Corresponding Author

(Lomolino, 2000a), area and distance still play primary roles in alternative theories (Heaney, 2000; Lomolino, 2000b). In general island area, and to a lesser degree isolation, can hardly be disputed as important determinants of insular species richness.

Area might influence species richness directly in two ways: larger islands present larger targets for dispersing individuals and they generally support larger populations. Thus, island area may influence species richness by its effect on colonization rates or on the outcomes of several mechanisms that determine vulnerability to extinction (MacArthur & Wilson, 1967). Area might also influence species richness indirectly via its correlation with other factors that affect diversity directly. Among the most plausible of such potentially confounding variables is habitat diversity, which is often presumed to increase in direct relation to island area (Kohn & Walsh, 1994). The negative correlation between island isolation (distance from either the mainland and/or the large islands) and species richness, although not as strong, is also well documented. Since species differ in the maximum distance over which they can disperse, islands that are near the mainland will potentially receive propagules from more species than will distant islands (Rosenzweig, 1995).

During the last decade, ecologists and biogeographers have devoted increasing attention to the pattern of nested species assemblages in insular habitats. Nestedness occurs where assemblages in depauperate sites are comprised of species that constitute subsets of species that occur in successively richer sites. In nested biotas, common species tend to occur in all sites while rare species tend to occur only in the richest sites. This pattern indicates a high level of non-random organization of assemblages and has important implications for conservation (Patterson & Atmar, 1986; Patterson, 1990; Patterson & Brown 1991; Fleishman et al., 2007). Nestedness has been interpreted as a measure of biogeographic order in the distribution of species (Atmar & Patterson, 1993). This pattern indicates a high level of non-random organization of assemblages and has important implications for maintaining or maximizing species diversity in ecosystems threatened by anthropogenic effects (Maron et al., 2004; Fleishman et al., 2007).

Diverse biotic and abiotic processes are believed to generate nested distributions, including selective extinction (Atmar & Patterson 1993; Wright et al., 1998), differential colonization (Kadmon, 1995), nested habitats (Wright et al., 1998; Honnay et al., 1999), and differential environmental tolerances among species (Fleishman et al., 2007). Differences in environmental tolerances among species may interact with nested habitats to produce nestedness. According to this hypothesis, species-rich sites are those that contain the greatest habitat heterogeneity and/or have environmental conditions tolerable to the largest number of species (Cook, 1995; Honnay et al., 1999). Differential nestedness among groups of species (e.g., taxonomic groups or guilds) that vary in sensitivity to a particular environmental variable may determine how that variable contributes to the general pattern of species nestedness.

Nestedness has important implications for conservation, when species assemblages on an archipelago or habitat fragments show nestedness, it is more efficient to protect large islands or fragments than smaller islands or fragments (Patterson, 1987). Others have suggested that the management of colonization processes might also be important for the long-term maintenance of diversity (Lomolino, 1994; Cook, 1995).

On the arid archipelagoes, environmental features such as salinity, aridity, habitat diversity, elevation and human disturbance may interact with life history characteristics of plant species in determining local extinctions or colonization. The islands and archipelagos of Red Sea attracted less attention about their pattern of vegetation distribution and dynamics, compared to the Mediterranean Sea (Panitsa & Tzanoudakis, 1998, 2001; Panitsa et al., 2006; Médail & Vidal, 1998; Khedr & Lovett-Doust, 2000; Bergmeier & Dimopoulos, 2003; El-Bana, 2009).

Here we explore the patterns exhibited by plant species richness and nestedness on 20 islands of the Farasan archipelago in the Red Sea (Saudi Arabia) to identify possible effects of island size, elevation, number of habitats and distance from species pool. We also examine the best fit model for the total species richness, as well as the special patterns exhibited by certain important taxonomic and ecological subgroups of plant species.

## 2. Materials and methods

### 2.1 Study area

The Farasan archipelago consists of more than 36 vegetated islands and extends between longitudes $41^0$ 20' and $42^0$ 25' E and latitudes $16^0$ 20' and $17^0$ 10' N along the southern Red Sea (Figure 1). The islands, with elevation in the order of tens of metres, range in size from very small, a few m², to the very large island of Farasan Alkabir, about 319.5 km². All islands are an uplifted coral reef that formed during the Pleistocene on a foundation of salt diapirs (i.e. domes of salt rocks from the Miocene; Dabbagh et al., 1984). There is some variation in geomorphology among the islands despite their similar origin. The shore may rise gently to be followed by salt marshes and sandy plains, or be marked by small cliffs emerging from the coralline plateau and covered by coral rubble, and some islands feature a rugged structure of hillocks and outcrops. Some islands such as Zifaf and Sasu islands are hilly. Large boulders, gravels and small stones are found in the steep runnels of these islands.

The islands are an important habitats for both local and migrating birds. In addition, the islands home for the threatened and endemic Arabian gazelle and other mammals (Masseti, 2010). Most of the islands are subjected to heavy human activities such as overgrazing and wood cutting. Furthermore, the exotic and invasive tree *Prosopis juliflora* was introduced for greening landscape along roadsides in Farasan Alkabir island. It has escaped the cultivated sites and invaded the rich natural habitats such as Wadi Mattar.

Unfortunately, there are no climatic records available for Farasan Islands. The climate at Jizan city (42 km from Farasan Islands) is hot and humid with a maximum daily temperature in the range of 35–40°C during July. The overriding influence on the islands is the high year-round humidity, mitigated by winds. The mean annual rainfall is about 70 mm at Jizan. As in other arid regions, the condensation of dew is very important for the growth of vegetation on these islands (Osborne, 2000).

### 2.2 Data collection

Vegetation surveys were commenced in 2009 and 2010 during the rainy season from January to April. Random sampling was used in selecting 20 islands to represent an array of sizes, which ranged in area from 0.081 km² to 319.5 km² (Figure 1). Area (km²), distance (km) to the

Fig. 1. Farasan archipelago showing the location of the 20 studied islands (Abkar, Abu Shawk Umm Hawk, Ad Dissan, Al Hindiyah, Aslubah, At Targ, Dumsuk, Dushak, Farasan Alkabir, Kayyirah, Manzar Abu Shawk, Manzar Sajid, North Reefs, Rayyak Al Kabir, Safrah, Sajid, Shura, South Reefs, Sulayn and Zufaf.

nearest large island, and elevation (m) of each surveyed island were calculated by the program (Arc*GIS, 2008 USA). Two hundred and ten stands were selected to represent the main habitats on each island. Seven main habitat types were recognized: wet saline marshes, dry saline marshes, sand plains, mobile sand dunes, wadi channels, and coral rocky crevices and runnels. The stand size was about 10 m × 10 m in all habitats, except for the salt marshes and the rocky crevices and runnels where vegetation appeared as strips; the shape was modified to 5 m × 20 m. In each stand, shoot presence/absence of all vascular plant species was recorded. The position of each sampled stand was georeferenced using GARMIN GPS map 276.

All plant species were identified in each island following Chaudhary (1989, 2000); Collenette (1999). Plant species were categorized in terms of their life-forms (therophytes, hemicryptophytes, geophytes, chamaephytes and phanerophytes), salt tolerance (halophytes and glycophytes) and succulence (succulents and non-succulents). Life-forms of the plants were determined according to Raunkiaer classification (Raunkiaer, 1934). This classification is of special importance for the vegetation in arid regions. These categories reflect adaptation and tolerance of vegetation to the main environmental factors such as drought and salinity. Furthermore, this classification was used as the processes and factors that underlie species richness in these groups differ, resulting in different richness patterns (Khedr &Lovett- Doust, 2000; Panitsa et al., 2006; El-Bana, 2009).

## 2.3 Statistical analyses

To identify factors that were important in determining the distribution of plant species and their ecological subgroups, simple linear regression was performed on the species/ecological group richness and biogeographical variables to characterize the functional relationships between the variables, as well as to generate predictive values from empirically fitted regression models. Stepwise multiple regression analysis also was used to identify the best predictor of total species richness and the partitions of the data set of ecological subgroups, using area, elevation, shortest distance from the nearest large island and number of habitats as predictor variables. It is not always clear which measure of geographical isolation to use, i.e. distance from the mainland, the nearest large island, or just the nearest island, and usually a different measure might be necessary for different islands (Turchi et al., 1995; Sfenthourakis, 1996; Morand, 2000; Brose, 2003). In the present case, we chose distance from the nearest large island (Farasan Alkabir) because this island is the most likely candidate for serving as species pools for the other islands examined here. The regressions were run using both logarithmic and arithmetic values for all variables and the best functions according to the behaviour of residuals and the total variance explained ($R^2$) were chosen. All regressions and the estimations of parameters were carried out with SPSS v.16. We calculated Cole and Mao- Tau sample-based rarefaction curves (Colwell et al., 2004) using EstimateS software (Colwell, 2005, version 7.5).

## 2.4 Nested analyses

The data was prepared by constructing presence/absence matrices (1= present, 0 = absent) where columns and rows represented species and islands, respectively. The islands (rows) were rank ordered in relation to decreasing number of species and the species (columns) were rank ordered in relation to decreasing number of sites occupied. We then conducted nestedness analyses at two different spatial scales (entire species richness) and the ecological subgroup scales. To determine nestedness of assemblages we used the Nested Temperature Calculator computer program (Atmar & Patterson, 1995). This program calculates a temperature value (T) for the matrix ranging from 0 to 100, based on its presence/absence structure. A temperature of 0, indicates maximum order (maximum nestedness) and 100, indicates disorder (complete lack of nestedness) (Atmar & Patterson, 1993). To determine the significance of T (observed temperature) it is compared with the distribution of simulated temperatures produced by randomization of the matrix in Monte Carlo simulations (500 iterations). This method was used because of its statistical properties and because it can be directly compared among different taxonomic and ecological groups (Wright et al., 1998).

The effects of island area, number of habitats, isolation, and elevation on the degree of nestedness were evaluated by correlating the ranking order of islands in the observed matrix (arranged to maximize nestedness, Atmar & Patterson, 1995) with the order of islands after re-arranging the matrix in relation to the aforementioned factors using Spearman rank correlation. A significant relationship indicates that species are packed in a predictable order owing to the influence of a given factor (Atmar & Patterson, 1995). This procedure has proven useful for indicating possible mechanisms involved in nested structure (Atmar & Patterson, 1995; Kadmon, 1995; Honnay et al., 1999).

## 3. Results

### 3.1 Species richness

We detected a total of 191 species among 129 genera and 53 families on the surveyed islands. Most species occurred on relatively few islands (Figure 2a). About 95.5% (183 of 191) of the species occurred on ≤ 10 islands. Likewise, most islands contained relatively few species (Figure 2b). About 80% (16 of 20) of the islands contained less than 60 species. Rarefaction curves of Cole and Mao-Tau for species richness (Figure 3) reached the asymptote before 18 islands, indicating that the sampling effort was sufficient to fully capture the richness and diversity of plant species assemblages.

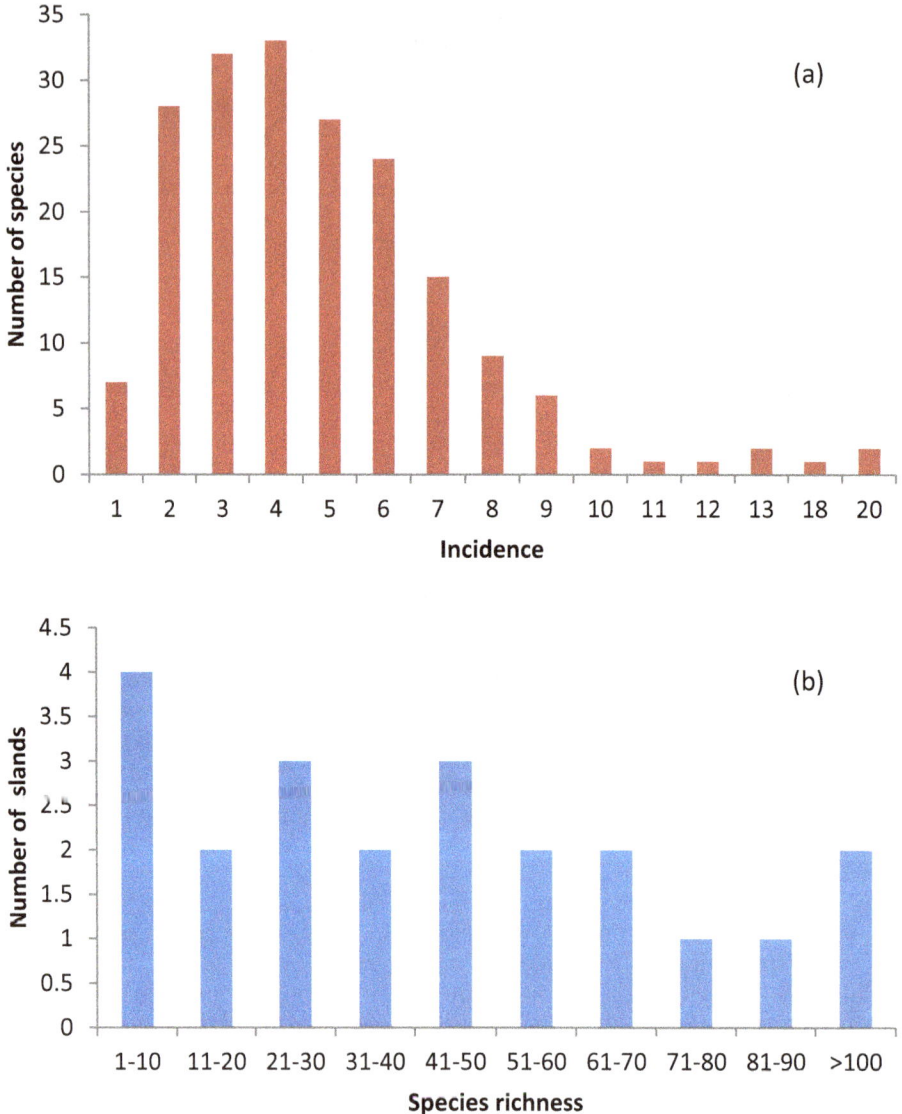

Fig. 2. Frequency distributions of incidence (i.e., the number of islands on which a species occurred) (a) and species richness (i.e. the number of species on an island) (b) for the toal flora of the Farasan archipelago.

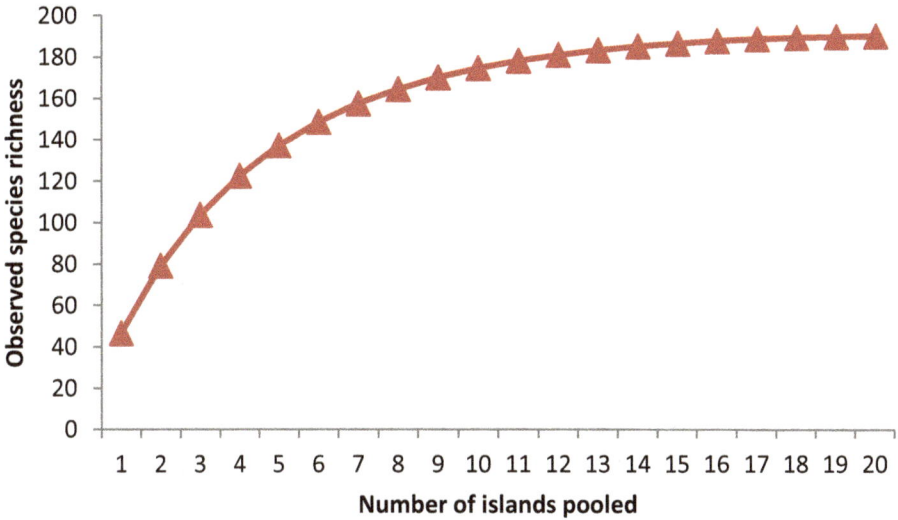

Fig. 3. Relationship between the number of islands pooled and the observed species richness of Farasan archipelago by rarefaction analysis. The asymptotic shape of the curve indicates that analysis of 18 islands provided sufficient sampling to fully capture the richness and diversity of plant species assemblages.

There was a significant positive relationship between island area and total plant species (Figure 4) with $r^2 = 0.732$ and $Z = 0.491$, $P < 0.0001$. Moreover, when the flora of each island was classified into different ecological groups and logS/logA was constructed, it appeared that each group had significantly different regressions. There were positive relationships between island area and each of perennials ($r^2 = 0.735$ and $Z = 0.312$, $P < 0.0001$) and annuals ($r^2 = 0.691$ and $Z = 0.168$, $P < 0.0001$) (Figure 4). Similarly, island area showed positive relationships with halophytes ($r^2 = 0.426$ and $Z = 0.049$, $P < 0.041$) and glycophytes (($r^2 = 0.737$ and $Z = 0.439$, $P < 0.0001$) (Figure 5). For succulence ecological groups, island area related positively with succulents (($r^2 = 0.669$ and $Z = 0.056$, $P < 0.0001$) and non-succulents ($r^2 = 0.73$ and $Z = 0.434$, $P < 0.0001$) (Figure 5). For the different growth forms, island area showed positive relationships with shrubs (($r^2 = 0.673$ and $Z = 0.087$, $P < 0.0001$), herbs ($r^2 = 0.729$ and $Z = 0.189$, $P < 0.0001$), trees ($r^2 = 0.816$ and $Z = 0.055$, $P < 0.0001$) and grasses ($r^2 = 0.684$ and $Z = 0.069$, $P < 0.0001$) (Figure 6).

The number of habitats was related positively with the island area ($r^2 = 0.516$, $P < 0.001$) (Figure 7a). In addition, the total number of species had a positive relationship with the number of habitats ($r^2 = 0.847$, $P < 0.0001$) (Figure 7b), and elevation ($r^2 = 0.366$, $P < 0.003$, data not shown). However, the distance from the largest island (Farasan Alkabir) has no effect on the species richness ($r^2 = -0.061$, $P < 0.887$).

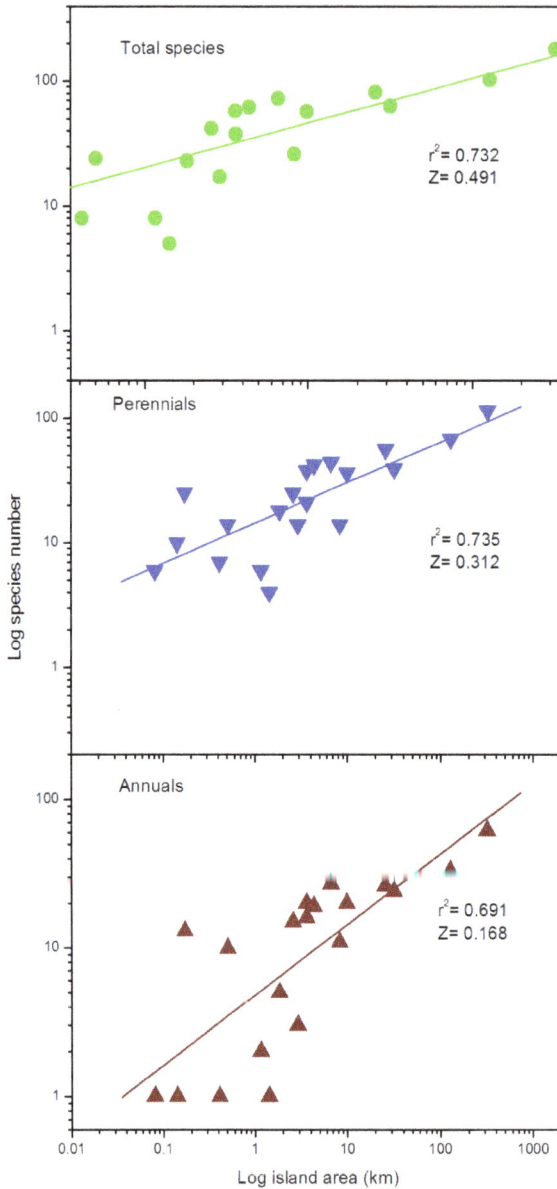

Fig. 4. Relationships of total species richness, number of perennials and annuals with island area of Farasan Archipelago

Influences of Island Characteristics on Plant Community Structure of Farasan Archipelago, Saudi Arabia: Island Biogeography and Nested Pattern

37

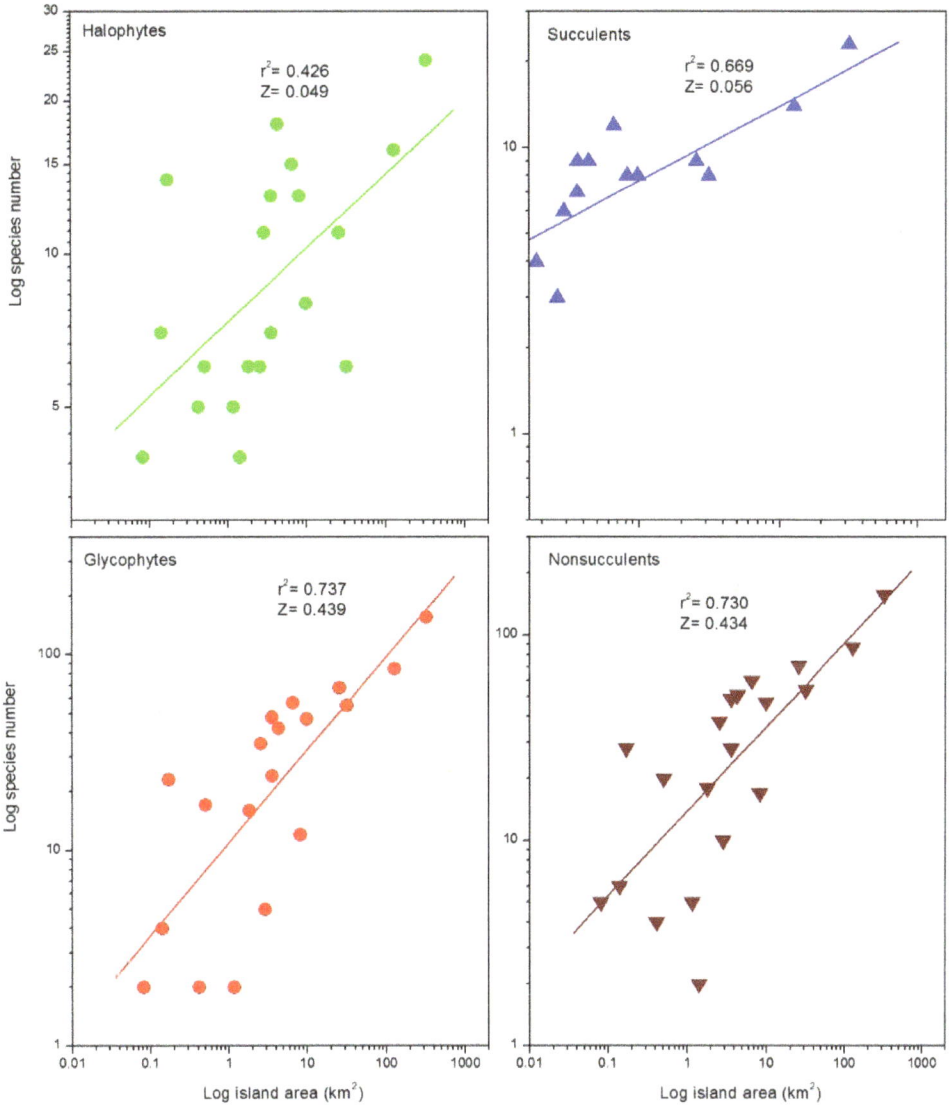

Fig. 5. Relationships of ecological groups (halophytes, glycophytes; succulents and non-succulents) with island area of Farasan Archipelago.

According to the stepwise regressions (Table 1), both island area and number of habitats affect species richness. When the same analyses were applied separately for each ecological groups, elevation was also significant parameter entering the model for perennials and annuals. Area, number of habitats and elevation explained a high percentage (88.7%) of total variance for annuals, while they explained about 72.3% of variance for the perennials. On the other hand, the number of habitats was not entering the model for shrubs, trees, non-succuelnts and halophytes (Table 1). Area and number of habitats entered the models of grasses, herbs, succulents, and glycophytes. Area and elevation were the only variables that entered the model for both trees and non-succulents, while area alone counted for shrubs (89.2%) and halophytes (76.2%). Distance from nearest large island (Farsan Alkabr) did not affect either the total species richness or any ecological groups.

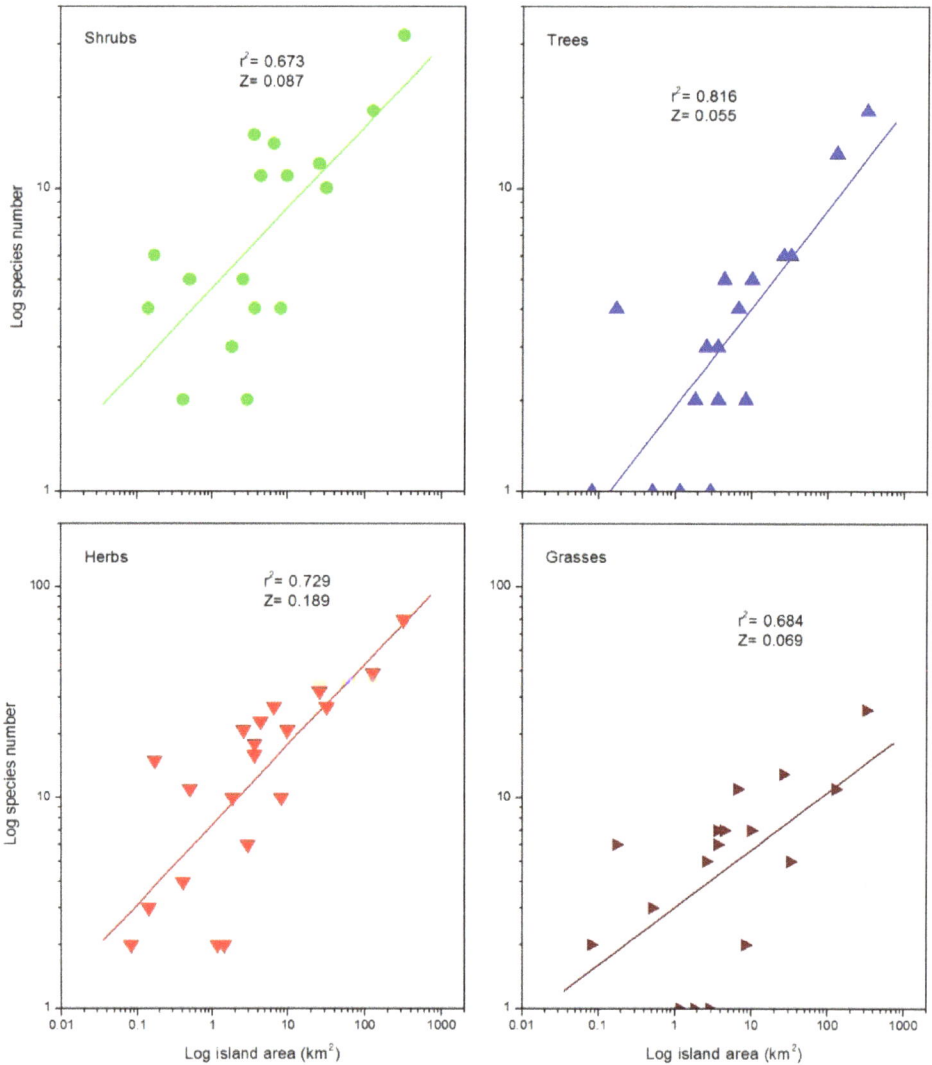

Fig. 6. Relationships of ecological groups (growth forms) with island area of Farasan Archipelago.

Influences of Island Characteristics on Plant Community Structure of Farasan Archipelago, Saudi Arabia: Island Biogeography and Nested Pattern

39

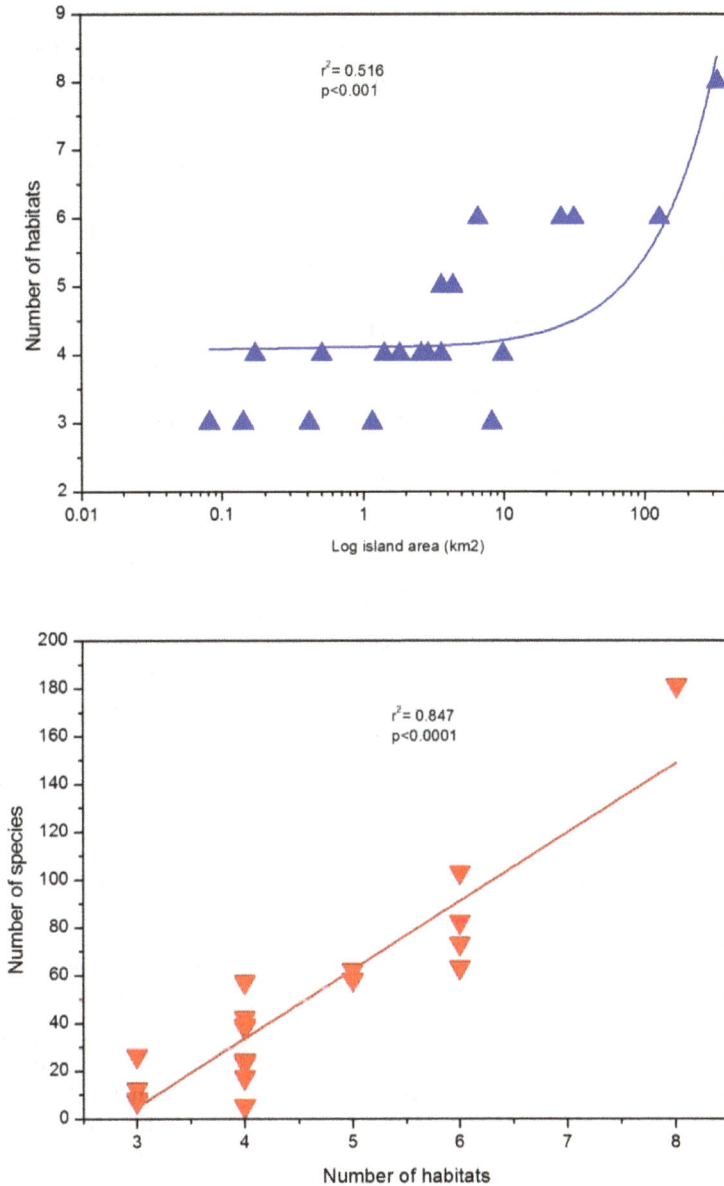

Fig. 7. Relationships of the number of habitats with island area (top) and with the total number of species (bottom) of Farasan Archipelago.

## 3.2 Nestedness pattern

The temperature nestedness calculator detected a high degree of nestedness for the entire flora as well as for each of the ecological subgroups (Table 2). The temperature of the maximally packed matrix ($T_{matrix}$ = 12.87$^0$) for the entire flora was significantly lower than the mean temperature of the random matrices generated by the Monte Carlo-derived null model ($T_{random}$ = 63.06$^0$, P<0.0001). Therefore, the plant communities were significantly nested.

| Data set | Function | Adjusted $R^2$ | $P$-value |
|---|---|---|---|
| All species | S = 0.41 + 4.16 A + 6.55 H | 0.856 | < 0.001 |
| Life span | | | |
| Annuals | S = 6.12 + 4.61A +2.66 H+ 3.52 E | 0.887 | < 0.001 |
| Perennials | S = 7.71 + 8.39 A + 9.85 H + 1.32 E | 0.723 | < 0.001 |
| Growth form | | | |
| Grasses | S = 3.67 + 5.05 A + 8.43 H | 0.849 | < 0.003 |
| Shrubs | S = 4.58 + 3.19 A | 0.892 | < 0.001 |
| Herbs | S = 2.45 + 2.31 H + 1.78 A | 0.715 | < 0.001 |
| Trees | S = 4.28 + 2.35 A + 5.38 E | 0.921 | < 0.000 |
| Succulence | | | |
| Succulents | S = 3.25 + 6.23 A + 2.12 H | 0.733 | < 0.007 |
| Non-succulents | S = 6.22 + 14.12A + 1.45 E | 0.832 | < 0.003 |
| Salt tolerance | | | |
| Halophytes | S = 3.59 + 1.16 A | 0.762 | < 0.016 |
| Glycophytes | S = 7.64 + 4.93 A + 14.73H | 0.899 | < 0.004 |

Table 1. Stepwise linear regressions of total species number and species number by ecological subgroup. Only variables that enter the model are shown, with the total variance explained and the statistical significance of the respective model. S abbreviates to species richness, A to island area, H to number of habitats and E to elevation.

When each ecological group was analyzed separately, the species distributions were significantly nested for all subgroups (Table 2). For the life span subgroups, the mean matrix temperatures for perennials and annuals were 13.36° and 12.69° that significantly different from the mean matrix temperatures of 62.64° and 58.92° generated randomly by Monte Carlo simulations, respectively (Table 2). The life-form distributions were significantly nested for all forms. The mean matrix temperatures were more strongly nested for therophytes, geophytes and chamaephytes with 11.14°, 13.35° and 13.63° compared to random temperatures of 58.44°, 59.87°and 58.92°, respectively (P<0.0001 for all). The mean matrix temperatures of hemicryptophytes, and phanerophytes were 29.48° and 17.27°, respectively. While, their random temperatures recorded 55.48° and 45.18°, respectively. For the salt tolerance subgroups, glycophytes were more nested with a matrix temperature of 13.35° compared to the random temperature of 59.78° generated by Monte Carlo simulations. On the other hand, the matrix temperature of halophytes was 22.33° which significantly different from the random temperature of 61.43°.

The ordered accumulation of species was affected mainly by island area and number of habitats, and to a lesser degree by elevation (Spearman's rank correlation, Table 3). Island area and number of habitats were also correlated for the different ecological groups. This indicates such that species appeared to accumulate in orderly fashion with increasing area and number of habitats. However, isolation was correlated neither to the total species richness nor to the ecological groups.

| Data set | Total number of species | Matrix temperature (°C) | Random temperature (°C) | P (T<T$_{Observed}$) |
|---|---|---|---|---|
| All species | 190 | 12.87 | 63.06 | <0.0001 |
| **Life span** | | | | |
| Perennials | 123 | 13.36 | 62.64 | <0.0001 |
| Annuals | 68 | 12.69 | 58.92 | <0.0001 |
| **Life-forms** | | | | |
| Therophytes | 56 | 11.14 | 58.44 | <0.0001 |
| Geophytes | 19 | 13.35 | 59.87 | <0.0001 |
| Hemicryptophytes | 27 | 29.48 | 55.48 | <0.0001 |
| Chamaephytes | 50 | 13.63 | 58.92 | <0.0001 |
| Phanerophytes | 21 | 17.27 | 45.18 | <0.0001 |
| **Growth forms** | | | | |
| Trees | 19 | 19.47 | 45.82 | <0.0001 |
| Shrubs | 35 | 14.62 | 57.03 | <0.0001 |
| Grasses | 29 | 15.48 | 54.07 | <0.0001 |
| Herbs | 74 | 12.5 | 60.97 | <0.0001 |
| **Succulence** | | | | |
| Succulents | 25 | 16.33 | 58.77 | <0.0001 |
| Non-Succulents | 166 | 12.14 | 61.78 | <0.0001 |
| **Salt tolerance** | | | | |
| Halophytes | 26 | 22.68 | 61.43 | <0.0001 |
| Glycophytes | 165 | 13.35 | 59.78 | <0.0001 |

Table 2. Results of the nestedness analyses as calculated by the nestedness temperature calculator for total plant species and the ecological subgroups.

## 4. Discussion

The equilibrium theory of island biogeography (MacArthur & Wilson, 1967) identifies island size and distance from the mainland as the two most important factors affecting species richness. In the present study, there was no effect of isolation from the largest island (Farasan Alkabir) on total species richness, or on richness of the ecological subgroups. However, all categories of plants increase in richness with island size. This shows that (a) Farasan islands adhere to the species-area relationship; and (b) this relationship exists across ecological groups despite differences in the processes and factors that govern diversity for these groups. It has been suggested that the value of the exponent Z should vary between 0.2 and 0.4 (MacArthur & Wilson, 1967; Rosenzweig, 1995). In the present study, the value of the exponent Z for the total species richness is larger than 0.4. However, this is in agreement with the reported values larger than 0.4 for the exponent Z in several other studies of plants on islands (Rydin & Borgegåd, 1988; Médail & Vidal, 1998; Panitsa et al., 2006; El-Bana, 2009). For example, the Z value of the log-log model for the Mediterranean arid islands is 0.56 (El-Bana, 2009). Rydin & Borgegåd (1988) recorded values varying between 0.36 and 0.56. The strong correlation of species richness with island area, number of habitats and elevation suggests that these quite steep slopes would not be due to the existence of a small island effect (Gentile & Argano, 2005).

| Data set | Area | Number of habitats | Isolation | Elevation |
|---|---|---|---|---|
| All species | 0.84** | 0.65** | -0.28 | 0.49* |
| **Life span** | | | | |
| Perennials | 0.73** | 0.58* | -0.19 | 0.53* |
| Annuals | 0.92** | 0.63** | -0.08 | 0.60** |
| Life-forms | | | | |
| Therophytes | 0.88** | 0.71** | -0.06 | 0.57* |
| Geophytes | 0.79** | 0.63** | -0.22 | 0.47* |
| Hemicryptophytes | 0.68** | 0.53* | -0.3 | 0.38 |
| Chamaephytes | 0.63** | 0.48 | -0.04 | 0.35 |
| Phanerophytes | 0.93** | 0.70** | -0.18 | 0.61* |
| **Growth forms** | | | | |
| Trees | 0.86** | 0.61* | -0.02 | 0.49* |
| Shrubs | 0.64** | 0.55* | -0.22 | 0.22 |
| Grasses | 0.68** | 0.47 | -0.34 | 0.33 |
| Herbs | 0.71** | 0.57* | -0.28 | 0.50* |
| **Succulence** | | | | |
| Succulents | 0.63** | 0.39 | -0.04 | 0.28 |
| Non-Succulents | 0.82** | 0.72** | -0.31 | 0.52* |
| **Salt tolerance** | | | | |
| Halophytes | 0.54* | 0.32 | -0.23 | 0.31 |
| Glycophytes | 0.87** | 0.67** | -0.12 | 0.55* |

* and ** indicate the values are significant at < 0.05 and 0.001, respectively.

Table 3. Spearman's rank correlations between the ranking order of islands in the observed matrix and the islands were ranked by area, number of habitats, isolation and elevation for the entire plant assemblage and their ecological groups.

In the present dataset the division of island flora into different ecological groups revealed that the slopes of the species area regressions are significantly different for each subgroup. For example, the slope of the log S/log A regression of glycophytes growing on the interior rocky and sandy habitats was higher than that of halophytes growing on the shorelines of islands. Similarly, the slope regression of succulents of saline habitats is lower than those of non-sucuulents. A similar pattern has been recognized by other studies of island and islet floras (Rydin & Borgegåd, 1988; Panitsa et al., 2006; El-Bana, 2009). Buckley (1985) divided the floras of small coastal islands on the basis of geographical origin. He found that the slope of log S/log A curves was smallest for the salt flat group growing on the coastlines of the islands (Z= 0.18) and greatest for the sand ridge group (Z= 0.6) which only occurred at the center of each island. Panitsa et al. (2006) found a difference in Z value between halophytes, therophytes, leguminosae and graminae. El-Bana (2009) reported that the slope of log A /log S regression for the halophytes was smaller than that of psammophytes (Z = 0.48 vs. Z= 0.64).

Nestedness appears to be a common phenomenon of insular flora (Kadmon, 1995; Wright et al., 1998; Honnay et al., 1999; Koh et al., 2002). Similarly, the present study detected a high degree of nestedness for the entire flora and for each ecological group. Wright et al. (1998) suggested that four filters operate to screen species occurrence in insular habitats and produce nested biotas. Among these were area and distance effects, passive sampling and

Influences of Island Characteristics on Plant Community Structure of Farasan Archipelago, Saudi Arabia: Island Biogeography and Nested Pattern

43

habitat nestedness. The area filter appears to be the most important in Farasan archipelago. Species-specific resource requirements and differential minimal area requirements result in different patterns of incidence on the islands.

Area- and species-dependent extinction rates have been suggested to play important roles for species richness of oceanic islands (MacArthur & Wilson, 1967), species composition structure (Nekola & White, 1999) and nestedness in land-bridge islands and in habitat fragments (Patterson & Atmar, 1986; Cutler, 1991; Simberloff & Martin, 1991; Wright et al., 1998). Also, differential immigration may be important in producing nestedness (Simberloff & Martin, 1991; Kadmon & Pulliam, 1993). In the current study, there was a lack of several species on smaller but not on larger islands. The reason could be area-dependent extinction and/or differential immigration, and, if so, one or both of these mechanisms may be influencing nestedness in the Farasan archipelago. The largest and the smallest islands surveyed differ in area by 3 orders of magnitude. The large islands are over 319 km² and the small islands <0.5 km² in area. For the entire flora and each ecological group, the distance has no effect on either species richness or nested pattern. This may suggest that the distance is short enough for recurrent colonization (the rescue effect, Brown & Kodric-Brown, 1977), which may affect nestedness (Cook, 1995; Hecnar et al., 2002). Taking into account that most of the recorded species are wind- and bird dispersed species. This dispersal mode with the short distances from the mainland and large island can explain the absence of isolation in the nestedness pattern (Butaye et al., 2001). Therefore, rescue effects (Brown & Kodric-Brown, 1977) and/or intra-island dispersal (King, 1988) may commonly operate but would be masked considering the wide range of areas and low isolation of the islands in the current study.

Habitat nestedness could induce nested structure in species assemblages because certain habitat specialists will be restricted to less common habitats found only on large islands (Wright et al., 1998; Honnay et al., 1999). The habitats among the islands of Farasan are not distributed randomly as the vegetation is characterized by clear zonation from the shorelines to the centre of islands resulting from both chemical and hydrophysical processes (El-Demerdash, 1996). Smaller islands tend to be salty with halophytic vegetation, while larger islands often have a combination of shoreline types (salt marsh, sand formations) and their interiors are usually rocky and have shrubs and trees. Furthermore, the positive and highly significant relationship of island area with number of habitats and elevation indicates that habitats accumulate in an orderly fashion as area increases.

Although all the ecological groups were significantly nested, there were differences in the degrees of nestedness among groups- halophytes and glycophytes, succulents and non-succulents, and plants corresponding to different life-forms. Despite the fact that halophytes and glycophytes share some similarities as xero-halophytic groups, they also have important differences (Danin, 1999). For example, halophytes are relatively more aquatic and tolerant to water logging and salt spray. On the other hand, glycophytes are more terrestrial and tolerant to sand burial (El-Bana et al., 2007). Therefore, it is not surprising that glycophytes were more highly nested than halophytes. This is can be explained by the increased representation of salt habitats in which halophytes tolerate, but which other plants cannot tolerate. Most of the surveyed shorelines of islands are exposed to the effects of seawater, thus sustaining more halophytes. These factors may enable halophytes to dominate the plant communities of shorelines (El-Demerdash, 1996), also taking the fact into account that halophytes are not affected by human disturbance, such as wood cutting and grazing.

Another mechanism which has been suggested for nested pattern is passive sampling whereby, larger islands capture more dispersing individuals than do smaller islands (Lomolino, 1990; Wright et al., 1998), and common species are more likely to be encountered than rare species. In the current study, passive sampling may account for nestedness. The result of the rarefaction suggests larger islands are capturing more richness and diversity of plant species assemblages. Consistent with this is the suggestion that those species most likely to occur on islands already are widely distributed regionally (King, 1988). For example, *Cyperus conglomerates*, *Arthrocnemum macrostachyum*, *Halopeplis perfoliata*, *Limonium axillare*, *Aeluropus lagopoides Zygophyllum coccineum* and *Zygophyllum simplex* have the highest incidence on the islands and they are also the species having the highest incidence on the coast of Saudi Arabia and southern Yemen (El-Demerdash et al., 1994; Hegazy et al., 1998; Kürschner et al., 1998). This suggests that a sampling filter (sensu Cutler, 1994) also may be operating in Farasan archipelago.

As suggested by Wright et al. (1998), many factors act as filters influencing the distribution of species on islands, and this differs by taxon and geographic setting (Atmar & Patterson, 1995). In this particular case, the nestedness of habitats, the tendency of common species to be widely distributed, rare species and habitats to be restricted to large islands and the differences in scale between large and small islands likely contribute jointly to nested pattern in Farasan archipelago.

## 5. Conclusion

In the current study, the high level of nestedness, the strong effect of area on total plant species richness and ecological groups, and the similarity of vegetation composition on the islands has several implications for conservation. First, the large and richest islands in Farasan archipelago such as Farasan Alkabir conserve higher diversity than an equivalent area of several smaller islands. This island also includes rare habitats like coral rocks and rare species. Second, the invasion of the unique habitats such as wadi channels and water catchments in this island by the exotic tree *Prosopis juliflora* should be managed to conserve the native biodiversity. Third, the current anthropogenic expansion on this island should be managed to conserve the existence of the rare habitats such as mangal vegetation where *Avicennia marina* and *Rhizophora mucronata* co-occur. Fourth, the protection of such critical mangal habitat is important on the other large island (e.g. Zufaf), due to its limited distribution in the country (Mandura, 1997; El-Juhany 2009, Zahran 2010).

## 6. Acknowledgment

We would like to thank Prince **Bander Bin Saud Bin Mohammad,Secretary General of the Saudi National Commission for Wildlife Conservation and Development** (NCWCD) for his assistance and access to the facilities at the Farasan Protected Area. The authors extend their appreciation to the Deanship of Scientific Research at King Saud University for funding the work through the research group project No RGP-VPP-031. We are also indebted to anonymous reviewers for fruitful comments on the manuscript. Many thanks are also given to Prof. Dr. A. Assaeed for his kind support and encouragement.

# 7. References

Atmar, W. & Patterson, B.D. (1993). The measure of order and disorder in the distribution of species in fragmented habitat. *Oecologia* Vol. 96, No. 3 (June 1993), pp. 373-382, ISSN 00298549

Atmar, W. & Patterson, B.D. (1995). The Nestedness Temperature Calculator: visual basic program, including 294 presence absence matrices. AICS Research, Inc., University Park, NM and the Field Museum, Chicago

Bergmeier, E. & Dimopoulos, F. (2003). The vegetation of the islets in the Aegean and the relation between the occurrence of the islet specialists, island size, and grazing. *Phytocoenologia*, Vol. 33, No. 2 (January 2003), pp. 447-474, ISSN 0340-269X

Brose, U. (2003). Island biogeography of temporary wetland carabid beetle communities. *Journal of Biogeography*, Vol. 30, No. 6 (June 2003), pp. 879-888, ISSN 03050270

Brown, J.H. & Kodric-Brown, A. (1977). Turnover rates in insular biogeography: effect of immigration on extinction. *Ecology*, Vol. 58, No. 2 (March 1977), pp. 445-449, ISSN 0012-9658

Buckley, R.C. (1985). Distinguishing the effects of area and habitat types on island plant species richness by separating floristic elements and substrate types and controlling for island isolation. *Journal of Biogeography*, Vol. 12, No. 6. (November 1985), pp. 527-535, ISSN 03050270.

Butaye, J., Jacquemyn, H. & Hermy, M. (2001) Differential colonization causing non-random forest plant community structure in a fragmented agricultural landscape. Ecography, Vol. 24, No. 4 (August 2001), pp. 369-380, ISSN 09067590

Chaudhary, S, (2000). *Flora of the Kingdom of Saudi Arabia*, In Ministry of Agriculture & Water, ISBN 9960-18-013-1, Riyadh, KSA

Chaudhary, S. (1989). *Grasses of Saudi Arabia*, In Ministry of Agriculture & Water, ISBN 89-60345 Riyadh, KSA

Collenette, S. (1999). Wildflowers of Saudi Arabia, In National Commission for Wildlife Conservation and Development, ISBN 9960614093, Riyadh, KSA

Colwell, R. K. (2005). EstimateS: Statistical estimation of species richness and shared species from samples. Version 7.5. User's Guide and application, 20.02.2011, Available from http://purl.oclc.org/estimates

Colwell, R.K., Mao, C.X. & Chang, J. (2004). Interpolating,extrapolating, and comparing incidence-based species accumulation curves. *Ecology*, Vol. 85, No. 10 (October 2004), pp. 2717-2727, ISSN 0012-9658

Cook, R.R. (1995) The relationship between nested subsets, habitat subdivision and species diversity. Oecologia, Vol. 101, No. 2 (February 1995), pp. 204-210, ISSN 00298549

Cutler, A. (1991). Nested faunas and extinction in fragmented habitats. Conservation Biology, Vol. 5, No. 4 (December 1991), pp. 496-505, ISSN 08888892

Dabbagh, A., Hotzl, H. & Schnier, H. (1984). Farasan Island, In: *Quanternary Periods in Saudi Arabia*, Jado, A. & Zotl, I. (Eds.) Springer, 212-232, ISBN 10-0387814485, New York, USA

Danin, A. (1999). Desert rocks as plant refugia in the Near East. *Botanical Review*, Vol. 65, No. 2 (April 1999), pp. 93-170, ISSN 00068101

El-Bana, M.I. (2009). Factors affecting the floristic diversity and nestedness in the islets of Lake Bardawil, North Sinai, Egypt: implications for conservation. *Journal of Coastal Conservation*, Vol. 13, No. 1 (March 2009), pp. 25-37, ISSN 1400-0350

El-Bana, M.I., Li, Z.Q. & Nijs, I. (2007). Role of host identity in effects of phytogenic mounds on plant assemblages and species richness on coastal arid dunes. *Journal of Vegetation Science*, Vol. 18, No. 5 (October 2007), pp. 635-644, ISSN 1654-1103

El-Demerdash, M.A. (1996).The vegetation of the Farasan Islands, Red Sea, Saudi Arabia. *Journal of Vegetation Science*, Vol. 7, No. 1 (February 1996), pp. 81-88, ISSN 1654-1103

El-Demerdash, M.A., Hegazy, A.K. & Zilay, A.M. (1994). Distribution of the plant communities in Tihamah coastal plains of Jazan region, Saudi Arabia. *Vegetatio*, Vol. 112, No. 2 (July 1994), pp. 141-151, ISSN 1385-0237

El-Juhany, L. (2009). Present Status and Degradation Trends of Mangrove Forests on the Southern Red Sea Coast of Saudi Arabia. *American-Eurasian Journal of Agricultural and Environmental Sciences*, Vol. 6, No. 3 (February 2009), pp. 328-340, ISSN 1818-6769

Fleishman, E., Donnelly, R., Fay, J. & Reeves, R. (2007). Applications of nestedness analyses to biodiversity conservation in developing landscapes. *Landscape and Urban Planning*, Vol. 81, No. 4 (July 2007), pp. 271–281, ISSN 0169-2046

Gentile, G., Argano, R. (2005). Island biogeography of the Mediterranean Sea: the species relationship for terrestrial isopods. . *Journal of Biogeography*, Vol. 32, No. 10 (October 2005), pp. 1715-1726, ISSN 03050270

Gilbert, F.S. (1980). The equilibrium theory of island biogeography, fact or fiction? *Journal of Biogeography*, Vol. 7, No. 3 (September 1980), pp. 209–235. ISSN 03050270

Heaney, L.R. (2000) Dynamic disequilibrium: a long-term, large-scale perspective on the equilibrium model of island biogeography. *Global Ecology and Biogeography Vol. 9*, No. 1 (January 2000), pp. 59–74, ISSN 1466822X

Heatwole, H. (1991). Factors affecting the number of species of plants on islands of the Great Barrier Reef, Australia. *Journal of Biogeography*, Vol. 18, No. 2 (March 1991), pp. 213–221. ISSN 03050270

Hecnar, S.J., Casper, G.S., Russell, R.W., Hecnar, D.R. & Robinson, J.N. (2002). Nested species assemblages of amphibians and reptiles on islands in the laurentian great lakes. *Journal of Biogeography*, Vol. 29, No. 4 (June 2002), pp. 475–485, ISSN 03050270

Hegazy, A.K., El-Demerdash, M.A. & Hosni, H.A. (1998). Vegetation, speciesdiversity and floristic relations along an altitudinal gradient insouth-west Saudi Arabia. *Journal of Arid Environments*, Vol. 38, No. 1 (January 1998), pp. 3-13, ISSN 0140-1963

Honnay, O., Hermy, M. & Coppin, P. (1999). Nested plant communities in deciduous forest fragments: species relaxation or nested habitats. *Oikos*, Vol. 84, No. 1 (January 1999), pp. 119–129, ISSN 00301299

Johnson, M.P. & Simberloff, D.S. (1974). Environmental determinants of island species numbers in the British Isles. *Journal of Biogeography*, Vol. 1, No. 3 (September 1974), pp. 149–154, ISSN 03050270

Kadmon, R. (1995). Nested species subsets and geographic isolation: acase study. *Ecology*, Vol. 76, No.2 (March 1995), pp. 458–465, ISSN 00129658

Kadmon, R. & Pulliam, H. R. (1993). Island biogeography: effect of geographical isolation on species composition. *Ecology*, Vol. 74, No. 4 (June 1993), pp. 977-981, ISSN: 0012-9658

Khedr, A.A. & Lovett-Doust, J. (2000). Determinants of floristic diversityand vegetation composition on the islands of Lake Burollos,Egypt. *Applied Vegetation Science*, Vol. 3, No.2 (December 2000), pp. 147–156, ISSN 14022001

King, R.B. (1988). Biogeography of reptiles on islands in Lake Erie, In: *The biogeography of the islands region of western Lake Erie*, J.F. Downhower, (Ed.), 125–133, Ohio State University Press, ISBN 0814204481, Columbus, Ohio

Koh, L.P., Sodhi, N.S., Tan, H.T.W. & Peh, K.S.H. (2002). Factors affecting the distribution of vascular plants, springtails, butterflies and birds on small tropical islands. *Journal of Biogeography*, Vol. 29, No. 1 (January 2002), pp. 93–108, ISSN 03050270

Kohn, D.D. & Walsh, D.M. (1994). Plant species richness-the effect of island size and habitat diversity. Journal of Ecology, Vol. 82, No 2, (June 1994) pp. 367–377, ISSN 00220477

Kürschner, H., Al-Gifri, A. N., Al-Subai, M. Y. & Rowaished, A. K. (1998). Vegetational patterns within coastal salines in southern Yemen. *Feddes Repertorium*, Vol. 109, No. 1/2 (April 1998), pp. 147-159, ISSN: 1522-239X

Lomolino, M. V. (1990). The Target Area Hypothesis: The Influence of Island Area on Immigration Rates of Non-Volant Mammals. *Oikos*, Vol. 57, No. 3 (April 1990), pp. 297-300, ISSN: 00301299

Lomolino, M.V. (1994) Species richness patterns of mammals inhabiting nearshore archipelagoes: area, isolation, and immigration filters. *Journal of Mammalogy*, Vol. 75, No. 1 (February 1994), pp. 39–49, ISSN 00222372

Lomolino, M.V. (2000a) A call for a new paradigm of island biogeography. *Global Ecology and Biogeography*, Vol. 9, No. 1 (January 2000), pp. 1–6, ISSN 1466-8238

Lomolino, M.V. (2000b) A species-based theory of insular zoogeography. *Global Ecology and Biogeography*, Vol. 9, No. 1 (January 2000), pp. 39-58, ISSN 1466-8238

MacArthur, R.H. & Wilson, E.O. (1967). *The theory of island biogeography*, In Princeton University Press, ISBN: 9780691088365, Princeton, USA

Mandura, A. S. (1997). A mangrove stand under sewage pollution stress: Red Sea. *Mangroves and salt Marshes*, Vol. 1, No. 4 (March 1997), pp. 255-262, ISSN 1386-3509

Maron, J.L., Vil, M., Bommarco, R., Elmendorf, S. & Beardsley, P. (2004). Rapid evolution of an invasive plant. *Ecological Monographs*, Vol. 74, No. 2 (May 2004), pp. 261–280, ISSN 00129615

Masseti, M. (2010). The mammals of the Farasan archipelago, Saudi Arabia. Turkish Journal of Zoolgy, Vol. 34, No. 3 (July 2010), pp. 359-365, ISSN 1300-0179

Médail, F. & Vidal, E. (1998). Organisation de la richesse et de la composition floristiques d'îles de Méditerranée occidentale (S.E. France). *Canadian Journal of Botany*, Vol. 76, No. 2 (February 1998), pp 321-331, ISSN 1916-2790

Morand, S. (2000). Geographic distance and the role of island area and habitat diversity in the species–area relationships of four Lesser Antillean faunal groups: acomplementary note to Ricklefs & Lovette. *Journal of Animal Ecology*, Vol. 69, No. 6 (December 2000), pp. 1117–1119, ISSN 1365-2656

Nekola J. C. & White P. S. (1999). The distance decay of similarity in biogeography and ecology. *Journal of Biogeography*, Vol. 26, No. 4 (July 1999), pp. 867–878, ISSN 03050270

Osborne, P.L. (2000). *Tropical ecosystems and ecological concepts*, In Cambridge University Press, ISBN 10-0521645239, Cambridge, UK

Panitsa, M. & Tzanoudakis, D. (1998). Contribution to the study of the Greek flora: flora and vegetation of the E Aegean islands Agathonisi and Pharmakonisi. *Wildenowia*, Vol. 28, No. 1 (December 1998), pp 95–116, ISSN 05119618

Panitsa, M. & Tzanoudakis, D. (2001) A floristic investigation of the islet groups Arki and Lipsi (East Aegean Area, Greece). *Folia Geobotanica,* Vol. 36, No. 3 (June 2001), pp 265-279, ISSN 12119520

Panitsa, M., Tzanoudakis, D., Triantis, K.A. & Sfenthourakis, S. (2006). Patterns of species richness on very small islands: the plants of the Aegean archipelago. *Journal of Biogeography,* Vol. 33, No. 7 (July 2006), pp. 1223-1234, ISSN 03050270

Patterson, B.D. (1987). The principle of nested subsets and its implications for biological conservation. *Conservation Biology,* Vol. 1, No. 4 (December 1987), pp. 323-334, ISSN 08888892

Patterson, B.D. (1990). On the temporal development of nested subsets patterns of species composition. *Oikos,* Vol. 59, No. 3 (December 1990), pp. 330-342, ISSN 00301299

Patterson, B.D. & Atmar, W. (1986). Nested subsets and the structure of insular mammalian faunas and archipelagos. *Biological Journal of the Linnean Society,* Vol. 28, No 1 (May 1986), pp. 65-82, ISSN 1095-8312

Patterson, B.D. & J.H. Brown. 1991. Regionally nested patterns of species composition in granivorous rodent assemblages. *Journal of Biogeography,* Vol. 18, No. 4 (July 1991), pp. 395-402, ISSN 03050270

Rafe, R.W., Usher, M.B. & Jefferson, R.G. (1985). Birds on reserves: the influence of area and habitat on species richness. *Journal of Applied Ecology,* Vol. 22, No. 2 (August 1985), pp. 327-335, ISSN 00218901

Raunkiaer, C. (1934). *The life forms of plants and statistical plant geography,* In Oxford University Press, ISBN 0-405-10418-9, Oxford, UK

Rosenzweig, M.L. (1995). *Species Diversity in Space and Time,* In Cambridge University Press, ISBN 0-521-499952-6, Cambridge, UK

Rydin, H. & Borgegård, S.O. (1988). Plant species richness on islands over a century of primary succession in Lake Hjälmaren. Ecology, Vol. 69, No. 4 (August 1988), pp. 916–927, ISSN 0012-9658

Sfenthourakis, S. (1996). A biogeographic analysis of terrestrial isopods (Isopoda, Oniscidea) from central Aegean islands (Greece). *Journal of Biogeography,* Vol. 23, No. 5. (September 1996), pp. 687–698, ISSN 03050270

Simberloff, D. & Martin, J.L. (1991). Nestedness of insular avifaunas:simple summary statistics masking complex species patterns. Ornis Fennica, Vol. 68, No. 4 (June 1991), pp. 178-192, ISSN 00305685

Turchi, G. M., Kennedy, P. L. , Urban, D. &Hein, D. (1995). Bird species richness in relation to isolation of aspen habitats. *Wilson Bulletin,* 107, No. 3 (June 1995), pp. 463-474, ISSN 00435643

Whittaker, R.J. (2000). Scale, succession and complexity in island biogeography: are we asking the right questions? *Global Ecology and Biogeography,* Vol. 9, No. 1 (January 2000), pp. 75–85, ISSN 1466822X

Wright, D.H. (1983). Species–energy theory: an extension of species–area theory. *Oikos,* Vol. 41, No. 3 (December 1983), pp. 496–506, ISSN 00301299

Wright, D.H., Patterson, B.D., Mikkelson, G.M., Cutler, A. & Atmar, W. (1998) A comparative analysis of nested subset patterns of species composition. Oecologia, Vol. 113, No. 1 (June 1998), pp. 1–20, ISSN 00298549

Zahran, M.A. (2010). *Climate-vegetation: Afro-Asian Mediterranean and Red Sea Coastal Lands,* In Springer, ISBN 978-90-481-8594-8, London, UK

# 3

# Passive Long-Distance Migration of Apterous Dryinid Wasps Parasitizing Rice Planthoppers

Toshiharu Mita[1], Yukiko Matsumoto[2],
Sachiyo Sanada-Morimura[3] and Masaya Matsumura[3]
*[1]Laboratory of Entomology, Faculty of Agriculture,*
*Tokyo University of Agriculture, Atsugi,*
*[2]National Institute of Agrobiological Sciences, Owashi, Tsukuba,*
*[3]National Agricultural Research Center for Kyushu Okinawa Region, Koshi,*
*Japan*

## 1. Introduction

The wasp family Dryinidae comprises predator and parasitoid wasps of leaf- and planthoppers (Hemiptera: Auchenorrhyncha). This family is morphologically distinct from other wasps. Females in most subfamilies of Dryinidae have forelegs that are modified into a chela, with an enlarged claw (Fig. 1: green) and 5th tarsomere (Fig. 1: red) that aid in grasping the host insect. The enlarged claw moves widely when the chela opens. Such foreleg morphology is not always the case, such as for females of the subfamily Aphelopinae R.C.L. Perkins and Erwiniinae Olmi & Guglielmino (Olmi & Guglielmino, 2010) that have simple forelegs. Dryinid wasps often show distinctive sexual dimorphism, such as the presence or absence of chela. The subfamily Gonatopodinae Kieffer is one of the extreme cases of sexual dimorphism, but in this case it is because females of most of the species are apterous. The pterothorax becomes so slender in the apterous form (Fig. 2: A, C) that they look like ants. In contrast, males have well developed wings and a pterothorax (Fig. 1: B).

Since apterous females cannot disperse very far by themselves, the current distribution of these wasps should have been caused by historical events, such as terrestrial immigration and local extinction. Immigration over large distances, especially across geographical barriers (e.g. open water) seems unlikely. Recent studies revealed the annual long-distance migration of rice planthoppers, which are host to apterous dryinids. If rice planthoppers carry dryinid larvae to distant locations, immigrant individuals of the Dryinidae should be recognizable from the destination locality. This behavior can be viewed as passive long-distance migration. Confirming the presence or absence of such behavior and determining the degree of influence that this has on a population at the destination locality will broaden our understanding on the current biogeographical distribution of the Dryinidae. In this chapter, we introduce a brief summary about apterous species of *Haplogonatopous* R.C.L. Perkins, which are common in Asian rice paddies. We then describe the life cycle of one of these species that inhabits other vegetation, discuss the possible mechanism for settling in a new environment, and show an important insight indicated from our recent phylogeographical approach using two species of *Haplogonatopus*.

## 2. Common species of *Haplogonatopus* in Asian rice paddies

This section briefly summarizes previous studies about *Haplogonatopus* species that parasitize rice planthoppers in monsoon Asia. There are two common species in rice paddies. The present understanding of the morphological differences, host ranges, and distribution of the two species are described.

### 2.1 Taxonomy

Like many other taxa of the subfamily Gonatopodinae, females of *Haplogonatopus* are wingless. The small palpal formula, 2/1 is one of the most distinctive apomorphies of this genus. The presence of "rhinarium" on the antennae, the large subapical tooth and a single row of lamellae on the enlarged claw, and a single spur on the hind tibia are also diagnostic characters for the female (Olmi, 1984, 1999). However, the combination of those latter character states also occurs within *Gonatopus* Ljungh. Among the four species of *Haplogonatopus* distributed in the Palaearctic and Oriental regions, *H. apicalis* R.C.L. Perkins, 1905 and *H. oratrius* (Westwood, 1833) are common in Asian rice paddies. Their appearance is very similar to each other, but they are morphologically distinguishable by the female coloration and male genitalia (Fig. 2).

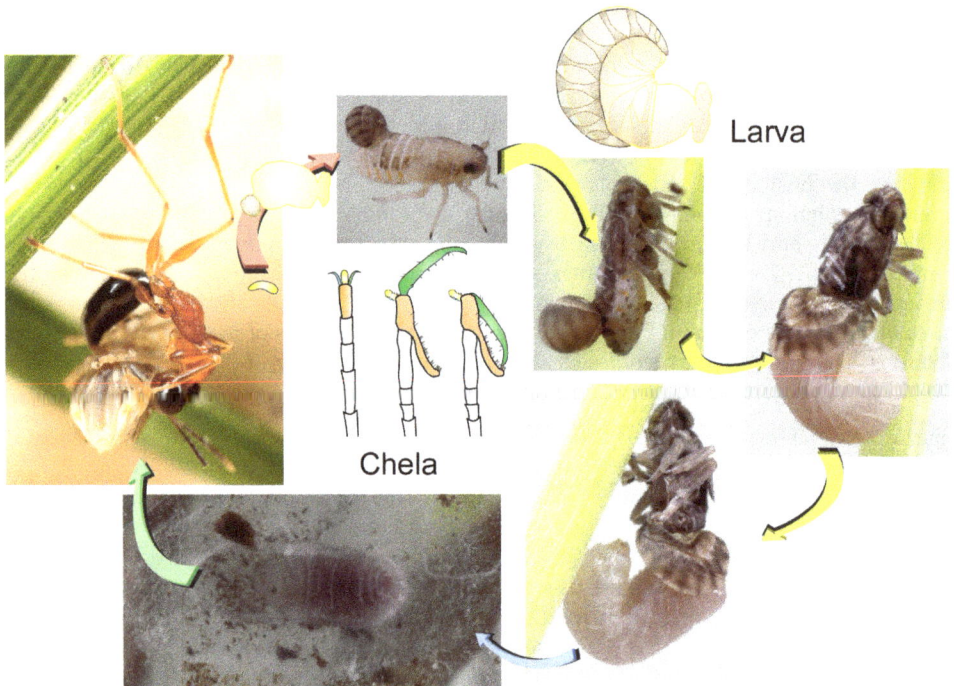

Fig. 1. Life cycle of *Haplogonatopus oratrius*.

### 2.2 Biology

Although the biology of most dryinid species is not well studied, species parasitizing rice planthopper are relatively well understood. *H. apicalis* attacks the white-backed planthopper,

*Sogatella furcifera* (Horváth, 1899), and *H. oratrius* attacks the small brown planthopper, *Laodelphax striatellus* (Fallén, 1826) (Fig. 1). The host preference shows an important contrast. *H. apicalis* only parasitizes *S. furcifera* under natural conditions; although there are other host records (Guglielmino & Olmi, 1997), they are apparently exceptional. *H. oratrius* parasitizes many species (Guglielmino & Olmi, 1997), although the dominant host species in Asia is *L. striatellus*. The large biomass of *L. striatellus* could be advantageous for *H. oratrius*, as the wasp hibernates as a larva in the nymphs of this planthopper (Nishioka, 1980; Kitamura, 1987). Nishioka (1980) reported that a female *H. apicalis* was reared from an overwintering nymph of *L. striatellus* in Kochi Prefecture, southern Shikoku. However, despite an intensive field survey, parasitism of *H. apicalis* on *L. striatellus* has not been observed in Shimane, western Honshu (Kitamura, 1987, 1989; Kitamura & Nishikata, 1987). In view of the above, the hibernation of *H. apicalis* in *L. striatellus* is unlikely to be common in the temperate region.

The general life cycle of *Haplogonatopus* can be summarized as follows (Fig. 1). An adult female captures a host using its chela (= distal apex of modified foreleg) and oviposits an egg into the posterior part of abdomen. The larva develops in the abdomen of the host. The mature larva consumes the entire content of the host, and then emerges from the host's body. The larva spins a cocoon on plant tissue.

Fig. 2. General habitus and male genital organs of *Haplogonatopus* that are common in rice paddies. A *H. oratrius*, female; B ditto, male; C *H. apicalis*, female; D male genitalia, *H. oratrius* (left), *H. apicalis* (right). Scale = 1.0 mm (A–C), 0.1 mm (D).

## 2.3 Distribution

The distribution of the two *Haplogonatopus* species overlaps with the main distribution of their dominant hosts. *H. oratrius* is common from Europe to the temperate region of Asia, including Taiwan. Although there is no record from South East Asia, they have been recognized from the Mariana Islands (Olmi, 1999). Recently, they were found in the oceanic and tropical Bonin Islands (Mita, unpublished data). Many organisms have been accidentally introduced to new regions by human activity, and the distribution of *H. oratrius* needs additional evaluation. *H. apicalis* was originally described from Northern Australia. However, it is rather common in Monsoon Asia. There is no record from oceanic islands, but its host, *S. furcifera*, is distributed in Fiji, Micronesia and New Hebrides (Asche & Wilson, 1990).

## 3. Seasonal host shift of *Haplogonatopus oratrius*

Adult *L. striatellus* often moves a short distance from a rice paddy, but the destination locality may not be always suitable. This section will show recent results of field research conducted on vegetation unsuitable for *L. striatellus*.

As it is too cold for winter rice cropping in Japan, *L. striatellus* needs to hibernate on other vegetation. They can be found at paddy-side levees occupied with grasses. They may also be found in dry river beds occupied by *Eragrostis curvula* (Schrad.) Nees in autumn. Another planthopper, *Hosunka hakonensis* (Matsumura, 1935) is the predominant species in the bush of *E. curvula*. They also successfully reproduce and thrive on *E. curvula*. However, *L. striatellus* do not appear to reproduce on this plant as nymphs are not found on it. Figure 3 shows the seasonal fluctuation of *H. hakonensis* collected from a colony of *E. curvula* at Sagami-gawa River (Kanagawa, Central Honshu) by net-sweeping. The field research was conducted four times a month from April to October on 2006. The sweeping-netting was undertaken for 20 minutes each sampling. Nymphs were collected from April to September but they were also found in winter (data not shown). Conversely, all collected individuals of *L. striatellus* collected from *E. curvula* were adults and nothing was collected from April to June (Fig. 4). *H. oratrius* and *Gonatopus dromedarius* (A. Costa, 1882) inhabit the same environment. When nymphs parasitized by Dryinidae were collected, they were kept in a plastic tube with a grass leaf until the dryinid larvae emerged. The identification of dryinid species was undertaken on adults or, when they did not become adults, by coloration of the larval sac. The two dryinid species are easily identified by the color of the larval sac: in *H. oratrius* it is ash gray (Fig. 1), whereas in *G. dromedarius* it is black. *H. hakonensis* were found to be parasitized by both dryinid species (Figs. 3, 5). The rate of parasitism was highest in June (Fig. 5), and most dryinid species parasitizing *H. hakonensis* in June were *G. dromedarius* (*H. hakonensis*: *L. striatus* = 34: 11). Because parasitism of overwintering nymphs by both dryinid species was observed in April, *H. hakonensis* is regarded as a winter host.

The seasonal fluctuation of *L. striatellus* at Sagami-gawa River is similar to that of a paddy-side levee (Figs. 4, 6), except that the latter contains adults and nymphs (data not shown). At Sagami-gawa River, parasitism was observed in September (Fig. 5). All of the dyrinids were *H. oratrius* except for one female *G. dromedarius* (*L. striatus*: *H. hakonensis* = 23: 1). The parasitism of *G. dromedarius* toward *L. striatellus* appeared to be highly restricted. In a paddy-side levee at Atsugi, *L. striatellus* was not collected from April through May (Fig. 6). However, this result may have resulted from low density. Kitamura (1989) reported the presence of overwintering nymphs and a relatively high ratio of parasitism (9–34%) by *H. oratrius* in a similar environment from winter to spring in Shimane.

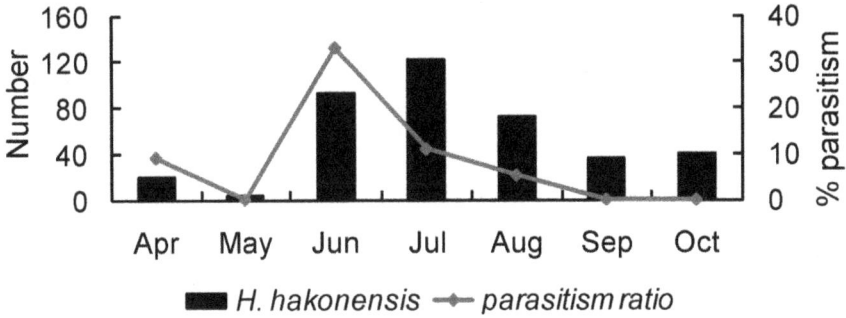

Fig. 3. Seasonal fluctuation of *Hosunka hakonensis* and parasitism by Dryinidae at Sagami-gawa River in 2006.

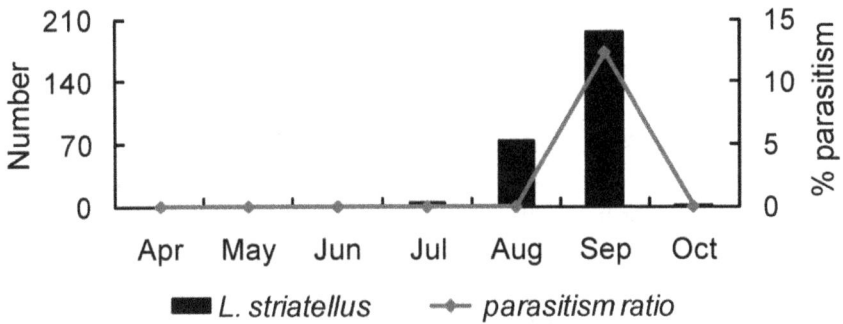

Fig. 4. Seasonal fluctuation of *Laodelphax striatellus* and parasitism by *Haplogonatopus oratrius* at Sagami-gawa River in 2006.

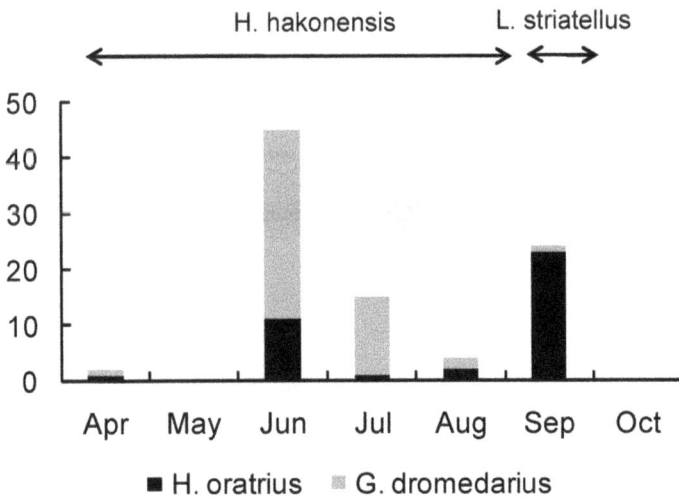

Fig. 5. Seasonal fluctuation of dryinid wasps that emerged from *Hosunka hakonensis* and *Laodelphax striatellus* at Sagami-gawa River in 2006.

The abundance of parasitizing dryinid individuals indicates that *G. dromedarius* had one population peak in June in 2006. Although the host species was different from that of *G. dromedarius*, *H. oratrius* displays two population peaks, one each in June and September (Fig. 5). Kitamura's (1983) developmental model based on physiology and ambient temperature indicated that *H. oratrius* can have five generations a year in Matsue, Shimane Prefecture. As the climate in Kanagawa is similar to that of Shimane, the two observed population peaks are considered to be caused by a lack of host resource abundance (Figs. 3, 4).

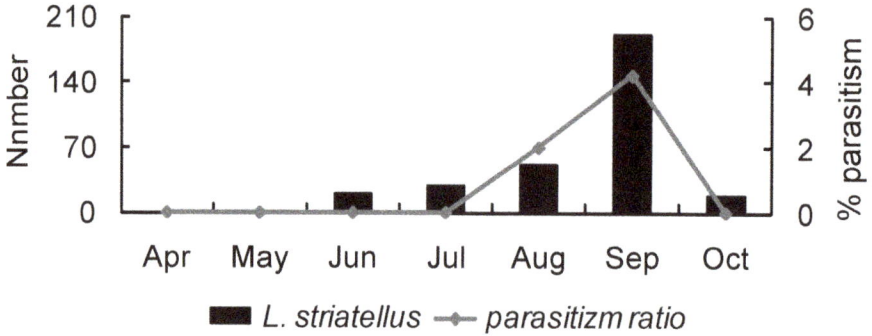

Fig. 6. Seasonal fluctuation of *Laodelphax striatellus* and parasitism by *Haplogonatopus oratrius* at a paddy-side levee in Atsugi in 2006.

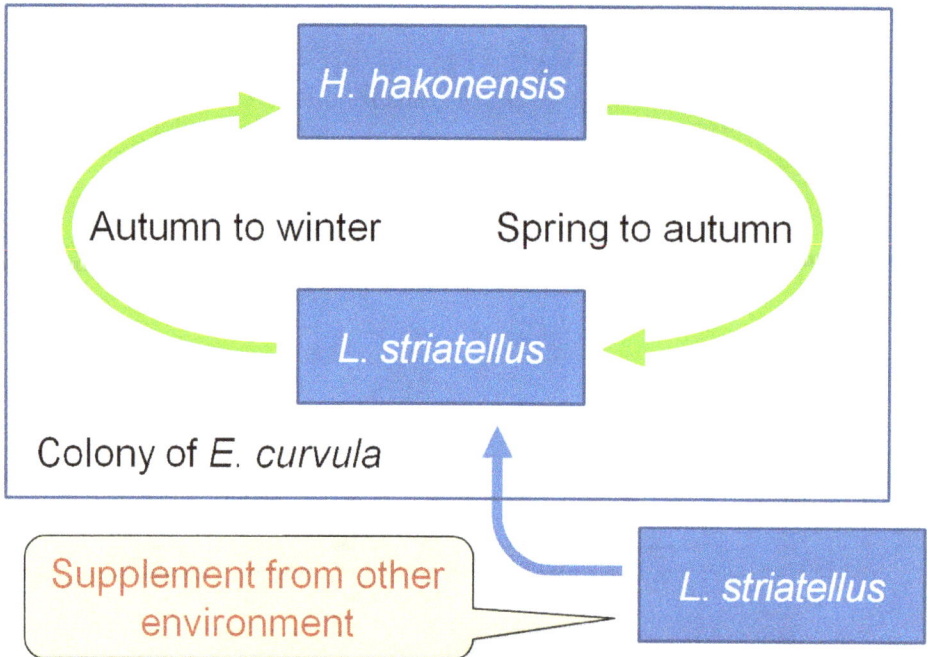

Fig. 7. Host-shift cycle of *Haplogonatopus oratrius* occurring on a colony of *Eragrostis curvula*. *Hosunka hakonensis* hibernates on *E. curvula* whereas *Laodelphax striatellus* dies in winter.

Adults of immigrant *L. striatellus* could be already parasitized by *H. oratrius*, as indicated in Fig. 6. Therefore, *H. oratrius* both the original inhabitants and immigrants, coexist in the same locality during autumn. In winter, overwintering nymphs of *H. hakonensis* are considered to be parasitized by both species. The seasonal host shift and life cycle of *H. oratrius* at Sagami-gawa River are summarized in Fig. 7. Although *H. oratrius* parasitizes only *L. striatellus* and *H. hakonensis* at Sagami-gawa River, many planthopper species can be parasitized by *H. oratrius* (Guglielmino & Olmi, 1997). We conclude that the relatively wide host range allows *H. oratrius* to settle in new environments when the parasitized hosts relocate.

## 4. Long-distance migration of rice planthoppers

*S. furcifera* and *L. striatellus*, together with the brown planthopper, *Nilaparvata lugens* (Stål, 1854) are the dominant planthopper pests of the rice plant in Asia. They are known as long-distance migrant insects (Kisimoto, 1975). Their migration ability in Asia has been demonstrated by temporal biotype changes (Sogawa, 1992, 1993) and migration analysis (Otuka et al., 2005a, 2005b, 2008, 2010).

*S. furcifera* is a tropical to subtropical species. Those found in Japan in the rainy season (June to July) originated from northern Vietnam via southern China (Sogawa, 1993; Otuka et al., 2008). As *S. furcifera* cannot hibernate in temperate regions, including Japan and Korea, it is thought that they immigrate from overseas each year. Conversely, *L. striatellus* is a temperate species. It is distributed mainly in the temperate to subarctic regions of Asia to Europe, but also in some subtropic or tropic regions (Taiwan, southern China, the high altitude areas of Southeast Asia). They hibernate as fourth instar nymphs (Kisimoto, 1957). Planthoppers captured at a weather ship on the East China Sea include many *L. striatellus* (Kisimoto, 1983), but this species is not distributed in the lower altitude areas of the tropic region, as are the other two planthoppers. Recently, a mass migration of *L. striatellus* from around Jiangsu Province of China to Kyushu, Japan, was strongly indicated by trap catches and source estimation using backward trajectory analysis (Otuka et al., 2010).

## 5. Parasitized rice planthoppers collected over the ocean

Several authors have reported that the parasitism of dryinid larval sacs and stylops on rice planthoppers was probably a direct result of migration from overseas. Kisimoto (1975) reported parasitized *S. furcifera* and *N. lugens* collected by a monitoring net trap for migrant insects. Kitamura & Nishikata (1987) reported the seasonal changes of parasitism ratio on leaf- and planthoppers probably migrated from China. The above two observations were conducted on land. The parasitizing dryinid larvae and stylops were also recognized from individuals collected over the ocean. In 1967 and 1968, a mass flight of *S. furcifera* and *N. lugens* was observed around the weather ship, "Ojika" in the Pacific Ocean (N29°, E135°) (Kisimoto, 1983; Kisimoto & Sogawa, 1995). Three females of *H. apicalis* were reared from *S. furcifera* collected from the ship in 1968. These species have also been found over the East China Sea. *S. furcifera* parasitized by stylops and dryinids were collected on the weather ship, "Keifu-maru", on the East China Sea (N31°, E126°) in 1984 (Noda, 1986). The stylops was identified as *Elenchus japonicus* (Esaki & Hashimoto, 1931) (Kifune & Maeta, 1986). The dryinid species was not identified, but was likely *H. apicalis*.

## 6. Genetic variation among East Asian populations

The phylogeographical analysis was conducted using the two species of *Haplogonatopus* based on 807-bp mitochondrial COI sequence data obtained from many localities in East Asia (Fig. 8). The monophyly of the species was strongly supported by cladistic analysis. In the parsimonious network of *H. oratrius*, a combination of three star-like core haplotype groups (Fig. 9, A–C) was indicated. The largest haplotype, group A, was composed from China, Taiwan and Kyushu elements. Haplotype group C was regarded as the Japanese endemic population. Haplotype group B was composed from mainly Japanese elements, but a few foreign elements were also included. The network structure was moderately equivalent to the geographical distribution. On the other hand, the network of *H. apicalis* is complicated. A single star-like structure and two circular structures were recognized. Most of the others show multimodal distribution and the structure is very different from the geographical distribution. Compared to *H. oratrius*, *H. apicalis* has high genetic variation over all localities sampled. However, they could not been isolated by geographical distribution. This could imply that all sampling sites should be regarded as the same population, which would be consistent with the annual long-distance migration of *S. furcifera*. However, *S. furcifera* showed no genetic variation among the East Asian populations sampled (Mun et al., 1999). Because the present study investigated the original and the secondary migration direction, the similar genetic structure of *H. apicalis* might be retrieved from the other localities in East

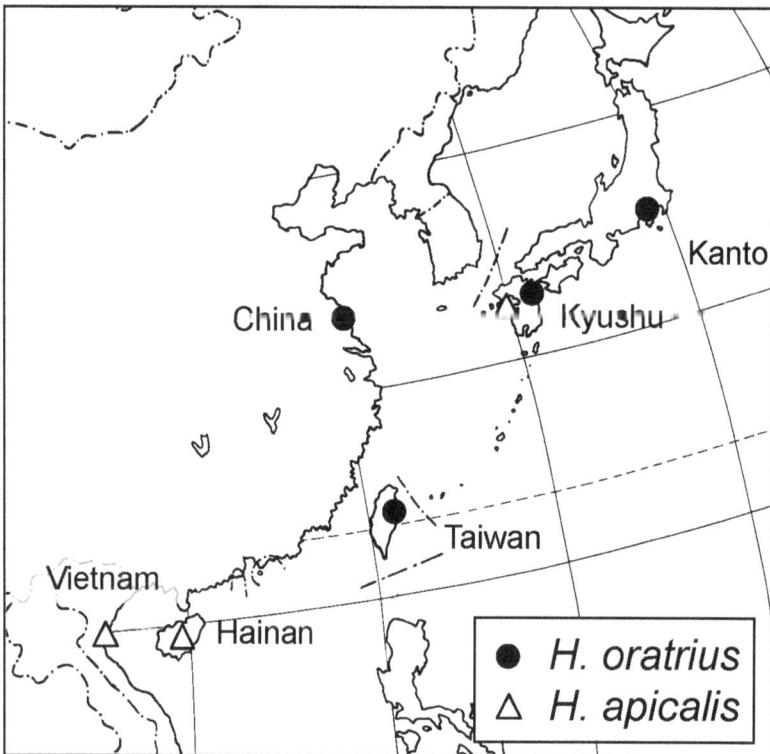

Fig. 8. Sampling localities.

Asia. However, East Asia is only part of the distribution. The genetic structure in South East Asia and South Asia is disputed.

The haplotype network of *H. oratrius* moderately reflects geographical distances. This pattern is similar to *E. japonicus* (Matsumoto et al., 2011). The stylops can parasitize all three rice planthoppers and other species. Furthermore, host preference is considered to differ with region (Maeta et al., 2007; Chandra, 1980). Conversely, *H. oratrius* mostly relies on *L. striatellus* as a host resource, and thus the geographical difference in host preference, excluding *L. striatellus*, could be a minor effect. The genetic structure of the Kyushu population sampled was intermediate between the Kanto and Taiwan-China populations. Haplotypes are seemingly composed of two elements: the more-variable part directly connected with Kanto, and the less-variable part connected with Taiwan and China. The natural populations of Kyushu, Taiwan and China probably share many haplotype components. However, the dominant haplotypes among the three populations are observed in haplotype group A (Fig. 9). This indicates a current large gene flow caused by the migration of parasitized *L. striatellus*. The haplotype network of Taiwan and China is somewhat simpler than the Japanese populations. They are geographically distant, but genetically indistinguishable. The current migration of *L. striatellus* between Taiwan and China should be tested. Otherwise, it is possible that *H. oratrius* has experienced past fragmentation.

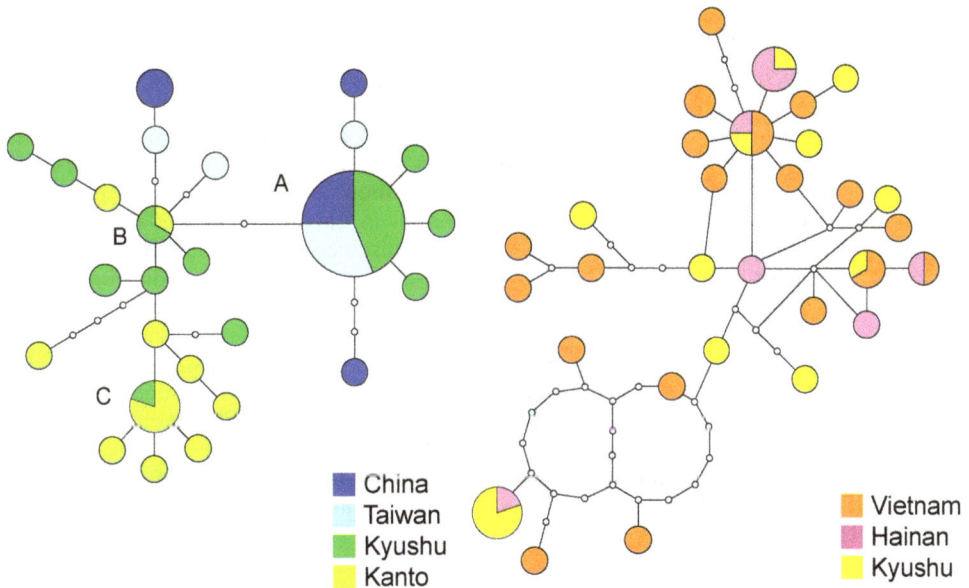

Fig. 9. Parsimonious network of two species of *Haplogonatopus* (left: *H. oratrius*, right: *H. apicalis*). Each circle indicates a different haplotype. The frequency of a geographical population is indicated by a different color.

## 7. Discussion

Host range and hibernation ability contrasts between *H. apicalis* and *H. oratrius*. However, both species have long-distance passive migration ability in association with their hosts. The hibernation ability of their main hosts and the host-shifting ability of dryinids at the destination locality may influence their genetic structure. According to the Palaearctic distribution of *L. striatellus* and their biology, *H. oratrius* could have been distributed widely without long-distance migration of the host. The overseas mass migration of *L. striatellus* demonstrated by Otuka et al. (2010) and genetic structure of *H. oratrius* suggests the presence of current gene flow among domestic populations of both species in China, Taiwan, and Kyushu. *H. apicalis* migrates into northern temperate regions as larvae parasitizing *S. furcifera*. This host probably becomes extinct each winter (Kitamura, 1987). Before the origin of rice culture in monsoonal Asia, heavy outbreaks of rice planthoppers should have occurred rarely. Gene flow within *H. apicalis* may have been accelerated by human activity. Another important dryinid species is *G. fluvifermur* (Esaki & Hashimoto, 1935). It parasitizes *N. lugens* in rice paddies. The observation of insecticide resistance and virulence to resistant rice varieties implies that the East Asian population of *N. lugens* has different traits from the South East Asia population (Sogawa, 1992, Matsumura et al., 2008, unpublished data). During field research at Sagami-gawa River, some females of *G. fluvifemur* were reared from other delphacid species collected on reed bush (*Phragmites australis* (Cav.) Trin. ex Steud.) habitat and grasslands in Kanto, central Honshu (Mita, unpublished data). Because few individuals of *N. lugens* reach there, these dryinids might be considered to adapt to other delphacids and establish in the new locality. It is important to compare many traits of different species for further discussion on the significance of passive migration (Mita et al., in preparation).

Dryinid wasps, including apterous taxa like *Haplogonatopus*, can be passively transported to distant localities by their hosts. This is similar to the dispersal capability of "aerial plankton" arthropods (Richter, 1970; Bowman et al., 1978; Mound, 1983). Insect migration is often considered to be an active behavior (Drake et al., 1995); however, passive aerial dispersal of minute apterous arthropods has been reported (Washburn & Washburn, 1984; Jung and Croft, 2001). Such species may actively launch themselves into air drafts, and are passively carried by the wind. On the other hand, the long-distance dispersal of dryinid wasps completely depends on hosts' activity. The long-distance migration of dryinid wasps together with stylops is an interesting example of passive behavior. The concept of "passive migration" is an important element in the historical biogeography of the Dryinidae. In this chapter, we reported the possibility of a local host shift by *H. oratrius* caused by the immigration of *L. striatellus* from elsewhere. So far, *H. apicalis* have not adapted to their planthopper host's destination localities. However, the expansion of the distribution caused by host dispersal is a highly probable event in certain taxa, perhaps together with host change. Consequently subsequent allopatric speciation and/or secondary contact might have occurred. The long-distance migration of their host has greatly influenced the distribution of the Dryinidae in monsoon Asia, not only in the past, but also as a progressive event.

## 8. Acknowledgments

The figure plate of the life cycle of *H. oratrius* was kindly provided by Y. Tanaka. The field research at Sagami-gawa River was conducted in collaboration with R. Watanabe and M. Oishi. We thank the following individuals for their assistance in the molecular experiment: M. Maruyama (Kyushu University Museum), F. Ryu (Kyushu University), Y. Ando, and C. Horie. We are also much indebted to the following individuals for their support on the material and assistance in field surveys: S. Yoshimatsu (National Institute of Agro-Environmental Sciences); S. Okajima (Tokyo University of Agriculture); A. Sakai (ditto); K.

Watanabe (Kobe University); S. Shobu (Saga Prefectural Agriculture Research Center); R. Otsu (Nagasaki Prefectural Government, Plant Protection Office); M. Kajisa (Miyazaki Plant Protection and Fertilizer Inspection); and H. Inoue (Kagoshima Prefectural Institute for Agricultural Development). This study is supported by KAKENHI 20-4137 and 22880035.

## 9. References

Asche, M. & Wilson, M. (1990) The delphacid genus *Sogatella* and related groups: a revision with special reference to rice-associated species (Homoptera: Fulgoroidea). *Sys. Entmol.*, 15, 1-42

Bowman, J., Cappuccino, N. & Fahrig, L. (2002) Patch size and population density: the effect of immigration behavior. *Conserv. Ecol.*, 6, 9

Chandra, G. (1980) Taxonomy and bionomics of the insect parasites of rice leafhoppers and planthoppers in the Philippines and their importance natural biological control. *Philipp. Entomol.*, 4, 119-139

Drake, V., Gatehouse, A. & Farrow, R. (1995) Insect migration: a holistic conceptual model, In: *Insect Migration: Tracking resources through space and time*, V. Drake & A. Gatehouse (Eds.), 427-459, Cambridge University Press, UK.

Guglielmino, A. & Olmi, M. (1997) A host-parasite catalog of world Dryinidae (Hymenoptera: Chrysidoidea). *Cont. Entomol. Internat.*, 2(2), 165-298

Jung, C. & Croft, B. (2001) Aerial dispersal of phytoseiid mites (Acari: Phytoseiidae): estimating falling speed and dispersal distance of adult females. *Oikos*, 94, 182-190

Kennedy, J. (1961) A turning point in the study of insect migration. *Nature*, 189, 785-791

Kifune, T. & Maeta, Y. (1986) New host records of *Elenchus japonicus* (Esaki et Matsumoto, 1931) (Strepsiptera, Elenchidae) from Japan and the East China Sea. *Kontyu*, 54, 359-360

Kisimoto, R. (1957) Studies on the diapause in the planthoppers effect of photoperiod on the induction and completion of diapause in the fourth larval stage of the small brown planthopper, *Delphacodes striatella* Fallèn. *Appl. Ent. Zool.*, 2, 128-134

Kisimoto, R. (1975) *Transoceanic migration of planthopper*. Chuokoronsha, Tokyo, Japan (in Japanese)

Kisimoto, R. (1983) Long-distance migration of planthoppers. *Bull. Facul. Agr. Mie Univ.*, 67, 17-29 (in Japanese)

Kisimoto, R. & Sogawa, K. (1995) Migration of the Brown Planthopper *Nilaparvata lugens* and the White-backed Planthopper Sogatella furcifera in East Asia: the role of weather and climate, In: *Insect Migration: Tracking resources through space and time*, V. Drake & A. Gatehouse (Eds.), 67-90, Cambridge University Press, UK.

Kitamura, K. (1983) Comparative study on the biology of dryinid wasps in Japan (2) Relationship between temperature and the developmental velocity of *Haplogonatopus atratus* Esaki et Hashimoto (Hymenoptera: Dryinidae). *Bull. Fac. Agr. Shimane Univ.*, 17, 147-151.

Kitamura, K. (1987) Seasonal changes in percentage parasitism of the parasitoids of leaf- and planthoppers in Shimane Pref. (Homoptera: Auchenorrhyncha). *Bull. Fac. Agr. Shimane Univ.*, 21, 155-170

Kitamura, K. (1989) Comparative studies on the biology of dryinid wasps in Japan 6. Hibernation and development of *Haplogonatopus atratus* Esaki et Hashimoto (Hymenoptera: Dryinidae) on overwintering leaf- and planthoppers (Homoptera: Auchenorrhyncha). *Jpn. J. Appl. Entomol. Zool.*, 33, 24-30 (in Japanese, English summary)

Kitamura, K & Nishikata, Y (1987) A monitor-trap survey of parasitoids of the leaf- and planthoppers supposedly migrated from the mainland China (Homoptera:

Auchenorrhyncha). *Bull. Fac. Agr. Shimane Univ.*, 21, 171–177 (in Japanese, English summary)

Maeta, Y., Machita, Y. & Kitamura, K. (2007) Studies on the biology of *Elenchus japonicus* (Esaki & Hashimoto) (Strepsiptera, Elenchidae). *Jpn. J. Entomol.*, 10, 33-46 (in Japanese, English summary)

Matsumoto, Y., Matsumura, M., Hoshizaki, S., Sato, Y. & Noda, H. (2011) The strepsipteran parasite *Elenchus japonicus* (Strepsiptera, Elenchidae) of planthoppers consists of three genotypes. *Appl. Entomol. Zool.*, 46, 435–442

Matsumura, M., Takeuchi, H., Satoh, M., Sanada-Morimura, S., Otuka, A., Watanabe, T. & Thanh, DV. (2008) Species-specific insecticide resistance to imidacloprid and fipronil in the rice planthoppers *Nilaparvata lugens* and *Sogatella furcifera* in East and South-east Asia. *Pest Manag. Sci.*, 64, 1115-1121

Mound, L. (1983) Natural and disrupted patterns of geographical distribution in Thysanoptera (Insecta). *J. Biogeogr.* 10, 119–133

Mun, JH., Song, YH., Heong, KL. &, Roderick, GK. (1999) Genetic variation among Asian populations of rice planthoppers, *Nilaparvata lugens* and *Sogatella furcifera* (Hemiptera: Delphacidae): mitochondrial DNA sequences. *Bull. Entomol. Res.*, 89, 245–253

Nishioka, T. (1980) Biological notes on *Haplogonatopus atratus* Esaki & Hashimoto. *Gensei*, 38–39, 9–19 (in Japanese, English summery)

Olmi, M. (1999) Hymenoptera Dryinidae-Embolemidae. *Fauna d'Italia*, 37, Edizioni Calderini, Bologna

Olmi, M. & Guglielmino, A. (2010) Description of Erwiniinae, new subfamily of Dryinidae from Ecuador (Hymenoptera: Chrysidoidea). *Zootaxa*, 2605, 56–62

Otuka, A., Watanabe, T., Suzuki, Y. & Matsumura, M. (2005a) Estimation of the migration source for the white-backed planthopper *Sogatella furcifera* (Horváth) (Homoptera: Delphacidae) immigrating into Kyushu In June. *Jpn. J. Appl. Entomol. Zool.*, 49, 187–194 (in Japanese, English summary)

Otuka, A. Watanabe, T., Suzuki, Y., Matsumura, M., Furuno, A. & Chino M. (2005b) A migration analysis of the rice planthopper *Nilaparvata lugens* from the Philippines to East Asia with three-dimensional computer simulations. *Popul. Ecol.*, 47, 143-150.

Otuka, A., Matsumura, M., Sanada-Morimura, S., Takeuchi, H., Watanabe, T., Ohtsu, R. & Inoue, H. (2010) The 2008 overseas mass migration of small brown planthopper, *Laodelphax striatellus*, and subsequent outbreak of rice stripe disease in western Japan. *Appl. Entomol. Zool.*, 45, 259–266

Otuka, A., Matsumura, M., Watanabe, T. & Dinh, T. V. (2008) A migration analysis for rice planthoppers, *Sogatella furcifera* (Horváth) and *Nilaparvata lugens* (Stål) (Homoptera: Delphacidae), emigrating from northern Vietnam from April to May. *Appl. Entomol. Zool.*, 43, 527–534

Richter, C. (1970) Aerial dispersal in relation to habitat in eight wolf spider species (Pardosa, Araneae,Lycosidae). *Oecologia* 5, 200–214

Sogawa, K. (1992) A change in biotype property of brown planthopper populations immigrating into Japan and their probable source areas. *Proc. Assoc. Plant. Prot. Kyushu*, 38, 63–68

Sogawa, K. (1993) Source estimation of brown planthopper based upon biotype. *Japan Agriculture Technology*, 37, 36–40 (in Japanese)

Washburn, J. & Washburn, L. (1984) Active aerial dispersal of minute wingless arthropods: exploitation of boundary-layer velocity gradients. *Science*, 223, 1088-1089

# 4

# Phylogenetic Systematics and Biogeography: Using Cladograms in Historical Biogeography Methods

Raúl Contreras-Medina[1] and Isolda Luna-Vega[2]
[1]Escuela de Ciencias, Universidad Autónoma "Benito Juárez" de Oaxaca (UABJO),
[2]Laboratorio de Biogeografía y Sistemática, Departamento de Biología Evolutiva,
Facultad de Ciencias, Universidad Nacional Autónoma de México (UNAM),
México

## 1. Introduction

Phylogenetic systematics (or cladistics) was proposed by the German entomologist Willi Hennig (1966). Since its formulation it has had a great impact on taxonomy and other biological disciplines such as biogeography, paleontology, and evolutionary biology. In the case of biogeography, phylogenetic systematics has been fundamental and the basis for several historical biogeography approaches, playing a crucial role in the current status of this biological discipline (Crisci, 2001). The term cladistics was first used by authors such as Camin and Sokal or even Ernst Mayr (Schuh, 2000) and was applied to phylogenetic systematic studies that followed Hennig (1966). Notwithstanding that the term cladistics is currently in common use (even a scientific journal has that name), the word cladist was initially used as pejorative, to refer to those authors who used the methods of Willi Hennig (Schuh, 2000).

The methodology of phylogenetic systematics is mainly comparative (Espinosa & Llorente, 1993) and results in a dendrogram called cladogram (Nelson & Platnick, 1981), which represents a hypothesis of phylogenetic relationship between the members of the biological group studied. This taxonomic approach proposes sister group relationships among species by common ancestry through the evaluation of character states, avoiding descendant-ancestry hypothesis, thus eliminating the search of missing links (Espinosa & Llorente, 1993). Thus cladograms became a powerful way to represent the phylogeny of organisms and communicate these hypotheses to other biologists (Crisci, 2001). From biological and historical perspectives, phylogenetic relationships between taxa and their geographical distribution are considered to be intimately linked to part of the evolutionary process; for this reason it is assumed that a cladogram includes potentially useful information to elucidate the distributional history of organisms and data about the relationships among the areas inhabited by them (Crisci, 2001).

Historical biogeography studies the distribution of organisms, emphasizing processes occurring over millions of years and generally at great spatial resolutions, many times at a worldwide level. In historical biogeography, the proliferation of competing disciplines has generated a great number of approaches (Crisci, 2001; Morrone 2009); to this end, several methodologies have been proposed, and the use of taxonomic cladograms is a basic tool in many of them. Among the best-known approaches are: ancestral areas, phylogenetic biogeography, cladistic biogeography, comparative phylogeography, and event-based methods (Crisci et al., 2000).

In this chapter, we briefly discuss the different methods of historical biogeography in which cladograms play an important role, and compare them in the light of the processes that affect the distribution of organisms. For this, we used case studies of animals and plants of Latin America.

## 2. Methods of historical biogeography

### 2.1 Phylogenetic biogeography

This was the first historical biogeographical method that used cladograms as a basic tool to infer biogeographic histories (Crisci, 2001). The approach was proposed by Brundin (1966) and Hennig (1966), and consists of interpreting the biogeographic history of the taxonomic cladogram obtained for a particular taxon, applying two methodological rules: the progression rule and the deviation rule. The first methodological rule assumed that the basal members of a monophyletic group are found closer to/or in the center of origin than those apomorphic members, which are located on the periphery. The deviation rule implies that in any speciation event the apomorphic species accumulate more advanced character states (apomorphies) than the basal species, and are considered more deviated from the ancestor (Morrone et al., 1996).This approach assumes possibilities of dispersal and extinction, and its main concern is to interpret the distributional history of individual taxa. The center of origin could be identified as the area inhabited by the taxon located in the most basal position of the cladogram.

Two studies applying this approach are relevant. The study of Dávila-Aranda (1991) had the main goal of obtaining the phylogeny of a group of species of *Sorghastrum* (Poaceae) represented in Mexico, and as a secondary objective to propose a biogeographic hypothesis to explain the presence of this genus in Mexico. With this study, Dávila-Aranda separated the most plesiomorphic species of *Sorghastrum* from the most apomorphic. Reynoso and Montellano-Ballesteros (2004) worked with the desert tortoise genus *Gopherus*, distributed in northern Mexico and the southern United States. At first, the authors obtained the phylogenetic analysis of tortoises using extant and fossil species; the cladogram obtained was used to reconstruct the biogeographic history of the genus *Gopherus*. Reynoso and Montellano-Ballesteros (2004) considered that the origin of *Gopherus* can be traced back to the Oligocene on the Central Plains of North America (where *G. laticuneus* was found, for them the oldest and most primitive known species of the genus), and later it extended southward from eastern Arizona to Florida (where *G. polyphemus* inhabits) and from northern Texas to Aguascalientes, Mexico during the Plio-Pleistocene (where *G. flavomarginatus* occurred); successful expansion of *Gopherus* during the Pleistocene was followed by a series of extinctions (mainly in Texas and eastern Mexico) and the reduction of the range affecting most of the tortoise species.

This approach is considered to be an eclectic one, because it tries to explain the general patterns of distribution through vicariance and exceptional cases through dispersal. It also intended to find the centers of origin of the groups by analyzing the cladograms, and proposed probable routes of long distance dispersal through a dynamic Earth.

## 2.2 Ancestral areas

This approach was developed by Bremer (1992) and it was used for the recognition of an ancestral area of a monophyletic group from the information of its cladogram; this method has been considered the formalization of a cladistic procedure based on a dispersalist approach (Crisci, 2001). It is based on two assumptions: (1) the area located in the basal position of the cladogram (the most plesiomophic) has a high probability of being considered the ancestral area for a particular taxon, in relation to those located in other positions (apomorphics); (2) an area represented in several branches of the same cladogram has a high probability of representing the ancestral area, in relation to those located in few or one branches.

The first step to carry out an analysis of ancestral areas is to construct an area-cladogram, which is obtained from the substitution of terminal taxa by the area or areas where each taxon inhabits; this area-cladogram is analyzed and each area is considered as a binary character with two states (present or absent) and optimized on the cladogram (Crisci et al., 2000). From a comparison of the number of gains and losses, it is possible to estimate which area is considered the ancestral area for the taxon under study, from the highest values observed in gain/loss quotients. This method of historical biogeography is based on dispersal principles and its main concern is the distributional history of individual taxa.

Katinas and Crisci (2000) offered one of the main studies applying this method in South America, based on the flowering plant sister genera *Moscharia* and *Polyachyrus* (Asteraceae). In this study, the areas of endemism analyzed were the Coastal Desert, Cardonal, North Central Chile, and South Central Chile provinces. Applying the ancestral areas method, the analysis showed that the most probable area identified as the ancestral area is North Central Chile, which had the highest gain/loss quotient value. According to their results, Katinas & Crisci (2000) hypothesized that the ancestor of *Moscharia* and *Polyachyrus* may have inhabited a part of the area of North Central Chile, and during humid climate periods, the biota of this region increased its range both to the south (South Central Chile) and to the north (Coastal Desert), with the high Andean slopes (Cardonal) being the last area to be occupied.

## 2.3 Cladistic biogeography

This approach was proposed by Rosen (1978), and Nelson & Platnick (1981); it combines the method of cladistics with theoretical aspects of panbiogeography (Crisci et al., 2000; Espinosa & Llorente, 1993). Its basic premise is the search for patterns of relationships among areas of endemism (Humphries & Parenti, 1999). The central axis of this method supposes a relationship between the history of life and history of Earth (Espinosa & Llorente, 1993).

The first step in cladistic biogeography is to construct area cladograms from taxonomic cladograms, which are obtained by replacing their terminal taxa by the areas of endemism where they occur (Morrone & Crisci, 1995); from the information of two or more area cladograms, we can apply one or more of the methods that have been proposed in cladistic biogeography (see Luna-Vega & Contreras-Medina in this book), in order to obtain the general area cladogram (Morrone, 2005). The general area cladogram is the final result of any analysis of cladistic biogeography and represents a hypothesis of relationships among areas of endemism analyzed and also reflects vicariance events that occurred in the biogeographic history of the biota analyzed (Contreras-Medina, 2006; Morrone, 1997). One problem detected is related to the basal position of some areas in the general area cladogram, influenced by low diversity or underrepresentation of the biological group studied (e. g. Contreras-Medina & Luna-Vega, 2002).

The study of the gymnosperm genera *Ceratozamia, Dioon* and *Pinus* by Contreras-Medina et al. (2007) is among the studies applying this method in Mesoamerica. In this study, the areas of endemism analyzed were the 19 Mexican floristic provinces proposed by Rzedowski (1981), and the areas of endemism proposed by Morrone (2001) for Central America and by Takhtajan (1986) for North America (see Luna-Vega & Contreras-Medina in this book). Two methods of cladistic biogeography were applied: Brooks Parsimony Analysis and Paralogy-free Subtrees; the consensus cladogram was obtained from each method. Only two clades were consistent in both consensus cladograms; one clade formed by the Sierra Madre Occidental plus Sierra Madre Oriental-Altiplano provinces, and another clade formed by the Great Basin and Mojavean provinces. These authors considered that both peninsulas of Mexico have a different history in relation to the continental portion of the country.

## 2.4 Event-based methods

This approach creates explicit models of biogeographic processes that affect the geographical distribution of organisms (Crisci, 2001; Morrone, 2009). This approach includes some proposals, one of them being the dispersal-vicariance analysis (or DIVA) proposed by Ronquist (1997). This last method reconstructs the biogeographic history of individual taxa, and also allows reconstruction of biogeographic scenarios that include the possibility of reticulate relationships that do not necessarily follow a hierarchical pattern, as occurs in other methods of historical biogeography (Crisci et al., 2000; Morrone, 2009).

This biogeographic reconstruction is based on a cost matrix, which is constructed according to certain premises (Crisci et al., 2000; Morrone, 2009): (1) vicariance events have a null cost of 0, which implies that speciation is due to vicariance; (2) duplication events have a null cost of 0, which is assumed due to sympatric speciation; (3) dispersal events have a cost of 1 per area unit added to a distribution, and (4) extinction events have a cost of 1 per unit area deleted from a distribution.

Among studies applying the DIVA method in South America, we found the study based on several genera of weevils (Curculionidae) by Posadas & Morrone (2003). In this study, the areas of endemism analyzed were the Maule, Valdivian Forest, Magellanic Forest, Magellanic Moorland, and Falkland Islands provinces. The dispersal-vicariance analysis showed that the most frequent dispersal event involved the Maule-Valdivian Forest (21.4%), whereas the most frequent vicariance event involved the separation of the Falkland Islands from the Magellanic Forest-Magellanic Moorland set.

The DIVA has some advantages over other event-based methods, allowing reconstructing biogeographic scenarios, which can include a reticulate area history; colonizations are treated as integral components of evolution of organisms; and additionally, analysis with co-occurring taxa can also be used to explore general biogeographic events (Kodandaramaiah, 2010; Posadas & Morrone, 2003). Although DIVA is an approach to the event-based methods, this method has a low probability of invoking extinctions, inability to distinguish between contiguous range expansions and dispersal across a barrier, and has problems when events of speciation due to dispersal are being erroneously considered as vicariance (Kodandaramaiah, 2010).

## 2.5 Comparative phylogeography

Phylogeography studies the principles and processes governing the geographical distribution of genealogical lineages at intraspecific level using sequences of mitochondrial DNA in animals and chloroplast in plants (Crisci, 2001); it was originally proposed by Avise et al. (1987). Several individuals of the target species are examined along their distribution range, in order to obtain DNA sequences. The sets of similar sequences are recognized as haplotypes and all the information is represented in a phylogeographic tree; the localities (geography) where each specimen was collected are related with the phylogeographic pattern (tree). Generally, results obtained with this approach are based on dispersal principles and dubious clock calibrations (Heads, 2005); its main concern is the distributional history of one species or related species.

The algorithms used to construct taxonomic cladograms, such as parsimony or maximum likelihood are also used to construct phylogeographic trees; the genealogy of haplotypes presents a branched hierarchical structure as observed in taxonomic cladograms, which can be used for a historical biogeographic analysis applying the same principles of the cladistic biogeography (Contreras-Medina, 2006), but at an intraspecific level when comparing two or more phylograms. In this way, the application of comparative phylogeography approach (Arbogast & Kenagy, 2001) implies the comparison of phylogeographic studies of two or more species that are co-distributed (sympatric), in order to search for common historical patterns of distribution (Zink, 2002; Morrone, 2005).

Among studies applying comparative phylogeography in Mesoamerica, we found the study of Sullivan et al. (2000) based on highland rodents (*Peromyscus aztecus/Peromyscus hylocetes* complex and *Reithrodontomys sumichrasti*). The areas considered in this study included several mountain ranges located in Central and Southern Mexico and northern Central America; these mountain chains are the Sierra Madre Oriental, Trans-Mexican Volcanic Belt, Sierra Madre del Sur, Oaxaca Highlands, and the mountains of Chiapas and northern Central America (see Fig. 2 of Luna-Vega & Contreras-Medina in this book). The results showed that these rodents presented certain common phylogeographic patterns, as well as areas of incongruence. A vicariant pattern between the Oaxaca Highlands and part of the Sierra Madre del Sur was noted, as well as the separation of all mountain ranges in relation to the Chiapas and Central American Highlands, where the Isthmus of Tehuantepec (a lowland region) acted as a barrier and played a relevant role (located in the basal position) of these southern areas. The Sierra Madre Oriental (SMOR) is an example of incongruence, because in the case of *Reithrodontomys sumichrasti* the SMOR is the sister area of the clade Oaxaca Highlands-Sierra Madre del Sur, while in the *Peromyscus aztecus/Peromyscus hylocetes* complex the SMOR is sister to the area of the Trans-Mexican Volcanic Belt.

## 3. Comparison of methods

A comparison of the approaches mentioned in this chapter includes the following aspects: the process involved, reconstruction of biotas or individual histories, and the taxonomic level used in the analysis (Table 1).

Dispersal, vicariance and extinction are of major or minor importance in some of these approaches; dispersal is used mainly in ancestral areas and phylogenetic biogeography, while in the event-based methods, cladistic biogeography and comparative phylogeography dispersal and vicariance are assumed (Crisci et al., 2000); extinction is implemented in all approaches (Crisci et al., 2000). Some methodologies give more importance to the biogeographic history of a particular taxon (ancestral areas, phylogenetic biogeography and phylogeography), while others emphasize the historical relationships among areas of endemism (cladistic biogeography and comparative phylogeography). In analyses based on only one taxon, the concept of center of origin is maintained, e.g. ancestral areas, phylogenetic biogeography and phylogeography. The taxonomic level used in biogeographic analysis is different among these approaches: phylogeography is applied only at species level, while in phylogenetic biogeography, cladistic biogeography and event-based methods are applied at species level or supraspecific taxa; the ancestral areas method is applied at any taxonomic level (Crisci et al., 2000). The historical relationship between areas and the search of common patterns of distribution are the main objectives in cladistic biogeography and comparative phylogeography.

|  | Minimum number of taxa to work with | Main biogeographic process used | Use of center of origin concept | Taxonomic level used | Explanation of individual histories |
|---|---|---|---|---|---|
| Phylogenetic biogeography | 1 | Dispersal | Yes | Species or genera | Yes |
| Ancestral areas | 1 | Dispersal | Yes | Species or genera | Yes |
| Cladistic biogeography | 2 | Vicariance | No | Species or genera | No |
| Event-based methods | 1 | Dispersal and vicariance | No | Species or genera | Yes & No |
| Comparative phylogeography | 2 | Vicariance | No | Infra-specific | No |

Table 1. Main characteristics of the different historical biogeography approaches commented in this chapter.

The development of historical biogeography has been driven by the confrontation of two main biogeographic processes, dispersal and vicariance. Wegener (1929) drew attention to plant distributions, especially of the Southern Hemisphere, where related genera and even congeneric species were separated by vast oceans, representing the living evidence of continental drift (Contreras-Medina & Luna-Vega, 2002). This distributional pattern has two different historical explanations, which are dispersal and vicariance. The former process involves a common ancestor that originally occurred in one area and later dispersed into another, where its descendants survived until the present day; vicariance implies an ancestor that was originally widespread in a larger area that became fragmented, leaving descendants that have survived in the fragments until now (Morrone & Crisci, 1995) (Fig. 1).

The dispersalist program began with Darwin (1859), especially with his two chapters on geographic distribution, in which the main axis of the Darwinian conception was noted: a random dispersal on a stable geography (Bueno & Llorente, 1991). This point of view was maintained for more than a century and influenced biogeographical thinking for many decades. This influence is reflected in several methods of biogeography, v. gr. ancestral areas method and phylogenetic biogeography (Morrone, 2005); notwithstanding that these methods used cladograms, the center of origin concept is implemented in both approaches.

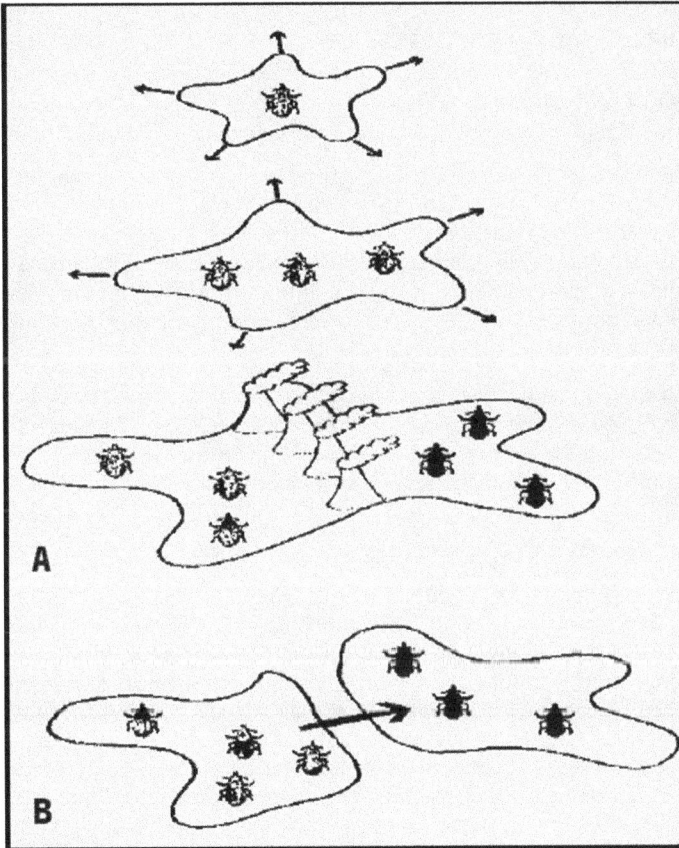

Fig. 1. Historical explanations of disjunct distributions: (A) vicariance, and (B) dispersal. Redrawn from Contreras-Medina et al. (2001).

Vicariance was first considered to be an important component of biogeography after the studies of Croizat (1958, 1964); his conception of space as part of the evolutionary process was later included in the cladistic biogeographic approach, which is also known as vicariant biogeography (Espinosa & Llorente, 1993). This point of view began at the middle of the XX century and has influenced biogeographical thinking in recent decades. This influence is reflected mainly in several methods of biogeography, e.g. cladistic biogeography and comparative phylogeography. The use of phylogenies in cladistic biogeography led this approach being considered the most robust method in historical biogeography (Contreras-Medina, 2006; Humphries, 2000).

## 4. Conclusions

The use of cladograms in all the approaches mentioned above is essential and emphasizes the relevance of phylogenetic evidence in historical biogeographic studies. These approaches differ in the number of taxa used for analysis, the interpretation of area cladograms, and in that they give different importance and weight to events that modify the geographic distribution of organisms. In this sense, the effects of extinction can be profound, but they are not commonly considered, many times ignored, in biogeographic studies (Lieberman, 2002). Local extinctions are probably as important as dispersal and vicariance, but unfortunately they are never inferred in biogeographic analyses (Kodandaramaiah, 2010).

It is important to consider that a cladogram represents only a hypothesis of the phylogeny of certain biological group and not necessarily the truth of how the evolutionary history of organisms occurred. If this first hypothesis is contrasted with new evidence, it is possible that the previous topology of the cladogram may change and, in consequence these changes might affect our biogeographic analysis results. Notwithstanding, cladograms are the main source of evidence on phylogeny for all the methods mentioned above and represent the basis for their implementation.

Crisci (2001) considered that the history of life on Earth is complex and we will probably never see it totally revealed. Notwithstanding, historical biogeography is part of the scientific challenge that attempts to resolve the relationship between the history of Earth and the evolution of life; cladograms represent an essential tool for addressing this difficult task.

## 5. Acknowledgments

The first author gives thanks for the honor that the Symposium Committee on Applications of Phylogenies in Botany of the XVIII Mexican Botanical Congress (Guadalajara, Mexico, November 2010) did him by inviting him to present the conference that originated part of this contribution. David Espinosa and Othón Alcántara made constructive comments to the manuscript. RCM dedicates this chapter to his son José Arturo Contreras Córdoba on occasion of his first year of life. Funds for the publication of this contribution were provided by the Secretaría de Planes y Programas Estratégicos directed by Josefina Aranda Bezaury of the Universidad Autónoma "Benito Juárez" de Oaxaca (UABJO). Financial support was given by PAPIIT 221711.

## 6. References

Arbogast, B. S. & Kenagy, G. J. (2001). Comparative phylogeography as an integrative approach to historical biogeography. *Journal of Biogeography* Vol. 28, pp. 819−825, ISSN 0305-0270

Avise, J.C., Arnold, J., Ball, R.M., Bermingham, E., Lamb, T., Neigel, J.E., Reeb, C.A. & Saunders, N.C. (1987). Intraspecific phylogeography: The mitochondrial DNA bridge between population genetics and systematics. *Annual Review of Ecology and Systematics* Vol. 18, pp. 489–522, ISSN 0066-4162

Bremer, K. (1992). Ancestral areas: a cladistic reinterpretation of the center of origin concept. *Systematic Biology* Vol. 41, pp. 436–445, ISSN 1063-5157

Brundin, L. (1966). Transantarctic relationships and their significance, as evidenced by chironomid midges. *Kungliga Svenska vetenskapsakadamiens handlingar* Vol. 11, No. 1, pp. 437–472.

Bueno, A. & Llorente, J. (1991). El centro de origen en la biogeografía: historia de un concepto. In: *Historia de la biogeografía: centros de origen y vicarianza*. Llorente, J. (ed.), México, D. F., Ciencias Servicios Editoriales UNAM, pp. 1–33. ISBN 968-36-2156-2

Contreras-Medina, R. (2006). Los métodos de análisis biogeográfico y su aplicación a la distribución de las gimnospermas mexicanas. *Interciencia*, Vol. 31, No. 3, pp. 176–182, ISSN 0378-1844

Contreras-Medina, R. & Luna-Vega, I. (2002). On the distribution of gymnosperm genera, their areas of endemism and cladistic biogeography. *Australian Systematic Botany* Vol. 15, No. 2, pp. 193–203, ISSN 1030-1887

Contreras-Medina, R., Luna-Vega, I. & Morrone, J.J. (2001). Conceptos biogeográficos. *Elementos* Vol. 8, No. 41, pp. 33–37, ISSN 0187-9073

Contreras-Medina, R.; Luna-Vega, I. & Morrone, J.J. (2007). Gymnosperms and cladistic biogeography of the Mexican Transition Zone. *Taxon* Vol. 56, No. 3, pp. 905–915, ISSN 0040-0262

Crisci, J.V. (2001). The voice of historical biogeography. *Journal of Biogeography* Vol. 28, No. 2, pp. 157–168, ISSN 0305-0270

Crisci, J. V., Katinas, L. & Posadas, P. (2000). *Introducción a la teoría y práctica de la biogeografía histórica*. Buenos Aires, Sociedad Argentina de Botánica (ISBN 987-97012-4-0) (English translation: 2003, *Historical biogeography: An introduction*. Cambridge, Mass. Harvard University Press).

Croizat, L. (1958). Panbiogeography. Published by the author. Caracas. 1731 p. ISBN 978-0854860340

Croizat, L. (1964). *Space, time, and form: The biological synthesis*. Published by the author. Caracas. ISBN 978-0854860364

Darwin, C. (1859). *El origen de las especies*. Planeta-Agostini, Barcelona, Spain. (Spanish version 1992). ISBN 84-395-2172-3

Dávila-Aranda, P. (1991). Consideraciones filogenéticas y biogeográficas preliminares del género *Sorghastrum* (Poaceae). *Acta Botanica Mexicana* Vol. 14, 59–73, ISSN-0187-715

Espinosa, D. & Llorente, J. (1993). *Fundamentos de biogeografías filogenéticas*. Universidad Nacional Autónoma de México-CONABIO. México, D. F. ISBN 968-36-2984-9

Heads, M.J. (2005). Toward a panbiogeography of the seas. *Biological Journal of the Linnean Society* Vol. 84, No. 4, pp. 675–723, ISSN 1095 8312

Hennig, W. (1966). *Phylogenetic systematics*. University of Illinois Press, Urbana IL, 280 p. ISBN 978-025-2068-140

Humphries, C. J. (2000). Form, space and time: which comes first? *Journal of Biogeography* Vol. 27, No. 1, pp. 11–15, ISSN 0305-0270

Humphries, C. J. & Parenti, L. R. (1999). *Cladistic biogeography*. Oxford University Press, New York. ISBN 019-854818-4

Katinas, L. & Crisci, J.V. (2000). Cladistic and biogeographic analyses of the genera *Moscharia* and *Polyachyrus* (Asteraceae, Mutisieae). *Systematic Botany* Vol. 25, No. 1, pp. 33–46, ISSN 0363-6445

Kodandaramaiah, U. (2010). Use of dispersal-vicariance analysis in biogeography – a critique. *Journal of Biogeography* Vol. 37, No. 1, pp. 3-11, ISSN 0305-0270

Lieberman, B.S. (2002). Phylogenetic biogeography with and without the fossil record: gauging the effects of extinction and paleontological incompleteness. *Palaeogeography, Palaeoclimatology, Palaeoecology* Vol. 178, No. 1, pp. 39-52, ISSN 0031-0182

Luna–Vega, I. & Contreras-Medina, R. (2012). Contributions of cladistic biogeography to the Mexican Transition Zone. In: *Global advances in Biogeography*,L. Stevens (Ed.). InTech, Rijeka, Croatia. ISBN 979-953-307-415-2

Morrone, J.J. (1997). Biogeografía cladística: conceptos básicos. *Arbor* Vol. 158, pp. 373 – 388, ISSN 0210-1963

Morrone, J. J. (2001). *Biogeografía de América Latina y el Caribe*. SEA y M & T Tesis, Vol. 3, Zaragoza, Spain. ISBN 84-922495-4-4

Morrone, J. J. (2005). Cladistic biogeography: identity and place. *Journal of Biogeography* Vol. 32, pp. 1281 – 1286, ISSN 0305-0270

Morrone, J. J. (2009). *Evolutionary biogeography: An integrative approach with case studies*. Columbia University Press, New York. ISBN 978-0-231-14378-3

Morrone, J. J. & Crisci, J. V. (1995). Historical biogeography: Introduction to methods. *Annual Review of Ecology and Systematics* Vol. 26, pp. 373 – 401, ISSN 0066-4162

Morrone, J.J., Espinosa, D. & Llorente, J. (1996). *Manual de biogeografía histórica*. Ciencias Servicios Editoriales, Universidad Nacional Autónoma de México, México, D. F. ISBN 968-36-4842-8

Nelson, G. & Platnick, N.I. (1981). *Systematics and biogeography: Cladistics and vicariance*. Columbia University Press, New York. ISBN 0-231-04574-3

Posadas, L. & Morrone, J.J. (2003). Biogeografía histórica de la familia Curculionidae (Coleoptera) en las regiones Subantártica y Chilena Central. *Revista de la Sociedad Entomológica Argentina* Vol. 62, No. 1-2, pp. 75 – 84, ISSN 0373-5680

Reynoso, V.H. & Montellano-Ballesteros, M. (2004). A new giant turtle of the genus *Gopherus* (Chelonia: Testudinidae) from the Pleistocene of Tamaulipas, Mexico, and a review of the phylogeny and biogeography of gopher tortoises. *Journal of Vertebrate Paleontology*, Vol. 24, No. 4, pp. 822 – 837, ISSN 0272-4634

Ronquist, F. (1997). Dispersal-vicariance analysis: a new approach to the quantification of historical biogeography. *Systematic Biology* Vol. 46, pp. 195-203, ISSN 1063-5157

Rosen, D.E. (1978). Vicariant patterns and historical explanation in biogeography. *Systematic Zoology* Vol. 27, pp. 159 – 188, ISSN 0039-7989

Rzedowski, J. (1981). *Vegetación de México*. Limusa. Mexico, D.F. ISBN 968-18-0002-8

Schuh, R.T. (2000). *Biological systematics, principles and applications*. Cornell University Press, New York. ISBN 0-8014-3675-3

Sullivan, J., Arellano, E. & Rogers, D. S. (2000). Comparative phylogeography of Mesoamerican highland rodents: concerted versus independent response to past climatic fluctuations. *The American Naturalist* Vol. 155, No. 6, pp. 755 – 768, ISSN 0003-0147

Takhtajan, A. (1986). *Floristic regions of the world*. University of California Press, Berkeley. ISBN 0520040279

Wegener, A. (1929). El origen de los continentes y océanos. Planeta-Agostini, Barcelona, Spain (Spanish version 1992). ISBN 84-395-2237-1

Zink, R.M. (2002). Methods in comparative phylogeography, and their application to studying evolution in the North American aridlands. *Integrative and Comparative Biology*, Vol. 42, No. 1, pp. 953 – 959, ISSN 1540-7063

# Part 2

# Regional Biogeography of Individual Taxa

# 5

# Establishment of Biogeographic Areas by Distributing Endemic Flora and Habitats (Dominican Republic, Haiti R.)

Eusebio Cano Carmona[1] and Ana Cano Ortiz[2]

[1]*Dpto. Biología Animal, Biología Vegetal y Ecología, Área de Botánica, Universidad de Jaén,*
[2]*Dpto. Sostenibilidad, INTERRA, Ingeniería y Recursos S.L.,*
*Spain*

## 1. Introduction

Despite the large number of botanical studies conducted on the flora of the Island of Hispaniola, some of which adopted a floristic or physiognomical approach (e.g., Zanoni et al., 1990; Höner and Jiménez 1994; Guerrero et al., 1997; May, 1997, 2000, 2001; Mejía and Jiménez 1998; Rivas-Martínez et al. 1999; May and Peguero, 2000; Mejía et al., 2000; Slocum et al., 2000; García and Clase 2002; García et al. 2002; Peguero and Salazar 2002; Veloz and Pequero 2002); and Cano et al., 2009a, 2009b, 2010a, 2010b, 2011, 2012 with a phytosociological methodology, only few have adopted phytogeographical or phytosociological approaches. Here we present a biogeographic profile on the flora of the Island of Hispanola.

## 2. Study area

Our study area is the Island of Hispaniola (Dominican Republic, Republic of Haiti); (Fig. 1). The island belongs to the so-called Greater Antilles (Cuba, Hispaniola, Jamaica and Puerto Rico), which are located approximately in the centre of the Antillean Arc. With an area of 76,484 km² Hispaniola is the second largest island of the group, second in size only to Cuba (110,861 km²). By comparison, the Bahamas (Grand Bahama, Andros, Mayaguana, Great Inagua and Grand Caicos) are a group of islands open to the Atlantic and extend from SE to NW. The arc of the Lesser Antilles, with the Virgin Islands, St Kitts Nevis, Antigua and Barbuda, Dominica, St Lucia, St Vincent and the Grenadines, Barbados, Grenada, Trinidad and Tobago, is located to the SE of Puerto Rico.

## 3. Methods

Our biogeographical approach for the vegetation of the Island of Hispaniola relies both on previous geological studies compiled by Liogier (2000), Mollat et al. (2004), Cano et al. (2009a, 2010a) and also on some geomorphological studies (A.R.N. 2004). We carried out a total of 300 vegetation relevês on the island These samples, together with the numerous floristic studies used references, allowed us to plot the different biogeographical areas of the island. For an accurate mapping of the floristic differences in the territory, we first studied the endemic plants growing not only all over the island but also on restricted sites of it. We

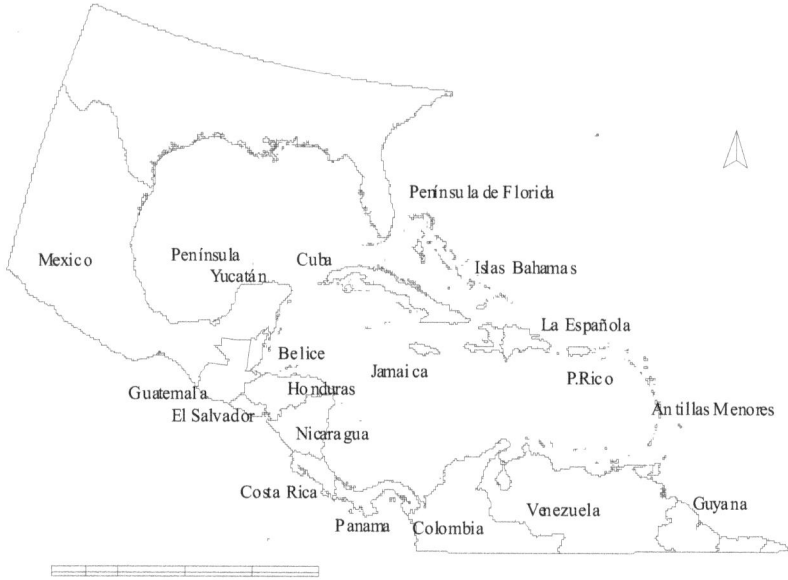

Fig. 1. Location of Hispaniola in the Caribbean area (Cano et al., 2009a)

provide here the number of endemic species per sampling unit. The area of these sample units ranges from 500 to 2,000 m², depending on the vegetation unit involved in each case, either of herbaceous, scrub or forest species. To check that the units defined in this manner were suitable as far as the differences recorded in the flora and vegetation are concerned, we simultaneously applied our own Jaccard's and Pearson's numerical analyses already published. Our approach relies heavily on the bioclimatic and biogeographical analyses up to the rank of biogeographical sector, as conducted by us in previous studies (Cano et al. 2009a, 2010a). With the presentation of some bioclimatic charts representing the study area in this paper we further extend this approach. For that purpose we also made use of the globalbioclimatic.org website Rivas Martínez (2009). Our biogeographical mapping is based on edaphological, bioclimatic, floristic, vegetational and historical criteria. Our analysis of the flora makes use of 1,582 endemic species recorded either in the references or in our field samples. We followed the taxonomy of Liogier (1996-2000) and Martín & Cremers (2007).

We determined the degree of kinship between each pair of 19 biogeographic areas (Fig. 2), and we consulted previous studies (Cano et al., 2010a). These references were further extended through application of the Jaccard's coefficient and the Pearson's index between the 19 areas according to the presence/absence of the species.

## 4. Results and discussion

### 4.1 Analysis of vegetation

Island of Hispaniola exhibits a wide altitudinal range, from 0 at sea level to 3,175 masl on Pico Duarte (Cordillera Central), a great variety of soils, and a rainfall gradient ranging from 400 to 4,600 mm. These three parameters, together with the isolation of the territories involved have been crucial for the emergence of the current vegetation.

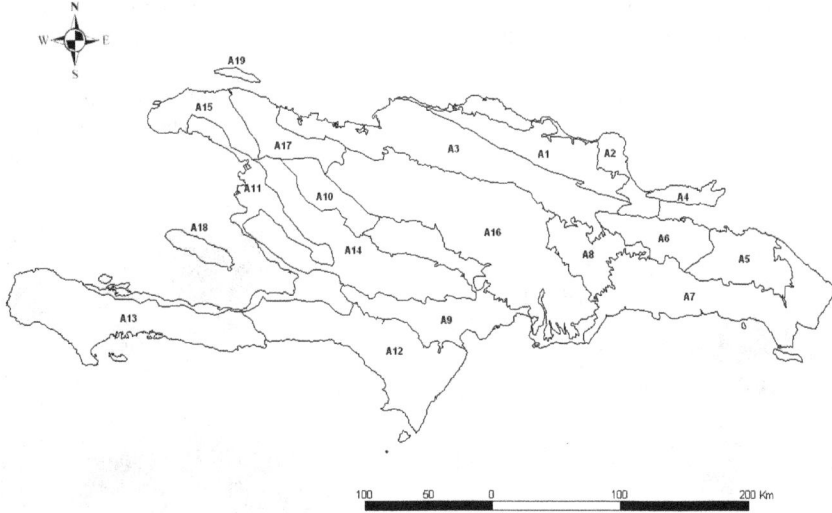

Fig. 2. Floristic areas on the Island of Hispaniola.

To study the vegetation we defined some large areas according to rainfall and temperature records. These areas include dry areas, subhumid areas, humid-hyperhumid areas and high mountain areas, as defined in Cano et al. (2009a, b).

Dry areas exhibit a tropical xeric macrobioclimate dominated by an infratropical semiarid and dry thermotype (Fig. 3 and 4). These areas also support a high richness of endemic species and correspond with our study areas A3, A9 and A12. From a physiognomical point of view, the vegetation is very similar in all the semiarid and dry areas, and is usually dominated by plants of the *Agavaceae* and *Cactaceae* families, among other, such as *Lemaireocereus hystrix* (Haw.) B.&R., *Cylindropuntia caribae* (B.&R,) Kunth, *Consolea moniliformis* (L.) Haw., *Leptochloopsis virgata* (Poir.) Griseb., *Pilosocereus polygonus* (Lam.) B.& R., *Opuntia dillenii* (Fer.- Gawl) Haw., *Leptocereus weingartianus* (Hartm.) Britt. & Rose, *Acacia skleroxyla* Tuss., *Agave antillarum* Descourt., *Pithecellobium unguis-cati* (L.) Mart.

In the southwest of the island (A12) we find the dry forest of Pedernales Ccitillan (Procurrente de Barahona; Photograph 1), developed on limestone dog-tooth substrates. Several endemic species include *Melocactus pedernalensis* (Ait.) M. Mejía & R. García, *Galactia dictyophylla* Urb., *Coccoloba incrassata* Urb., *Caesalpinia domingensis* Urb., and *Guettarda stenophylla* Urb. However, the dry forest growing in the area A9, with an ombrothermic index value Oi = 2.1, presents a slightly lower level of endemism. Here the most species differentiating the dry forest of Pedernales are *Melocactus lemairei* (Monv.) Miq. *Neoabbottia paniculata* (Lam.) Britt. & Rose, *Coccotrinax spissa* Bailey. Area A3, located in the northwest of the island supports a dry forest dissimilar to those previously mentioned in that it supports an endemic flora, including *Salvia montecristina* Urb. & Ekm., *Mosiera urbaniana* Borhidi, *Croton poitaei* Urb., *Croton sidaefolius* Lam., *Guettarda tortuensis* Urb. & Ekm. and *Coccoloba buchii* Urb. The most representative plant communities growing in dry areas belong to the following endemic habitats: *Lepotogono buchii-Leptochloopsietum virgatae* Cano, Velóz & Cano-Ortiz 2010, which is included in the endemic serpentinicolous alliance *Tetramicro canaliculatae-*

*Leptochloopsion virgatae* Cano, Veloz & Cano-Ortiz 2010; *Crotono astrophori-Leptochloopsietum virgatae* Cano, Veloz & Cano-Ortiz 2010, *Melocacto pedenalensi-Leptochloopsietum virgatae* Cano, Velóz & Cano-Ortiz 2010, *Solano microphylli-Leptochloopsietum virgatae* Cano, Velóz & Cano-Ortiz 2010, which are included in the endemic alliance *Crotono poitaei-Leptochloopsion virgatae* Cano, Veloz & Cano-Ortiz 2010; and the pine forests of *Leptogono buchii-Pinetum occidentalis* Cano, Velóz & Cano-Ortiz 2011 on serpentine substrates, which we include in the endemic alliance Leptogono buchii-Pinion occidentalis ined.

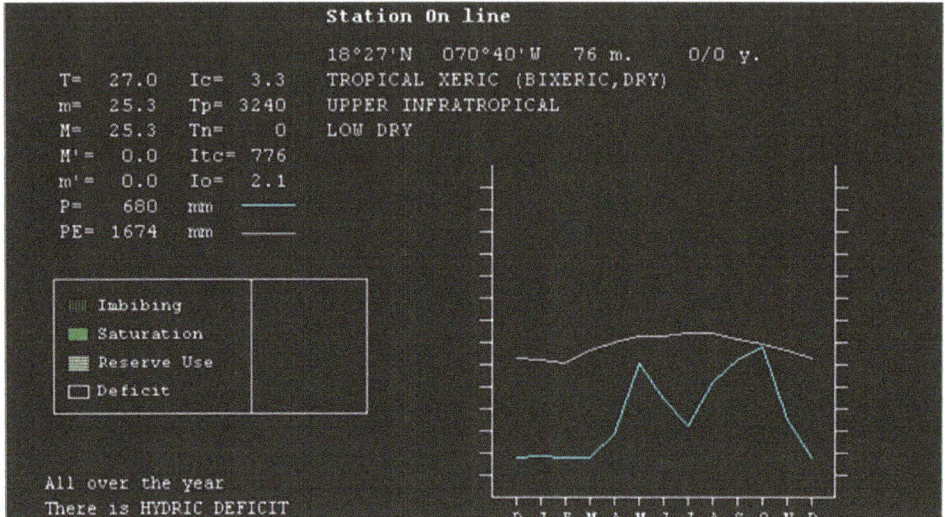

Fig. 3. Weather station of Azua: upper infratropical, lower dry bixeric, (A9).

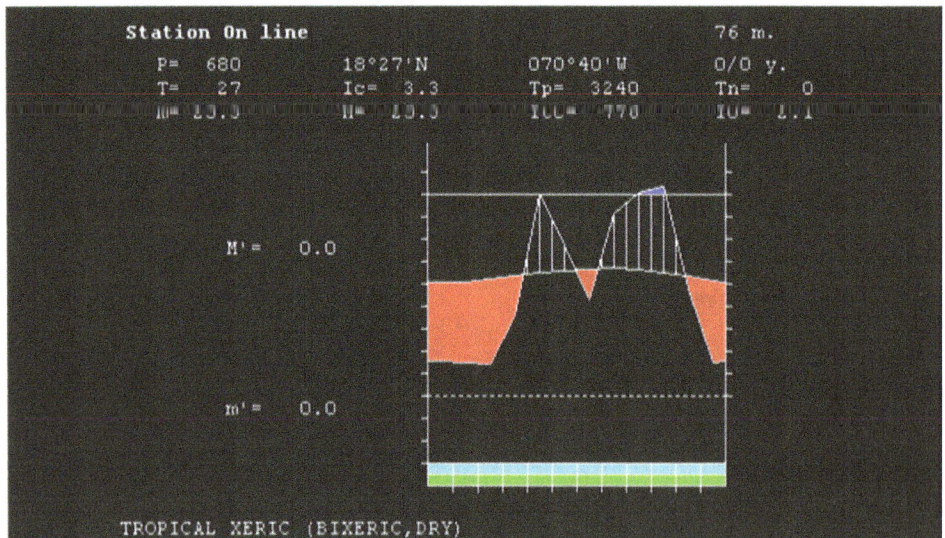

Fig. 4. Weather station of Azua: tropical xeric (bixeric), (A9).

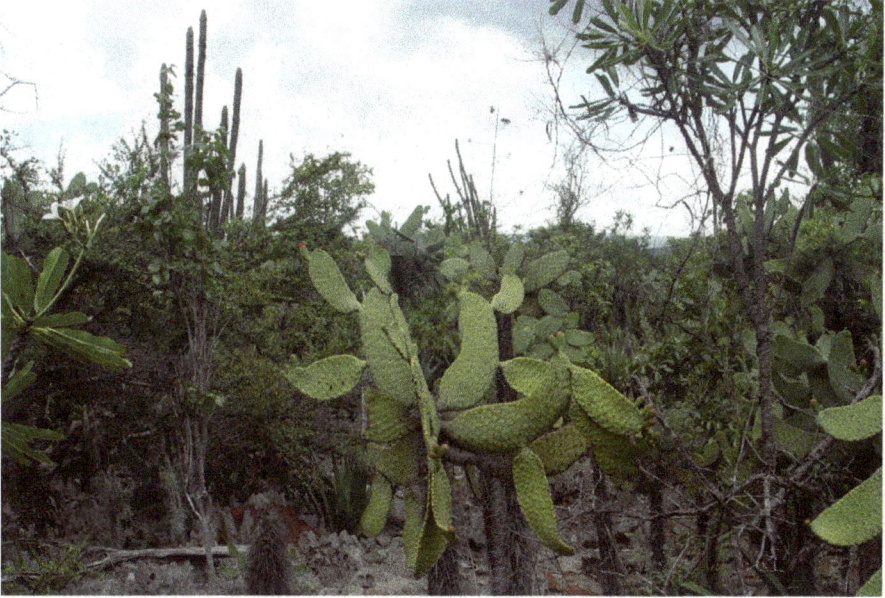

Photo 1. Dry forest of *Consolea moniliformis* of Procurrente of Barahona. Dominican Republic.

Most of the territory of Hispaniola presents a tropical pluviseasonal macrobioclimate (Figs. 5, 6) and dominance of the subhumid ombrotype, with rainfall rates ranging from 1,000 to 2,000 mm and Oi values ranging from 3.7-4.3 (Parque Nacional del Este), Oi = 4 (El Seibo), Oi = 6.2 (Miches), Oi = 5.9 (Mayaguana), to Oi = 9.3 (Jarabacoa). The dominant vegetation in these areas is a subhumid, broad-leaved forest undergoing a dry season from December to April. As a result of the water stress, this flora includes tree-like, deciduous species. This is the case of *Bursera simaruba* (L.) Sarg., *Swietenia mahagoni* (L.) Jacq. and other species, such as *Metopium toxiferum* (L.) Krug & Urb., *Krugidendron ferreum* (Vahl) Urb., *Acacia macracantha* H. & B. ex Willd., *Coccoloba diversifolia* Jacq. and *Bucida buceras* L. These formations include important endemic taxa, such as the climbing plant *Aristolochia bilobata* L., the tree-like *Melicoccus jimenezii* (Alain) Acev. Rodr. and scrub-like plants, such as *Lonchocarpus neurophyllus* Benth. There also are other dominant scrub formations playing the role of dynamic substitution stages, such as *Zamia debilis* L., which occus with the endemic taxa *Pereskia quisqueyana* Alain and *Goetzea ekmanii* O.E. Schulz.

If these subhumid forests are located on reef-perforated limestones, the territory adopts a dry profile as a result of the heavy water loss through the soil. Such settings include floristic elements such as *P. polygonus*, *P. unguis-cati*, *L. weingartianus* and *Hylocereus undatus* (Haw.) Britt. & Rose. The plant formations peculiar to A7 are associated with the dry forests of the southwest region of the island, with which they also share some physiognomical features. A similar situation occurs in in the rocky escarpments of Samaná, where *B. simaruba*, *Coccotrinax gracilis* Burret, *A. antillarum*, *L. weingartianum* and *O. dilleni* occur frequently. As a result of water stress, these habitats tend to exhibit deciduous species related to the

*Chrysophyllo oliviformi-Sideroxyletum salicifolii* Cano & Velóz 2011 and *Zamio debilis-Metopietum toxiferi* Cano & Velóz 2011 associations of Cano & Velóz (2012); (Photograph 2).

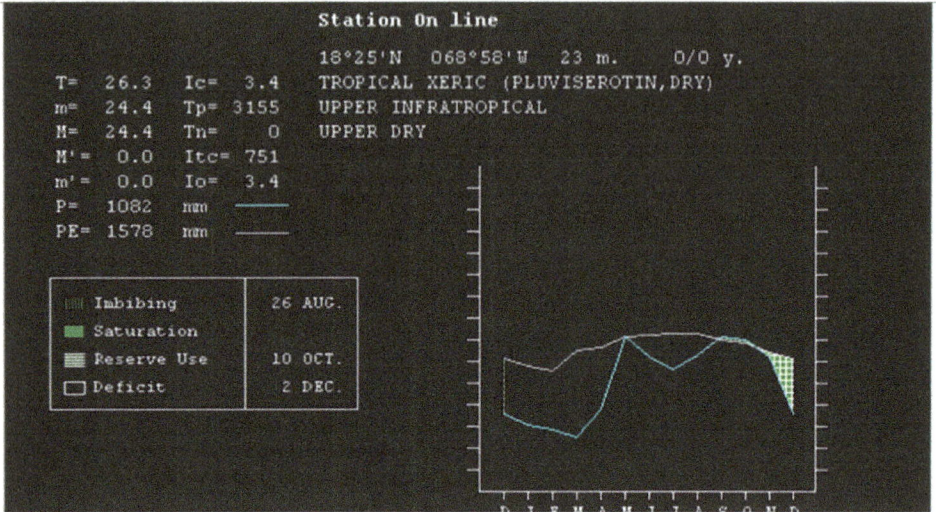

Fig. 5. Weather station of La Romana (40 years): upper infratropical, upper dry, (A7).

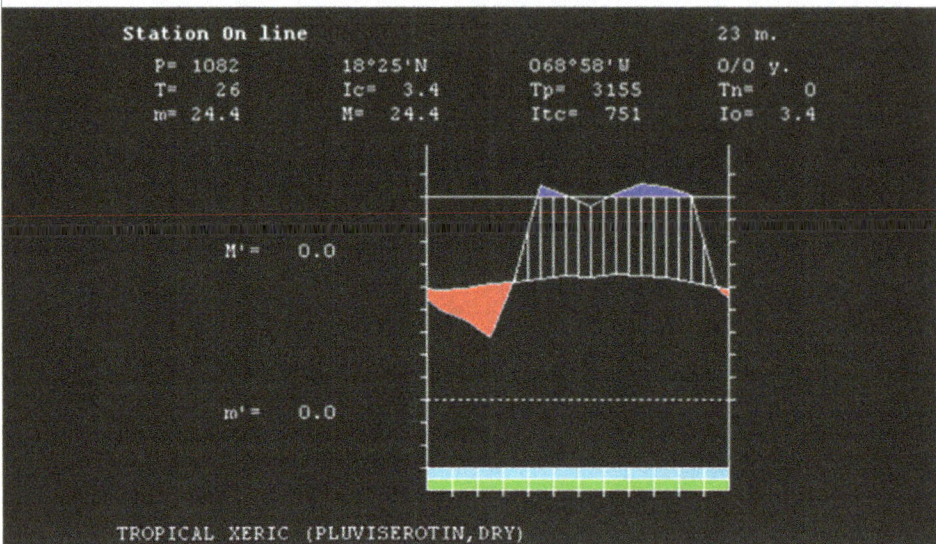

Fig. 6. Weather station of La Romana (40 years): tropical pluviseasonal, xeric, (A7).

Dry and subhumid areas with serpentinicolous vegetation belonging to the phytosociological classes of *Tabebuio-Bureserea* Knapp (1964) Borhidi 1991 and *Phyllantho-Neobracetea valenzuelanae* Borhidi & Muñiz in Borhidi et al. (1979) are of great interest.

In humid areas the macrobioclimate is tropical pluvial (Figs. 7, 8) with no dry season. Actually, rainfall rates are higher than 2,000 mm. These humid areas tend to be located on mountain ranges, such as the Cordillera Septentrional, Cordillera Central, Sierra de

Photo 2. Dry, edaphoxerophilous forest (Parque Nacional del Este, Dominican Republic) in Area A7.

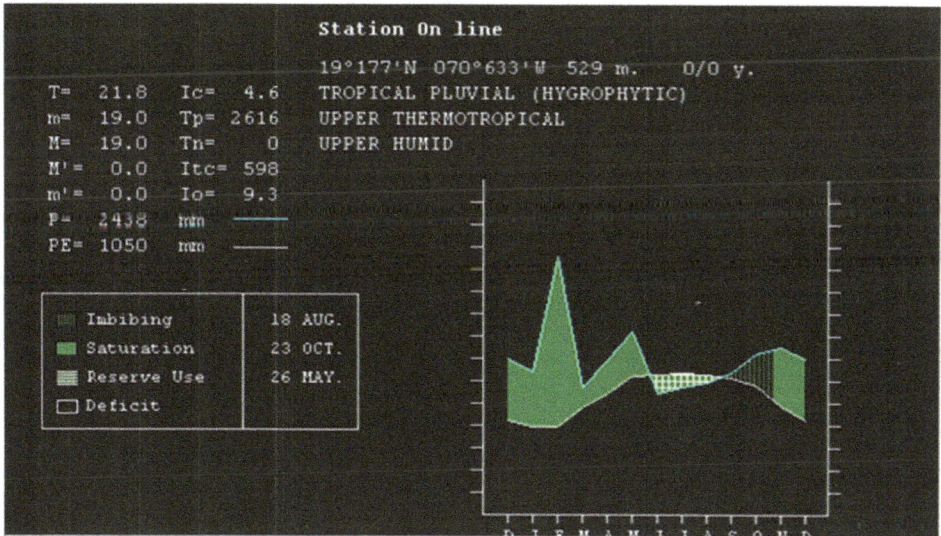

Fig. 7. Weather station of Jarabacoa (1931-2000): upper thermotropical, upper huimid, (A16).

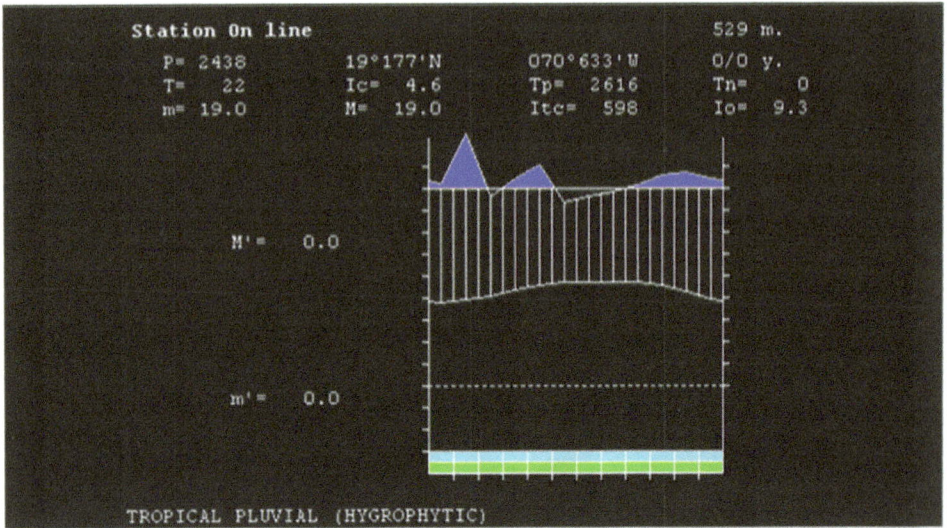

```
Station On line                                        529 m.
     P= 2438          19°177'N         070°633'W       0/0 y.
     T=   22          Ic=  4.6         Tp=  2616       Tn=   0
     m= 19.0          M=  19.0         Itc=  598       Io=  9.3


                M'=   0.0



                m'=   0.0




          TROPICAL PLUVIAL (HYGROPHYTIC)
```

Fig. 8. Weather station of Jarabacoa (1931-2000): tropical pluvial, hygrophytic, (A16).

Photo 3. Humid-hyperhumid forest in Mogotes of Haitises. Dominican Republic, (A6).

Bahoruco, Cordillera Oriental, Los Haitises (Photograph 3), and the Samana Peninsula. In these sites, pluvial, humid plant formations and broad-leaved, ombrophilous forests are usually found. The physiognomical profile of these forests is variable. In the Loma la Herradura (Cordillera Oriental) the dominant plants are *Sloanea berteriana* Choisy, *Ormosia krugii* Urb., *Didymopanax morototoni* (Aubl.) Dcne. & Planch., and *Oreopanax capitatus* (Jacq.) Dcne. & Planch. In stream bottoms, the manacla forest or *manaclar* of *Prestoea montana* (Grah.) Nichol, exists, along with associated *Guarea guidonia* (L.) Sleumer, *D. morototoni, Alchornea latifolia* Sw., and *Eugenia domingensis* Berg (Honer & Jiménez, 1994).

On the Cordillera Central (A16), in the Ebano Verde Scientific Reserve (Photograph 4), the ombrophilous forest is dominated by species in the genus *Magnolia* , which are endemic to the island. These are *Magnolia pallescens* Urb. & Ekm. and *Magnolia domingensis* Urb., together with the wind tree *Didymopanax tremulus* Krug & Urb., *Ocotea leucoxylon* (Sw.) Lanessan, *Persea oblongifolia* Kopp, *Cyrilla racemiflora* L., *Cecropia schreberiana* Miq., *Dendropanax arboreus* (L.) Dcne. & Planch., and other endemic species, such as *Myrsine nubicola* A. Liogier, *Odontadenia polyneura* (Urb.) Woods, *Marcgravia rubra* A. Liogier, *Pinguicula casabitoana* J. Jiménez, and *Tabebuia vinosa* A. Gentry. A similar situation is found in the Loma la Herradura with the *manaclar* or community of *P. montana*, which takes refuge in the most humid gullies. When these plant communities are altered and their cover decreased, the fern community or *calimetal* of *Dicranopteris pectinata* (Willd.) Underw. and *Gleychenia bifida* (Willd.) Spreng. (May, 2000) emerges immediately.

Photo 4. Humid forest of the Cordillera Central. Dominican Republic, (A16).

In the Loma Humeadora the wind tree cloud forest of *D. tremulus* is located at altitudes ranging from 1,100 to 1,315 m and occurs in combination with *Clusia clusioides* (Griseb.) D´Arcy, *C. racemiflora, Ocotea foeniculacea* Mez, *Lyonia alainii* W. Judd and *P. montana*. At lower altitudes, e.g. 850-1,100 m, in gullies and hillsides with slopes of 45-60 ° and a thick, water-retaining layer of fallen leaves, *P. montana* becomes dominant in association with *A. latifolia, O. leucoxylon, Bombacopsis emarginata* (A. Rich.) A. Robins., *S. berteroana, Mora abbottii* Rose & Leon., *Turpinia occidentalis* (Vent.) G. Don, *Bactris plumeriana* Mart., and *Ditta maestrensis* Borhidi (Mejía & Jiménez, 1998).

The relevés conducted both in the Cordillera Central and in Sierra de Bahoruco not only revealed different substrates but also clear dissimilarities in the broad-leaved forest. *M. pallescens* and *M. domingensis* are peculiar to the Cordillera Central, and *Magnolia hamorii* Howard only occurs in the Sierra de Bahoruco. The forests of *M. hamorii* and *D. tremulus* occur in combination with a large number of endemic species, such as *Lasianthus bahorucanus* Zanoni, *Psychotria guadalupensis* (DC.) Howard, *H. domingensis* Urb., *Mecranium ovatum* Cog. (a locally endemic plant), *Vriesea tuercheimii* (Mez.) L.B. Smith, *Macrocarpaea domingensis* Urb., *Cestrum daphnoides* Griseb., *Hypolepis hispaniolica* Maxon, *Columnea domingensis* (Urb.) Wiehler, and *Ilex tuerckheimii* Loes. Cano et al. (2009a) included this vegetation in the classes *Ocoteo-Magnolietea* Borhidi & Muñiz in Borhidi, Muñiz & Del Risco (1979) and *Weinmannio-Cyrilletea* Knapp (1964).

Photo 5. Cloud forest of Bahoruco. Dominican Republic, (A12).

Our study of high mountain areas was carried out in the Sierra de Bahoruco (A12) and the Cordillera Central (A16), which we crossed from Constanza to Sán José de Ocoa. The high mountain macrobioclimate is tropical pluviseasonal and mesophytic. From a physiognomical point of view, the plant formations sampled between 1,203 m (Sierra Bahoruco) and 2,383 m (Cordillera Central) are similar and are a pine forest of *Pinus occidentalis* Sw. In these territories precipitation rates are lower, as the sea of clouds carried by trade winds that supports broad-leaved forest at lower elevations never reaches these higher altitudes. Winter time temperatures can fall below 0 °C. Cold, xeric conditions in the high mountain support the *P. occidentalis* forest, which in the Cordillera Central is accompanied 8-10 endemic species per sampling relevé.. A similar scenario has been observed in Sierra de Bahoruco (Photograph 5), where pine forest support an average record of 20 endemic plants per relevé. The high level of endemism in these two mountain ranges is likely attributable to a lengthy period of isolation.

In the Cordillera Central these forests develop on siliceous substrates and exhibit a large number of endemic species, such as *I. tuerckheimii*, *Ilex fuertesiana* (Loes.) Loes., *Garrya fadyenii* Hooker, *Mikania barahonensis* Urb., *Myrica picardae* Krug & Urb., *Rubus eggersii* Rydberb., *Tetrazygia urbaniana* (Cogn. in Urb.) Croizat ex Moscoso, and *Fuchsia pringsheimii* Urb. Endemic parasitic species, such as *Pinus occidentalis*, *Dendropemon pycnophyllus* Krug & Urb., and *Dendropemon constantiae* Krug & Urb. play an important role in these pine forests. Meanwhile, in the underbrush of this pine forest the Gramineae *Isachne rigidifolia* (Poir.) Urb. is abundant, but where the pine forest becomes sparse it is replaced by a formation of tufted Gramineae dominated by *Danthonia domingensis* Hack. & Pilg., which covers large areas above 1,800 m in the Cordillera Central.

By contrast, the pine forest of *P. occidentalis* develops on limestone soils in Sierra de Bahoruco and includes a different floristic composition. As particularly interesting plants, special mention must be made of the endemic species *Coccotrinax scoparia* Becc. *Agave intermixta* Trel., *Senecio barahonensis* Urb., *Cestrum brevifolium* Urb., *Eupatorium gabbii* Urb., *Lyonia truncatula* Urb., *Sideroxylon repens* (Urb. & Ekm.) T. Pennigton, *Cordia selleana* Urb., *Narvalina domingensis* Cass., and *Galactia rudolphiodes* (Griseb.) Benth. & Hook. *var. haitiensis* Urb., together with other endemic grasses, including *Pilea spathulifolia* Groult, *Tetramicra ekmanii* Mansf., *Artemisia domingensis* Urb., *Gnaphalium eggersii* Urban, and *Polygala crucianelloides* DC. In our opinion, high mountain pine forest habitats endemic to Hispaniola (Cano et al., 2011a) include *Dendropemom phycnophylli-Pinetum occidentalis* Cano, Velóz & Cano-Ortiz 2011 and *Cocotrino scopari-Pinetum occidentalis* Cano, Velóz & Cano-Ortiz 2011 (Photograph 6).

### 4.2 Distribution analysis of endemic species

Endemic plant species are found in many of the 19 floristic areas of Hispaniola. The total number of endemic species of all 19 floristic areas combined is 2,094. Meanwhile, the total number of endemic taxa is 1,162. The difference of these two figures, 932 taxa, indicates that a large number of endemic species is widely distributed, occurring widely across the island. However, there are three areas that should be considered as hot spots of endemism, including areas A12, A16, A13 (Fig. 2). Second in levels of endemism, A4 and A9 exhibit endemicity levels far above average, and are particularly interesting because of their endemic species exclusive to the territory. Also, in A18 and A19 are found endemic plant genera.

Photo 6. Community of *Pinus occidentalis* and *Coccotrinax scoparia*. Sierra de Bahoruco. Dominican Republic. As. *Cocotrino scopari-Pinetum occidentalis* Cano, Velóz & Cano-Ortiz (2011), (A12).

## 4.3 Biogeographical analysis

The geological background of the island, the wide spectrum of bioclimatic thermotypes (which range from infratropical to supratropical standards) and ombrotypes (which range from semiarid to hyperhumid standards), the origin of the flora as a result of dispersal routes, and the intense isolation of *sierras* and mountains have generated a high level of endemic habitats and species.. The island supports 1,284 plant genera, of which 31 are endemic, including: *Zombia, Leptogonum, Arcoa, Neobuchia, Fuertesia, Sarcopilea, Salcedoa, Eupatorina, Vegaea, Coeloneurum, Theophrasta, Haitia, Stevensia, Samuelssonia, Hottea, Tortuella,* and *Anacaona*, among others. Some endemic genera are monotypic and have a fairly restricted distributional area. This is the case of *Vegaea pungens* Urb., *Zephyranthes ciceroana* M. Mejía & R. García, *Gautheria domingensis* Urb., *M. domingensis, Omphalea ekmanii* Alain, *Gonocalyx tetrapterus* A. Liogier, *G. ekmanii, Reinhardtia paiewonskiana* R.W. Read, *T. Zanoni &* M. Mejía, *Pseudophoenix ekmanii* Burret, and *Salcedoa mirabaliarum* F. Jiménez & L. Katinas. Others are endemic at a very local level, as is the case for *Pinguicola casabitoana, Fuertesia domingensis* Urb., *P. quisqueyana, M. jimenezii* Alain and *S. montecristina*.

According to Liogier (1996), the total richness of Hispaniola plants is 5,800 taxa, Including islands Beata, Saona, Gonave and Tortuga.. Subsequently, Mejía (2006) increased the figure to 6,000 vascular species encompassing 1,284 genera, including 2,050 endemic species (34.1% of all Hispaniola plant species). This high level of endemism makes Hispaniola one of the world's hot spots for the conservation of the flora, and is consistent with the high levels of endemism on other Carribean islands. For example, Cuba supports a total of 6,500 species, approximately 50% of the species of which are endemic, and including 66 endemic genera. In comparison, Madagascar, in the Palaeotropical Kingdom, African Subkingdom supports a total of 12,000 plant species, 80% of which are endemic, and distributed among 12 families and 350 genera (Costa, 1997).

Our biogeographical description of island of Hispaniola includes 1,582 endemic species distributed in 19 floristic areas (Cano et al., 2010a). This high proportion of endemic taxa, together with the existence of peculiar vegetation catenae supports ascription of the rank of biogeographical province to Hispaniola. The Hispaniola Province supports 154 endemic species that are broadly distributed across the island (Table 1; Fig. 9). Of these widely distributed species, 114 are in the family *Melastomataceae* (Cano et al., 2010a).

Despite the presence of a relatively large number of widely distributed endemic species, Pearson analysis produced low pairwise correlations between floristic areas A12 and A16 (r = 1.25), A16 and A13 (r = 1.17), and A12 and A13 (r = 1.23). We attribute these low scores todifferences in geological, edaphic, climatological, and land use factors. Low pairwise correlation. For example, in the latter case, both areas A12 and A13 have calcareous substrates, but A13 has sustainedmore intense human land use impacts. A16 and A17 are distinctively different floristically because the Massif du Nord (A17) is an extension of the Cordillera Central (A16). The common occurrence of calcareous outcrops in A17, and the high intensity of human use there increased the difference between these two areas, making A17 more similar to A15 (northwest of Haiti; Jaccard analysis revealed a dissimilarlity distance of 0.9 between areas A12 and A16 a 10% match and a 90% difference; (Fig. 10). We obtained similar results in the comparison of A16 and A17, and this analysis confirmed that A17 was more similar to A15. Also, Jaccard analysis for areas A12 and A13 corroborated the Pearson analysis (Cano et al., 2010 a).

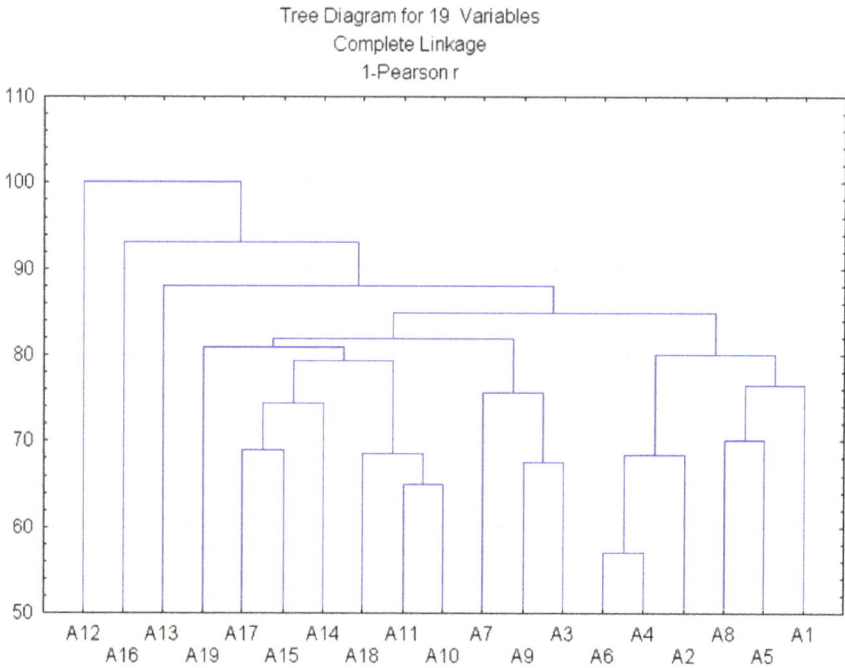

Fig. 9. Cluster diagram for 19 variables based on Pearson´ r. Correlation analysis of the 19 areas (X-axis) and the number and endemism per area (Y-axis).

|  | A1 | A2 | A3 | A4 | A5 | A6 | A7 | A8 | A9 | A10 | A11 | A12 | A13 | A14 | A15 | A16 | A17 | A18 | A19 |
|---|---|---|---|---|---|---|---|---|---|---|---|---|---|---|---|---|---|---|---|
| A1 | 0.00 | 0.99 | 0.95 | 0.93 | 0.96 | 1.00 | 1.03 | 0.95 | 0.99 | 0.96 | 1.02 | 1.13 | 1.06 | 0.99 | 1.01 | 1.05 | 1.03 | 0.95 | 0.97 |
| A2 | 0.99 | 0.00 | 1.03 | 0.73 | 0.97 | 0.86 | 0.97 | 0.91 | 1.02 | 1.02 | 1.04 | 1.09 | 1.07 | 1.03 | 1.02 | 1.05 | 1.04 | 1.02 | 0.99 |
| A3 | 0.95 | 1.03 | 0.00 | 0.99 | 1.02 | 1.04 | 0.95 | 0.97 | 0.85 | 0.97 | 0.96 | 1.09 | 1.04 | 0.96 | 0.98 | 1.09 | 0.95 | 1.00 | 1.02 |
| A4 | 0.93 | 0.73 | 0.99 | 0.00 | 0.81 | 0.71 | 0.93 | 0.98 | 0.98 | 1.02 | 1.06 | 1.17 | 1.10 | 1.04 | 1.04 | 1.05 | 1.06 | 1.03 | 1.00 |
| A5 | 0.96 | 0.97 | 1.02 | 0.81 | 0.00 | 0.91 | 1.02 | 0.88 | 0.98 | 1.01 | 1.02 | 1.03 | 1.03 | 1.01 | 1.02 | 0.93 | 1.02 | 1.01 | 1.01 |
| A6 | 1.00 | 0.86 | 1.04 | 0.71 | 0.91 | 0.00 | 1.04 | 0.95 | 1.01 | 0.96 | 1.02 | 1.11 | 1.05 | 1.02 | 1.04 | 1.07 | 1.02 | 1.02 | 1.02 |
| A7 | 1.03 | 0.97 | 0.95 | 0.93 | 1.02 | 1.04 | 0.00 | 1.01 | 0.92 | 0.97 | 1.01 | 1.03 | 1.06 | 1.00 | 1.03 | 1.12 | 1.01 | 1.02 | 1.02 |
| A8 | 0.95 | 0.91 | 0.97 | 0.98 | 0.88 | 0.95 | 1.01 | 0.00 | 1.02 | 1.01 | 1.01 | 1.04 | 1.02 | 1.01 | 1.01 | 1.00 | 1.01 | 1.01 | 1.01 |
| A9 | 0.99 | 1.02 | 0.85 | 0.98 | 0.98 | 1.01 | 0.92 | 1.02 | 0.00 | 0.91 | 0.97 | 1.00 | 1.08 | 0.99 | 1.01 | 1.09 | 1.02 | 1.00 | 0.99 |
| A10 | 0.96 | 1.02 | 0.97 | 1.02 | 1.01 | 0.96 | 0.97 | 1.01 | 0.91 | 0.00 | 0.81 | 1.05 | 0.97 | 0.95 | 0.84 | 1.01 | 0.93 | 0.86 | 1.01 |
| A11 | 1.02 | 1.04 | 0.96 | 1.06 | 1.02 | 1.02 | 1.01 | 1.01 | 0.97 | 0.81 | 0.00 | 1.11 | 0.92 | 0.92 | 0.89 | 1.09 | 0.97 | 0.83 | 0.99 |
| A12 | 1.13 | 1.09 | 1.09 | 1.17 | 1.03 | 1.11 | 1.03 | 1.04 | 1.00 | 1.05 | 1.11 | 0.00 | 1.23 | 1.05 | 1.14 | 1.25 | 1.16 | 1.09 | 1.09 |
| A13 | 1.06 | 1.07 | 1.04 | 1.10 | 1.03 | 1.05 | 1.96 | 1.02 | 1.08 | 0.97 | 0.92 | 1.23 | 0.00 | 1.02 | 1.04 | 1.17 | 1.01 | 1.02 | 1.03 |
| A14 | 0.99 | 1.03 | 0.96 | 1.04 | 1.01 | 1.02 | 1.00 | 1.01 | 0.99 | 0.95 | 0.92 | 1.05 | 1.02 | 0.00 | 0.93 | 1.04 | 0.91 | 0.97 | 1.01 |
| A15 | 1.01 | 1.02 | 0.98 | 1.04 | 1.02 | 1.04 | 1.03 | 1.01 | 1.02 | 0.84 | 0.89 | 1.14 | 1.04 | 0.93 | 0.00 | 1.07 | 0.86 | 0.94 | 0.89 |
| A16 | 1.05 | 1.05 | 1.09 | 1.05 | 0.93 | 1.07 | 1.12 | 1.00 | 1.09 | 1.01 | 1.09 | 1.25 | 1.17 | 1.04 | 1.07 | 0.00 | 1.04 | 1.04 | 1.06 |
| A17 | 1.03 | 1.04 | 0.95 | 1.06 | 1.02 | 1.02 | 1.01 | 1.01 | 1.02 | 0.93 | 0.97 | 1.16 | 1.01 | 0.91 | 0.86 | 1.04 | 0.00 | 0.99 | 0.99 |
| A18 | 0.95 | 1.02 | 1.00 | 1.03 | 1.01 | 1.02 | 1.02 | 1.01 | 1.00 | 0.86 | 0.83 | 1.09 | 1.02 | 0.97 | 0.94 | 1.04 | 0.99 | 0.00 | 1.01 |
| A19 | 0.97 | 0.99 | 1.02 | 1.00 | 1.01 | 1.02 | 1.02 | 1.01 | 0.99 | 1.01 | 0.99 | 1.09 | 1.03 | 1.01 | 0.89 | 1.06 | 0.99 | 1.01 | 0.00 |

Table 1. Results of the Pearson's pairwise.

Within the Hispaniola Province, we distinguished two subprovinces: the Central Subprovince and the Caribbean-Atlantic Subprovince, based on differences in their geological origins and bioclimatic, floristic, and vegetational profiles. Key elements from our previous work that supports this differentiation include: 1) the siliceous Central Subprovince, which includes only one sector containing one single area (A16). The Central subprovince (A16) includes only the central sector (1.1) Fig. 12; (Fig.9); and 2) the calcareous Caribbean-Atlantic Subprovince, with the following 5 biogeographical sectors 2.1.-Bahoruco-Hottense (A12, A13); 2.2.- Neiba-Matheux-Northwest (A14, A15, A17 and A19); 2.3.-Azua- Sán Juan-.Hoya Enriquillo-Port-au-Prince-Artiobonite-Gonaivës (A9, A10, A11 and A18); 2.4.-Caribeo-Cibense (A3, A7 and A8); and 2.5.-North (A1, A2, A4, A5 and A6), which comprises the other 18 areas (Fig. 11 and 12).

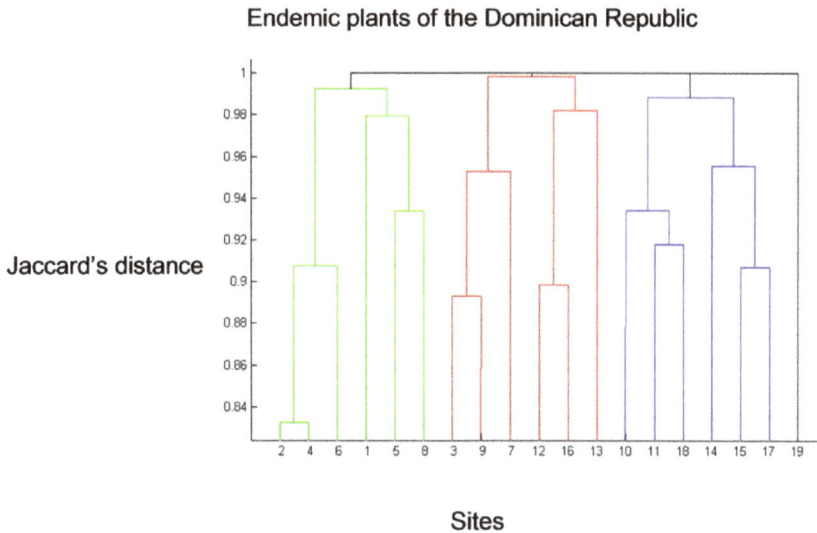

Endemic plants of the Dominican Republic

Jaccard's distance

Sites

Fig. 10. Cluster analysis of the Jaccard's dissimilarity distances between study areas.
1.- Green group in northern areas with calcareous substrates domain and serpentines.
2.- The red areas are grouped with south central siléceos substrates and dry areas.
3.- Blue are grouped in areas of highly altered Republic of Haiti.
(Jaccard's analysis)

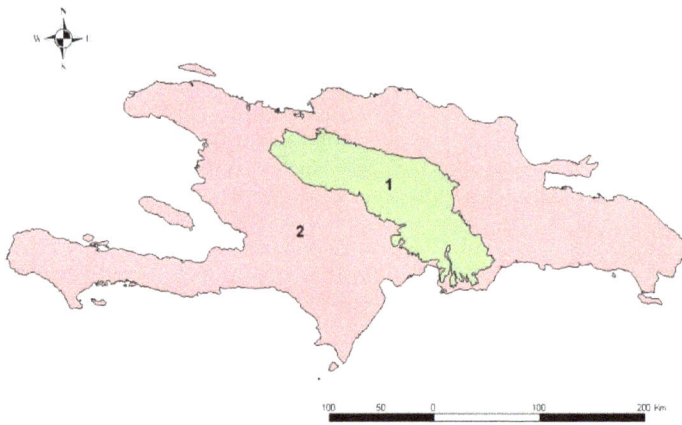

Fig. 11. Map of the biogeographical subprovinces of Hispaniola. from Cano et al. (2009a). 1.-Central (A16); 2.- Caribbean-Atlantic.

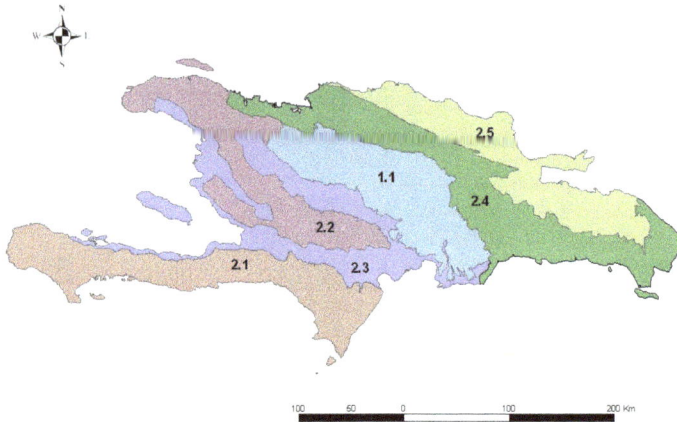

Fig. 12. Map of the biogeographical sectors of Hispaniola. 1.1.- Central (A16). 2.1 - Bahoruco-Hottense, 2.2.- Neiba-Matheux-Northwest, 2.3.-Azua - Sán Juan-. Hoya Enriquillo-Port-au Prince-Artiobonite-Gonaivës, 2.4.-Caribeo-Cibense, 2.5.-North. Redrawn from Cano et al. (2009a).

# 5. Conclusions

## 5.1 Floristic biogeography

Hispaniola had been previously considered as a single biogeographical sector (Rivas-Martínez et al., 1999) and included in the province of Antilles. However , we recommend promoting Hispaniola to the rank of a biogeographical province (Cano et al., 2009a) and contend that the province of Antilles should be promoted to a superprovince, namely that of Central-Eastern Antilles, with the inclusion of the islands of Jamaica and Hispaniola. Hispaniola, together with a group of some small nearby islands –Beata, Saona, Gonave and Tortuga– make up the Hispaniola Province. Our biogeographical proposal for the Hispaniola Province encompasses 5 biogeographical sectors and 19 areas within the two countries of the Dominican Republic and the Republic of Haiti. We distinguish the following biogeographic hierarchy: The Caribbean-Mesoamerican Region. Central-Eastern Antilles Superprovince. Hispaniola Province. 1.- Central Subprovince. 1.1.- Central Sector. A16.- Central-Eastern Area. 2.- Caribbean-Atlantic Subprovince. 2.1.- Bahoruco-Hottense Sector. A12.- Bahoruco-La Selle Area. A13.- Hottense Area. 2.2.- Neiba-Matheux-Northwest Sector. A14.- Neiba-Matheux Area. A15.- Northwest Area. A17.- Centre-West Area. A19.- Tortuga Island Area. 2.3.- Azua-Sán Juan-.Hoya Enriquillo-Port-au-Prince-Artiobonite-Gonaivës Sector. A9.- Azua-Sán Juan-Hoya Henriquillo Area. A10.- Central Plain Area. A11.- Port-au-Prince-Arbiobonite-Gonaivës Area. A18.- Gonave Island Area. 2.4.- Caribeo-Cibense Sector. A3.- Cibao Valley Area. A7.- Caribeo-Eastern Area. A8.- Yamasense Area. 2.5.-North. A1.- Cordillera Septentrional Area. A2.- Atlantic Coastal Area. A4.- Samanense Area. A5.- Eastern Area. A6.- Haitiense Area.

# 6. Acknowledgments

This paper is the result of a number of research projects granted by the AECI (*Ministerio de Asuntos Exteriores* of Spain, in cooperation with the Universidad INTEC and the Jardín Botánico Nacional Rafael Ma. Moscoso of Santo Domingo, Dominican Rep.). We thank the Universidad de INTEC and the Jardín Botánico Nacional for taxonomic support. We also express our gratitude to Dr. Francisco J. Esteban Ruiz for invaluable help with statistical analyses.

# 7. References

A.R.N. (2004). *Atlas de los Recursos Naturales de la República Dominicana*. Ed. Frank Moya Pons. Secretaría de Estado de Medio Ambiente y Recursos Naturales, ISBN, 99934-996-4-1, Santo Domingo.

Borhidi, A. (1991). *Phytogeography and vegetation ecology of Cuba*, Akadémiai Kiadó, ISBN, 963 05 5295 7, Budapest.

Cano, E., Veloz, A., García Fuentes, A., León, Y., Ruiz, L., Salazar, C., Tores, J.A., Cano-Ortiz, A. & Montilla, R.J. (2006b). Caracterización preliminar y biodiversidad del bosque seco en República Dominicana, *Proceedings of* IX Congreso Latinoamericano de Botánica, Santo Domingo, 18-25 Junio, 2006.

Cano, E., García Fuentes, A., León, Y., Veloz, A., Ruiz, L., Salazar, C., Torres, J.A., Cano-Ortiz, A. & Montilla, R.J. (2006a). Ensayo biogeográfico de La Española, *Proceedings of* IX Congreso Latinoamericano de Botánica, Santo Domingo, 18-25 Junio, 2006.

Cano, E. & Veloz Ramirez, A. (2012). Contribution to the knowledge of the plant communities of the Caribean-Cibensean sector in the Dominican Republic. *Acta Botanica Gallica* (in press), ISSN: 1253-8078

Cano, E., Veloz Ramirez, A., Cano-Ortiz, A. & Esteban, F. J. (2009a). Distribution of Central American *Melastomataceas*: A Biogeographical analysis of the Islands of the Caribbean. *Acta Botanica Gallica*, Vol. 156 (4): 527-558, ISSN: 1253-8078.

Cano, E., Veloz Ramírez, A., Cano-Ortiz, A. & Esteban, F. J. (2009b). Analysis of *Pterocarpus officinalis* forest in the Gran Estero (Dominican Republic). *Acta Botanica Gallica*, 156 (4): 559-570, ISSN: 1253-8078.

Cano, E., Veloz Ramírez, A. & Cano-Ortiz, A. (2010a). Contribution to the biogeography of the Hispaniola (Dominican Republic, Haiti). *Acta Botanica Gallica*, 157 (4): 581-598, ISSN: 1253-8078.

Cano, E., Veloz Ramírez, A. & Cano-Ortiz, A. (2010b). The habitats of *Leptochloopsis virgata* in the Dominican Republic. *Acta Botanica Gallica*, 157 (4): 645-658, ISSN: 1253-8078.

Cano, E., Veloz Ramírez, A. & Cano-Ortiz, A. (2011). Phytosociological study of the *Pinus occidentalis* forests in the Dominican Republic. *Plant Biosystems*, 145(2): 286-297, ISSN: 1126-3504.

Cano, E., Cano-Ortiz, A., Velóz, A., Alatorre, J. & Otero, R. (2012). Comparative analysis between the mangrove swamps of the Caribbean and those of the State of Guerrero (Mexico). *Plant Biosystems* (in press), ISSN: 1126-3504.

Costa, M. (1997). Biogeografía in Izco et al. (eds). McGraw-HILL-INTERAMERICANA DE ESPAÑA, Madrid.

García, R., Mejía, M. & Zanoni, TH. (1994). Composición florística y principales asociaciones vegetales en la reserva científica de Ébano Verde, Cordillera Central, República Dominicana. *Moscosoa*, 8:86-130, ISSN: 0254-6442, ISSN: 0254-6442.

García, R., Mejía, M., Peguero, B & Jiménez, F, (2001). Flora endémica de la Sierra de Bahoruco, República Dominicana. *Moscosoa*, 12:9-44, ISSN: 0254-6442.

García, R. & Clase, T. (2002). Flora y vegetación de la zona costera de las provincias Azua y Barahona, República Dominicana. *Moscosoa*, 13:127-173, ISSN: 0254-6442.

García, R., Mejía, M., Peguero, B., Salazar, J. & Jiménez, F. (2002). Flora y vegetación del Parque Natural del Este, República Dominicana. *Moscosoa*, 13:22-58, ISSN: 0254-6442.

Guerrero, A., Jiménez, F., Höner, D. & Zanoni, T. (1997). La flora y la vegetación de la Loma Barbacoa, Cordillera Central, República Dominicana. *Moscosoa*, 9:84-116, ISSN: 0254-6442.

Höner, D. & Jiménez, F. (1994). Flora vascular y vegetación de la Loma la Herradura (Cordillera Oriental, República Dominicana). *Moscosoa*, 8:65-85, ISSN: 0254-6442.

Liogier, A.H. (2000). *Diccionario Botánico de nombres vulgares de la Española.* Jardín Botánico Nacional Dr. Rafael Ma. Moscoso, Santo Domingo.

Liogier, A. H. (1996-2000). *La Flora de la Española. Vol. I-IX.* Jardín Botánico Nacional Dr. Rafael Ma. Moscoso, ISBN, 84-8400-217-9, Santo Domingo.

May, TH. (1997). Fases tempranas de la sucesión en un bosque nublado de *Magnolia pallescens* después de un incendio (Loma de Casabito, Reserva Científica Ébano Verde, Cordillera Central, República Dominicana). *Moscosoa*, 9: 117-144, ISSN: 0254-6442.

May, TH. (2000). Respuesta de la vegetación en un calimetal de *Dicranopteris pectinata* después de un fuego, en la parte oriental de la Cordillera Central, República Dominicana. *Moscosoa*, 13: 113-132, ISSN: 0254-6442.

May, TH. (2001). El endemismo de especies de plantas vasculares en República Dominicana, en relación con las condiciones ambientales y los factores biogeográficos. *Moscosoa*, 12:60-78, ISSN: 0254-6442.

May, TH. & Peguero, B. (2000). Vegetación y flora de la Loma El Mogote, Jarabacoa, Cordillera Central, República Dominicana. *Moscosoa*, 11: 11-37, ISSN: 0254-6442.

Mejía, M. & Jiménez, F. (1998). Flora y vegetación de Loma La Humeadora, Cordillera Central, República Dominicana. *Moscosoa*, 10: 10-46, ISSN: 0254-6442.

Mejía, M., García, R. & Jiménez, F. (1998). *Gaussia attenuata* (O.F.Cook) Becc. y *Coccothrinax barbadensis* (Lodd. Ex Mart) Becc. (Arecaceae). Dos nuevos registros para la Isla Española. *Moscosoa*, 10:3-9, ISSN: 0254-6442.

Mejía, M., García, R. & Jiménez, F. (2000). Sub-región fitogeográfica Barbacoa-Casabito: Riqueza florística y su importancia en la conservación de la flora de la Isla Española. *Moscosoa*, 11:57-106, ISSN: 0254-6442.

Mejía, M. (2006). Flora de la Española: conocimiento actual y estado de conservación. *Proceedings of* IX Congreso Latinoamericano de Botánica, Santo Domingo, Junio, 2006.

Mollat, H., Wagner, B.M., Cepek, P. & Weiss, W. (2004). *Mapa geológico de la República Dominicana 1:250.000*. Geologisches Jahrbuch, ISBN, 3-510-95927-2, Hannover.

Peguero, B. & Salazar, J. (2002). Vegetación y flora de los cayos Levantado y La Farola, Bahía de Samaná, República Dominicana. *Moscosoa*, 13: 234-262, ISSN: 1126-3504.

Rivas Martínez, S., Sánchez Mata, D. & Costa, M. (1999). North American boreal and western temperate forest vegetation. Syntaxonomical synopsis of the potential natural plant communities of North America, II. *Itinera Geobotanica*,12:5 326, ISSN: 0213-8530.

Salazar, J., Peguero, B. & Veloz, A. (1997). Flora de la Península de Samaná, República Dominicana. *Moscosoa*, 8:133-188, ISSN: 0254-6442.

Samek, V. (1988). Fitorregionalización del Caribe. *Revista del Jardín Botánico Nacional (Cuba)*, 9: 25-38, ISSN: 0253-5696.

Slocum, M., Mitchell, T., Zimmerman, J.K. & Navarro, L. (2000). La vegetación leñosa en helechales y bosques de ribera en la Reserva Científica de Ébano Verde, República Dominicana. *Moscosoa*, 11:38-56, ISSN: 0254-6442.

Takhtajan, A. (1986). *Floristic Regions of the World.* Transl. by T.J. Crovello and ed. by A. Cronquist. University of California Press, ISBN, 0520040279, Berkeley.

Tolentino, L. & Peña, M. (1998). Inventario de la vegetación y uso de la tierra en la República Dominicana. *Moscosoa*, 10:179-203, ISSN: 1126-3504.

Trejo-Torres, J.C. & Ackerman, J.D. (2001). Biogeography of the Antilles based on a parsimony analysis of orchid distributions. *Journal of Biogeography*, 28:775-794, ISSN: 1365-2699.

Velóz, A. & Peguero, B. (2002). Flora y vegetación del Morro de Montecristi, República Dominicana. *Moscosoa,*13: 81-107, ISSN: 0254-6442.

Zanoni, Th., Mejía, M. Pimentel, J.D. & García, R. (1990). La flora y vegetación de los Haitises, Republica Dominicana. *Moscosoa*, 6:46-98, ISSN: 0254-6442.

# 6

# Biogeographic Insights in Central American Cycad Biology

Alberto S. Taylor B.[1], Jody L. Haynes[2], Dennis W. Stevenson[3],
Gregory Holzman[4] and Jorge Mendieta[1]
[1]*Universidad de Panamá, Departamento de Botánica,*
[2]*IUCN/SSC Cycad Specialist Group,*
[3]*The New York Botanical Garden,*
[4]*Pacific Cycad Nursery,*
[1]*Panamá*
[2,3,4]*USA*

## 1. Introduction

Cycads (Cycadophyta) are dioecious, palm-like, gymnosperm (non-flowering seed plants) trees and shrubs found broadly across the tropical belt of the world (Fig. 1). Cycads are veritable living fossils once thought to date from the Mesozoic as early as 280 million years ago (mya); however, recent reevaluation of the fossil evidence indicates they originated in the upper Paleozoic, more than 300 mya (Pott et al., 2010). They predated, and were contemporaneous with, the dinosaurs and have survived to the present. As a group, the cycads have been systematically restructured, making use of morphological and molecular characters and cladistic analysis (Caputo et al., 2004). Three families are currently considered valid (Cycadaceae, Stangeriaceae, and Zamiaceae), with a world total of about 330 described species (Osborne et al., 2012). *Cycas* and *Zamia* are the most widely distributed and researched genera.

Cycads are solely represented in Panama by the genus *Zamia*, with 16 described species and a new one soon to be described, of these 17 species, 12 are endemic. Although these 17 species are many more than were previously known and/or described by earlier workers (Stevenson, 1993; Schutzman et al., 1998), still many others are likely to be found, especially in the forested areas near the Colombian border. *Zamia*, the most widely distributed genus of cycads after *Cycas* (Fig. 2), extends the length of the isthmus, stretching from Darién Province in the east, abutting on Colombia, to Bocas del Toro and Chiriquí provinces in the west, next to the Costa Rican border (Fig. 3). Only the south-central provinces of Herrera and Los Santos lack cycad populations, and this is likely due to habitat destruction (deforestation for agricultural and cattle ranching endeavors) during past centuries, rather than a natural deficiency. Conversely, on the Atlantic slope there are many cycad populations, including *Z. pseudoparasitica*, the only known obligate gymnosperm epiphyte.

The main objective of this paper is to present data justifying the study of cycads as biogeographical subjects, taking into account their evolutionary antiquity, as well as their extinct and extant distribution, general biology, and reproductive strategy. In this context, we present an example of tropical cycad biology, with the Isthmus of Panama as our main biogeographical framework. Much work on the biology and conservation of cycads has been carried out in this region for over 100 years, starting with many of the noteworthy European plant and animal collectors of the 19th century, up to the Panamanian national workers of today. The cycads serve as a living case of the result of island biogeography, with current distribution being limited by both biotic (pollinators, for example) and abiotic (climate, geography/orography, and habitat restrictions) factors.

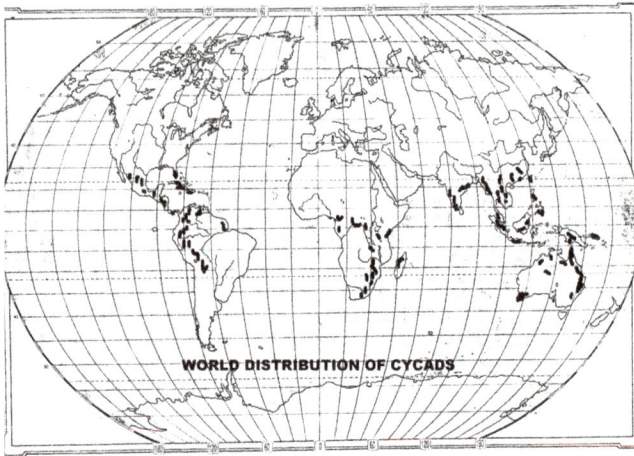

Fig. 1. World distribution of extant cycads.

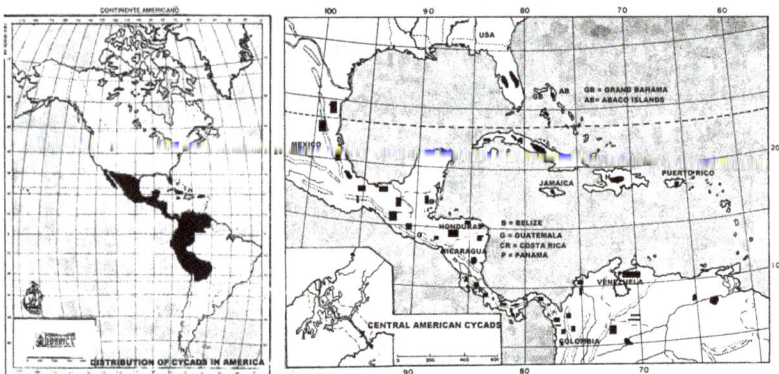

Fig. 2. Distribution of cycads (mostly *Zamia*) on the North and South American continents and in Central America and the Caribbean.

The Panama Land Bridge—definitely closed about 3-2.7 mya (Coates & Obando, 1996; Webb, 1997)—served as a physical connection to the northern and southern continents and allowed rapid and almost explosive species radiation. The Isthmus of Panama consists of about 75,500 km² of land mass (Wikipedia, 2011) and harbors about two-thirds the number of cycad species as Mexico, which is over 20 times larger in area. Panama has almost the same number of species as Colombia, which also is nearly 20 times larger. Rising and falling sea levels and the closing of the land bridge made possible the paleontological "Great American Biodiversity Interchange," involving the movement of animals and plants in two directions (north-south and south-north), including cycads from North and South America, into the isthmus (Fig. 4. & Table 1), where they colonized new habitats and expanded into new niches, giving rise to new species that, together with the existing flora, make this the richest cycad flora per unit area in the neotropics (Stevenson, 1993). Even though only one genus is represented, it is by far the most diverse among the cycads and has the widest distribution after *Cycas* (Fig. 2).

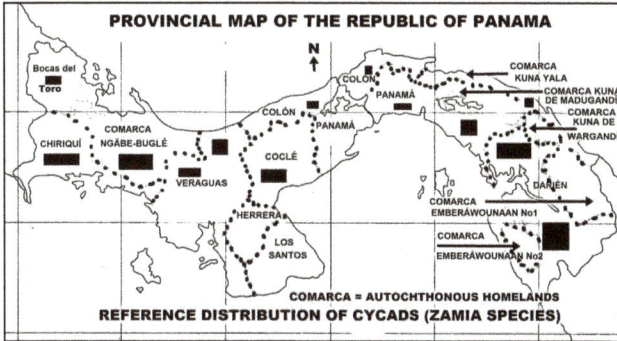

Fig. 3. Generalized distribution of cycads (*Zamia*) across the Isthmus of Panama.

Fig. 4. Interchange of *Zamia* species on the Isthmus of Panama and new types that evolved there. 1-4. *Zamia* species from Colombia; 5-7. *Zamia* species from Central America.

*Zamia* includes the only known obligate gymnosperm epiphyte (*Z. pseudoparasitica*). There are types with underground stems (e.g., *Z. ipetiensis, Z. cunaria, Z. dressleri*) and others with large, above-ground trunks (e.g., *Z. elegantissima, Z. nesophila, Z. obliqua*); types with leaflets ranging from quite small (*Z. acuminata*) to very large (e.g., *Z. hamannii, Z. imperialis*); lowland types sporadically inundated with seawater (*Z. nesophila, Z. hamannii*) to upland types above 1,500 masl (e.g., *Z. lindleyi, Z. pseudomonticola*); and types pollinated by only one pollinator (e.g., *Pharaxonotha* in *Z. manicata*) to those pollinated by two different genera of pollinators (e.g., *Pharaxonotha* and *Rhopalotria* [or *Rhopalotria*-like weevil] in *Z. obliqua* and *Z. fairchildiana*).

| CYCAD SPECIES | COUNTRY OF ORIGIN | | |
|---|---|---|---|
| | PANAMA | COSTA RICA | COLOMBIA |
| Z. acuminata* | Yes | Yes | No |
| Z. cunaria | Yes | No | No |
| Z. dressleri | Yes | No | No |
| Z. elegantissima | Yes | No | No |
| Z. fairchildiana | Yes | Yes | No |
| Z. hamannii | Yes | No | No |
| Z. imperialis | Yes | No | No |
| Z. ipetiensis | Yes | No | No |
| Z. lindleyi | Yes | No | No |
| Z. manicata | Yes | No | Yes |
| Z. nesophila | Yes | No | No |
| Z. neurophyllidia | Yes | Yes | No |
| Z. obliqua | Yes | No | Yes |
| Z. pseudomonticola | Yes | Yes | No |
| Z. pseudoparasitica | Yes | No | No |
| Z. skinneri | Yes | No | No |
| Zamia cf. elegantissima | Yes | No | No |

* Considered not synonymous with the described species from Nicaragua.

Table 1. Relationship between the cycad flora of Costa Rica, Panama, and Colombia.

## 2. Methods

More than 40 populations including all 17 species of isthmian cycads were studied between 1998 and 2010. We collected data on plant size and reproductive strategy, including coning cycles, sex ratios, time of dehiscence and receptivity of cones, pollination and pollinators, ripening of seed cones, and disarticulation of cones. Efforts were made to find dispersal agents of cycads, and data on herbivory were collected on almost every visit. Data were also collected pertaining to biogeographic and taxonomic issues of the plicate-leaved species, which have puzzled botanists for more than a century. A concise phylogeny of the genus, distribution among forest types, and conservation threats rounded out the data gathered. Trips were made on foot, by car, by special all-terrain vehicles, by train, boat, and plane, and even on horseback. Different detection methods for pollinating beetles were employed, including exclusion experiments, the use of greased microscope slides tied to petioles near pollen and ovulate cones, and baiting pollinators outside of the normal coning period using dehiscing cones from garden plants.

### 2.1 Study area

The Isthmus of Panama occurs in the tropics of the western hemisphere with the following coordinates: 7°12'07"to 9°38'46" North and 77°09'24"to 83°03'07" West. It is limited to the west by Costa Rica, to the north by the Caribbean Sea, to the east by Colombia, and to the south by the Pacific Ocean. The isthmus emerged through volcanic activities and elevation of the sea floor (Coates, 2003), and is presently traversed by two cordilleras: one extending from Nicaragua to the western side of the canal (the Talamanca range), and the other from the east side of the canal into Colombia (which is related to the Andean system of South America; Wikipedia, 2012). The result is a typically montane and irregular topography with peaks to 3,300 masl. Moreover, there are narrow coastal plains and, in certain zones, contracted alluvial terraces. The cordillera rocks are volcanic in origin. In the central part of the isthmus where the Panama Canal is found, the topography is sinuous, being formed by rocks from the rising sea beds (Instituto Geográfico Nacional Tommy Guardia, 2007).

Because of its geography, Panama has a typical tropical climate. Average temperature is 21.0°C at the lowest and 34.5°C at the highest, with an annual average of 27.7°C. April has the highest temperature readings, and the lowest are found in December and March. Rainfall varies according to region, with values from 1,000-7,000 mm annually, and there are two distinct seasons: a rainy season from the end of April until November, and a dry season from December to April. Panama's rainfall pattern is governed by the seasonal displacement of tropical Pacific air pockets and subtropical ones of the Atlantic and is modified by local orography, although there is a fairly uniform pattern of rainfall distribution throughout the year on the Caribbean slope (ANAM, 2006).

Climatic conditions in Panama are adequate for forest development. However, at present only 44% of the surface has a natural forest covering. Deforested sites are being used for agriculture and cattle ranching (ANAM, 2006). Extant forests are found north and east in Panama, and the process of forest destruction began with the arrival of the first settlers and has henceforth continued. At present, the continued destruction of natural forests is due to the advancing agricultural and cattle ranching frontier in different regions and to the development of surface mining activity (ANAM, 2006).

## 2.2 Forest types & cycad populations

During the Great American Biodiversity Interchange, a great many species of plants migrated across the isthmus, some of which have become adapted to the local climate, forming the structure of the natural vegetation observed today. In the case of forests, species biodiversity is from different sources. High-altitude forests have Mesoamerican influence, while in the lowlands, conditions have been influenced by neotropical species and migration from the Chocó region (Louis Berger Group, 2003). The differences between lowland vegetation and those from the highlands are due principally to temperature changes concordant with changes in altitude.

Panamanian cycads are associated with different forest types, and can be found alongside natural montane vegetation and with localized forests in the lowlands. Moreover, it is possible to find cycads in association with human-altered forests. In general, Panamanian cycads grow in three general types of forests: lowland evergreen forest, lowland semideciduous forest, and submontane evergreen forest (although the geographic distribution of most species coincides with the lowland forest types) (Fig. 5, Table 2).

**Lowland evergreen forests:** These are found below 700 masl; annual rainfall is above 3,000 mm, and temperature varies between 24-30°C. Tree tops bear leaves year-round, which leads to the understory being relatively sparse. These forests have densities of around 600 individual trees per hectare, representing approximately 100 distinct species (Mendieta, unpubl.). Thirteen species of native cycads are found here, representing 76% of described isthmian species (Fig. 5, Table 2).

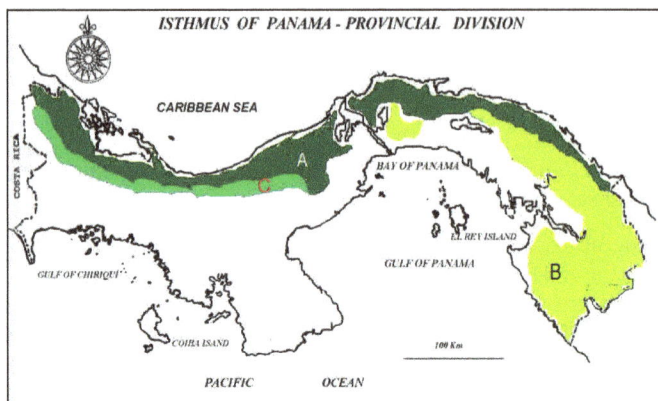

Fig. 5. Distribution of isthmian cycads according to forest type. A. Lowland evergreen forest;
B. Lowland semideciduous forest; C. Submontane evergreen forest.

| Species | Lowland evergreen forest | Evergreen submontane forests | Lowland semideciduous forests |
|---|---|---|---|
| Z. acuminata | X | | |
| Z. cunaria | X | | |
| Z. dressleri | X | | |
| Z. elegantissima | X | | |
| Z. fairchildiana | X | | |
| Z. hamannii | X | | |
| Z. imperialis | X | | |
| Z. ipetiensis | X | | X |
| Z. lindleyi | | X | |
| Z. manicata | X | | |
| Z. obliqua | X | | X |
| Z. pseudomonticola | | X | |
| Z. pseudoparasitica | | X | |
| Z. nesophila | X | | |
| Z. neurophyllidia | X | | |
| Z. skinneri | X | | |
| Z. cf. elegantissima | | | X |
| **Totals** | **13** | **3** | **3** |

Table 2. Presence of isthmian cycads in natural forest types.

**Lowland semideciduous forests.** These are found below 700 masl, with annual rainfall between
2,000-3,000 mm, and with temperatures ranging between 24-30°C. Under these conditions, 30-
40% of tree species are deciduous, and understory species are generally evergreen year round.
Tree densities are on the order of 430  individual trees per hectare, representing approximately
100 distinct species (Mendieta, unpubl.). Three native cycad species are found in this forest type,
representing 17% of described isthmian species (Fig. 5, Table 2).

**Submontane evergreen forests.** These are found between 700-1,200 masl, with annual rainfall above 3,000 mm. Altitude modifies these forests, so the temperature varies between 21-24°C during the year. Tree tops are evergreen year round. Tree densities are on the order of 700 individual trees per hectare, representing 110 distinct species (Louis Berger Group, 2000). Three native cycad species are found in this forest type, representing 17% of described isthmian species (Fig. 5, Table 2).

## 3. Isthmian cycad flora

Cycads abound throughout the neotropics and even in warm temperate areas of the globe, but they are much more abundant in tropical and subtropical sites with high humidity and temperatures (Fig. 1). Considering its relatively small size and abundance of cycad species, the Isthmus of Panama is an ideal locale for their study, and cycad populations across the isthmus have been examined by the authors over the past three decades. The following is a brief summary of the Panamanian cycad flora:

1.  *Zamia cunaria* (Fig. 6A) – a plant with no above-ground stem and with very few leaves in mature specimens; its habit and morphology are similar to that of *Z. ipetiensis*; it is a narrow Panamanian endemic, found only along a road encompassing national land and part of the Kuna or Dule homeland in eastern Panama, where it grows at 400-650 masl.
2.  *Z. dressleri* (Fig. 6B) – another species with no above-ground stem and with very few leaves in mature specimens (1-3 generally). This is a Panamanian endemic found on the Atlantic side of the isthmus in Colón Province. It is very hardy and will withstand rough handling, drought, and some direct sunlight.
3.  *Z. obliqua* (Fig. 6C) – a shrub to large tree with distinctive leaflets indicative of its specific epithet, although there are variations on the theme; it is found in separate sites in Darién Province, and there are disjunct, remnant populations abutting on the Atlantic coast, near the Panama Canal area, and in northwestern Panamá Province, growing in extended sympatric association with *Z. imperialis* and *Z. pseudoparasitica* in north-central Panama. It almost rivals *Z. pseudoparasitica* in distribution and is the only widely-disjunct cycad in Panama.

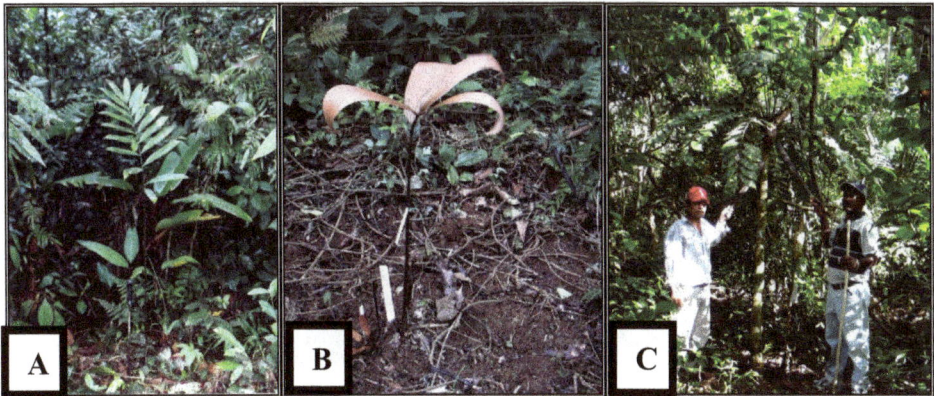

Fig. 6. Biodiversity of Panamanian cycads. A. *Zamia cunaria*, with underground stem; B. *Z. dressleri*, idem, with pollen cones near soil level; C. *Z. obliqua* with huge, arborescent trunk.

4.  *Z. pseudoparasitica* (Fig. 7A) – this obligate tree-dwelling cycad, while a Panamanian endemic, is perhaps the widest ranging species in Panama, being found on high, and sometimes lower, tree branches from Bocas del Toro Province in the west to Colón Province near the northern end of the Panama Canal; germinated seeds have been found on the ground, but these apparently are eaten by unknown animals, and neither juvenile nor adult plants are ever seen growing terrestrially.

5.  *Z. acuminata* (Fig. 7B) – a plant with no above-ground stem in habitat (potted plants may show aboveground stems). It is found above 500 masl in east-central Coclé Province. This species is small, with many long-acuminate leaves; it has been treated as synonymous with another plant along the Río San Juan in southern Nicaragua, but it likely is a different species (Schutzman, 2004; Lindström, pers. comm.). The long-acuminate leaves are distinctive in the genus.

6.  *Z. imperialis* (Fig. 7C-D) – previously known as the "red leaf *skinneri*," it is quite attractive with crimson-colored leaves seen all over the forest when the leaves are emerging. The plants are generally short-stemmed with a few large leaves (1-3 generally) and huge leaflets. It is an endemic species of the central Atlantic versant.

Fig. 7. Biodiversity of Panamanian cycads (cont'd.): A. Specimens of *Zamia pseudoparasitica* hanging from tree branches in north-central Panama; B. *Z. acuminata*, a small-leaved species with underground stems; C-D. *Z. imperialis,* one of the large-leafleted, plicate-leaved species.

7. *Z. hamannii* (Fig. 8A) – a shrubby to large, arborescent, island-dwelling plant that often grows near the sea in Bocas del Toro Province. Emergent leaves are rosy red to orange. This species is similar to *Z. imperialis* in overall appearance, but has a silver tomentum on emerging leaves, bears many more leaves, and has different seed cones.

8. *Z. nesophila* (Fig. 8B) – shrubby to arborescent, with leaves green (emergent and mature) and leaflets and cones differing from those of other plicate-leaved species. It is found near beaches, withstanding sea spray and occasional inundation by seawater. Like *Z. hamannii*, it is an island-dweller in the Bocas del Toro region.

9. *Z. skinneri* – long confused with other plicate-leaved taxa but recently recharacterized (Taylor et al., 2008; 2012), it grows large, arborescent trunks, bears green-emergent leaves, and is widely distributed throughout coastal mainland Bocas del Toro Province.

10. *Z. neurophyllidia* – similar to *Z. skinneri* (and similarly confused for many years), but much smaller and with a different form of seed cone. It is distributed in extreme northwestern Panama and may occur in Costa Rica.

11. *Z. elegantissima* – a small shrub to very large tree (Fig. 8C), it is found near eastern and northern Colón Province on the Atlantic versant and is sympatric with *Z. cunaria*, where it grows a stem to almost 3 m high. Populations are very small, numbering less than 50 plants, and emergent leaves are cream-colored.

12. *Z.* cf. *elegantissima* – known among cycad workers as "blanco," or white, because of the snow-white appearance of its new leaves, this species appears to be a diminutive variant of *Z. elegantissima*. However, this is the only Panamanian species with this type of emergent leaf. The shape and size of the cones and seeds, as well as the overall size and number of leaves, and the diameter and height of the trunks of mature specimens also are smaller in every respect. Reproductive plants range from nearly acaulescent to 1.5 m high, and this species is endemic to Panamá Province.

13. *Z. pseudomonticola* – this species is very similar to *Z. fairchildiana* (see below) and is distributed in far western Panama.

14. *Z. fairchildiana* – described by Luis Diego Gómez, the same author as *Z. pseudomonticola*, plants apparently are also found in Costa Rica and are somewhat similar in overall appearance to *Z. elegantissima*, but the pointed leaflets have a raised surface on the upper or adaxial side; the size and form of cones and the emergent leaves also differ. It is found in western to southwestern Panama.

15. *Z. lindleyi* – a shrubby to arborescent, narrow-leafleted plant in the high-altitude, humid forests of Chiriquí Province, northwestern Panama (Fig. 8D).

16. *Z. ipetiensis* – a small to medium-sized plant with branching underground stems and leaves to approximately 1 m tall (Fig. 8E). It is endemic to eastern Panama in one of the homelands of the Emberá tribal group. This is the only species in Panama in which the pollen cones in natural populations can number up to 13 on an individual plant.

17. *Z. manicata* – found only in Darién Province, it is the only Panamanian *Zamia* (and one of only two zamias, the other of which is in Peru) having a petiolule, or leaflet stalk, near the base of which is a small, gland-like protuberance (Fig. 8F-G).

## 4. Reproductive biology, sex ratio & herbivory in isthmian cycads

### 4.1 Reproductive biology

Most Panamanian cycads in natural populations, when coning, do so only once a year. Four months are required for cone maturation, 1-2 months for pollen dehiscence and receptivity,

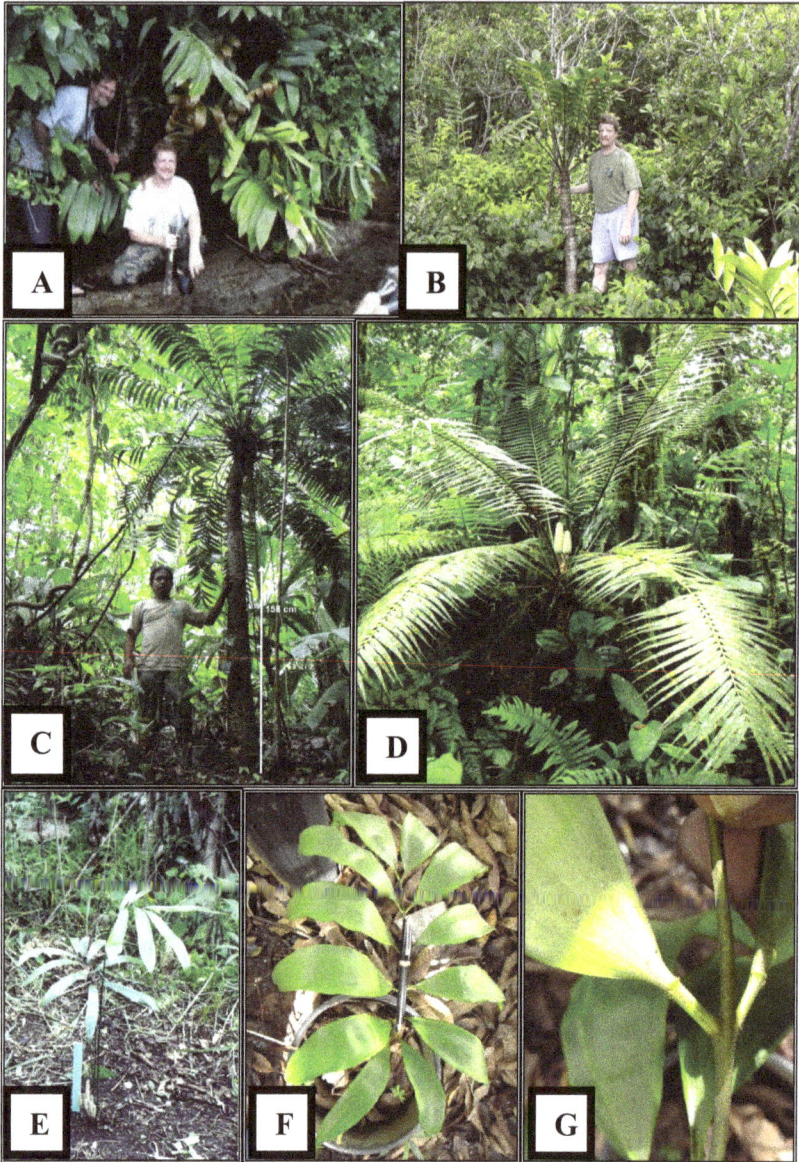

Fig. 8. Biodiversity of Panamanian cycads (cont'd.). A. *Zamia hamannii*, a species sometimes inundated by seawater; B. *Z. nesophila*, a species mostly growing on sand and sometimes inundated by seawater; C. *Z. elegantissima*; D. *Z. lindleyi*, a high mountain species, with two pollen cones; E. *Z. ipetiensis*, a species similar to *Z. cunaria* but which produces many more pollen cones per plant; F-G. *Z. manicata*, showing the petiolules and associated gland at the base of the leaflets.

and 1-2 years for seed ripening. There are exceptions (e.g., *Zamia acuminata* seeds ripen in about eight months after pollination), but this is the general rule for almost all of the populations studied. Outside the natural environment (gardens, landscapes, or isolation due to habitat destruction), coning can occur at any time and sometimes occurs many times during a year. We have not observed any stress-triggered reproduction—coning after disturbance as a do-or-die strategy—in Panamanian *Zamia* species, although examples of having mature cones many times a year in natural populations have been observed in *Z. nesophila* in the Bocas del Toro archipelago and in *Z. hamannii* somewhere far away in the same political jurisdiction. *Zamia nesophila* has been observed having dehiscing pollen cones and receptive ovulate cones three times during a given year and *Z. hamannii* twice.

The grandfather of cycad studies, Charles Joseph Chamberlain, thought that cycads were wind pollinated just like other gymnosperms, even knowing that insects were found in, upon, and about pollen cones at the time of pollination (Chamberlain, 1919; 1935), and that others (Pearson, 1906; Rattray, 1935) had, indeed, observed beetles on pollen cones and written about the possibility of insect pollination in the South African cycads *Encephalartos altensteinii* and *E. villosus*. Pollination by insects is now strongly supported for cycads as a group (Norstog & Nicholls, 1997; Wilson, 2002; Proches et al., 2009), and pollinators have been found for all populations studied on the Isthmus of Panama with the exception of *Zamia neurophyllidia*. Pollinators represent weevils that are morphologically similar to *Rhopalotria*, as well as the erotylid pleasing fungus beetle, *Pharaxonotha* (Fig. 9D). *Pharaxonotha* is always found in dehiscing pollen cones and is sometimes accompanied by *Rhopalotria*, suggesting that *Pharaxonotha* is the primary pollinator of isthmian zamias. The insects enter the pollen cones and ovulate cones guided by aromatic chemicals emitted by the cones (Stevenson et al., 1998; Taylor, 2002).

The exact form of pollination and the reward offered to pollinators have been variously discussed by others, and Terry et al. (2007) put forth an odor-mediated "push-pull" pollination mechanism in which the volatiles emitted by dehiscing pollen cones and receptive ovulate cones are increased at certain times in a diel cycle. When the dehiscing cone temperature becomes quite high above ambient, the pollinators are "pushed" out of the cone and move toward available receptive ovulate cones. When the temperature is reduced, the volatile concentration also declines, and the insects are once more "pulled" into the pollen cones. In a preliminary experiment accompanied by the senior author in 2006, Terry and Roemer (unpublished) observed only a slight increase in pollen cone temperature above ambient in an excised cone in the lab that was maintained in a water-absorbent material. The extremely rainy period and no available assistance and infrastructure to safeguard the instruments, however, made it impossible to undertake the field experiment; this will be taken up in the future. Suinyuy et al. (2012) have confirmed the relationship between volatile emissions and insect attraction to dehiscent pollen cones and receptive ovulate cones of *Encephalartos villosus* in South Africa. Thermogenesis exists, but the confirmation of the "push-pull mechanism" of Terry et al. (2007) still remains to be supported in other cycad species, including those of *Encephalartos*.

Another question to consider is the mutualistic reward offered to the plant by pollinator destruction of pollen cones, which, according to Marler (2010), is not only the conveying of pollen to receptive ovulate cones, but also a cryptic benefit of cone tissue disposal which translates into an increase in ultimate lifetime reproductive effort or events. In fact, any event, natural or accidental, that rapidly disposes of cycad cones in a given reproductive cycle will hasten the onset of the next reproductive cycle. This has been observed by the senior author in at least one of the isthmian species (*Zamia* cf. *elegantissima*) in which the destruction of an almost mature ovulate cone by larvae of the lepidopteran *Eumaeus godarti* brought about the onset of a new reproductive cycle in a short time (about a year). When this does not happen, the reproductive cycle takes many years to be repeated in this species. A point should be made here that cone disposal by non-pollinating insects or natural destruction and abscission of pollen cones in isthmian cycads takes place even without the benefit of the pollinators, both in natural populations and in garden-grown plants.

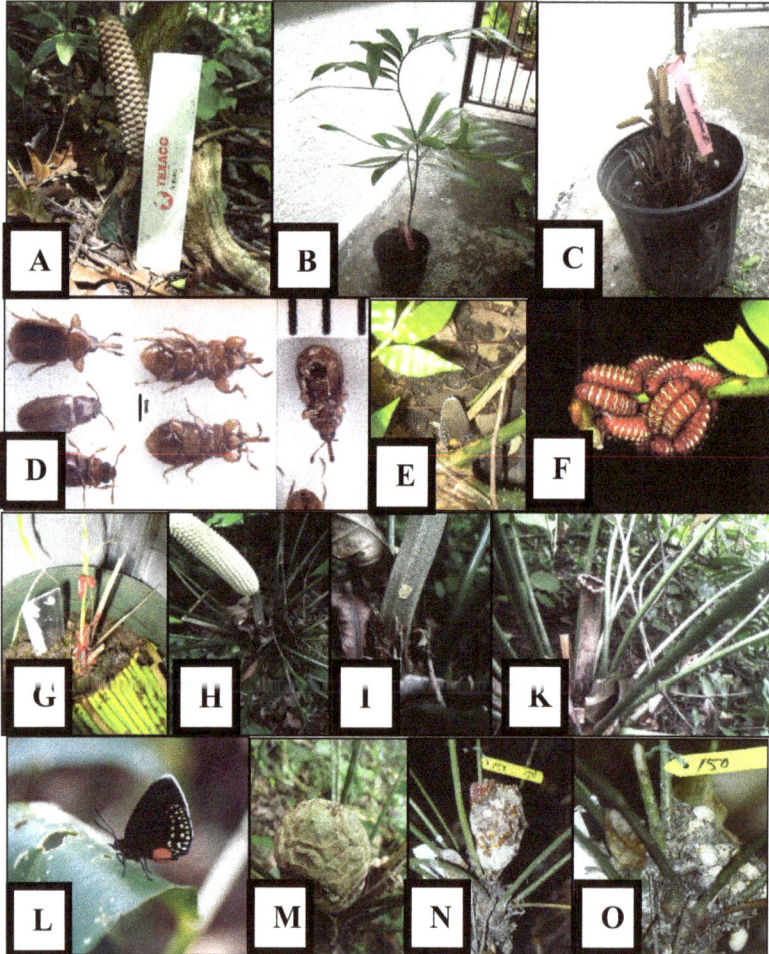

Fig. 9. Pollination and herbivory in Panamanian cycads. A. Pollen bait cone of *Zamia* cf. *elegantissima*; B. Small potted pollen plant of *Z. ipetiensis* used as bait; C. Pollen cones of B; D. Cycad pollinators: *Pharaxonotha* beetles (left middle and left bottom) and *Rhopalotria*-like weevils (center and right); E. *Eumaeus godarti* butterfly laying eggs; F. *E. godarti* larvae on *Z. lindleyi* leaf; G. *Zamia* germling destroyed by *E. godarti* larvae; H-K. Destruction of *Z. elegantissima* pollen cones by *E. godarti* larvae; L. *E. godarti* mature butterfly; M-O. Attacked seed cone of *Z.* cf. *elegantissima* with liberation of seeds (these seeds later germinated).

A method devised by the senior author known as the "pollen bait method" makes use of *ex situ* dehiscing pollen cones of the same or related species taken to a natural population, with the pollen cone peduncle in a water-absorbent material and held in an appropriate receptacle (Fig. 9A). Small potted plants with dehiscing pollen were also used (Fig. 9B-C). After two days, the bait cone is retrieved and, if it was effective (bait cones of the same species are always effective), pollinators (weevils and/or beetles) are captured even outside of the normal coning period for the species. This, of course, means that the pollinators in some form (perhaps as dormant pupae) are present in the populations and are always ready for a dehiscent or receptive cone. Because cones of different species have been effectively used as bait, it follows that some pollinators are either not species-specific or different cycad species emit the same or similar aromatic attractants. This inter-specific attraction is not found in all isthmian species. Pollinators are found in large numbers throughout the Isthmus of Panama. However, because the same insect species sometimes pollinates different cycad species, and different pollinators can pollinate the same species, it appears to be relatively easy for isthmian cycads to extend their ranges. This probably explains the broad distribution of *Zamia obliqua* and *Z. pseudoparasitica* across the isthmus.

## 4.2 Coning cycle & sex ratio

Timing of coning is different for the various species of *Zamia* in Panama; in most populations, individuals do not cone every year, or those that do generally represent a small percentage of the total population (Table 3). However, in the case of *Z. ipetiensis* and two populations of *Z. cunaria*, most individuals that appeared mature, judging from leaf size, bore cones. In populations of other species (e.g., *Z.* cf. *elegantissima*), most individuals observed in the population studied bore cones, but in one coning cycle (year 2000) most plants did not cone. One must consider that ovulate plants use more resources in cone building than pollen plants. While most species appear to have a nearly 1:1 sex ratio, the ratio may vary markedly from one coning cycle to the next. In the case of *Z. ipetiensis*, it is almost certain that there is a bias toward ovulate plants. Pollen plants, however, produce many more cones per plant (up to 13) than other species, and this apparently results in the population being quite vigorous, for there is always enough pollen during a coning cycle to account for the greater number of ovulate plants.

## 4.3 Herbivory

*Eumaeus godarti* (Lepidoptera: Lycaenidae) (Figs. 9E & L) is the principal cycad herbivore in Panama, but only in a very small number of cases has it adversely affected the populations infested, and most plants recover rapidly after being attacked. The life cycle of the insect has been worked out by the senior author for isthmian populations. Even in cases in which the larvae of the butterfly eat most of the mature ovulate cone parenchyma, the seeds are left intact, and these germinate sooner than if the cone is left to disintegrate naturally (Fig. 9M-O). In this case, the herbivore can be considered an opportunistic semi-mutualist *Zamia* symbiont, at least part of the time (Taylor, 1999), and an aid to limited seed dispersal, especially if seed cleaning occurs on plants near natural embankments or cliffs.

| Species & Year | Pop. Size (approx.) | Plants Observed | # Coning ♀ | # Coning ♂ | % Coning | Sex Ratio (♀/♂) |
|---|---|---|---|---|---|---|
| Z. acuminata 1999-2002 | >400 | 170 | 65 | 73 | 81.2 | 0.9 |
| Z. acuminata 2002 | >400 | 170 | 24 | 22 | 27.0 | 1.1 |
| Z. cunaria 2007 | >100 | 31 | 3 | 5 | 25.8 | 0.6 |
| Z. cunaria 2007 | >100 | 28 | 14 | 2 | 57.1 | 7.0 |
| Z. cunaria 2008 | >100 | 36 | 32 | 3 | 97.2 | 10.7 |
| Z. cf. elegantissima 1998-2000 | >100 | 60 | 3 | 14 | 28.3 | 0.2 |
| Z. cf. elegantissima 1999 | >300 | 223 | 9 | 17 | 11.7 | 0.5 |
| Z. cf. elegantissima 1999-2001 | >300 | 223 | 57 | 42 | 44.4 | 1.4 |
| Z. cf. elegantissima 2000-2001 | >700 | 362 | 32* | 61** | 25.7 | 0.52 |
| Z. cf. elegantissima 1999-2001 | >700 | 362 | 90 | 110 | 55.2 | 0.8 |
| Z. ipetiensis 1999-2001 | >300 | 150 | 75 | 29 | 69.3 | 2.6 |
| Z. ipetiensis 2000 | >300 | 55 | 33 | 21 | 98.2 | 1.6 |

Table 3. Coning events in Panamanian cycads during different coning cycles. (* Plants that had coned the year before were still bearing old cones [19]; ** some of these [24] had also coned the year before).

## 5. Dispersal of isthmian cycads

Cycad seeds are formed by two layers: an outer pulpy layer known as the sarcotesta – which harbors toxins in some cycads and is attractive to many biotic dispersers (Fig.10A-B) – and an inner hard layer known as the sclerotesta (Fig. 10C). Cycad seeds range from relatively large to relatively small, according to species (Fig. 10), varying in isthmian species from 2.8-3.0 cm long and 1.6-1.8 cm wide in *Zamia elegantissima* (on the large side), to 1-1.5 cm long and 0.5-0.8 cm wide in *Z. manicata* (medium-sized), down to 1.6-1.9 cm long and 0.4-0.5 cm wide in *Z. acuminata* (the smallest). Most cycad seeds require an after-ripening period, after seed cone disarticulation, to germinate. This period can range from a few weeks, as in most *Zamia* species – including those from Panama – to more than a year, as in some *Encephalartos* species in South Africa. In a year-long experiment at the University of Panama (unpublished), the germination rates for *Z.* cf. *elegantissima*, *Z. acuminata*, and *Z. ipetiensis* seeds decreased during the year, and germination was negligible after that.

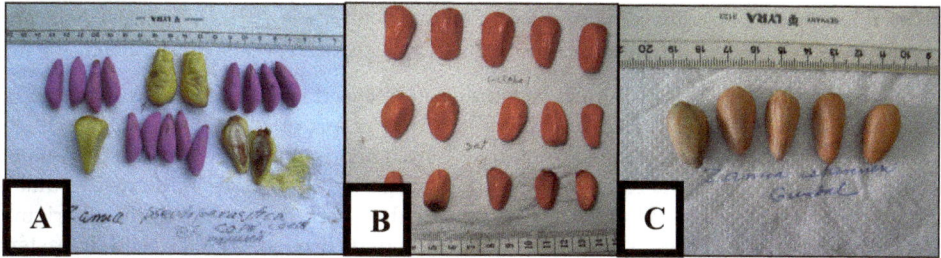

Fig. 10. Istmian cycad seeds. A. *Zamia pseudoparasitica* with yellow sarcotesta (violet color is due to antifungal solution on the sclerotesta); B. *Z. imperialis* (above) and *Z. nesophila* (below) with red sarcotesta; C. *Z. imperialis,* showing the stony sclerotesta.

Palatability of the sarcotesta is known for many cycad species (Norstog & Nicholls, 1997; Jones, 2002), and this should also stand for isthmian species such as *Zamia pseudoparasitica*, the sarcotesta of which exudes a gummy substance when ripe (Fig.10A). The sarcotesta of all isthmian species is red, except for that of *Z. pseudoparasitica*, which is yellow. It generally becomes soft and quite pulpy during after-ripening. At the moment, there are no known long-distance dispersal agents for any cycad species in Panama, except gravity when seeds fall near cliff sides or on the sides of steep hills. Ants and other insects and also crabs have been seen moving seeds over short distances. Perhaps the animals that dispersed seeds no longer exist, or perhaps there has just not been enough monitoring time or effort put into studying and discovering the dispersers. In the case of *Z. pseudoparasitica*, bats and toucans have been reported by a few of our field supporters in western Panama to transport seeds, but definitive proof is still needed. Two of us have seen the toucans flying in the area where the species is growing, and there are bats in the area and lots of anecdotal statements, but no real proof—although the seeds have been seen germinating on the branches of different trees not containing cycads, and the most likely dispersers would be birds, bats, or mammals. Traps with mature seeds were set in preliminary trials and will be repeated, because of both negative results and the field observance of the senior author and one of his field assistants of the overnight consumption of one third of a mature *Z. pseudoparasitica* cone in north-central Panama in an area with toucans and bats.

Janzen (1981a; 1981b), in his seminal studies of the *guanacaste* tree in neighbouring Costa Rica (*Enterolobium ciclocarpum*: Mimosoideae), hypothesized that equids the size of contemporary Costa Rican range horses were present in Costa Rica in the late Pleistocene. Removed by evolutionary forces for about $10^6$ years, they were again introduced as modern horses (*Equus caballus*) by the Spaniards after the European colonization of the Americas. Examples of these horses are Costa Rican free ranging horses that eat and then excrete the *guanacaste* (also known as Corotu or curutu in Panama) seeds in their dung, with most of these germinating later. The equids, then, are the ideal dispersors of the tree, although they were absent for so many thousand years. Other approaches to this problem of apparent missing dispersers (Hunter, 1989) consider that no animal interaction is necessary to explain the spread and germination of *E. ciclocarpum* seed, and that hidrochory (dissemination by water) is suggested as the preferred dispersal strategy. In the case of cycad seeds, their toxicity and knowing that almost all with viable embryos sink in water, hydrochory (except in known cases of a few *Cycas* species, such as *C. litoralis*, *C. micronesica*, and *C. rumphii*) is ruled out in most cases, and dispersal by animals has been reported as actually being oberserved in many cycads in Africa, Southeast Asia, and Australia, but very little information has been given for American cycads (Jones, 2002).

There are anecdotal reports of human consumption of cycad seeds in the aboriginal region of eastern Panama (Taylor, 2012), and there is a first report of the pre-Columbian use of *Zamia* starch grains on cutting utensils in an unknown aboriginal group in western Panama during preceramic times (Dickau, 2007). But in none of these cases is there irrefutable evidence of human beings as actual dispersers of the species. At present, the aboriginal groups that make use of these plants usually take the parts needed from the wild. In eastern Panama, however, we have seen a few plants that have been sown around the Embera village, but still near the original natural population of *Z. ipetiensis*.

# 6. "Island" biogeography & isthmian cycads

As mentioned above, the immigration of cycads from North and South America into the newly forming isthmus and subsequent speciation events—which have occurred from late Pliocene to recent geologic time—account for most of the cycad species on the isthmus based on congruence of observed patterns of phylogenetic relationships combined with the current geographical distribution of species (Caputo et al., 2004). Conversely, the plicate-leaved zamias of northwestern Panama—specifically those in the Bocas del Toro region—may have originated on an ecological time scale, and the current distribution of this unique assemblage may best be explained by a modification of the "dynamic habitat" hypothesis of Gregory and Chemnick (2004), which states that *Dioon* populations in Mexico migrated up and down in elevation, and north and south, in response to warming and cooling climatic cycles corresponding to glaciation events (see Taylor et al., 2008).

## 6.1 The plicate-leaved zamias & their distributions

The plicate-leaved zamias are distributed from western Central America into northern South America. There is evidence that leaflet "plication" evolved independently within the genus, and that the South American plicate-leaved taxa (e.g., *Zamia roezlii* and *Z. wallisii*) are phylogenetically unrelated to those in Central America (Caputo et al., 2004). The Central American representatives are largely restricted to the Atlantic slope of the continental divide and include *Z. dressleri*, *Z. hamannii*, *Z. imperialis*, *Z. nesophila*, *Z. neurophyllidia*, and *Z. skinneri* in Panama (the latter two may also occur in coastal southeastern and south-central Costa Rica, respectively), as well as one or more undescribed species in Costa Rica and southern Nicaragua. All of the named species, except *Z. dressleri*, occur in the Bocas del Toro region, with at least three of these species—*Z. hamannii*, *Z. imperialis*, and *Z. nesophila*—being Panamanian endemics (Taylor et al., 2008).

## 6.2 Evolutionary hypothesis

The Bocas del Toro region of northwestern Panama was a site of intense volcanic activity from about 20 mya to less than 2.8 mya when the final closure of the isthmus took place (Coates et al., 2005; Fig. 11). During this period, there was ample time for the immigration of plant and animal species from the northern and southern continents into the isthmus. Sea level rise and continental submergence over the past few thousand years have created numerous islands and peninsulas in the Bocas del Toro region (Anderson & Handley, 2002) and have resulted in marked morphological differentiation in insular mammals (Anderson & Handley, 2001; 2002) and plants (Taylor et al., 2008) compared to their mainland counterparts. Surprisingly, the processes that have led to this striking level of divergence may have occurred in as little as 5,000-10,000 years, according to recent estimates of the ages of the islands in this region (Anderson & Handley, 2002).

Taylor et al. (2008) hypothesized an explanation for the biogeographical and evolutionary patterns of the plicate-leaved zamias of northwestern Panama in their "dynamic habitat" hypothesis, which is similar to that proposed by Gregory and Chemnick (2004) for the Mexican species of *Dioon*. In Mexico, movement to higher elevations during inter-glacial periods resulted in colonization of new river valleys by *Dioon*, followed by secondary isolation within and between river drainages and subsequent rapid vicariance of the majority of extant species. During ice age phases, however, populations are forced downslope and southward, possibly resulting in genetic remixing. These processes have, then, been repeated numerous times over the past two million years (Gregory & Chemnick, 2004).

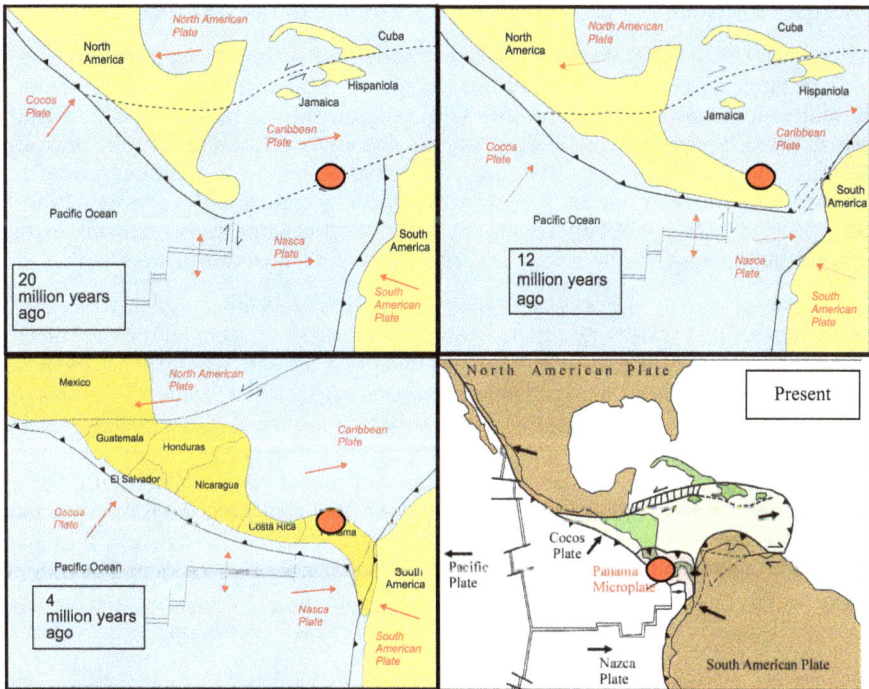

Fig. 11. Timeline of formation of the Isthmus of Panama. Maps kindly donated by Dr. Anthony Coates of the Smithsonian Tropical Research Institute (STRI) in Panama City. (The red dot indicates the approximate location of the Bocas del Toro region of Panama.)

The primary difference between Gregory and Chemnick's (2004) hypothesis and that proposed by Taylor et al. (2008) for the plicate-leaved zamias of Bocas del Toro is that, rather than (or in addition to) temperature fluctuations being the primary driving factor, the current distribution of taxa is primarily due to the periodic rise and fall of sea levels that accompanied the dramatic

Pleistocene changes in climate. Higher sea levels during warmer inter-glacial periods created geographical barriers that isolated cycad populations and allowed them to evolve independently, whereas lowered sea levels during the periodic ice ages allowed for renewed genetic interchange and introgression as populations merged back together into newly exposed land areas. This sea-level change cycle has been repeated several times, following each ice age, resulting in relatively brief but intense periods of isolation.

Following this logic, isolation onto separate islands combined with unique environmental pressures associated with an insular existence likely contributed to allopatric speciation of *Zamia hamannii* and *Z. nesophila* from their mainland counterparts, possibly over just the past 5,000-10,000 years. Similar evolutionary processes likely played a role in the vicariance of *Z. imperialis* and *Z. neurophyllidia*, except that instead of being restricted to different islands, the propensity of these latter species to grow inland away from the coast was likely an important factor in maintaining their isolation from populations nearer the coast, thus allowing for their independent evolutionary pathways (Taylor et al., 2008). Because the isolation processes have occurred several times in the past million years, genetic evidence will be needed to distinguish the most recent event from previous sea-level increases.

As Taylor et al. (2008) suggested, *Zamia skinneri* is unique within this species complex with regard to its biogeographical and evolutionary patterns because it is widespread throughout coastal mainland Bocas del Toro Province (and possibly into southeastern Costa Rica) and because it exhibits surprisingly high levels of phenotypic variation within and among populations, thereby suggesting that the entire coastal region of northwestern Panama represents a zone of repeated, episodic convergence/hybridization events resulting from regular cycles of rising and falling sea levels. If so, this area represents an extremely important genetic refuge for the plicate-leaved zamias of northwestern Panama.

The plicate-leaved zamias of Bocas del Toro are members of a convoluted, actively evolving species complex that has undergone numerous cycles of geographic isolation and admixture/introgression during inter-glacial periods and ice ages, respectively (Schutzman, 2004; Taylor et al., 2008). The coastal mainland region, which is home to *Zamia skinneri*, represents the center of diversity of this complex, and *Z. hamannii*, *Z. imperialis*, *Z. nesophila*, and *Z. neurophyllidia* are incipient species that have evolved allopatrically on the periphery as a result of geographic isolation (in the case of *Z. hamanni* and *Z. nesophila*) and/or other as yet unknown abiotic and/or biotic factors (in the case of *Z. imperialis* and *Z. neurophyllidia*) perhaps during the current inter-glacial period (Taylor et al., 2012). As such, the 'Skinneri Complex' represents another example (in addition to the Mexican species of *Dioon*) of rapid evolution in this ancient plant group, effectively countering the assumption of gradual divergence of taxa over evolutionary time that is often ascribed to such long-living organisms.

## 7. Phylogeny of isthmian cycads

All of the described species of isthmian cycads have acquired species status by the usual method of Adansonian discrimination, based on overall similarity of morphologic characteristics considered important. However, newer methods in phylogenetics that combine morphology and molecular data are opening new avenues for understanding speciation in cycads. Morphological and morphometric character sets in cycad identification and systematics should be used with caution, however, because under *ex situ* conditions (gardens, homes, landscapes), modification of those characters can cause confusion if not taken into account and clarified. Conversely, data from morphology and morphometrics, combined with a molecular approach and analyzed using cladistic methods, can define monophyletic groups (clades) of taxa (Schuh and Brower, 2009; Wiley and Lieberman, 2011).

Within the Zamiaceae, *Zamia* Linnaeus, with 71 currently recognized species (Osborne et al., 2012) distributed from southeastern USA through Central America and the Caribbean to Brazil, is the least-studied genus in terms of phylogenetics. However, there are two previous studies on the phylogeny of *Zamia* (Caputo et al., 1996; 2004), with the latter using a data matrix that combined molecular data with morphological data. Caputo et al. (2004) demonstrated that *Zamia* is divided into four major clades, which are also supported by geography (Fig. 12). The backbone of the consensus tree was composed of various species found in Mega-Mexico (see Nicolade-Morejon et al. [2009] for a thorough discussion of *Zamia* in Mega-Mexico) and is herein referred to as the Mega-Mexico Grade (Fig. 12). While the pattern of phylogenetic relationships uncovered by Caputo et al. (2004) was broadly congruent with geographical distribution, it was not so with morphological resemblance within the genus. An example of the latter is the independent evolution of leaflet "plication" within the genus, as mentioned above and discussed again below.

The current phylogenetic research on *Zamia* has expanded the sampling of species by 30%, from 23 species in the Caputo et al. (2004) paper to 36 species in this paper. All data were obtained and analyzed exactly as in Caputo et al. (2004), and the results produced one fully resolved tree whose topology has remained the same as the consensus tree of Caputo et al. (2004), including the nesting of *Chigua* within *Zamia* (Fig. 12). Not too surprisingly, Lindström (2009) recently synonymized *Chigua* with *Zamia*. The phylogenetic and biogeographic implications of this expanded study led to some surprising conclusions as summarized below.

The results of this investigation offer new insights into the phylogeny of *Zamia* and provide fertile ground for further testing the addition of new species. One of the most surprising results is the relationship of the Panamanian endemic *Z. ipetiensis* in the Panama Clade with the Colombian endemic *Z. wallisii* in the South American Clade. These two species show little or no resemblance to each other, other than belonging to the same genus. However, given that *Z. manicata* is found only in Colombia and the Darién region of Panama — and as such represents a South American element in Colombia — the same may well be true for *Z. ipetiensis*, which, along with *Z. cunaria*, may represent a relic. This issue should be more rigorously tested by inclusion of the latter in future studies.

Within the Panama Clade (the "*pseudoparasitica* clade" of Caputo et al. [2004]), there is a subclade of plicate-leaved zamias, consisting of *Z. dressleri*, *Z. neurophyllidia*, and *Z. skinneri*, with *Z. obliqua* (without plicate leaflets) nested inside. Interestingly, *Z. obliqua* was taken by Caputo et al. (2004) to be an exception in their inclusive clade of Central American taxa because its distribution was then thought to be extreme eastern Panama to the southern Chocó of Colombia; however, the known range of this species actually extends to at least west-central Panama. As mentioned previously, the presence of plicate leaflets has clearly evolved at least twice within *Zamia*, and the relationship of plicate leaflets to habitat and other factors is enigmatic, as is the apparent loss of leaflet plication in *Z. obliqua*.

The distribution of *Zamia obliqua* also warrants further field work, with more samples taken across its range included in phylogenetic studies. *Zamia obliqua* has a disjunct distribution from central Panama to the Darién and adjacent Chóco of Colombia and then to Cabo Corrientes, Colombia. At issue here is whether or not *Z. obliqua* represents a disjunct

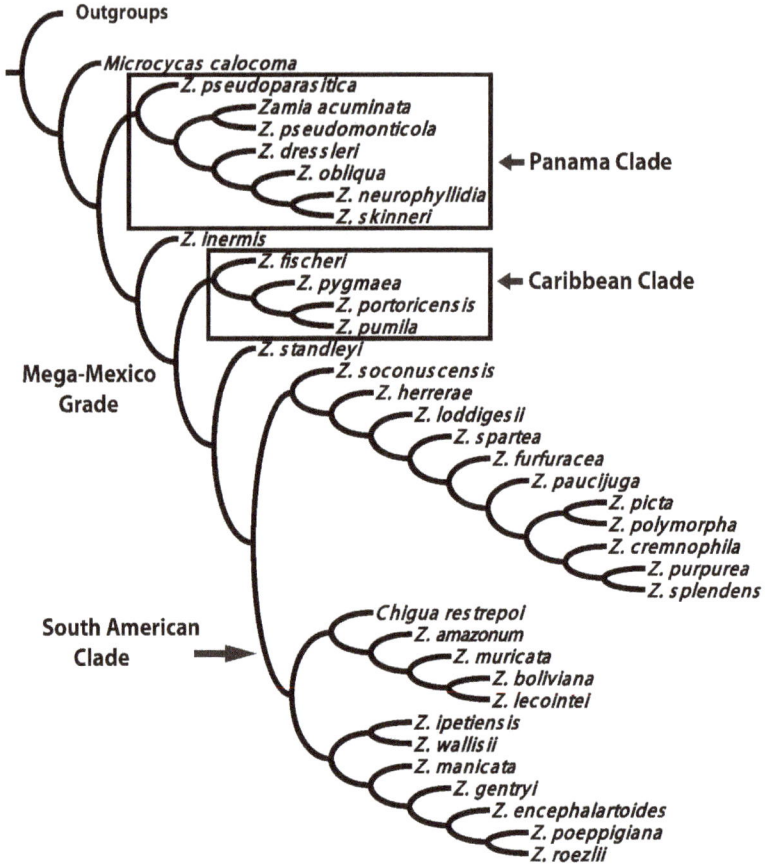

Fig. 12. Resolved tree of the genus *Zamia* using molecular and morphological data (Caputo et al., 2004; current paper). 1. Caribbean Clade, with the Mexican *Z. fischeri* as its sister species; 2. South American Clade; 3. Panama Clade; and 4. Mega-Mexico grade.

distribution of a single species or the existence of more than one related species. Regardless if it is one species or a clade of sister species, it appears that *Z. obliqua* is a Panamanian taxon that also occurs in Colombia. If, however, *Z. obliqua* represents two species in different clades (i.e., one in the Panama Clade and one in the South America Clade), then the common morphology of plants in the various disjunct localities must have evolved independently in response to similar environmental pressures.

Another interesting disjunct distributional pattern is that of *Zamia chigua* and *Z. lindleyi*. These two species are currently recognized as distinct taxa (Osborne et al., 2012) but have been considered conspecific in previous treatments of *Zamia* for Panama (Stevenson, 1993) and for Colombia (Stevenson, 2001). However, recent work has shown that these do, indeed, represent two distinct species based on morphological features of leaves, cones, and spines (Calonje et al., 2011). These two entities are strikingly disjunct, with *Z. chigua* known only from the Cabo Corrientes area of the Chóco region of Colombia, and *Z. lindleyi* known only from a small area of Chiriquí Province in northwestern Panama. Given that these two species are obviously very similar morphologically, the question arises whether they are sister species or are each related to species in other clades. Regardless, the ecology and biology of these organisms must be better known to understand why they are so widely disjunct but yet currently so restricted in range.

Some striking aspects concerning the distribution of *Zamia* in Central America, excluding Mega-Mexico, is the high endemism and morphological diversity in western Panama and adjoining Costa Rica. The Panama Clade contains one of the smallest *Zamia* species known (*Z. acuminata*), the only epiphytic cycad (*Z. pseudoparasitica*), and the majority of the largest trunking species in the genus (*Z. hamannii, Z. nesophila, Z. obliqua,* and *Z. skinneri*). Habitats also vary greatly, ranging from high-level cloud forest to sea-level island beaches. The recent elucidation of the variability within the previously described *Z. neurophyllidia* and *Z. skinneri* resulted in the recognition of three new distinct species, *Z. hamannii, Z, imperalis,* and *Z. nesophila,* all of which are plicate-leaved (Taylor et al., 2008). These discoveries await detailed phylogenetic and ecological analyses, but clearly show adaptations to varied conditions. In contrast to the other plicate-leaved species in Panama that occur primarily in montane forests, *Z. nesophila* and *Z. hamannii* are island-dwellers that are exposed to, and often inundated by, seawater. Interestingly, *Z. roezlii,* which is endemic to the Buenaventura area of the Chóco, Colombia, also has plicate leaves and can live in mangrove swamps.

Finally, it should be noted that *Zamia acuminata* needs to be better defined, because the originally described species grows along the San Juan River in Nicaragua, and is reportedly morphologically dissimilar to the plants bearing the same (misapplied) name in Panama (Schutzman, 2004; Lindström, pers. comm.).

## 8. Conservation efforts & isthmian cycads

Cycads in Panama face an ongoing condition of destruction and illegal extraction in almost all natural populations visited. Populations near newly opened highways are readily disposed of, either because the people working nearby do not know the academic or commercial value of the plants or because they become readily available to poachers. On many occasions a few colleagues have addressed the issue, and an international conference on cycad biology took place in Panama City in 2008, in which all delegates signed a public petition for the government of Panama through the auspices of the University of Panama to take under its responsibilities the extension of cycad biology works and conservation and to make the public aware of the problem at issue. However, outside the support of the university for the senior author's basic research on isthmian cycad biology and conservation as a full time research faculty, including minimal support for a fledging *ex situ* cycad garden, very little has been done to insure the health of cycad populations throughout the isthmus. Furthermore, basic funding for cycad research is lacking.

### 8.1 Conservation status of natural forests & cycad survival

The Isthmus of Panama has a natural forest surface of approximately 3 million ha (44% of the land surface), although conditions of climate and soil are adequate for almost all the available land surface to be forested. The loss of forest surface is associated with anthropogenic activities occurring at different times, and deforestation of natural vegetation is a threat to the survival of isthmian cycads because they are almost all understory species. The history of the isthmus shows that deforestation began with the first human settlements, approximately 7 kya (Cooke,

2003). Since then, destruction of the forests has been continual, with periods of lesser activities. More recently, although the rate of deforestation has been reduced, there is still pressure to use forest land for different types of human activities, such as the establishment of human centers and for the production of foodstuffs. To face this threat of losing natural forests and their biodiversity, a National System of Protected Areas (SINAP for the Spanish acronym for Sistema Nacional de Áreas Protegidas) has been established, which covers 36% of the land mass of the isthmus. Although there are efforts to preserve forests, the habitat of cycads, there are still natural populations of cycads outside of protected areas. In certain cases, forests outside protected areas still have large populations of cycads, such as *Zamia pseudoparasitica*, which survives primarily on very high branches of trees in Veraguas, Coclé, and Colón provinces of north-central Panama. Because of this, it is vital to conduct research that points to forests with extant cycad populations for these forests to be included in SINAP.

## 8.2 Local conservation initiatives

The best approach for cycad conservation in Panama is to meet with local people and to form a team, with communication between researchers and locals at fixed intervals. Donating cell phones, in certain areas, and even small digital cameras, can help volunteers locate and relay situations in the field to a center with better infrastructure. At all times, Panamanian National Environmental Authority (ANAM) personnel should be kept abreast of what is being done. Thousands of plants and many populations have been destroyed for road building, creation of pastures for cattle grazing, or just because the plants had no immediate use for the community. ANAM states that there is no money to save cycads. This being the case, the job is left to academic institutions and the private sector. As a result, all of the Panamanian species should have at least the following Red List category rating: Critically Endangered (CR B1, I, iii-v) (IUCN, 2001). Land owners in most provinces with cycads have been approached, and the discussion of conservation has been positive. However, the problem remains quite unsettled (see also Taylor et al., 2007). In fact, many populations that were quite healthy and with sizeable numbers of individuals, are now down to few and scattered plants, and even places in which the owners had given hope of conservation have not been left unscathed, basically because new roads have been opened during the intervening years and new settlements have been made. Populations in northwestern Panama along river embankments are at the moment safe because of the difficulty of getting to those places and the very aggressive attitude of aboriginals in their homelands to defend their environment. Alas, even here, at present, there is danger of deforestation due to the construction of dams and the human use of the surrounding land in which the cycad populations survive.

## 9. Conclusion

Cycads arrived from both South and North America into the Isthmus of Panama, and new species evolved there, the original progenitors having existed globally for over 300 million years and predating the breakup of Pangaea. Today's extant cycads occupy mainly tropical and subtropical regions of the world. Panama supports at least 17 species, making the isthmus a cycad biodiversity hotspot because there are more representatives per unit land mass than in any other region of the neotropics.

Endemism has been the rule, with most cycads in the Isthmus of Panama forming relatively small, endemic populations, and with most populations numbering only a few hundred individuals. Morphological diversity is rampant within the Panamanian cycad flora—from those with underground stems to those with huge trunks and a wide range of other divergent characters, such as cone morphology and morphometrics, and leaf type and number. We still cannot answer the question of why cycads have survived so long with relatively similar vegetative and reproductive structures of millions of years ago, but it could be linked to their individual longevity, production of toxic substances, and absence of herbivores as a result of those substances. Cycad survival is challenged by human alteration of habitat, including clearing of forests and converting "cycad land" to roads, development, and cattle ranching, as well as illegal removal of plants for black market sale, which is taking many rare species to the brink of extinction.

We describe cycad distribution and evolution in relation to the origin of the Isthmus of Panama and the Great American Biodiversity Interchange, which was completed beginning about 3 mya (although there were temporary closings and openings of the seaway for millions of years before definite closure), and during which period cycads likely made their way from both the north and the south towards the isthmus. We also summarize our hypothesis for the current biogeography and ongoing evolution of the plicate-leaved 'Skinneri Complex' in northwestern and north-central Panama.

A complete phylogeny is still lacking for the genus *Zamia*, but what is known provides insight and testable hypotheses into how the genetic makeup of this genus relates to geologic changes and the present-day geographic distribution of the member species. Future work with more species examined and using other genome sets and/or molecular markers will provide a better understanding of this problem, and more species surveys and descriptions are needed to address the question of synonymy of many entities, with *Z. cunaria* vs. *Z. ipetiensis* and *Z. acuminata* in Panama vs. *Z. acuminata* from Central America being noteworthy examples. More work is also needed in the 'Skinneri Complex', and the arduous work already conducted (Taylor et al., 2008) should be addressed when speaking of *Z. skinneri* in a general way.

The study of cycad pollination biology in Panama is under way, and the genera (or similar/related genera) of some insect pollinators have been determined. This ongoing work includes a phylogeny of the pollinators as well as descriptions of putative new species. Also, new work will soon begin on heat production in *Zamia* cones and the volatile emissions that are associated with pollinator activity.

Last but not least, we address the problem of disappearing forests and the clearing of natural populations of *Zamia* on the isthmus. A great fear is that populations not yet studied will disappear in the immediate future, due to the clearing of land for human use and the inability of the understaffed and inadequate organization of the Natural Environmental Authority of Panama. In this respect, just trying to save what is left of these populations is a great challenge.

## 10. Acknowledgment

We acknowledge the support by the administration of the University of Panama (President, Vice-President for Research and Graduate Studies, and various deans of the Faculty of Natural and Exact Sciences and Technology) of the senior author as full-time research faculty and also for infrastructure where possible (e.g., space for a cycad garden—with over 2,000 plants, including germlings, young plants, and mature, non-coning and coning individuals—to grow native and naturalized species of cycads and a reservoir of specimens for all of the known populations of isthmian cycads, plus representatives of the world cycad flora). Partial financial support came from NSF grants (BSR-8607049, EF 0629817, IOS 0421604) to D.W.S. We are also grateful for the partial support of the National Environmental Authority of Panama (ANAM) for granting us permission to carry out research on isthmian cycads in national parks and especially in the homelands of the aboriginal people in almost every province of the republic. The senior author is also indebted to his wife, Isabel Debora Herrera Antaneda, for her moral support during the writing of this paper. Our thanks go to Mr. Eduardo Sánchez, who has been working as a field assistant on the ecology of *Zamia elegantssima*, and to Dr. Anthony Coates of the Smithsonian Tropical Research Institute in Panama, who provided valuable suggestions and permission to use the maps in Fig. 11.

## 11. References

ANAM. (2006). *Indicadores Ambientales de la República de Panamá (Environmental Meters of the Republic of Panama)*, ANAM, Panamá, Quebecor World Bogotá, S.A., Colombia

Anderson, R.P. & Handley, Jr., C.O. (2001). A new species of three-toed sloth (Mammalia: Xenarthra) from Panama, with a review of the genus *Bradypus. Proceedings of the Biological Society of Washington,* Vol. 114, No. 1, (April 2001), pp. 1-33

Anderson, R.P. & Handley, Jr., C.O. (2002). Dwarfism in insular sloths: Biogeography, selection, and evolutionary rate. *Evolution,* Vol. 56, No. 5, (May 2002), pp. 1045-1058

Calonje, M.A., Taylor, A.S., Stevenson, D.W., Holzman, G. & Ramos, Y.A. (2012). *Zamia lindleyi*: A misunderstood species from the highlands of western Panama, In: Proceedings of the 8th International Conference on Cycad Biology, Panamá, Panamá, January 2-8, 2008, *Memoirs of the New York Botanical Garden*, Vol. 106, in press

Caputo, P., Cozzolino, S., Gaudio, L., Moretti, A. & Stevenson, D.W. (1996). Karyology and phylogey of some Meso-American species of *Zamia* (Zamiaceae). *American Journal of Botany,* Vol. 83, No. 11, (November 1996), pp. 1513-1520

Caputo, P., Cozzolino, S., De Luca, P., Moretti, A. & Stevenson, D.W. (2004). Molecular phylogeny of *Zamia* (Zamiaceae), In: *Cycad Classification: Concepts & Recommendations*, Walters, T. & Osborne, R. (Eds.), pp. 149-157, CABI Publishing, ISBN 0-85199-741-4, Cambridge, MA, USA

Chamberlain, C.J. 1919. *The Living Cycads*, University of Chicago Press, Chicago, IL, USA

Chamberlain, C.J. 1935. *Gymnosperms: Structure & Evolution*, University of Chicago Press, Chicago, IL, USA

Coates, A. (2003). Forja de Centroamérica (The Rise of Central America), In: *Una Historia de la Naturaleza y Cultura de Centroamérica: Paseo Pantera*, Coates, A. (Ed.), pp. 1–40, Smithsonian Books, Washington, D.C., USA

Coates, A.G., McNeill, D.F., Aubry, M-P., Berggren, A. & Collins, L.S. (2005). An introduction to the geology of the Bocas del Toro Archipelago, Panama. *Caribbean Journal of Science,* Vol. 41, No. 3, (2005), pp. 374-391

Coates, A. & Obando, J. (1996). The geologic evolution of the Central American isthmus, In: *Evolution & Environment in Tropical America*, Jackson, J., Budd, A. & Coates, A. (Eds.), pp. 21-56, University of Chicago Press, Chicago, IL, USA

Cooke, R. (2003). Los pueblos indígenas de Centroamérica durante las épocas precolombinas y colonial, In: *Una Historia de la Naturaleza y Cultura de Centroamérica: Paseo Pantera*, Coates, A. (Ed.), pp. 153–196, Smithsonian Books, Washington, D.C., USA

Dickau, R.A., Ranere, A.J. & Cooke, R.G. (2007). Starch grain evidence for the preceramic dispersals of maize root crops into tropical dry and humid forests in Panama. *Proceedings of the National Academy of Sciences*, Vol. 104, No. 9, (2007), pp. 3651-3656

Gomez, L.D. (1982). Plantae Mesoamericanae Novae. II. *Phytologia*, Vol. 50, No. 6, (1982), pp. 401-404

Gregory, T.J. & Chemnick, J. (2004). Hypotheses on the relationship between biogeography and speciation in *Dioon* (Zamiaceae), In: *Cycad Classification: Concepts & Recommendations*, Walters, T. & Osborne, R. (Eds.), pp. 137-148, CABI Publishing, ISBN 0-85199-741-4, Cambridge, MA, USA

Instituto Geográfico Nacional Tommy Guardia. (2007). *Atlas Nacional de la República de Panamá*, Ministerio de Obras Públicas, Panamá, Panamá

IUCN. (2001). *IUCN Categories & Criteria, ver. 3.1*, IUCN Species Survival Commission, Gland, Switzerland

Lindström, A.J. (2009). Typification of some species names in *Zamia* L. (Zamiaceae), with an assessment of the status of *Chigua* D.W. Stev. *Taxon*, Vol. 58, No. 1, (February 2009), pp. 265-270

Louis Berger Group. (2000). *Mapa de vegetación de Panamá. Corredor Biológico Mesoamericano del Atlántico Panameño*, Autoridad Nacional del Ambiente (ANAM), Panamá, Panamá

Louis Berger Group. (2003). *Informe: Recopilación y Presentación de Datos Ambientales y Culturales en la Región Occidental de la Cuenca del Canal de Panamá*, Louis Berger Group, Inc., Panamá, Panamá

Janzen, D.H. (1981a). Guanacaste tree seed-swallowing by Costa Rican range horses. *Ecology*, Vol. 62, No. 3, (June, 1981), pp. 587-592

Janzen, D.H. (1981b). *Enterolobium cyclocarpum* seed passage rate and survival in horses, Costa Rican Pleistocene seed dispersal agents. *Ecology*, Vol. 62, No. 3 (June, 1981), pp. 593-601

Jones, D.L. (2002). *Cycads of the World. Ancient Plants in Today's Landscape* (2nd edition), Smithsonian Institution Press, ISBN 1-58834-043-0, Washington, D.C., USA

Nicolade-Morejón, F., Vovides, A.P. & Stevenson, D.W. (2009). Taxonomic revision of *Zamia* in Mega-Mexico, *Brittonia*, Vol. 61, (2009), pp. 1-38

Norstog, K.J. & Nicholls, T.J. (1997). *The Biology of the Cycads*, Cornell University Press, ISBN 0-8014-3033-X, Ithaca, NY, USA

Osborne, R., Stevenson, D., Hill, K. & Stanberg, L. (2012). The World List of Cycads/La Lista Mundial de Cícadas, In: Proceedings of the 8th International Conference on Cycad Biology, Panamá, Panamá, January 2-8, 2008, *Memoirs of the New York Botanical Garden*, Vol. 106, in press

Pearson, H.H.W. (1906). Note on the South African cycads. I. *Transactions of the South African Philosophical Society*, Vol. 16, (1906), pp. 341-354

Pott, C., McLaughlin, S. & Lindström, A. (2010). Late Palaeozoic foliage from China displays affinities to Cycadales rather than to Bennetitales necessitating a re-evaluation of the Palaeozoic *Pterophyllum* species. *Acta Palaeontologica Polonica*, Vol. 55, No. 1, (2010), pp. 157-168

Proches, W. & Johnson, S. (2009). Beetle pollination of the fruit-scented cones of the South African cycad *Stangeria eriopus, American Journal of Botany*, Vol. 96, No. 9, (2009), pp. 1722-1730

Rattray, G. (1913). Notes on the pollination of some South African cycads, *Transactions of the Royal Society of South Africa*, Vol. 3, (1913), pp. 259-270

Schuh, R. & Brower, A. (2009). *Biological Systematics: Principles & Applications* (2nd edition), Cornell University Press, ISBN 978-0-8014-4799-0, Ithaca, NY, USA

Schutzman, B. (2004). Systematics of Meso-American *Zamia* (Zamiaceae), In: *Cycad Classification: Concepts & Recommendations*, Walters, T. & Osborne, R. (Eds.), pp. 159-172, CABI Publishing, ISBN 0-85199-741-4, Cambridge, MA, USA

Schutzman, B., Vovides, A.P. & Adams, R.S. (1998). A new *Zamia* (Zamiaceae, Cycadales) from central Panama. *Phytologia*, Vol. 85, No. 3, (September 1998), pp. 137-145

Stevenson, D.W. (1993). The Zamiaceae in Panama with comments on phytogeography and species relationships, *Brittonia*, Vol. 45, No. 1, (March 4, 1993), pp. 1-16

Stevenson, D.W., Norstog, K.J. & Fawcet, P.K.S. (1998). Pollination biology of cycads, In: *Reproductive Biology*, Owens, S.J. & Rudall, P.J. (Eds.), pp. 277-324, Royal Botanic Gardens, Kew, UK

Stevenson, D.W. (2001). Cycadales, In: *Flora de Colombia*, Bernal, R. & Forero, E., (Eds.), pp. 1-92, Instituto de Ciencias Naturales, Universidad Nacional de Colombia, ISSN 0120-4351, Bogotá, Colombia

Suinyuy, T.N., Donaldson, J.S. & Johnson, S.T. (2012). Role of cycad cone volatile emissions and thermogenesis in the pollination of *Encephalartos villosus* Lehm: Preliminary findings from studies of plant traits and insect responses, In: Proceedings of the 8th International Conference on Cycad Biology, Panamá, Panamá, January 2-8, 2008, *Memoirs of the New York Botanical Garden*, Vol. 106, in press

Taylor, A.S. (1999). Natural reproductive population structure and pollination in Panamanian *Zamia*. Abstract, XVI International Botanical Congress, St. Louis, MO, USA

Taylor, A.S. (2002). Irrefutable proof of insect pollination in *Zamia elegantissima* Schutzman, Vovides & Adams, In: *Proceedings of the International Conference on Tropical Ecosystems Tropical Forests: Past, Present & Future (Association for Tropical Biology Annual Meeting & Smithsonian Tropical Research Institute)*, Panamá, Panamá, July 30-August 2, 2002

Taylor, A.S., Haynes, J.L. & Holzman, G. (2008). Taxonomical, nomenclatural, and biogeographical revelations in the *Zamia skinneri* species complex (Cycadales: Zamiaceae) of Central America, *Botanical Journal of the Linnean Society*, Vol. 158, (2008), pp. 399–429

Taylor, A.S., Haynes, J.L. & Holzman, G. (2012). The *Zamia skinneri* (Cycadales: Zamiaceae) complex in Panama, In: Proceedings of the 8th International Conference on Cycad Biology, Panamá, Panamá, January 2-8, 2008, *Memoirs of the New York Botanical Garden*, Vol. 106, in press

Taylor, A.S., Haynes, J.L., Holzman, G. & Mendieta, J. (2007). Variability of natural populations and conservation issues facing plicate-leaved *Zamia* species in Central America, In: Proceedings of the 7th International Conference on Cycad Biology, pp. 557-577, ISSN 0077-8931, Xalapa, Mexico, January 8-12, 2005, *Memoirs of the New York Botanical Garden*, Vol. 97, pp. 557-577

Taylor, A.S. (2012). Notes on the ethnobotany of Panamanian cycads, In: Proceedings of the 8th International Conference on Cycad Biology, Panamá, Panamá, January 2-8, 2008, *Memoirs of the New York Botanical Garden*, Vol. 106, in press

Terry, I., Walter, G., Moore, C., Roemer, R. & Hull, C. (2007). Odor-mediated push-pull pollination in cycads, *Science*, Vol. 318, (2007), p. 70

Webb, D.S. (1997). The great American faunal interchange, In: *Central America, a Natural and Cultural History*, Coates, A.G. (Ed.), pp. 97-122, Yale University Press, New Haven, CT, USA

Wikipedia. (2011). Panamá, *Wikipedia La Encyclopedia Libre*, 5 December 2011, Available from <http://es.wikipedia.org/wiki/Panama>

Wikipedia. (2012). Geography of Panamá, *Wikipedia*, 7 January 2012, Available from <http://en.wikipedia.org/wiki/Geography_of_Panama>

Wiley, E.O. & Lieberman, B.S. (2011). *Phylogenetics: The Theory of Phylogenetic Systematics* (2nd edition), Wiley-Blackwell, 978-0-470-90596-8, Hoboken, NJ, USA

Wilson, G. (2002). Insect pollination in the cycad genus *Bowenia* Hook ex Hook. f. (Stangeriaceae), *Biotropica*, Vol. 34, No. 3 (September, 2002), pp. 438-441

# 7

# Biogeography of Intertidal Barnacles in Different Marine Ecosystems of Taiwan – Potential Indicators of Climate Change?

Benny K.K. Chan[1,2] and Pei-Fen Lee[2,*]
*[1]Biodiversity Research Center, Academia Sinica, Taipei,*
*[2]Institute of Ecology and Evolutionary Biology, National Taiwan University, Taipei,*
*Taiwan*

## 1. Introduction

The Indo-Pacific region contains the highest global marine biodiversity including fishes, molluscs, benthic crustaceans and corals (Tittensor et al., 2010). Understanding the biogeography and ecology of marine species in such biodiversity hot spot is essential to protect our marine resources because this can allow us to understand the present biodiversity status and predict changes under the effect of global warming (Firth et al., 2009; Tittensor et al., 2010).

The life cycle of common intertidal organisms includes a planktonic larval phase and sessile or benthic adult phase. As a result, the biogeography of intertidal species is driven by the supply of larvae, their dispersal range, settlement, and subsequent recruitment into the adult population. In most cases, larvae with longer developmental periods have longer dispersal distance (Scheltema, 1988), resulting in a wider geographical distribution. Recent studies (e.g., Barber et al., 2000, 2006; Zakas et al., 2009), however, showed that the interplay between larval supply, local oceanographic currents and geographical isolation can result in retention of larvae in both regional or local scales, leading to distinct assemblage structure and geographic distribution (see Scheltema, 1988; Kojima et al., 1997; Pannacciulli *et al.*, 1997; Wethey, 2002; Dawson *et al.*, 2010; Reece *et al.*, 2010). In addition to the effects of ocean current on larval dispersal, temperature and recruitment also can affect the survival and sustainability of both cold and warm marine species, thus affecting their biogeography (Herbert et al., 2007; Jones et al., 2009).

On rocky shores, barnacles often are used as model and representative species to study the biogeography of intertidal species, as they are the major space occupiers and often have wide geographical distribution (Darwin, 1854; Southward & Crisp, 1956; Crisp & Southward, 1958; Crisp et al., 1981; Southward, 1991; Chan et al. 2007a, b; Tsang et al. 2008a, b, 2011). The larval (planktonic) phase of intertidal thoracican barnacles consists of six naupliar stages and a single cypris stage prior to settlement (Walker et al., 1987). The complete larval development period of intertidal barnacles including the genus *Tetraclita* and *Chthamalus* in the Pacific ranges from 14-21 days for dispersal in the ocean (Chan, 2003; Yan & Chan, 2004).

---

*[*]* Corresponding Author

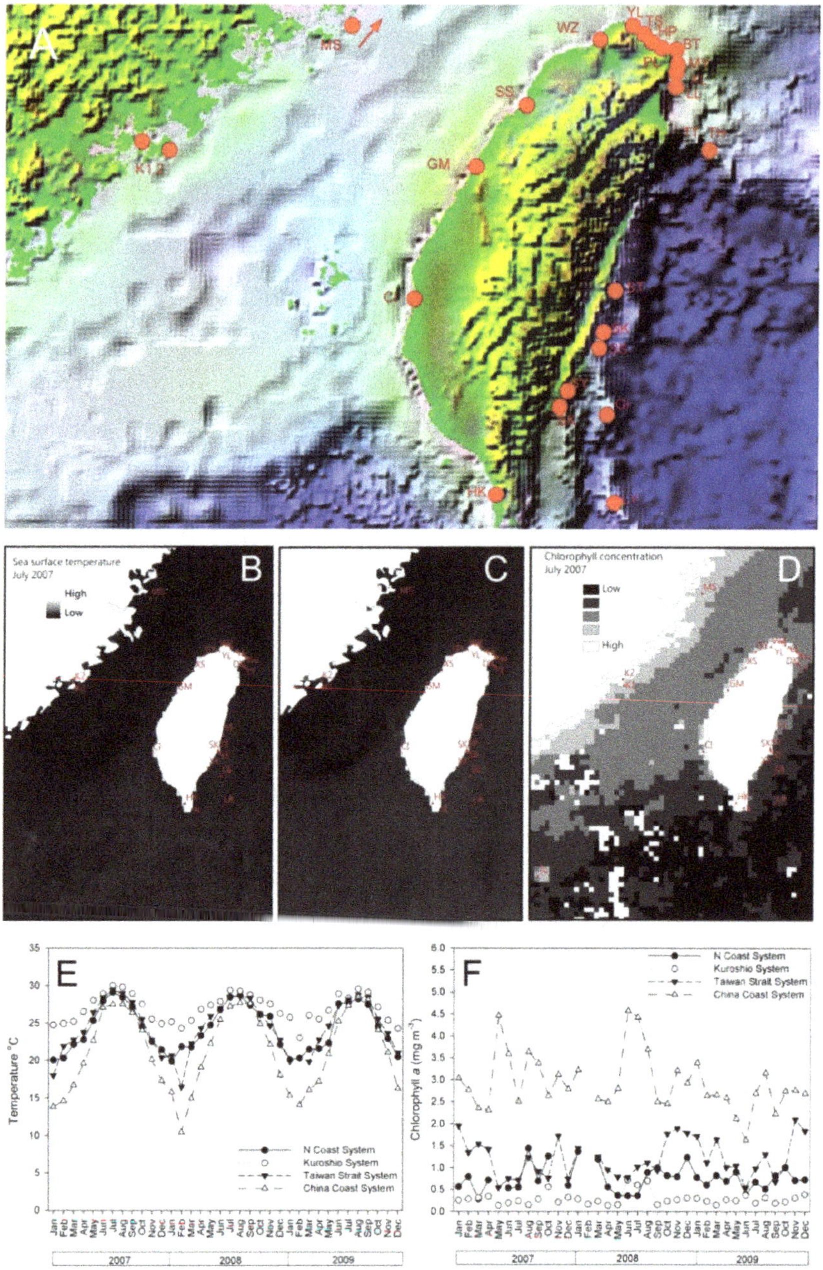

Fig. 1. A. Taiwan marine topography map showing sampling sites located in different ecosystems. For sites abbreviations, refer to Table 1. B. Sea surface temperature derived from satellite images at July 2007. C. Sea surface temperature derived from satellite images at December 2007. D. Sea surface Chlorophyll *a* concentration obtained from satellite images at

July 2007. E. Average sea surface temperature and F. Chlorophyll *a* concentration (detected from Giovanni Satellite System) from 2007-2009. N Coast Ecosystem – averaged from YL, HP, TI, Kuroshio Ecosystem, averaged from ST, XG, HK, Taiwan Strait Ecosystem, averaged from SS, GM and CI , China Coast Ecosystem, averaged from MT and K1 (site abbreviation, see Table 1). Sea surface temperature map in Fig. 1C, D. was derived from NOAA/AVHRR SST images. Sea surface temperature and chlorophyll *a* data in Fig.1E, F was produced with the Giovanni online data system (developed and maintained by the NASA GES DISC).

| Site name | Latitude | Longitude | Marine Ecosystem | Abbreviated site name |
|---|---|---|---|---|
| Tiu-shi | 25.16.43N | 121.36.57E | N Coast | TS |
| Yeliu | 25.12.39N | 121.41.48E | N Coast | YL |
| He-Ping Dao | 25.09.49N | 121.45.48E | N Coast | HP |
| Ping-Long Qiu | 25.08.07N | 121.48.20E | N Coast | PL |
| Bitou | 25.07.40N | 121.55.15E | N Coast | BT |
| Mei-Yan-Shan | 25.04.15N | 121.55.27E | N Coast | MY |
| Xiang Lan | 25.01.33N | 121.58.22E | N Coast | XL |
| Dai Xiang Lan | 25.01.31N | 121.58.42E | N Coast | DX |
| Lai Lai | 25.00.08N | 122.00.08E | N Coast | LL |
| Turtle Island - tail | 24.50.39N | 121.56.41E | N Coast | TT |
| Turtle Island - head | 24.50.10N | 121.57.35E | N Coast | TH |
| Shi Ti Ping | 23.29.28N | 121.30.45E | Kuroshio | ST |
| Siu Kang | 23.09.16N | 121.24.16E | Kuroshio | SK |
| San San Tai | 23.07.25N | 121.25.06E | Kuroshio | SS |
| Shi Yu Sha | 23.10.31N | 121.24.11E | Kuroshio | SY |
| Xiao Yeliu | 22.48.56N | 121.11.45E | Kuroshio | XY |
| Jia Ler Shui | 21.59.55N | 120.52.23E | Kuroshio | JL |
| Hai Kou | 22.05.54N | 120.42.57E | Kuroshio | HK |
| Lanyu | 22.03.32N | 121.30.24E | Kuroshio | LA |
| Green Island | 22.38.49N | 121.28.39E | Kuroshio | GI |
| Tung Sha Island | 20.40.15N | 116.43.01E | Kuroshio | TU |
| Wa Zhi Wei | 25.10.14N | 121.24.42E | Taiwan Strait | WZ |
| Xiang Shan | 24.45.36N | 120.53.42E | Taiwan Strait | XS |
| Gao Mei | 24.18.57N | 120.32.26E | Taiwan Strait | GM |
| Ciqu | 23.06.49N | 120.02.32E | Taiwan Strait | CI |
| Guan Yan | 25.03.44N | 121.05.29E | Taiwan Strait | GY |
| Ma Su Island | 26.13.20N | 119.59.59E | China Coastal | MS |
| Kinmen-Xi-Bin | 24.24.32N | 118.26.17E | China Coastal | K1 |
| Kinmen-site2 | 24.31.34N | 118.25.03E | China Coastal | K2 |

Table 1. Sampling locations with their correspondence marine ecosystems for the intertidal barnacles in the present study. Sites were abbreviated in figures for clarity.

In the northwestern Pacific, Taiwan is a large island located between the Philippines and Japan, and is a mixing zone of several major oceanographic currents. This current mixing has produces a great diversity of physical characteristics (including sea surface temperatures and salinities) in the marine ecosystems there. The north and north east coasts are mainly rocky shores and their hydrography is primarily subject to the East China Sea Large Marine Ecosystem (Tseng et al., 2000). The East China Sea is the largest marginal sea in the northwestern Pacific region. The East China Sea is bounded to the east by the Kuroshio Current (Chen et al., 1995), to the west by mainland China, and receives much surface runoff from Chanjiang River (Tian et al., 1993). The northeast coast of Taiwan is the eastern boundary of the East China Sea, separating from the Kuroshio Current, with its steep continental slope (> 1000 m depth, Fig. 1A) in which the Kuroshio Current flows. The Kuroshio Current intrudes from the East China Sea to the boundary region in northeastern Taiwan (Tseng et al., 2000). Between 2007 and 2009, average summer (July-September) sea surface temperature in the northeastern coast varied from 27-28°C (Giovanni database, NASA, USA; see section 2.2). In winter (December–February) the average surface seawater temperature in northeastern coast varied from 20-21°C (Figs. 1 B, C, E).

The eastern and southeastern coastlines of Taiwan include rocky shores and coral reefs, with hydrography affected by the warm Kuroshio Current, which has a low chlorophyll *a* concentration (Figs. 1D, F). The Kuroshio Current originates in the Philippines, flowing northward to pass through Taiwan and reaches Honshu in Japan (Muromtsev, 1970). The northward velocity of the current can reach 4 knots (Nitani & Shoji, 1970). Compared to the northeastern coast of Taiwan, the eastern and southeastern coasts sustain only small variation in annual seawater temperature. From 2007-2009, variation in the annual seawater temperature along the eastern and southeastern coasts of Taiwan ranged from 24-25°C in winter and 29°C in summer (see section 2.2; Figs. 1 B, C, E).

The western coastlines of Taiwan facing the Taiwan Strait are mainly sandy shores and mangroves. The Taiwan Strait is a wide strait between the mainland China and Taiwan, with an average depth of about 60 m (Jan et al., 2002; Fig. 1A). The hydrography of the Taiwan Strait adjacent to the Taiwan main island is affected by South China Sea Surface Current and the Kuroshio Branching Current, which arrives from the diversion of the Kuroshio Current in Basi Channel in southern Taiwan. In summer, the South China Sea Surface Current flows northward along the Taiwan Strait and enters the East China Sea, becoming the Taiwan Warm Current (Jan et al. 2010). In winter, due to the strong monsoon coming from the East China Sea Ecosystem from the north, the northward flow of Kuroshio Branch in the Taiwan Strait is blocked in the Penghu waters, because of the shallower depths in the northern part of the strait (Jan et al., 2002). Average seawater temperatures range from 19°C in winter to 28°C in summer (Figs. 1B, C, E; see section 2.2). Kinmen and Matsu are small islands located in the Taiwan Strait near the mainland China coastlines and their hydrography is mainly affected by the South China Sea Warm Current and the China Coastal Current (Tseng et al., 2000). In winter, the average seawater temperature can dropped to 14-15°C in 2007-2009 (Figs. 1B, C, E; see section 2.2). Such low temperature is due to the southward flowing of the cold China Coastal Current. The China Coastal Current also has high nutrients and productivity, resulting the waters affected by this current has a very high chlorophyll *a* concentration (Figs. 1D, F).

In addition to having high diversity of marine ecosystems, Taiwan sustains a latitudinal gradient of tropical to subtropical climate from south to north. The Tropic of Cancer bisects

Taiwan's main island approximately at the middle portion of the island. Northern Taiwan has a subtropical climate, with cold winters (average air temperature 15°C) and hot summers, while the southern part of Taiwan is tropical, having warmer winters (average air temperature 19°C) and very hot summers.

Under such diverse marine ecosystems and temperature regimes, biogeographic zones of intertidal species are distributed in relation to these different marine influences and climate zones. A gradient of warm water and cold water species appears to exist from the eastern to the northern coastlines of Taiwan. Chan et al. (2009) reviewed and reported on the distribution of 94 thoracican barnacles in Taiwan, suggesting that their distribution was related to marine currents and climatic patterns. For example, the intertidal barnacle *Tetraclita* spp. exhibited distinct distribution among different marine ecosystems. *T. kuroshioensis* and *T. japonica formosana* are distributed along the north, northeastern, and southeastern coasts of Taiwan, their presences related to the Kuroshio Current (Chan et al., 2007a, b; 2008a). *Tetraclita squamosa* is present along the coasts of the mainland China and the distribution is affected by the South China Sea Surface Current and the China Coastal Current (Chan et al., 2007a, 2008a). Along the southeastern and eastern coasts are both northern (cold water) and southern (warm water species) barnacle species. *Chthamalus moro*, *C. malayensis* and *T. chinensis* which are more abundant in the southern waters of Taiwan, are reduced in abundance towards the north (Tsang et al., 2008, Chan et al., 2009). The cold water species, *C. challengeri* are only found in Matsu and Kinmen Islands, which are influenced by the cold China Coastal Current, and these species are absent from the Taiwan main island (Chan et al., 2009).

With the anthropogenic increase in carbon dioxide production, the temperature of the earth is increasing and the recent global climate is changing, affecting the ecosystems of the world (Poloczanska et al. 2008; Hawkins et al., 2009). Among different ecosystems, the rocky intertidal zone is a harsh habitat strongly influenced by both physical (e.g., thermal stress) and biological (e.g. competition) factors. Rocky intertidal species therefore are sensitive to environmental changes, and can show rapid responses (Thompson et al., 2002, Rivadeneira & Fernandez, 2005; Helmuth et al., 2006; Hawkins et al., 2009). Increased seawater temperature has resulted in range shifts of intertidal species in northern temperate Atlantic waters (Southward *et al.* 2005; Helmuth et al. 2006; Lima *et al.*, 2006, 2007a, b; Mieskowska et al., 2006, 2007; Herbert et al., 2007; Hawkins et al., 2008). In the northwestern Pacific at present there are no studies concerning the range shift of intertidal organisms, because the baseline biogeographic pattern species is still poorly known. To understand the effects of climatic change on intertidal species distributions, it is essential to understand in detail the biogeography of multiple intertidal species, particularly those in the northwestern Pacific. A comparative multispecies approach can reveal broad biogeographic patterns because different species have different reproductive seasons and larval periods, thus seasonal variation in larval survival and development will be strongly affected by current patterns and thus geographic distribution.

In this chapter, we report on the biogeography of 21 intertidal barnacles in different marine ecosystems and climatic zones in Taiwan to provide quantitative baseline biogeographic data for future temporal comparative uses. We hypothesize that the biogeography of intertidal species in Taiwan is being affected by the marine ecosystems, which differ in environmental factors, including water temperature, salinity and chlorophyll concentration (an indication on the primary production of the marine ecosystem), and along the climatic zones in southeastern and eastern coastal Taiwan.

## 2. Materials and methods

### 2.1 Study sites and sampling

Barnacles were surveyed at 29 locations distributed around the Taiwan main island and on outlying islands in 2007 and 2008 (Fig. 1A). Sampling locations covered all the marine ecosystems of Taiwan (Fig. 1). Eleven sites were located on the north and northeastern coasts of Taiwan, hereafter named as North Coast (NC) ecosystem for clarity (Table 1). Nine sites were located in the eastern and southeastern coasts of Taiwan, including the outlying Turtle, Lanyu, and Green Islands, which are influenced by the warm Kuroshio Current, hereafter referred to as the Kuroshio Ecosystem (KS; Table 1). One site was selected as a reference site: Pratas Island (Tung Sha or Dongsha), which is located in the South China Sea and close to the Luzon Strait, and which receives the Kuroshio Branch and the South China Sea Currents. Five sites were selected on the west coast of Taiwan and are part of the Taiwan Strait Marine Ecosystem. Three sites were selected in Kinmen and Matsu Islands, which are located in close proximity to the coastline of mainland China, and are affected by the South China Sea Warm Current and the China Coastal Current, hereafter named as China Coastal Ecosystem.

At each site, 30-50 m stretches of the shoreline were chosen based on accessibility and covered high diversity of habitats including sloping platforms, vertical rocks and large boulders. All intertidal barnacles species distributed from the highest tidal level to the lowest shore stages (1 m above Chart Datum, C.D.) were identified and recorded. Additionally, specific searching was conducted on shaded rocks and in cracks to detect stalked barnacles, including *Capitulum* and *Ibla* species, and the Tetraclitid barnacle *Tetraclitella* species. The abundance of each species of barnacles was scored using a semi-quantitative scale, modified from Herbert et al., 2007 (Table 2).

| Frequency | Description | Abundance Score |
|---|---|---|
| None | None found | 0 |
| Rare | <10% cover on the shore, only a few found in 30 minutes searching. | 1 |
| Occasion | >10% to < 30% cover on the shores | 2 |
| Common | 30 - 50% cover on the shore. | 3 |
| Abundant | Rocks well covered (>50% cover on the barnacle zone) | 4 |

Table 2. Semi-quantitative scale to measure the abundance of barnacles, modified from Hebert et al., 2007.

### 2.2 Environmental factors

Variation in monthly seawater temperature and chlorophyll *a* concentration were obtained from the Giovanni database, NASA, USA (see web link in reference section) at Yeliu, He

Ping Dao, Turtle Island (located in the N Coast Ecosystem; site details, see Table 1), Shi Ti Ping, Xiu Gang, Hai Kou (located in the Kuroshio Ecosystem), Sheung Shan, Gaomei and Ci Qu (located in the Taiwan Strait Ecosystem) and Kinmen and Matsu (located in the China Coastal Ecosystem).

## 2.3 Variation in barnacle assemblages in different marine ecosystems

Variations in the barnacle species semi-quantitative abundance among different marine ecosystems in Taiwan (N Coast Ecosystem, Kuroshio Ecosystem, Taiwan Strait Ecosystem and the China Coastal Ecosystem) were investigated using multivariate analysis (PRIMER 6, Plymouth Routine in Multivariate Analysis, PRIMER-E Ltd; Clarke, 1993). The species abundance data were square root transformed prior to analysis and the matrix of similarity between each pair of sites was calculated using Bray-Curtis similarity index (Bray & Curtis, 1957). Non-metric multidimensional dimension scaling (nMDS) was used and we plotted the two dimensional ordinations of the ranked orders of similarity among the species composition at sites (Clarke, 1993). One-way analysis of similarity test (ANOSIM; Factor: marine ecosystem) and global $R$ test (Clarke & Green, 1988) was calculated to test the significant differences in the assemblage structure among the marine ecosystems in Taiwan. SIMPER analysis was conducted to examine the diagnostic species between each pair of marine ecosystems (Clarke, 1993).

## 2.4 Variation in barnacle species composition and environmental factors

For representative species that exhibit diagnostic distribution among the marine ecosystems, the distribution and semi-quantitative abundance of the species was plotted on the satellite image ocean sea surface temperature and chlorophyll $a$ concentration (NOAA/AVHRR SST images, 7/2007 and 12/2007). Through this approach we attempted to detect environmental factors responsible for the variation in species abundance among different marine ecosystems

## 3. Results

### 3.1 Species composition in relation to marine ecosystems

A total of 21 species of intertidal barnacles were recorded from all marine ecosystems in Taiwan. Fifteen species were recorded in the N Coast Ecosystem. The Kuroshio Ecosystem includes 17 species. Overlap of species occurred between the N Coast Ecosystem and the Kuroshio Ecosystem, with 13 species (except *Pseudoctomeris sulcata* which is only present in the N Coast Ecosystem) detected in the N Coast Ecosystem also recorded in the Kuroshio Ecosystem; however, abundance values varied between the two ecosystems (Table 3). The barnacle *Octomeris brunnea* and *Yamaguchiella coerulescens* were only recorded in the Kuroshio Ecosystem. The mangrove barnacle *Fistulobalanus albicostatus* was recorded only in the Taiwan Strait Ecosystem: this species is absent from all other marine ecosystems. In the China Coastal Ecosystem, 10 species were recorded, of which 3 species were unique to the China Coast Ecosystem: including *Chthamalus challengeri*, *Tetraclita squamosa* and *Tetraclita japonica japonica* (Table 3).

| Species | N coast | Kuroshio | Taiwan Strait | China Coastal |
|---|---|---|---|---|
| Tetraclita kuroshioensis | + | + | - | - |
| Tetraclita japonica formosana | + | + | - | - |
| Chthamalus challengeri | - | - | - | + |
| Tetraclita squamosa | - | - | - | + |
| Tetraclita japonica japonica | - | - | - | + |
| Chthamalus malayensis | + | + | - | - |
| Chthamalus moro | + | + | - | - |
| Hexechamaesipho pilsbryi | + | + | - | - |
| Megabalanus volcano | + | + | - | - |
| Tetraclitella divisa | + | + | - | - |
| Tetraclitella karandei | + | + | - | - |
| Tetraclitella chinensis | + | + | - | - |
| Capitulum mitella | + | + | - | + |
| Pseudoctomeris sulcata | + | - | - | - |
| Octomeris brunnea | - | + | - | - |
| Newmanella sp. | + | + | - | - |
| Fistulobalanus albicostatus | - | - | + | + |
| Chthamalus sp. | - | + | - | - |
| Ibla cumingi | - | + | - | - |
| Yamaguchiella coerulescens | + | + | - | - |
| Megabalanus tintinnabulum | + | + | - | - |

Table 3. Summary of presence and absences of intertidal barnacle species surveyed in the marine ecosystems of Taiwan

From nMDS ordination plots of the species abundance, the ordinations of the sites in each marine ecosystem formed distinct clusters (Fig. 2A). The clusters of the N Coast Ecosystem and the Kuroshio Ecosystem are closer together (Figs. 2B, 3A) and, except the Tung Sha Island, which is located in the South China Sea, were distinct from all main clusters. The clusters of Taiwan Strait Ecosystem and the China Coastal Ecosystem separated clearly from the other ecosystems (Figs. 2A, 2B, 3A). The pattern of the nMDS ordination plot was further supported from the ANOSIM analysis, which showed significant differences in the species abundance among the marine ecosystems ($R = 0.78$, $P < 0.001$). From cluster analysis, the similarity of the assemblages between the East China Sea and the Kuroshio Ecosystems was 60%, showing these two ecosystems have similar species composition (Fig. 2B). Similarity of the China Coastal Ecosystem from the East China Sea and Kuroshio Ecosystems was 20%, showing strong differences in assemblage composition. The Taiwan Strait Ecosystem had < 10% similarity from other ecosystems, with little overlap of species from the Taiwan Strait Ecosystems to the other ecosystems (Fig. 2B). From SIMPER analysis, *Tetraclita kuroshioensis*, *Chthamalus moro*, *Yamaguchiella coerulescens*, *Chthamalus malayensis* and *Hexechamaesipho pilsbryi* contributed a total of 60% differences between the N Coast and Kuroshio Ecosystems. *T. kuroshioensis* is more abundant in the N Coast Ecosystem and is reduced abundance in the Kuroshio Marine Ecosystem. *Chthamalus moro*, *C. malayensis*, *Hexechamaesipho pilsbryi* are more abundant in the Kuroshio ecosystem, and a gradient of decreasing abundance to the north existed (Figs. 3B, C, D). Comparing the China Coastal

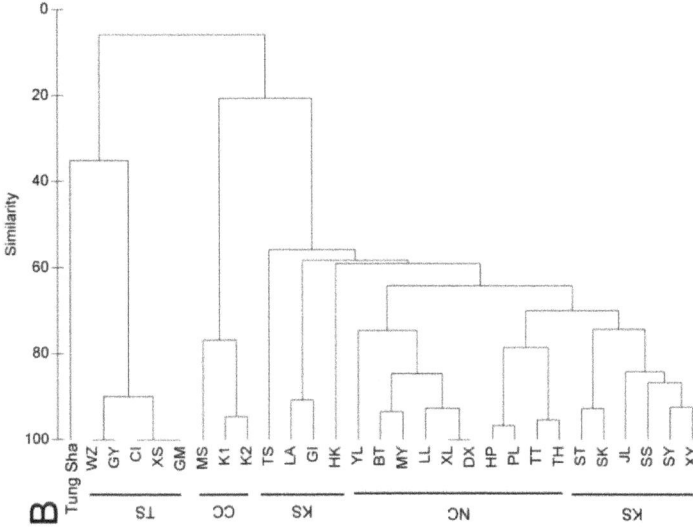

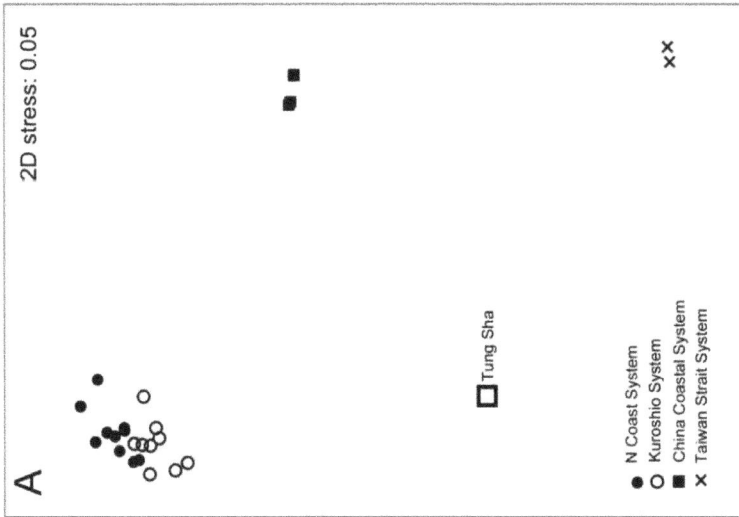

Fig. 2. A. Non-metric nMDS plots on the species composition in all sites from the N Coast Ecosystem, Kuroshio Ecosystem, Taiwan Strait Ecosystem and the China Coastal Ecosystem. B. Cluster plots on the species composition on all sites from the N Coast Ecosystem (NC), Kuroshio Ecosystem (KS), Taiwan Strait Ecosystem (TS) and the China Coastal Ecosystem (CC). Tung Sha island was not grouped in the ecosystem because it is located in the South China Sea. The Taiwan Strait Ecosystems involved 4 sites, but 2 of them overlap in the ordination plot.

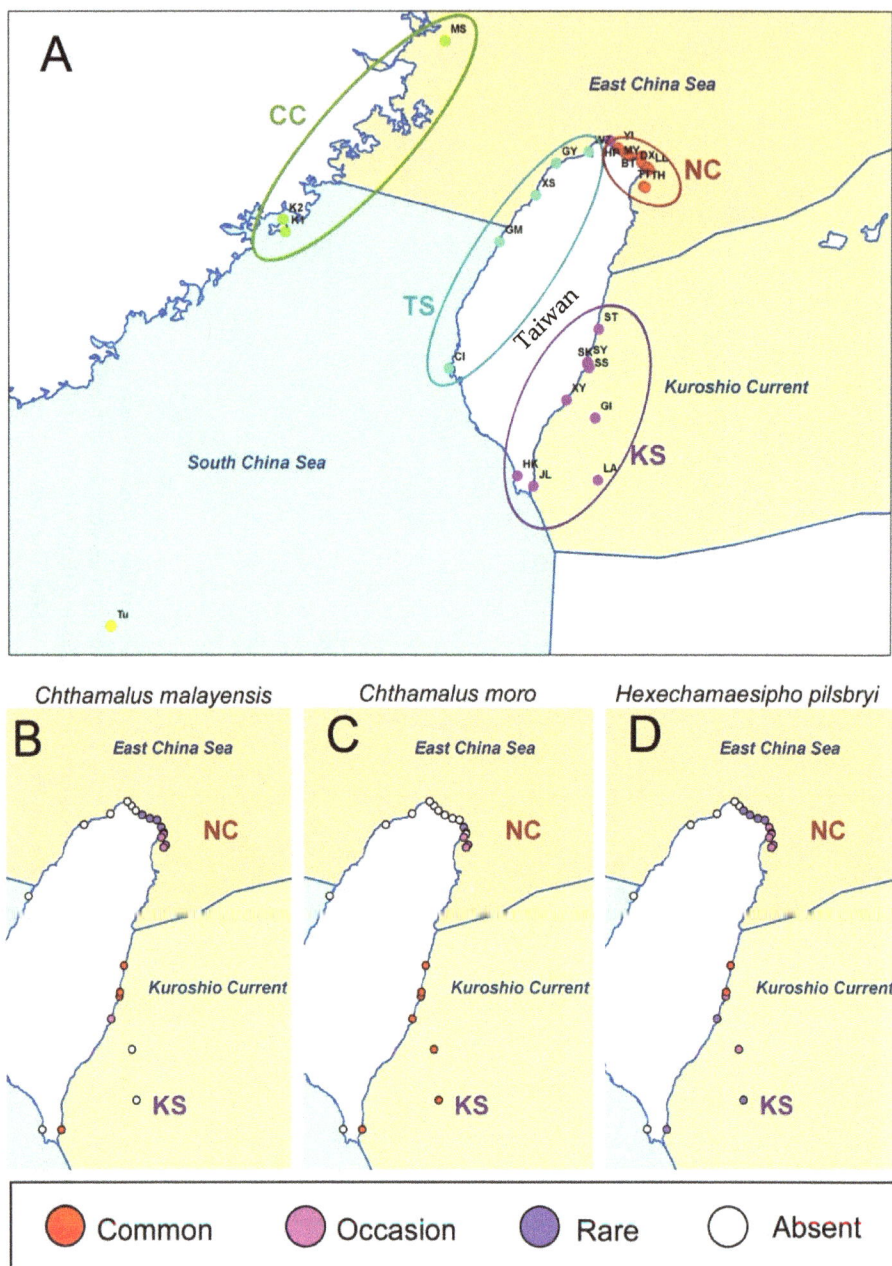

Fig. 3. A. Distribution of the large marine ecosystems of Taiwan, corresponding to the cluster distribution in the nMDS plots in Fig. 2A. Semi-quantitative distribution of B. *Chthamalus malayensis*, C. *C. moro* and D. *Hexechamaesipho pilsbryi* from the S to N coast of Taiwan main island. For details of the semi-quantitative scale, refer to Table 2.

ecosystem and the N Coast ecosystem, *Chthamalus challengeri*, *Tetraclita squamosa*, *Tetraclita japonica japonica*, *Tetraclita japonica formosana*, *Tetraclita kuroshioensis* and *Hexechamaesipho pilsbryi* contributed a total of 60% difference between these two ecosystems (Figs. 2B, 3A). *C. challengeri*, *T. squamosa* and *T. j. japonica* are only present in the China Coastal ecosystem, while the remaining representative species are absent from the China Coastal ecosystem.

## 4. Discussion

### 4.1 Biogeography, marine ecosystems and environmental factors

In this study, the species composition of intertidal barnacles in Taiwan is related to the individual marine ecosystems and to obvious biogeographical zones among the marine ecosystems. Barnacle species composition is similar between the N Coast Ecosystem and the Kuroshio Ecosystem, but abundance patterns differed between these two ecosystems. Species composition the Taiwan Strait Ecosystem and the China Coast Ecosystem are clearly distinct from the other ecosystems.

Comparing species composition between the N Coast Ecosystem and Kuroshio Ecosystem, almost all species collected in the N Coast Ecosystem were present in the Kuroshio Ecosystem, but the Kuroshio Ecosystem has additional diagnostic species. Similar species composition between the N Coast Ecosystem and the Kuroshio Ecosystem can be attributed to the intrusion and mixing of the Kuroshio Current at the N coast of Taiwan. In summer, the Kuroshio sub-surface water in the Pacific Ocean intrudes into the East China Sea in the waters around N and NE Taiwan, resulting in cold eddies in those regions (Tseng et al., 2000). In winter, the intrusion of the Kuroshio Current further shifts westwards onto the N coast of Taiwan (Tseng et al., 2000). Under the mixing of the Kuroshio and the waters in the East China Sea, the larvae of the intertidal barnacle can probably disperse between ecosystems, resulting in overlap of a considerable proportion of species. Although there are common species between the N Coast Ecosystem and the Kuroshio Ecosystem, latitudinal variation of species abundance occurs from the Kuroshio Ecosystem to the N Coast Ecosystem. *Chthamalus malayensis* and *Chthamalus moro* are warm water species common in the South China Sea region, including Hong Kong, the Philippines, and SE Asian locations (Yan & Chan, 2001; Rosell, 1972, 1986; Southward & Newman, 2003; Tsang et al., 2008). These two species had higher abundance in the southern locations of Taiwan main island and their abundance decreased from the E coast to the NE coast and were absent from the N coast locations. The decrease in abundance when approaching northwards in the east coast of Taiwan may be related to the change in seawater temperature. The seawater temperature in the N Coast Ecosystem is colder than the Kuroshio Ecosystem due to the strong NE monsoon in the East China Sea (Figs. 1B, C, E). *C. malayensis* and *C. moro* were not recorded from geographical locations (e.g., Japan) further north than Taiwan (Chan et al., 2008b; Southward & Newman, 2003), suggesting the NE coast of Taiwan may be their northern geographical limits. Decreasing abundance in cooler water species along a latitudinal gradient in Taiwan (Figs. 3B, C) is similar to barnacle distribution patterns reported in the British Isles. There, the warm water species *Chthamalus montagui* and *C. stellatus* reach range limits in the Central English Channel and fail to enter the North Sea due to reduced water temperature. By comparing the distribution pattern of *Chthamalus* in the UK from 1950 to 2004, *Chthamlaus* showed range extension in the English Channel and with extensive recruitment in warmest years (Herbert et al., 2007). Under the effect of global warming, warm water barnacle species along the east coast of Taiwan can expand northwards. The E to NE coast in Taiwan can provide excellent sites for monitoring range extension of intertidal species under the influence of the global climatic changes. Monitoring abundance and recruitment of warm water barnacle species at the NE Coast where their populations sharply decrease in abundance will clarify temporal and spatial scales of change and allow future comparative studies.

*Hexechamaesipho pilsbryi* is a high-shore barnacle recorded from Honshu and Okinawa in Japan and also in Taiwan (Chan et al., 2008b). The abundance of *Hexechamaesipho pilsbryi* decreased from south to north along the eastern coast (Fig. 3D). *Hexechamaesipho pilsbryi* is also recorded in the Philippines and Borneo waters (B. K. K. Chan unpublished data). From molecular analysis, it appears there are northern and southern populations of *Hexechamaesipho,* and the Taiwan populations belong to the latter (B.K.K. Chan, unpublished data). This suggests that *Hexechamaesipho* is distributed along the Kuroshio Current, but there is a physical boundary at Taiwan that blocks gene flow between the northern and southern populations.

The tetraclitid barnacles *Yamaguchiella coerulescens* and *Octomeris brunnea* were present only in the Kuroshio Ecosystem and were entirely absent from the N Coast Ecosystem. *Y. coerulescens* is reported common in the waters of the Philippines (Rosell, 1972). This suggests *Y. coerulescens* is distributed by the Kuroshio Current and its larvae do not intrude into the East China Sea waters. *Octomeris brunnea* was only recorded in Hai Kou, southern Taiwan in an extensive barnacle survey by Hiro (1939). In the present study, *O. brunnea* was also recorded in Hai Kou, but was absent from all other locations in Taiwan. *O. brunnea* live in shaded habitats and the rock formations in Hai Kou contain many large crevices. High habitat specificity in *O. brunnea* may limit its distribution in Taiwan.

The barnacle species composition of Kinmen and Matsu Islands, is distinct from that on the Taiwan main island. *Tetraclita squamosa, Tetraclita japonica japonica* and *Chthamalus challengeri* are only present this ecosystem. *T. squamosa* and *T. j. japonica* also have been recorded from South China to the East China coastline. The distribution of these two barnacle taxa is likely related to the flow of the China Coastal Current along the mainland China continent. The influence of the China Coastal Current is pronounced as little as 50 m offshore (Tseng et al., 2000). On the N Coast Ecosystem, *Tetraclita kuroshioensis* and *Tetraclita japonica formosana* become abundant, indicating no horizontal transfer of larvae across the Taiwan Strait. This likely is due to strong longitudinal flow along the strait. A similar distributional pattern also occurs for the lobster *Palinurus delagoa* in the East Africa and Madagascar. *P. delagoa* exhibits distinct genetic divergence between Madagascan and east African populations (Gopal et al., 2006), suggesting that gene flow across the Mozambique Channel is rather limited.

*Chthamalus challengeri* is a northern species that is abundant on temperate shorelines, including Yellow Sea, East China and the Japanese coastline. In Kinmen and Matsu, due to the low seawater temperature of the Chiangjian runoff and the China Coastal Current (Tseng et al., 2000), the environment may favor the survival of *C. challengeri. C. challengeri* is absent from Taiwan main island probably because the relatively higher seawater temperature on the N Coast Ecosystem and in the Kuroshio Ecosystem prevents its survival and recruitment. Further monitoring and examination of larval composition in Taiwan is needed to reveal the arrival of *C. challengeri* larvae into the waters around the Taiwan main island.

Most of the shorelines in the intertidal of the Taiwan Strait are occupied by mangroves or are open, sandy shores, and consequently, few intertidal barnacle species were recorded. In the present study, only *Fistulobalanus albicostatus* was collected on mangrove trunks. *Fistulobalanus* is a widespread estuarine barnacle species in Pacific waters, and can be abundant in brackish waters. In the present study, *Fistulobalanus* is abundant in the Taiwan Strait Ecosystem because of its association with mangrove and soft shore habitats. The absence of *F. albicostatus* from the other three marine ecosystems (except its rare occurrence in Kinmen) may be due to the rocky intertidal shorelines of those sites, and the lack of soft benthic habitats.

## 4.2 Conclusions

The marine ecosystems and associated barnacle species of Taiwan provide an excellent opportunity to study and monitor the processes of climate change. The present study demonstrates that the biogeography of intertidal barnacles in Taiwan is strongly influenced by oceanographic currents and water temperature differences among the several marine ecosystem. The Kuroshio and N Coast Ecosystems share a great proportion of species in common. However, we detected a gradient of decreasing abundance of the warm water species including *Chthamalus malayensis*, *C. moro* and *Hexechamaesipho pilsbryi* along the East coast of Taiwan, and with a sharper decrease in abundance to the NE coast of Taiwan. *Yamaguchiella coerulescens* was only detected on the east coast of Taiwan. Under the effects of global warming, warm water barnacle species are expected to expand their distributional ranges, similar to the case of *Chthamlaus* spp. in the English Channel (Herbert et al., 2007). The abundance of warm water barnacle species in the E and NE coast can be used as an indicator for global warming. Further studies should conduct regular monitoring to sample the abundance and recruitment of warm water barnacle species on the NE coast of Taiwan, and be designed to detect range extensions of intertidal species under the effects of climate change.

## 5. Acknowledgement

This study was partially funded by the Career Development Award from Academia Sinica to BKKC (AS-98-CDA-L15) and the grant from National Science Council of Taiwan to PFI, (101-2631-I I-002-005- and 100-2631-H-002-019-). Analyses and visualizations used in Fig 1C, D. were produced with the NOAA/AVHRR SST images, Fig. 1E, F were produced with the Giovanni online data system, developed and maintained by the NASA GES DISC.

## 6. References

Barber, P.H., Palumbi, S.R., Erdmann, M.V. & Moosa, M.K. (2000). A marine Wallace's line? *Nature*, Vol. 406, pp. 692-693.

Barber, P.H., Erdmann, M.V. & Palumbi, S.R. (2006). Comparative phylogeography of three codistributed stomatopods: origins and timing of regional lineage diversification in the Coral Triangle. *Evolution*, Vol. 60, pp. 1825-1839.

Bray, J.R. & Curtis, J.T. (1957). An ordination of the upland forest communities of Southern Wisconsin. *Ecological Monograph*, Vol. 27, pp. 325-349.

Chan, B.K.K. (2003). Studies on *Tetraclita squamosa* and *Tetraclita japonica* (Cirripedia: Thoracica) II: larval morphology. *Journal of Crustacean Biology*, Vol. 23, pp. 522-547.

Chan, B.K.K., Tsang, L.M. & Chu, K.H. (2007a). Cryptic diversity of *Tetraclita squamosa* complex (Crustacea, Cirripedia) in Asia: description of a new species from Singapore. *Zoological Studies*, Vol. 46, pp. 46-56.

Chan, B.K.K., Tsang, L.M., Ma, K.Y., Hsu, C.-H. & Chu, K.H. (2007b). Taxonomic revision of the acorn barnacles *Tetraclita japonica* and *Tetraclita formosana* (Crustacea: Cirripedia) in East Asia based on a combined molecular and morphological analysis. *Bulletin of Marine Science*, Vol. 81, pp. 101-113.

Chan, B.K.K., Murata, A. & Lee, P.F. (2008a). Latitudinal gradient in the distribution of the intertidal acorn barnacles *Tetraclita* species complex (Crustacea: Cirripedia) in NW Pacific and SE Asian waters. *Marine Ecology Progress Series*, Vol. 362, pp. 201-210.

Chan, B.K.K., Hsu, C.-H. & Southward, A.J. (2008b). Morphological variation and biogeography of an insular intertidal barnacle *Hexechamaesipho pilsbryi* (Crustacea: Cirripedia) in the western Pacific. *Bulletin of Marine Science*, Vol. 83, pp. 315-328.

Chan, B.K.K., Prabowo, R. & Lee, K.S. (2009). *Crustacea fauna of Taiwan: Barnacles, Vol. 1 – Cirripedia: Thoracica excluding the Pyrgomatidae and Acastinae*. National Taiwan Ocean University Press, Keelung, 297 pp.

Chen, C.T.A., Ruo, R., Pai, S.C., Liu, C.T. & Wong, G.T.F. (1995). Exchange of water masses between the East China Sea and the Kuroshio off northeastern Taiwan. *Continental Shelf Research*, Vol. 15, pp. 19-39.

Clarke, K.R. (1993). Non-parametric multivariate analyses of changes in community structure. *Australian Journal of Ecology*, Vol. 18, pp. 117-143.

Clarke, K.R. & Green, R.H. (1998). Statistical design and analysis for a 'biological effect' study. *Marine Ecology Progress Series*, Vol. 46, pp. 213-226.

Crisp, D.J. & Southward, A.J. (1958). The distribution of intertidal organisms along the coasts of the English Channel. *Journal of the Marine Biological Association of the United Kingdom*, Vol. 37, pp. 157–208.

Crisp, D.J., Southward, A.J. & Southward E.C. (1981). On the distribution of the intertidal barnacles *Chthamalus stellatus, Chthamalus montagui* and *Euraphia depressa*. *Journal of the Marine Biological Association of the United Kingdom*, Vol. 61, pp. 359–380.

Darwin, C. (1854). *A monograph on the sub-class Cirripedia with figures of all species. The Balanidae, Verrucidae, etc.* Ray Society, London.

Dawson, M.N., Grosberg, R.K., Stuart, Y.E. & Sanford, E. (2010). Population genetic analysis of a recent range expansion: mechanisms regulating the poleward range limit in the volcano barnacle *Tetraclita rubescens*. *Molecular Ecology*, Vol. 19, pp. 1585–1605.

Firth, L.B., Crowe, T.P., Moore, P., Thompson, R.C. & Hawkins, S.J. (2009). Predicting impacts of climate-induced range expansion: an experimental framework and a test involving key grazers on temperate rocky shores. *Global Change Biology*, Vol. 15, pp. 1413-1422.

Giovanni database (NASA, USA). 2011. Ocean Color Radiometry Online Visualization and Analysis. Global Monthly Products.
http://gdata1.sci.gsfc.nasa.gov/daac-bin/G3/gui.cgi?instance_id=ocean_month; download4

Gopal, K., Tolley, K.A., Groenveld, J.C. & Matthee, C.A. (2006). Mitochondrial DNA variation in spiny lobster *Palinurus delagoae* suggests genetically structured populations in the southwest Indian Ocean. *Marine Ecology Progress Series*, Vol. 319, pp. 191–198.

Hawkins, S.J., Moore, P., Burrows, M.T., Poloczanska, E., Mieszkowska, N., Herbert, R.J.H., Jenkins, S.R., Thompson, R.C., Genner, M.J. & Southward, A.J. (2008). Complex interactions in a rapidly changing world: responses of rocky shore species to recent climate change. *Climate Research*, Vol. 37, pp. 123-133.

Hawkins, S.J., Sugden, H.E., Mieszkowska, N., Moore, P.J., Poloczanska, E., Leaper, R., Herbert, R.J.H., Genner, M.J., Moschella, P.S., Thompson, R.C., Jenkins, S.R., Southward, A.J. & Burrows, M.T. (2009). Consequences of climate-driven biodiversity changes for ecosystem functioning of North European rocky shores. *Marine Ecology Progress Series*, Vol. 396, pp. 245-259.

Helmuth, B., Mieszkowska, N., Moore, P. & Hawkins, S.J. (2006). Living on the edge of two changing worlds: forecasting the responses of rocky intertidal ecosystems to climate change. *Annual Review of Ecology, Evolution, and Systematics*, Vol. 37, pp. 373-404.

Herbert, R.J.H., Hawkins, S.J., Sheader, M. & Southward, A.J. (2003). Range extension and reproduction of the barnacle *Balanus perforatus* in the eastern English Channel. *Journal of the Marine Biological Association of the United Kingdom*, Vol. 83, pp. 73–82.

Herbert, R.J.H., Southward, A.J., Sheader, M. & Hawkins, S.J. (2007). Influence of recruitment and temperature on distribution of intertidal barnacles in the English Channel. *Journal of the Marine Biological Association of the United Kingdom*, Vol. 87, pp. 487-499.

Hiro, F. (1939). Studies on the cirripedian fauna of Japan IV: cirripedes of Formosa (Taiwan) with some geographical and ecological remarks on the littoral forms. *Memoirs of the College of Science, Kyoto University Series B*, Vol. 15, pp. 245–284.

Jan, S., Wang, J., Chern, C.S. & Chao, S.Y. (2002). Seasonal variation of the circulation in the Taiwan Strait. *Journal of Marine Systems*, Vol. 35, pp. 249-268.

Jan, S., Tseng, Y-H., Dietrich, D.E. (2010) Sources of water in the Taiwan Strait. *Journal of Oceanography*, Vol. 66, pp. 211-221

Jenkins, S.R., Aberg, P., Cervin, G., Coleman, R.A., Delany, J., Hawkins, S.J., Hyder, K., Myers, A.A., Paula, J., Power, A.M., Range, P. & Hartnoll, R.G. (2001). Population dynamics of the intertidal barnacle *Semibalanus balanoides* at three European locations: spatial scales of variability. *Marine Ecology Progress Series*, Vol. 217, pp. 207-217.

Jones, S.J., Mieszkowska, N. & Wethey, D.S. (2009). Linking thermal tolerances and biogeography: *Mytilus edulis* (L.) at its southern limit on the east coast of the United States. *Biological Bulletin*, Vol. 217, pp. 73-85.

Kojima, S., Segawa, R. & Hayashi, I. (2000). Stability of the courses of the warm coastal currents along the Kyushu Island suggested by the population structure of the Japanese turban shell, *Turbo (Batillus) cornutus. Journal of Oceanography*, Vol. 56, pp. 601-604.

Lima, F.P., Queiroz, N., Ribeiro, P.A., Hawkins, S.J. & Santos A.M. (2006). Recent changes in the distribution of a marine gastropod, *Patella rustica* Linnaeus 1758, and their relationship to unusual climatic events. *Journal of Biogeography*, Vol. 33, pp. 812–822.

Lima, F.P., Ribeiro, P.A., Queiroz, N., Hawkins, S.J. & Santos, A.M. (2007a). Do distributional shifts of northern and southern species of algae match the warming pattern? *Global Change Biology, Vol. 13, pp. 2592–2604.*

Lima, F.P., Ribeiro, P.A., Queiroz, N., Xavier, R., Tarroso, P., Hawkins, S.J. & Santos, A.M. (2007b). Modelling past and present geographical distribution of the marine gastropod *Patella rustica* as a tool for exploring responses to environmental change. *Global Change Biology*, Vol. 13, pp. 2065–2077.

Mieszkowska, N., Kendall, M., Hawkins, S., Leaper, R., Williamson, P., Hardman-Mountford, N. & Southward, A. (2006). Changes in the range of some common rocky shore species in Britain – a response to climate change? *Hydrobiologia*, Vol. 555, pp. 241-251.

Mieszkowska, N., Hawkins, S.J., Burrows, M.T. & Kendall, M.A. (2007) Long-term changes in the geographic distribution and population structures of some near-limit populations of *Osilinus lineatus* in Britain and Ireland. *Journal of the Marine Biological Association of the United Kingdom*, Vol. 87, pp. 537–545.

Muromtsev, A.M. (1970). Some results of investigations of dynamics and thermal structure of the Kuroshio and adjacent regions, In: *The Kuroshio, A Symposium on the Japan Current*, Marr, J.C. (Ed), pp. 97-106, East-West Center Press, Honolulu.

Nitani, H. & Shoji, S. (1970). On the variability of the velocity of the Kuroshio. In: *The Kuroshio, A Symposium on the Japan Current*. Marr, J.C. (Ed), pp. 107-116, East-West Center Press, Honolulu.

Pannacciulli, F.G., Bishop, J.D.D. & Hawkins, S.J. (1997). Genetic structure of populations of two species of *Chthamalus* (Crustacea: Cirripedia) in the north-east Atlantic and Mediterranean. *Marine Biology*, Vol. 128, pp. 73–82.

Poloczanska, E.S., Hawkins, S.J., Southward, A.J. & Burrows, M.T. (2008). Modelling the response of populations of competing species to climate change. *Ecology*, Vol. 89, pp. 3138–3149.

Reece, J.S., Bowen, B.W., Joshi, K., Goz, V. & Larson, A. (2010). Phylogeography of two moray eels indicates high dispersal throughout the Indo-Pacific. *Journal of Heredity*, Vol. 101, pp. 391–402.

Rivadeneira, M.M. & Fernandez, M. (2005). Shifts in southern endpoints of distribution in rocky shore intertidal species along south-eastern Pacific coast. *Journal of Biogeography*, Vol. 32, pp.203-209.

Rosell, N.C. (1972). Some barnacles (Cirripedia Thoracica) of Puerto Galera found in the vicnity of the U.P. Marine Biological Laboratory. *Natural Applied Science Bulletin*, Vol. 24, pp. 143-285.

Rosell, N.C. (1986). *Philippine barnacles. Guide to Philippine Flora and Fauna Vol. VII.* Natural Resources Management Center, Ministry of Natural Resources and University of the Philippines, The Philippines.

Scheltema, R.S. (1988). Initial evidence for the transport of teleplanic larvae of benthic invertebrates across the East Pacific Barrier. *Biological Bulletin*, Vol. 174, pp. 145-152.

Southward, A.J. (1991). Forty Years of changes in species composition and population density of barnacles on a rocky shore near Plymouth. *Journal of the Marine Biological Association of the United Kingdom*, Vol. 71, pp. 495–513.

Southward, A.J. & Crisp, D.J. (1956). Fluctuations in the distribution and abundance of intertidal barnacles. *Journal of the Marine Biological Association of the United Kingdom*, Vol. 35, pp. 221–229.

Southward, A.J. & Newman, W.A. (2003). A review of some common Indo-Malayan and western Pacific species of *Chthamalus* barnacles (Crustacea: Cirripedia). *Journal of the Marine Biological Association of the United Kingdom*, Vol. 83, pp. 797-812.

Southward, A.J., Langmead, O., Hardman-Mountford, N.J., Aiken, J., Boalch, G.T., Dando, P.R., Genner, M.J., Joint, I., Kendall, M.A., Halliday, N.C., Harris, R.P., Leaper, R., Mieszkowska, N., Pingree, R.D., Richardson, A.J., Sims, D.W., Smith, T., Walne, A.W. & Hawkins, S.J. (2005). Long-term oceanographic and ecological research in the Western English Channel. *Advances in Marine Biology*, 47, pp. 1–105.

Thompson, R.C., Crowe, T.P. & Hawkins, S.J. (2002). Rocky intertidal communities: past environmental changes, present status and predictions for the next 25 years. *Environmental Conservation*, Vol. 29, pp. 168-191.

Tian, R.C., Hu, F.X. & Martin, J.M. (1993). Summer nutrient fronts in the Changjiang (Yangtze River) Estuary. *Estuarine Coastal and Shelf Science*, Vol. 37, No. 1, pp. 27–41.

Tittensor, D.P., Mora, C., Jetz, W., Lotze, H.K., Ricard, D., Berghe, E.V. & Worm, B. (2010). Global patterns and predictors of marine biodiversity across taxa. *Nature*, Vol. 466, pp. 1098-1102.

Tsang. L.M., Chan, B.K.K., Wu, T.H., Ng, W.C., Chatterjee, T., Williams, G.A. & Chu, K.H. (2008a). Population differentiation of the barnacle, *Chthamalus malayensis*: postglacial colonization and recent connectivity across Pacific and Indian Oceans. *Marine Ecology Progress Series*, Vol. 364, pp. 107-118.

Tsang, L.M., Chan, B.K.K., Ma, K.Y. & Chu, K.H. (2008b). Genetic differentiation, hybridization and adaptive divergence in two subspecies of the acorn barnacle *Tetraclita japonica* in the northwestern Pacific. *Molecular Ecology*, Vol. 17, pp. 4151–4163.

Tsang, L.M., Wu, T.H., Ng, W.C., Williams, G.A., Chan, B.K.K. & Chu, K.H. (2011). Comparative phylogeography of Indo-West Pacific intertidal barnacles. *Crustacean issue 19: Phylogeography and population genetics in Crustacea* (ed. S. Koenemann, C.D. Schubart and C. Held), CRC Press, Baton Raton.

Tseng, C., Lin, C., Chen, S. & Shyu, C. (2000). Temporal and spatial variation of sea surface temperature in the East China Sea. *Continental Shelf Research*, Vol. 20, pp. 373-387.

Walker, G., Yule, A.B. & Nott, J.A. (1987). Structure and function of balanomorph larvae. In: *Barnacle biology*, Southward, A.J. (ed), Crustacean Issues 5. pp. 307-328, A.A. Balkema, Rotterdam.

Wethey, D.S. (2002). Biogeography, competition, and microclimate: the barnacle *Chthamalus fragilis* in New England. *Integrative and Comparative Biology*, Vol. 42, pp. 872–880.

Yan, Y. & Chan, B.K.K. (2001). Larval development of *Chthamalus malayensis* (Cirripedia: Thoracica) reared in the laboratory. *Journal of the Marine Biological Association of the United Kingdom*, Vol. 81, pp. 623-632.

Yan, Y. & Chan, B.K.K. (2004). Larval morphology of a recently recognized barnacle *Chthamalus neglectus* (Cirripedia: Thoracica: Chthamalidae) from Hong Kong. *Journal of Crustacean Biology*, Vol. 24, pp. 519-528.

Zakas C., Binford, J., Navarrete, A. & Ware, J.P. (2009). Restricted gene flow in Chilean barnacles reflects an oceanographic and biogeographic transition zone. *Marine Ecology Progress Series*, Vol. 394, pp. 165-177.

# 8

# Biogeography of Chilean Herpetofauna: Biodiversity Hotspot and Extinction Risk

Marcela A. Vidal and Helen Díaz-Páez

*Departamento de Ciencias Básicas, Facultad de Ciencias, Universidad del Bío-Bío,*
*Departamento de Ciencias Básicas, Universidad de Concepción, Campus Los Ángeles,*
*Chile*

## 1. Introduction

The distribution of living organisms on our planet is not random: evidence accumulated since the eighteenth and nineteenth centuries by the pioneering work of European explorers and naturalists documented the existence of large differences in the number and types of species living in different places on the planet (Brown & Lomolino, 1998, Meynard et al., 2004). The importance and impacts of a geographical approach to the study of biodiversity are evident today, after more than two centuries, as the observations of these early naturalists are still under active investigation. In this biogeographical context, the study of the most biodiverse areas, and understanding of the mechanisms that operate to maintain diversity are fundamental to the development of conservation strategies. However, conservation strategies must be built on a solid understanding the biota, as well as clear identification of the life history, dispersal, and biogeographic and environmental factors that affect a region's biodiversity (Meynard et al., 2004).

Few prior studies are available to develop a dynamic synthesis of the variables influencing herpetofaunal biogeography in Chile. The lack of basic information about the herpetofauna and its biology, and the dispersed nature of existing information have impeded studies in this area of knowledge (Vidal, 2008). Biogeographical studies often are been based on understanding relationships between phylogeny and geographic distribution (e.g., Brooks & van Veller, 2001), but such studies have not been possible on the Chilean herpetofauna primarily because the phylogenetic relationships among many groups have not yet been resolved. A robust biogeographical analysis is needed to enhance opportunities for further evolutionary research and to frame conservation strategies.

## 2. Biogeography

Biogeography is the science of spatial pattern of biodiversity, both present and past, and how such patterns arise. The development of this branch of ecology addresses many questions (Vidal, 2008), including: Why are species or taxonomic groups (e.g., genera, families, orders) confined to current distributional ranges (García-Barros et al., 2002); what factors restrict a species to a particular place, and what prevents colonization of other areas (Teneb et al., 2004); how and to what extent do climate, topography and interactions with other organisms limit the distribution of species (Losos & Glor, 2003); and how do environmental events and processes (e.g., continental drift, Pleistocene glaciation, climate change) shape the current distribution of species (Brown & Lomolino, 1998; Hughes et al., 2002)?

In essence, biogeography investigates the relationships between patterns (non-random distribution and repetitive organization) and processes (pattern causality) that determine the geographical distributions of organisms. Although biogeographers attempt to summarize these patterns and processes from different perspectives of study (e.g., descriptive biogeography, ecological, historical, paleoecological), the emphasis of each is under constant discussion (Vidal, 2008). Historical biogeography has been particularly controversial. According to Nelson (1969), "the problem of historical biogeography" was the lack of methodology to uncover patterns of association between organisms and their geographical distribution, and the absence of a general explanation for these patterns. He concluded that the key elements that could solve the "problem" are the combination of information from phylogenetic systematics and Earth history. Nearly 50 years later, historical biogeography has been divided into at least two lines of research, fundamentally differing in their concepts and analytical techniques on distribution (Brooks & McLennan, 2002). These two approaches involve inductive/verification and hypothetic-deductive/falsacionist (Brooks et al., 2001). The former, commonly known as vicariance biogeography or cladistic biogeography (e.g., Nelson & Rosen, 1980), is based on the assumption that vicariant speciation is the most recurrent and link phylogeny to historical geology. The second approach on the other hand, originated from the proposal of Wiley (1981) using phylogenetic relationships between species and their geographic distributions to explore the contribution of different modes of speciation.

The ultimate objective of biogeography is an evolutionary perspective (e.g., Morrone, 2007) to understand the past, present, and future of the biota, and from the perspective of conservation biogeography to promote strategies appropriate for species stewardship (Myers, 1988; Álvarez & Morrone, 2004). In recent years biogeography has begun to play an important role in biodiversity conservation issues (Tognelli et al., 2008) since, as discussed below, these studies identify areas of high diversity or endemism that may be a high priority for conservation programs.

## 2.1 Biogeography of the Chilean herpetofauna

One of the main questions in historical biogeography is how to delimit the areas of greater richness or endemism within continents (Nelson & Platnick, 1981; Humphries & Parenti, 1986; Cardoso da Silva & Oren 1996). This question is usually analyzed by means of overlapping distribution maps of taxa, which can be used to detect areas with a high concentration of overlapping species ranges (Haffer, 1978; Cracraft, 1985). However, methodological difficulties have been reported when analyzing a large number of species (Morrone, 1994). It also is somewhat subjective, since there are no defined criteria for analyzing the inconsistencies (Linder, 2001). Such studies have been conducted in both plants and animals (e.g., Heyer, 1988; Benkendorff & Davis; 2002, García-Barros et al., 2002; Teneb et al., 2004), allowing the visualization of distribution patterns of many species, which are then contrasted with the geomorphological and bioclimatic history of the study area (Brown & Lomolino, 1998).

The inventory of Chilean herpetofauna has been in a state of flux due to taxonomic instability, especially among the reptiles (Donoso-Barros, 1966; Veloso & Navarro 1988; Núñez & Jaksic, 1992; Pincheira-Donoso & Núñez, 2005). However, the geographical distribution of many species recently has been clarified, improving information on taxa known only from type localities (Formas, 1995) and the fauna of undersampled areas (Mendez et al. 2005 ), as well as information from Chilean herpetological collections (e.g., Nuñez, 1992; e.g., Sepúlveda et al., 2006; Correa et al., 2007).

In Chile we now recognize 191 species of herpetozoans, including species of sea turtles and the island species, but excluding introduced species. Of this, 59 are amphibians, assigned to 14 genera among four families (Table 1).

| Family | Genus | Richness |
|---|---|---|
| Bufonidae | Rhinella | 4 |
| | Nannophryne | 1 |
| Cycloramphidae | Rhinoderma | 2 |
| | Alsodes | 16 |
| | Eupsophus | 9 |
| | Hylorina | 1 |
| | Insuetophrynus | 1 |
| Ceratophryidae | Atelognathus | 2 |
| | Batrachyla | 4 |
| | Chaltenobatrachus | 1 |
| | Telmatobius | 10 |
| Calyptocephalellidae | Calyptocephalella | 1 |
| | Telmatobufo | 4 |
| | Pleurodema | 3 |
| Total: 4 familes | 14 genera | 59 species |

Table 1. Families, genera, species number of amphibians found in Chile.

Reptiles include for 131 species, assigned to 17 genera among nine families (Table 2). Although the herpetofauna of Chile is low compared to other Neotropical countries, many authors have recognized Chile's high level of endemism. Formas (1979), Ortiz & Díaz-Páez (2006) and Vidal (2008) reported that 67%, 61% and 55% (respectively) of amphibians are endemic to Chile, while Veloso et al (1995) and Vidal (2008) indicate that 50% and 48%, respectively of reptiles are endemic. These authors considered different criteria to determine the endemism of particular species, illustrating the need for clear definition of the concept. High levels of endemism have been interpreted as the result of endogenic diversification (Figure 1) due to the existence of the natural barriers of the cold Pacific Ocean, the Andes, the Atacama Desert to the north, and extreme weather conditions in the south (Torres-Mura, 1994; Schulte et al., 2000; Díaz-Páez et al., (2002).

Although these natural isolating barriers have encouraged endemism, present-day herpetofaunal biogeography is the result of Pleistocene and earlier Cenozoic epochs, including the long and complex forest history of Patagonia. Glaciers were recently more extensive in Patagonia, covering most of southern continent. During episodes of glacial

| Family | Genus | Richness |
|---|---|---|
| Dermochelyidae | *Dermochelys* | 1 |
| Cheloniidae | *Chelonia* | 1 |
| | *Caretta* | 1 |
| | *Lepidochelys* | 1 |
| Colubridae | *Tachymenis* | 2 |
| | *Philodryas* | 4 |
| Elapidae | *Pelamis* | 1 |
| Teiidae | *Callopistes* | 1 |
| Scincidae | *Cryptoblepharus* | 1 |
| Leiosauridae | *Diplolaemus* | 4 |
| | *Pristidactylus* | 4 |
| Tropiduridae | *Liolaemus* | 94 |
| | *Phymaturus* | 6 |
| | *Microlophus* | 6 |
| Gekkonidae | *Homonota* | 2 |
| | *Lepidodactylus* | 1 |
| | *Phyllodactylus* | 1 |
| Total: 9 families | 17 genera | 131 species |

Table 2. Families, genera, species number of Chilean reptiles.

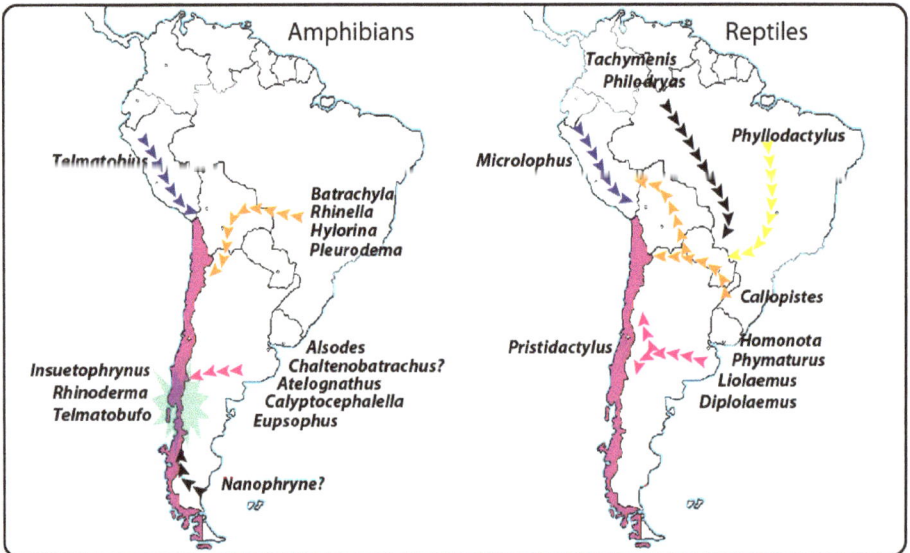

Fig. 1. Scheme representing the different origins of the genera of extant Chilean amphibians and reptiles, based on Vuilleumier (1968), Lynch (1978), Duellman (1979), and Cei (2000) and Basso et al (2011).

maximum herpetofaunal species associated with *Nothofagus* forests may have disappeared or, alternatively, may have been isolated in one or more refugia in southern South America, later expanding their ranges northward after the retreat of the glaciers (Vuilleumier, 1968; Lynch, 1978; Duellman 1979). However, according to Cei (2000), older parents of the fauna can be placed chronologically at the Cretaceous-Paleocene boundary, i.e. in the initial phase of the Andean uplift, which lead to the current configuration and topography of South America. Overall, some herpetofaunal species and genera have colonized the Andean biogeographic province from different parts of South America, while many taxa diversified *in situ* during late Cenozoic time (Figure 1).

The geographic distribution of both amphibians and reptiles shows the opposite distribution along to north-south axis in Chile. As shown in Figure 2, amphibians are found mainly in the central-southern Chile, while reptiles occupy the center-north. The genus *Telmatobius* is the only genus represented exclusively in the northern Chile, with species distributed mainly in elevation. Among the amphibians, *Rhinella* and *Pleurodema* have wide geographic ranges from 18 ° S to 49 ° S, but include few species. A few genera have restricted distributions, such as both *Atelognathus* (which has a few species) and *Insuetophrynus* (a monospecific genus). In contrast, other genera have wide geographic distributions, including *Hylorina* or *Calyptocephalella*, both of which are monospecific genera. Probably, the wide range of current distribution is due to the origin of these latter genera within the region (Duellman, 1979). Recently, Basso et al. (2011) reported a new genus in the family Ceratophryidae: *Chaltenobatrachus*, which has been described as monotypic genus (*C. grandisonae* = *A. grandisonae*) related to *Atelognathus*. The existence of *Chaltenobatrachus* in the region may be similar to the evolutionary history of *Atelognathus*, *Batrachyla* and *Hylorina*; however, given the recent description, it is difficult for us to explain its origin in the Argentinian-Chilean Patagonia.

Among the reptiles (Fig. 2), the genus *Liolaemus* has the largest range, while other genera, except *Microlophus*, *Phymaturus* and *Phyllodactylus* have intermediate sized distributions. Interestingly, when comparing the diversity of both groups, reptiles have a lower richness of genera than do amphibians. Moreover, within the reptiles no more than eight genera overlap in distribution, while among amphibians, up to 10 genera have overlapping ranges. This suggests that, at least for the reptiles, a few genera (e.g., *Liolaemus*, *Tachymenis*) have been able to adapt to a greater variety of environments, achieving greater diversification and breadth of geographic range (Vidal, 2008). On this point the geographic range of the genus *Liolaemus* may be related to the large number of species in Chile (Vidal et al., 2009). In contrast, the two *Tachymenis* colubrid species occur across a broad range, implying that it may be much more plastic than other reptile taxa.

## 3. Biodiversity hotspot

A biogeographic "hotspot" is a term was originally coined by Myers (1988, 1990) to refer to areas with elevated levels of species richness and endemism, and hotspots also often are areas that coincide with other human alterations. The term hot-spot was used by Prendergast et al. (1993) and Gaston & Williams (1996) to refer to areas of extreme taxonomic richness. While the initial definition contained restrictions, today this concept has been expanded and more overtly conceptualized, and from which, to contribute to new conservation strategies (Myers et al., 2000). For species richness and endemism, potential causal factors in the distribution patterns have been described, and which are associated with historical processes (Gaston, 2000; Allen et al., 2002).

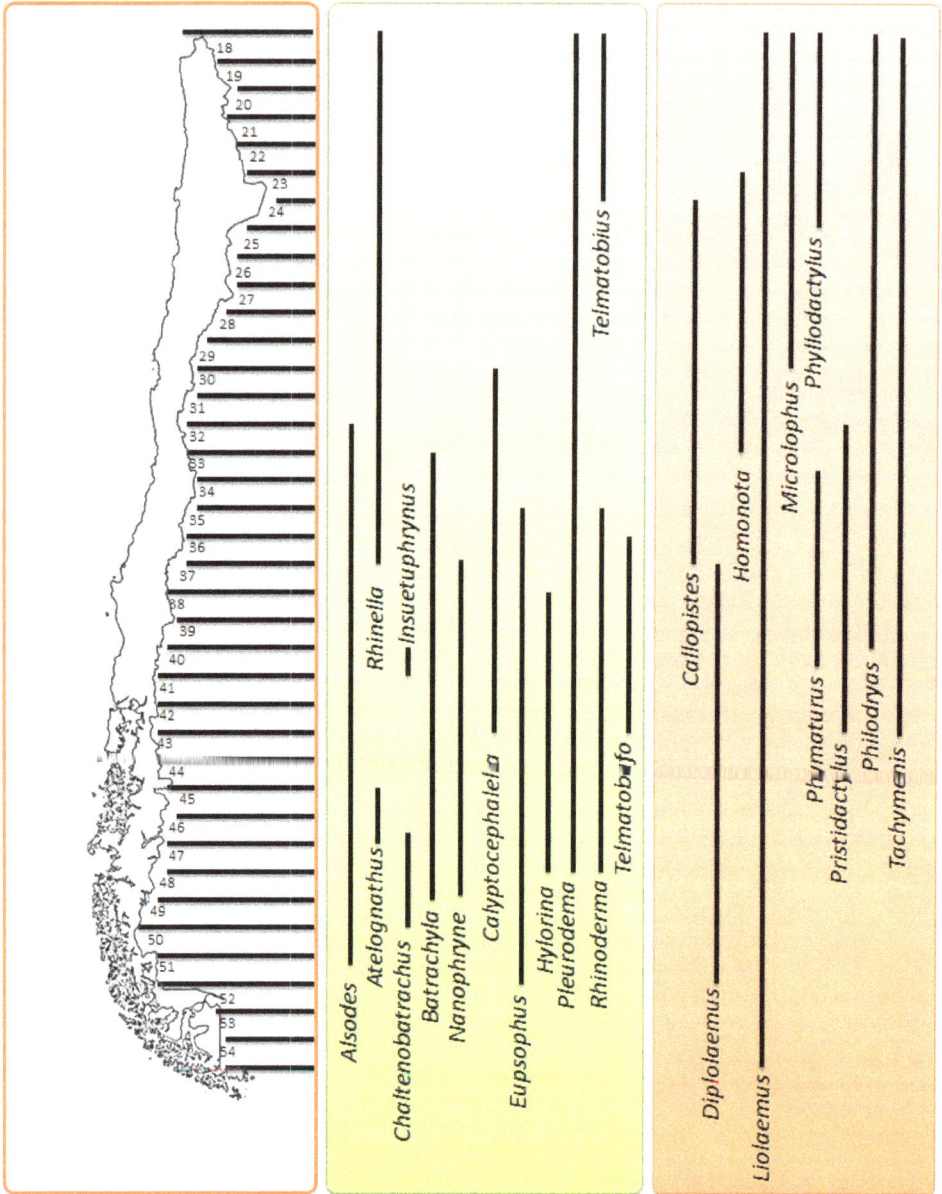

Fig. 2. Map of Chile showing diversity of amphibian and reptiles genera per degree of latitude.

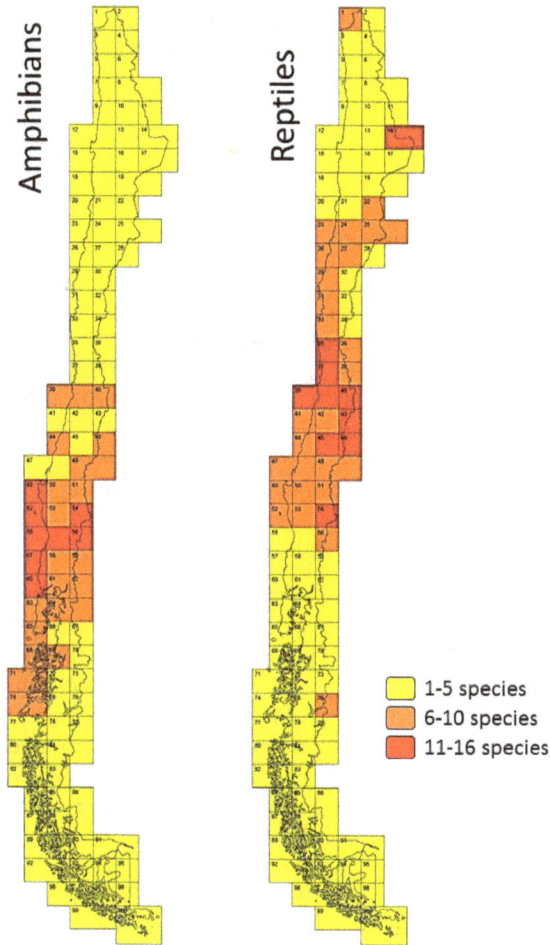

Fig. 3. Chilean amphibian and reptile species richness in 1° latitude by 1° longitude landscape quadrats (after Vidal, 2008). Quadrants in red show the highest species richness, with differences between amphibians (on the left) and reptiles (on the right).

In this context, many taxa are likely apomorphic species (apospecies; Moreno et al., 2006), which have not had sufficient time to move into other areas (e.g., *Eupsophus nahuelbutensis*, *Pristidactylus volcanensdis*), or correspond to ancestral forms (palaeospecies; Kirejtshuk, 2003;) that formerly occupied large areas (e.g., *Calyptocephalella gayi*, *Callopistes maculatus*) but now are restricted to small areas (Brown & Lomolino, 1998; Tribsch & Schönswetter, 2003; Cei, 2000). Thus, an area that concentrates many species (a hot-spot) may be an "evolutionary novelty", a site from which many new genera and species to emerge (Tribsch, 2004), whether remain endemic or not. Several potential hot spots have been reported in Chile, including the coastal range (Méndez et al., 2005; Smith-Ramírez, 2004; and in the Antofagasta region (Veloso & Núñez, 1998). In an analysis of endemism hotspots, Vidal (2008) considered the number of endemic species per degree latitude, finding hotspot located in north and central Chile (Fig. 3).

Interestingly, the distribution of herpetofauna are in direct relationship with its environment dependence, which would explain the presence of these proposed herpetofaunal hotspots coinciding with areas of higher winter rainfall in the Chilean-Valdivian forests (Chile Central), the hotspot proposed by Myers et al. (2000), and other in northern Chile. Both areas have the highest herpetofaunal species richness, but also more human intervention and fewer national parks that protect these species (Vidal et al., 2009).

## 4. Correlation between biological variables

Analysis of the conservation status of taxa in an area or country allows to link extinction risk with morphological, ecological and/or environmental variables. Studies focused on vertebrates have reported that several variables (e.g., body size) are positively associated with risk of extinction, ecological traits, phylogenetic and genetic features, and habitat degradation (Murray & Hose, 2005; Anderson et al., 2011). The loss of biodiversity of amphibians and reptiles has become an important global trend (Gibbons et al., 2000; IUCN, 2010). In this context, Corey & Waite (2008) suggest that threats to amphibians are concentrated in South and Central America, the Caribbean, and Australia. In addition, it has been suggested that some herpetozoan clades are especially prone to extinction by virtue of shared evolutionary histories (Lips et al., 2003; Case et al., 1998).

Body size among animals is directly related to physiological, morphological, ecological and evolutionary characteristics, as well as extinction risk. The relationship between body size and extinction risk recently has been a topic of interest to researchers because both variables are related to direct human influences (Fig. 4). As the body size of mammals increases so does the risk of extinction. However, similar studies of herpetozoa have not been conducted (Cardillo, 2003), nor have links between distribution, habitat conditions, and biological characteristics, such as body size. From our results, central Chile has a marked species concentration (Fig. 3). Biodiversity hotspots are biogeographic regions that are significant reservoirs of biodiversity and are threatened with destruction. Therefore, Chilean herpetozoa in this area are likely at increased extinction risk (Tribsch, 2004). Although the validity of this trend has been previously supported for herpetozoa, it has not yet been associated with other variables, such as body size, conservation status and extinction risk, as seen below

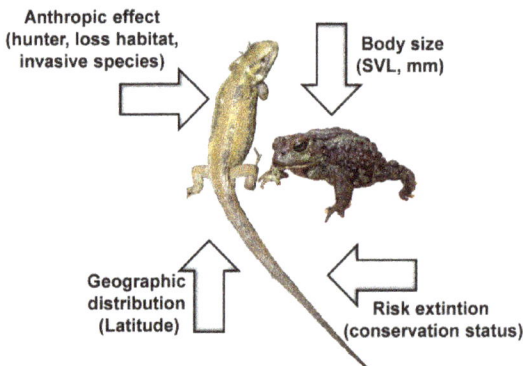

Fig. 4. Synergic effect of some variables involved in extinction risk among the amphibians and reptiles of Chile.

To evaluate extinction risk it is necessary to relate a species conservation status to variables realted to extinction. While Chilean herpetozoa are categorized at the species level as to conservation status, many taxa are categorized as Data Deficient (DD; IUCN, 2010). Here we consider species at risk of extinction, those species categorized as Critically Endangered (CE), Endangered (E) and Vulnerable (V), following categorizations for amphibians and reptiles as proposed by IUCN (2010), and Nuñez et al. (1997), respectively. In accordance to this are considered at risk only those species found within the categories mentioned above (EC, E, and V). By grouping them and observe their latitudinal distribution in which we can detect that Central Chile is an area with numerous species with elevated extinction risk. Of particular concern are reptiles in the north-central area from 25 ° to 44 ° S latitude, and amphibians in the south from 34 ° to 44 ° S latitude. This concentration of threatened and endangered species coincides with proposed biodiversity hotspot for herpetofauna in Chile (Fig. 3).

The scarcity of information on Chilean amphibians and reptiles prevents analysis of associative patterns: however, body size appears to be related of extinction risk for both classes (Meiri, 2008). In reptiles the risk of extinction increases with its frequency in quadrants, while in the case of amphibians restricted distribution is related to extinction risk (Fig. 5). These patterns appear related to human impacts on both classes because reptiles are

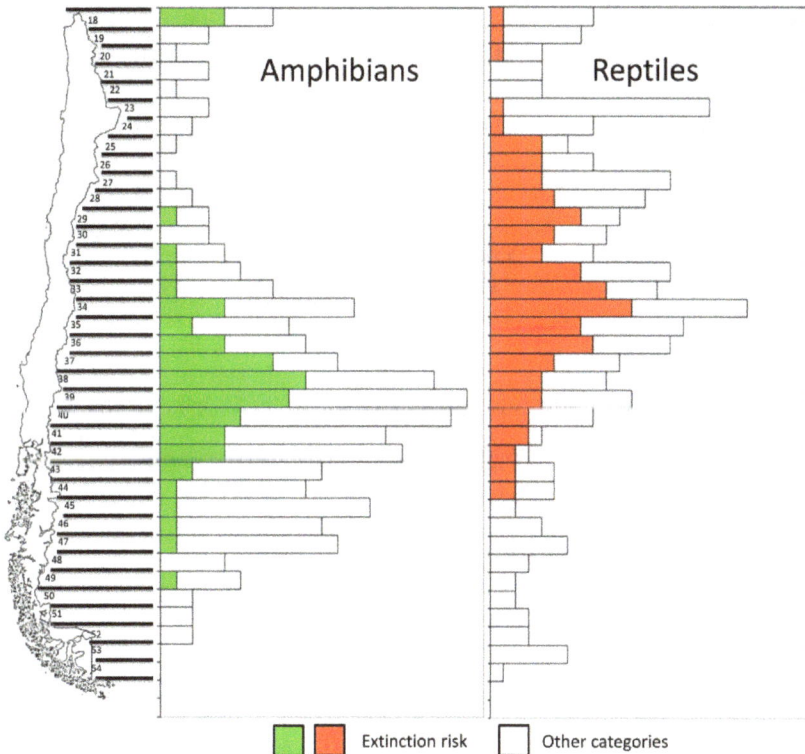

Fig. 5. Map of Chile showing herpetozoa taxa in relation to their extinction risk for each degree of latitude.

generally easier to observe and enjoy have greater interest on the part of man, whether due to aversive fear and beliefs, beauty (by virtue of colour, morphology, or as pets.) Furthermore, among amphibians the risk increases as frequency decreases, relating biological patterns and habitat dependence. It is intuitively obvious that the risk of extinction is greater for populations consisting of a few individuals than for those having many, but it also may be greater for populations undergoing greater flux than those with low temporal variability (Pimm et al., 1988).

Recent studies indicate that modern extinctions and declines of species have been phylogenetically selective (Cardillo, 2003). Thus, the habitat loss together the intrinsic traits some species make them particularly extinction-prone (Figure 4). Also, smaller-bodied species seem to be less vulnerable to decline and extinction than larger species (Gaston & Blackburn, 1995; Cardillo & Bromham, 2001). Furthermore, there may be tradeoffs between different traits; for example, smaller species may have an advantage in higher reproductive output and higher population densities, but larger species may have an advantage in greater mobility and energetic efficiency (Bielby, 2008; Sodhi et al., 2008). Nonetheless, larger vertebrates have a higher risk of extinction, and the explanation appears to be an inverse relationship between population size versus body size (Cardillo & Bromhman, 2001). In addition, the bigger the species the more vulnerable it may be to human persecution and hunting, while smaller species are generally more vulnerable to habitat loss due to anthropic activity (Cardillo, 2003; Sodhi et al., 2008). Similarly it has been established that smaller size confers greater protection.

We tested these extinction risk concepts using our Chilean herpetofaunal data. Among Chilean amphibians, the most important factor in risk is distribution (Figure 6, r Spearman = -0.52; $P<0.001$). For this class, size does not affect risk as much as habitat dependence; therefore, it appears that species with more limited ranges have the greatest risk of extinction. In contrast, body size among reptiles exacerbate extinction risk (Figure 6, r Spearman = 0.29; $P<0.05$), with many explanatory reasons. Thus, it is possible that body size directly determines a species' vulnerability: smaller species may, for instance, be less likely targets for human hunters, or less common prey items for invasive predators (Cardillo & Bromham, 2001). For example, Calyptocephalella gayi (Chilean Big frog) is the largest amphibian species in the country and is consumed due to their body size and good flavor of the meat. The species is broadly used for human consumption and an increase in the level of wild harvest has occurred since approximately 2000. The United States has been a significant commercial importer of wild-caught specimens of this species. From 2003 to 2007, 10,861 wild specimens were exported to the United States and were all traded for commercial purposes (Defenders of Wildlife, 2008). In the case of reptiles, Callopistes maculatus (Iguana Chilena) is the largest terrestrial reptile species and is negatively affected by traffic and trade (Auliya, 2003). According to Fitzgerald & Ortiz (1994), C. maculatus is "in danger" throughout its range due to habitat destruction, and in recent years by increasing harvest to meet international demand. During the years 1981-1991 this species sustained significant population loss from harvest, with the export of at least 2,400 live specimens (JNCC, 1993), which were sold as pets or used for the removal of skin (Díaz-Páez et al., 2008). Both of these large bodied species remain vulnerable to extinction due to harvest pressure.

Body size also may be correlated with other life-history or ecological traits that influence vulnerability, such as reproductive output, mobility, energy requirements or population density. Cardillo (2003) considered the additive impacts of environmental change on extinction risk. For example, the collapse of Pleistocene megafauna was exacerbated by environmental change, including deforestation (Sodhi et al., 2008).

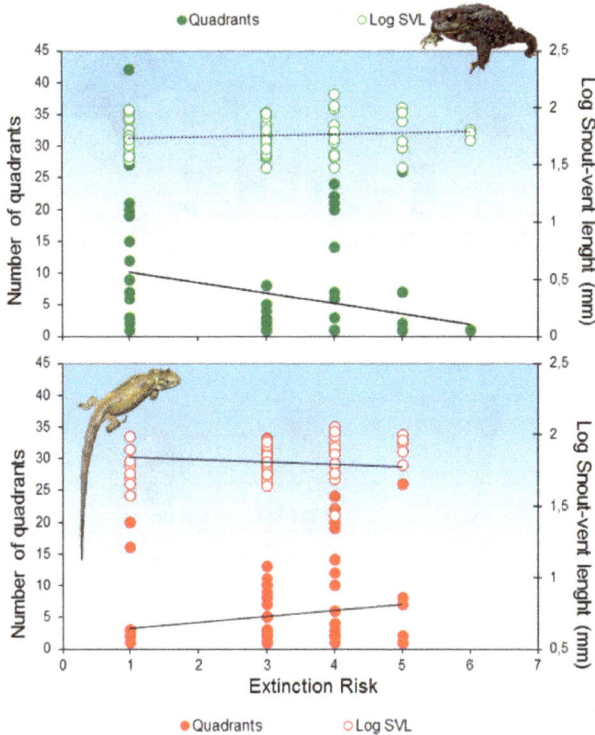

Fig. 6. Relationship between number of quadrants, snout-vent length (SVL) and extinction risk for Amphibians and Reptiles from Chile.

Behavioural, morphological, and physiological characteristics appear to make some species more susceptible than others to extinction. In general, large-sized species with restricted distributions and habitat specialization tend to be at greater risk of anthropogenic extinction than are others within their respective taxa (Sodhi et al., 2009). Our data support the findings of Cardillo (2003), who found that body size, distribution and ecological specialization increase the risk of anthropogenic extinction, especially in situations with rapid habitat loss.

## 5. Critical body size for conservation

Many studies suggest that larger bodied species are more susceptible to extinction than are smaller species (Cardillo & Bromham, 2001), while Murray & Hose (2005) reported no relationship. Our results show weak effect of body size on extinction risk in herpetozoans, although we note that snout–vent length was a predictor in the extinction risk (Figs. 5 and 6) based on IUCN status. There appear to be critical body size ranges in both classes: amphibians with small body sizes have a higher threat status, while increased body size in reptiles increases extinction risk. Similar analyses of Regional patterns from Australia and Central America corroborate our findings (Hero et al., 2005; Lips et al., 2003).

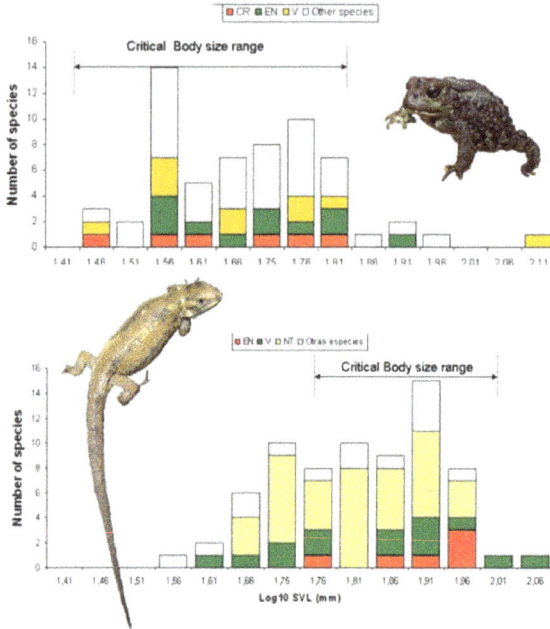

Fig. 7. Body sizes frequency distribution of Chilean herpetofauna, with status conservation according to IUCN (2010) for amphibians and Nuñez et al. (1997) for reptiles.

We used a combination of morphological and distribution data to elucidate extinction risk in Chile, finding that body size influenced extinction risk in opposite ways for the two classes (Fig. 7). Differences between our results and those of other analogous studies likely reflect different biogeographic realms (e.g., Australia Hero et al. 2005, Murray & Hose 2005, Williams & Hero 1998; or Central America Lips et al., 2003). Additionally, the biological traits underlying increased extinction risk/decline can and often do vary according to the particular threat involved, the environment of the location of study, and the species involved (Owens & Bennett, 2000). Our analyses explore the generalities that exist despite these differences, but will miss some of the specific regional correlates.

## 6. Conclusions

We present the first dynamic synthesis of the biogeographic variables affecting Chilean herpetozoa. Inadequate basic information has impeded or delayed studies in the Andean realm. Biogeographical knowledge plays a fundamental role in conservation because the relationship between geographic distribution and extinction risk can reveal new conservation issues and strategies. The herpetofauna of Chile has a lower richness relative to tropical and subtropical South America (Duellman, 1979) due to its prolonged geographical isolation (Armesto et al., 1995). This history of isolation has contributed to the uniqueness of Chilean herpetofaunal assemblage, with many endemic taxa (Arroyo et al., 1999; (Veloso et al., 1995). In Chile, amphibians and reptiles have the highest level of endemism of any vertebrate class, and endemism is focused in the hotspot in central Chile.

Different evolutionary processes are involved in anthropogenic extinction risk among the Chilean herpetofauna. Smaller species may have lower energy requirements and larger population sizes, making them more resilient to human disturbances. The higher reproductive potential of smaller species may reduce population recovery time following disturbance (Gaston & Blackburn, 1995). In contrast, larger species have lower reproductive rates and higher net energetic demands, requiring larger home ranges. We report a positive association between body size and extinction risk among reptiles. Larger species are affected more by harvest (Cardillo & Bromhan, 2001) and by habitat alteration, including the introduction of non-native species, as is happening in Chile today. Overall, we conclude that life-history traits influence extinction risk, with smaller-bodied amphibians affected by environmental changes, and larger bodied reptiles affected by harvest and habitat loss.

## 7. Acknowledgment

The authors were financed by CONICYT 79090026 and DIUC Project 210.412.045-1sp. We thank to Chilean Herpetological Network (RECH) for encouraging this work. We also thank the Department of Basic Sciences at the Universidad del Bío-Bío, and the Department of Basic Sciences at the Universidad de Concepcion for partial financial support of this study.

## 8. References

Anderson SC, RG Farmer, F Ferretti, AM Houde & JA Hutchings (2011). Correlates of vertebrate extinction risk in Canada. *BioScience,* Vol.61, N° 7, (August 2011), pp. 538-549, ISSN 0006-3568.

Allen A, JH Brown & JF Gillooly (2002). Global biodiversity, biochemical kinetics, and the energetic-equivalence rule. *Science,* Vol. 297, N° 5586, (August 2001), pp. 1545-1548, ISSN 1683-8831.

Álvarez E & JJ Morrone (2004). Propuesta de áreas para la conservación de aves de México, empleando herramientas panbiogeográficas e índices de complementariedad. *Interciencia* (Venezuela), Vol. 29 N°3, (March 2004), pp. 112-120, ISSN 0378-1844.

Armesto J, C Villagrán & MK Arroyo (1995). Ecología de los Bosques nativos de Chile. Editorial Universitaria. Chile. 469 pp.

Arroyo MTK, R Rozzi, J Simonetti, P Marquet & M Salaberry (1999). Central Chile. In: Mittermeier RA, N Myers & CG Mittermeier (eds) Hotspots: Earth's biologically richest and most endangered terrestrial ecosystems. CEMEX, México, Distrito Federal. Pp. 161-171.

Auliya M (2003). Hot Trade in Cool Creatures. A review of the live reptile trade in the European Union in the 1990s with a focus on Germany. A Traffic Europe Report. Published by Traffic Europe for IUCN, Brussels, Belgium. 105 pp, ISBN 978-2-96005-059-2.

Basso NG, CA Úbeda, MM Bunge & LB Martinazzo (2011). A new genus of neobatrachian frog from southern Patagonian forests, Argentina and Chile. *Zootaxa,* Vol. 3002, (August 2011), pp. 31–44, ISSN 1175-5334.

Benkendorff K & AR Davis (2002). Identifying hotspots of molluscan species richness on rocky intertidal reefs. *Biodiversity & Conservation,* Vol. 11, N°11, (November 2002), pp. 1959-1973, ISSN 0960-3115.

Bielby J (2008). Extinction Risk and Population Declines in Amphibians. A thesis submitted for the degree of Doctor of Philosophy at Imperial College London. 247pp.

Brooks DR & DA McLennan (2002). The Nature of Diversity: An Evolutionary Voyage of Discovery . University of Chicago Press, Chicago. 676 pp. ISBN, 9780226075907

Brooks DR, MGP Van Veller & DA McLennan (2001). How to do BPA, really. *Journal of Biogeography,* Vol. 28, N° 3, (March, 2001), pp. 345-358, ISSN: 1365-2699

Brown JM & MV Lomolino (1998). Biogeography. 2nd Edition. Sinauer Associates Sunderland MA 692 pp, ISBN 0878930736..

Cardillo M (2003). Biological determinants of extinction risk: why are smaller species less vulnerable? *Animal Conservation,* Vol. 6, N°1, (February 2003), pp. 63–69, ISSN 1469-1795.

Cardillo M & L Bromham (2001). Body size and risk of extinction in Australian mammals. *Conservation Biology* Vol. 15, N°5, (october 2001), pp. 1435-1440, ISSN 0888-8892.

Cardillo M, GM Mace, JL Gittleman, KE Jones, J Bielby & A Purvis (2008). The predictability of extinction: biological and external correlates of decline in mammals. *Proceeding the Royal Society Biological Science,* Vol. 275, N°1, (March 2008), pp. 1441–1448, ISSN 0962-8452.

Cardoso de Silva JM & DC Oren (1996). Application of parsimony analysis of endemicity in Amazonian biogeography: an example with primates. *Biological Journal of the Linnean Society,* Vol. 59, N°4, (December 1996), pp. 427-437, ISSN 1095-8312.

Case T J, Bolger AD & AD Richman (1998). Reptilian extinctions over the last ten thousand years. In: Fielder PL & PM Kareiva (eds.), Conservation biology for the coming decade, 2nd edition, Chapman & Hall, New York. Pp 157-186. ISBN 0-412-09661-7

Cei JM (2000). Centros de diversificación trans-ciscordilleranos y aislamiento por reducción de área como factores de la biodiversidad andino-patagónica XV Reunión de Comunicaciones Herpetológicas: San Carlos de Bariloche, 25 al 27 de octubre del 2000

Corey SJ & TA Waite (2008). Phylogenetic autocorrelation of extinction threat in globally imperilled amphibians. *Diversity and Distributions,* Vol. 14, N°4, (July 004), pp. 614–629, ISSN 1366-9516.

Correa CL, M Sallaberry, BA González, ER Soto & MA Méndez (2007). Amphibia, Anura, Leiuperidae, *Pleurodema thaul:* Latitudinal and altitudinal distribution extension in Chile. *Check List,* Vol. 5, N°4, (December 2009), pp. 267-270, ISSN 1809-127X.

Cracraft J (1985). Historical biogeography and patterns of differentiation within the South American avifauna: Areas of endemism. *Ornithological Monographs,* N°36, pp. 49-84. ISSN 0078-6594

Defenders of Wildlife (2008). Freedom of Information Act request to U.S. Fish and Wildlife Service. Information in LEMIS database.

Díaz-Paéz H, C Williams & RA Griffiths (2002). Diversidad y abundancia de anfibios en el Parque Nacional "Laguna San Rafael" (XI Región Chile). *Boletín Museo Nacional de Historia Natural* (Chile), N°51, pp. 135-145, ISSN 0027-3010.

Díaz-Páez H, J Núñez, H Núñez & Ortiz JC (2008). Estado de conservación de anfibios y reptiles. In: Herpetología de Chile. Vidal M & M Labra (eds), Science Verlag. Santiago, Chile.Pp. 233-267. ISBN 978-956-319-420-3.

Donoso-Barros R (1966). Reptiles de Chile. Ediciones Universidad de Chile, Santiago, Chile. cxliv + 458 pp, ISBN 0327-9375.

Duellman WE (1979). The South American herpetofauna: its origin, evolution and dispersal. *Monograph of the Museum of Natural History University of Kansas*, N°7, (December 1979), pp. 1-485, ISBN 0-89338-008-3.

Fitzgerald L & JC Ortiz (1994). Analyses of Proposals to Amend the CITES Appendices. IUCN Species Survival Commission Traffic Network, pp. 174-176, ISSN 1016-927X.

Formas JR (1979). La herpetofauna de los bosques temperados de Sudamérica. In: Duellman WE (ed), The South American herpetofauna: its origin, evolution and dispersal, pp. 341-369. *Museum of Natural History, University of Kansas, Kansas*, ISBN 0-89338-008-3.

Formas JR (1995). Anfibios. In: Simonetti JA, MTK Arroyo, AE Spotorno & E Lozada (eds), Diversidad biológica de Chile. Comisión Nacional de Investigación Científica y Tecnológica. Santiago, Chile. Pp. 314-325.

García-Barros E, P Gurrea, MJ Luciáñez, JM Cano, ML Munguira, JC Moreno, H Sainz, MJ Sanz & JC Simón (2002). Parsimony analysis of endemicity and its application to animal and plant geographical distributions in the Ibero-Balearic region (Western Mediterranean). *Journal of Biogeography* Vol. 29, N°1, (January, 2002), pp. 109-124, ISSN 1365-2699.

Gaston KJ (2000). Global patterns in biodiversity. *Nature*, Vol. 405, N° 6783, (May 2000), pp. 220-227, ISSN 0028-0836

Gaston K & TM Blackburn (1996). Rarity and body size: importance of generality. *Conservation Biology*, Vol. 10, N°4, (August 1996), pp. 1295-1298, ISSN 0888-8892

Gaston KJ & PH Williams (1996). Spatial patterns in taxonomic diversity. In: Gaston KJ (ed), Biodiversity: A biology of numbers and difference. Blackwell, Cambridge. Pp. 202-229.

Gibbons JW, DE Scott, TJ Ryan, KA Buhlmann, TD Tuberville, BS Metts, JL Greene, T Mills, Y Leiden, S Poppy & CT Winne (2000). The global decline of reptiles, déjà vu amphibians. *BioScience*, Vol. 50, N°8, (Augut 2000), pp. 653-666, ISSN 0006-3568

Hero JM, SE Williams & WE Magnusson (2005). Ecological traits of declining amphibians in upland areas of eastern Australia. *Journal of Zoology London*, Vol. 267, N°3, (November 2005), 221–232 ISSN 09528369

Haffer J (1978). Distribution of Amazon birds. *Bonner Zoologischen Beitragen* (German), Vol. 29, N°4, (June 1978), pp. 38-78, ISSN: 0006-7172.

Heyer WR (1988). On frog distribution patterns east of the Andes. In: Vanzolini PE & WR Heyer (eds), *Proceedings of a workshop on neotropical distributional patterns*, Academia Brasileira de Ciencias, Rio de Janeiro. Pp. 245-273.

Hughes TP, DR Bellwood & SR Connolly (2002). Biodiversity hotspots, centres of endemicity, and the conservation of coral reefs. *Ecology Letters*, Vol. 5, N°6, (November 2002), pp. 775-784, ISSN 1461-023X.

Humphries CJ & LR Parenti (1986). Cladistic biogeography. Oxford University Press, Oxford. xii + 98 pp, ISBN 019-854576-2.

IUCN (2010). IUCN Red List of Threatened Species. Version 2010.2. Downloaded in August 2010

Kirejtshuk AG (2003). Subcortical space as an environment for palaeoendemic and young groups of beetles, using mostly examples from sap-beetles (Nitidulidae, Coleoptera). *Proceedings of the second Pan-European conference on Saproxylic Beetles*, pp. 50-56. Royal Holloway, University of London. People's Trust for Endangered Species

JNCC (1993). Review of UK import of non CITES fauna from 1980 - 1991. Joint Nature Conservation Committee, Report N° 126. Compiled by the world Conservation Monitoring Centre in Collaboration with Traficc International. Joint Nature Conservation Committee, Peterborough.

Linder HP (2001). On areas of endemism, with an example from African Restionaceae. *Systematic Biology*, Vol. 50, N°6, (December 2001), pp. 892–912, ISSN 1063-5157

Lips KR, JD Reeve & LR Witters (2003). Ecological traits predicting amphibian population declines in Central America. *Conservation Biology*, Vol. 17, N°4, (August 2003), pp. 1078–1088, ISSN 0888-8892.

Losos JB & E Glor (2003). Phylogenetic comparative methods and the geography of speciation. *Trends in Ecology & Evolution*, Vol. 18, N°5, (May 2003), pp. 220-227, ISSN 0169-5347.

Lynch J (1978). A re-assessment of the telmatobiine Leptodactylid frogs of Patagonia. *Ocasional papers of the Museum of Natural History, The University of Kansas*, N° 72, pp. 1-57, *ISSN 0091-7958*.

Meiri Sh (2008). Evolution and ecology of lizard body sizes. *Global Ecology and Biogeography*, Vol. 17, N°6, (November 2008), pp. 724–734, ISSN 1466-8238.

Méndez MA, ER Soto, F Torres-Pérez & A Veloso (2005). Anfibios y reptiles de los bosques de la Cordillera de la costa (X Región, Chile). In: Smith-Ramírez C, JJ Armesto & C Valdovinos (eds), Historia, biodiversidad y ecología de los bosques costeros de Chile. Editorial Universitaria, Santiago, Chile. Pp. 441-451. ISBN 956-11-1777-0

Meynard C, H Samaniego & PA Marquet (2004). Biogeografía de aves rapaces de Chile. In: Munoz A, J Rau & J Yanez (eds.), Aves rapaces de Chile, Ediciones CEA, Valdivia, Chile. Pp. 129-140. ISBN 956-7279 08 X.

Moreno R, CE Hernández, MM Rivadeneira, MA Vidal & N Rozbaczylo (2006). Patterns of endemism in south-eastern Pacific benthic polychaetes of the Chilean coast. *Journal of Biogeography*, Vol. 33, N°4, (April 2006), pp. 750-759 ISSN 1365-2699

Morrone JJ (1994). On identification of areas of endemism. *Systematic Biology*, Vol 43, N°3, (May 1994), pp. 438–441 ISSN 1063-5157

Morrone JJ (2007). Hacia una biogeografía evolutiva. *Revista Chilena de Historia Natural*, Vol. 80, N°4, (December 2007), pp 509-520, ISSN 0716-078X.

Murray BR & GC Hose (2005). Life-history and ecological correlates of decline and extinction in the endemic Australian frog fauna. *Austral Ecology*, Vol. 30, N°5, (August 2005), pp. 564–571, ISSN 1442-9985.

Myers N (1988). Threatened biotas: hot-spots in tropical forest. *The Environmentalist*, Vol. 8, N°3, (September 1988), pp. 187-208, *ISSN 0251-1088*.

Myers N (1990). The biodiversity challenge: expanded hot-spots analysis. *The Environmentalist*, Vol. 10, N°4, (December 1990), pp. 243-256, ISSN *0251-1088*

Myers N, RA Mittermeier, CG Mittermeier, G Da Fonseca & J Kent (2000). Biodiversity hotspots for conservation priorities. *Nature*, Vol. 403, N° 6772, (February 2000), pp. 853-858, ISSN 0028-0836

Nelson G (1969). The problem of historical biogeography. *Systematic Zoology*, Vol. 18, N° 2, (June, 1969), pp. 243-246, ISSN 0039-7989.

Nelson G & N Platnick (1981). Systematics and biogeography: Cladistics and vicariance. Columbia University Press, New York. 567 pp, ISBN 0-231-04574-3.

Nelson G & N Platnick (1981). Systematics and biogeography: Cladistics and vicariance. Columbia University Press, New York. 567 pp.

Nelson G & DE Rosen (1980). Vicariance biogeography: A critique. Columbia University Press, New York. 593 pp.

Núñez H (1992). Geographical data of Chilean lizards and snakes in the Museo Nacional de Historia Natural, Santiago, Chile. *Smithsonian Herpetological Information Service* N° 91, pp. 1-29.

Núñez H & F Jaksic (1992). Lista comentada de los reptiles terrestres de Chile continental. *Boletín del Museo Nacional de Historia Natural* (Chile), Vol. 43, pp. 63-91, ISSN 0716-2537

Núñez H, V Maldonado & R Pérez (1997). Reunión de trabajo con especialistas de herpetología para categorización e especies según estados de conservación. *Noticiario Mensual Museo Nacional de Historia Natural* (Chile), Vol. 329, pp. 12-19, ISSN 0376-2041.

Ortiz JC & H Díaz-Páez (2006). Estado del conocimiento de los anfibios en Chile. *Gayana* (Chile), Vol. 70, N°1, (June 2006), pp 114-121. ISSN 0717-6538

Owens IPF & PM Bennett (2000). Ecological basis of extinction risk in birds: habitat loss versus human persecution and introduced predators. *Proceedings of the National Academy of Sciences* (U.S.A.), Vol. 97, pp, 12144-12148, ISSN 0027-8424

Pimm SL, HL Jones & JM Diamond (1988). On the risk of extinction. *The American Naturalist*, Vol. 132, N°6, (December 1988), pp. 757–785, ISSN 0003-0147.

Pincheira-Donoso D & H Nuñez (2005). Las especies chilenas del género *Liolaemus* Wiegmann, 1834 (Iguania: Tropiduridae: Liolaeminae). Taxonomía, sistemática y evolución. *Publicación Ocasional Museo Nacional de Historia Natural* (Chile), Vol. 59, pp. 1-486, ISSN 0716-0224.

Prendergast J, RM Quinn, JH Lawton, BC Eversham & DW Gibbons (1993). Rare species, the coincidence of diversity hotspots and conversation strategies. *Nature*, Vol. 365, N° 6444, (September 1993), pp. 335-337, ISSN 0028-0836.

Schulte JA, JR Macey, RE Espinoza & A Larson (2000). Phylogenetic relationships in the iguanid lizard genus *Liolaemus*: multiple origins of viviparous reproduction and evidence for recurring Andean vicariance and dispersal. *Biological Journal of the Linnean Society*, Vol. 69, N°1, (January 2000), pp. 75–102, ISSN 1095-8312.

Sepulveda M, MA Vidal & JM Fariña (2006). *Microlophus atacamensis*. Predation. *Herpetological Review*, Vol. 37, N°2, (October 2006), pp. 224-225, ISSN 0018-084X

Smith-Ramírez C (2004). The Chilean coastal range: a vanishing center of biodiversity and endemism in South American temperate rainforests. *Biodiversity & Conservation*, Vol. 13, pp. 373-393, ISSN 09060-3115.

Spotorno AE (1995). Vertebrados de Chile. In: Simonetti JA, MTK Arroyo, AE Spotorno & E Lozada (eds.). Diversidad Biológica de Chile, CONICYT, Santiago, Chile. Pp. 299-301.

Sodhi NS, D Bickford, AC Diesmos, TM Lee & LP Koh (2008). Measuring the Meltdown: Drivers of Global Amphibian Extinction and Decline. PLoS ONE 3(2): e1636. doi:10.1371/journal.pone.0001636, ISSN 1932-6203.

Sodhi NS, BW Brook & CJA Bradshaw (2009). Causes and consequences of species extinctions. In Levin SA (ed), The Princeton Guide to Ecology, Princeton University Press, Princeton, NJ. Pp. 514–520. ISBN 978-0-691-12839-9

Teneb EA, LA Caviares, MJ Parra & A Marticorena (2004). Patrones geográficos de distribución de árboles y arbustos en la zona de transición climática mediterráneo-templada de Chile. Revista Chilena de Historia Natural, Vol. 77, N°1, pp. 51-71, ISSN 0716-078X.

Tognelli MF, PI Ramirez & P Marquet (2008). How well do the existing and proposed reserve networks represent vertebrate species in Chile? Diversity and Distribution, Vol. 14, N° 1, (January 2008), pp. 148-158, ISSN 1366-9516.

Torres-Mura J (1994). Fauna terrestre de Chile. In: CONAMA (ed), Perfil ambiental de Chile, Santiago, Chile. Pp. 63-72.

Tribsch A (2004). Areas of endemism of vascular plants in the eastern Alps in relation to Pleistocene glaciation. Journal of Biogeography, Vol. 31, N°5, (May 2004), pp. 747-760, ISSN: 0305-0270.

Tribsch A & P Schönswetter (2003). In search for Pleistocene refugia for mountain plants: patterns of endemism and comparative phylogeography confirm palaeo-environmental evidence in the Eastern European Alps. Taxon, Vol. 52, N° 3, (August 2003), pp. 477–497, ISSN: 0040-0262

Veloso A & J Navarro (1988). Lista sistemática y distribución geográfica de anfibios y reptiles de Chile. Museo Regionale di Scienze Naturali Torino, Vol. 6, pp. 481-539. ISSN: 1590-6388.

Veloso A & H Núñez (1998). Inventario de especies de fauna de la región de Antofagasta (Chile) y recursos metodológicos para almacenar y analizar información de biodiversidad. Revista Chilena de Historia Natural, Vol. 71, N°4, (December 1998), pp. 555-569, ISSN 0716-078X

Veloso A, JC Ortiz, J Navarro, H Núñez, P Espejo & MA Labra (1995). Reptiles. In: Simonetti JA, MTK Arroyo, AE Opolorno & E Lozada (eds), Diversidad biológica de Chile, Comisión Nacional de Investigación Científica y Tecnológica, Santiago. Pp. 326-335.

Vidal MA (2008). Biogeografía de anfibios y reptiles. In: Vidal MA & A Labra (eds), Herpetología de Chile, Santiago, Science Verlag. Pp. 195-231. ISBN 978-956-319-420-3.

Vidal MA, E Soto & A Veloso (2009). Biogeography of Chilean herpetofauna: distributional patterns of species richness and endemism. Amphibia-Reptilia, Vol. 30, N°2, (April 2009), pp. 151-171, ISSN: 0173-5373

Vuilleumier F (1968). Origin of frogs of Patagonian forest. Nature, Vol. 219, N°5149, (July 1968), pp 87-89, ISSN 0028-0836.

Williams SE & JM Hero (1998). "Rainforest frogs of the Australian wet tropics: Guild classification and the ecological similarity of declining species. Proceedings of the Royal Society of London Series B Biological Sciences, Vol. 265, N° 1396, (April 1998), pp. 597-602, ISSN 1471-2954

Wiley RH (1981). Social structure and individual ontogenies: problems of description, mechanism, and evolution. In: Bateson PPG & PH Klopfer (eds), Perspectives in Ethology, vol. 4. Plenum Press, New York. Pp. 105-133. ISBN: 0-306-40511-3

# Part 3

# Biogeography of Complex Landscapes

# 9

# The Biogeographic Significance of a Large, Deep Canyon: Grand Canyon of the Colorado River, Southwestern USA

Lawrence E. Stevens

*Biology Department, Museum of Northern Arizona, Flagstaff, Arizona, USA*

## 1. Introduction

Mountains and uplifted areas occupy more than 10% of the Earth's surface, and their associated drainages rather commonly develop as constrained, canyon-bound channels. Therefore, large deep canyons (LDCs) are relatively frequently encountered, persistent landforms, occurring either as steeply dipping, V-shaped canyons that emerge from fold, fault-block, volcanic, and dome mountains, or occurring as drainages incised into uplifted plateaus. The latter type include the world's deepest canyons, including: the 3.35 km-deep Cotahuasi Canyon of the Rio Cotahuasi, a tributary of the Rio Ocona in southwestern Peru; the 2.44 km-deep Hells Canyon of the Snake River in Idaho; the >2.1 km-deep Barranca de Cobre in Chihuahua, Mexico; and the world-renowned 2.48 km-deep Grand Canyon in northern Arizona. LDCs also occur in submarine environments at the mouths of large rivers: the Nile and the Rhône Rivers have large submarine canyons at their mouths created during repeated Pliocene desiccation of the Mediterranean basin; the Indus and Ganges Rivers form lengthy submarine deltaic canyons; the Hudson River in northeastern North America has a substantial submarine canyon; and large canyons form in other tectonically active submarine areas. Whether terrestrial or subaqueous, LDCs support or generate strong ecological gradients, and periodically or perennially provide cascading deliveries of flow, sediments, nutrients, and biota to lower elevations (e.g., Gurnell and Petts 1995; Butman et al. 2006; Canals et al. 2009), processes that may influence the distribution and evolution of life around them.

Although both terrestrial and submarine LDCs are conspicuous landforms, their regional biogeographic significance has received remarkably little scientific attention, particularly in relation to that devoted to other major landforms, such as islands, lakes, and mountain ranges (Lomolino et al. 2010). Obviously functioning as barriers throughout human history and to many biota, terrestrial LDCs form complex habitat mosaics that also function as downstream and upstream corridors through higher elevation terrain, and LDCs contain an array of refugial habitats (Fig. 1; Stevens and Polhemus 2008). Consequently, LDCs may differentially facilitate or restrict gene flow, with ecological and evolutionary impacts on regional populations and assemblages across spatial and temporal scales, and in ways different from those of other conspicuous landforms. LDCs also are preferred sites for dam construction, and many LDCs and their associated assemblages have been altered by flow regulation throughout the world. Here I describe the biogeographical characteristics and significance of the world's most famous terrestrial LDC, Grand Canyon (GC) of the Colorado River in the American Southwest within its drainage basin, the Grand Canyon ecoregion (GCE). I compare LDC characteristics with those of other major landforms, and describe conservation issues.

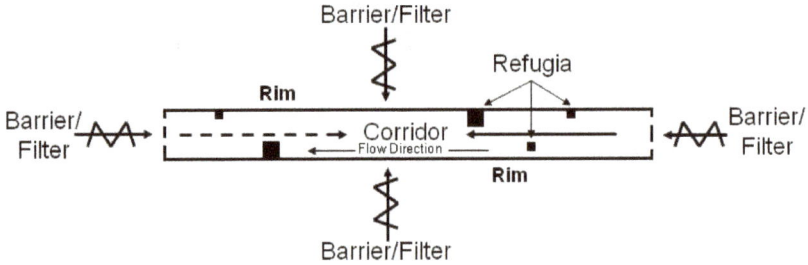

Fig. 1. Schematic diagram of the biogeographic functions of large, deep canyons.

## 2. Study area

The GCE includes 144,000 km² of the southern Colorado Plateau in northern Arizona, southern Utah, and western New Mexico (Fig. 2). This is a topographically complex region, ranging in elevation from 350 m AMSL on eastern Lake Mead to 3,850 m on the San Francisco Mountains near Flagstaff, Arizona, and with the highest point on the North Rim of GC at approximately 2,830 m. The climate of the GCE is continental and arid, with low elevation desert summertime high temperatures >45°C, and high elevation mountain winter low temperatures < -55°C. Precipitation is bimodal, with wintertime snowfall and rain, and mid-late summer monsoonal rains (Sellers et al. 1985). More than 90 percent of the GCE is managed by federal agencies or Indian Tribes, including: 12 National Park units, particularly including Grand Canyon National Park; 6 National Forest units; the Bureau of Land Management; and 8 Native America tribes. State lands include small tracts of state fish and wildlife management units, and private lands are few.

A major geologic province boundary divides Arizona in half, with the Basin and Range geologic province to the south and west, and the Colorado Plateau division of the Rocky Mountain geologic province to the north and east (Fig. 2). The boundary runs diagonally NW-SE through central Arizona, creating a lengthy escarpment known as the Mogollon Rim. This geologic province boundary forms a biogeographic ecotone between the Mexican-neotropical Madrean floristic region and the boreal Rocky Mountain floristic region, referred to as the Lowe-Davis line (Lomolino et al. 2010). Drainages that breach this geologic province boundary (e.g., the Virgin and Verde Rivers) are richer in species of aquatic invertebrates than are those that do not bridge the boundary (e.g., the Escalante River, the Little Colorado River, the Paria River, Kanab Creek, and Havasu Creek; Stevens and Polhemus 2008, Stevens and Bailowitz 2009).

The GCE lies at the intersection of 4 biomes, including the Madrean, Mohavean/Sonoran, intermountain, and cordilleran biomes. Low desert habitats to the south are occupied by Madrean, Sonoran, and Mohavean desert shrub vegetation, while middle elevations are

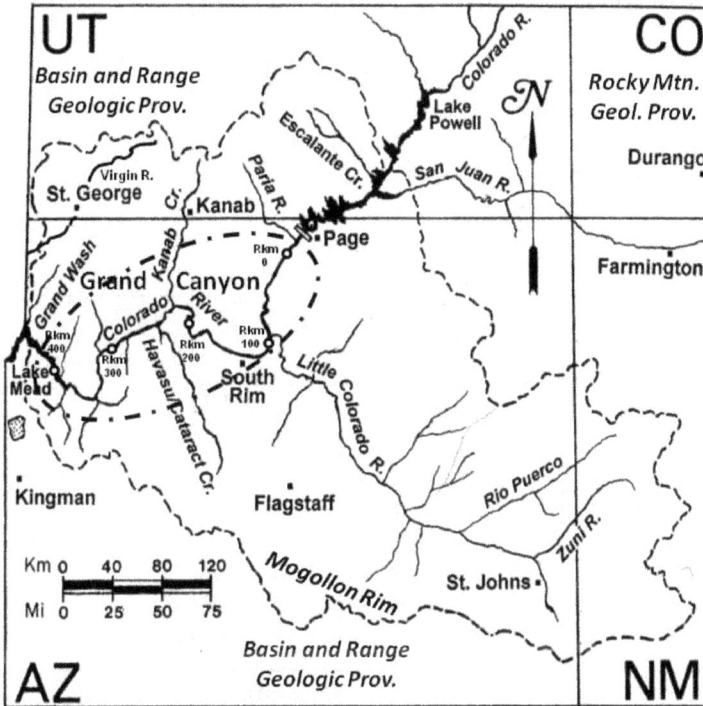

Fig. 2. Grand Canyon and the 144,000 km² Grand Canyon ecoregion (dashed line) on the southern Colorado Plateau, southwestern U.S.A.

dominated by intermountain Great Basin grasslands and pinyon-juniper woodlands. Upper elevation plateaus are occupied by cordilleran Rocky Mountain ponderosa pine (*Pinus ponderosa*) and mixed conifer forests, with large meadows, and highest elevations above 3600 m support 5 km² of Rocky Mountain alpine tundra habitat (Table 1). The GCE is strongly dominated by upland shrublands, woodlands, and ponderosa pine forests, which collectively occupy 95% of the land area, while deserts below 1000 m and montane to alpine habitats above 2800 m each occupy only about 2.5% of the land area.

The GCE is dominated by the large, deep canyons of the Colorado River and its many tributaries, and contains more than 75 ecosystems and habitats (Stevens and Nabhan 2002). Locations along the Colorado River are designated by distance from Lees Ferry, Arizona, and on the left (L) or right (R) side of the river looking downstream. Large tributaries include: the Paria River (Rkm 1.5R), the Little Colorado River (LCR; Rkm 98L), the Virgin River, Kanab Creek (Rkm 232R), Havasu/Cataract Canyon (253L), Diamond Creek/Peach Springs Wash (383L), and Grand Wash (479R; Figs. 2, 3). Hoover and Glen Canyon Dams on the Colorado River create the nation's two largest reservoirs, Lake Mead and Lake Powell, respectively, which overlap into the GCE. Not within GC, the Black, White, Eagle, and other tributaries of the upper Gila River form the southeastern drainage boundary of the GCE. Tonto Creek, a central Gila River tributary, drains the region west of the upper Gila River

| Elevation Range (m) | Vegetation Zone | Characteristic Flora | Comments |
|---|---|---|---|
| <1500 | Lower Sonoran | Desert shrubs, including cacti; Colorado River, springs, and reservoir riparian vegetation | GC floor and lower slopes |
| 1100-2300 | Upper Sonoran | Pinyon pine (*Pinus* spp.), *Juniperus* spp.; Great Basin shrublands; grasslands; stream and springs riparian vegetation | GC middle-upper slopes |
| 1600-2800 | Transition | Ponderosa pine (*Pinus ponderosa*), Gambel's oak (*Quercus gambelii*), Douglas fir; small streams and springs riparian vegetation | Both GC rims |
| 2500-3300 | Canadian | Mixed conifer: white fir (*Abies concolor*), Douglas fir (*Pseudotsuga mentsezii*), *Pinus* spp.; springs riparian vegetation | GC North Rim |
| 3000-3600 | Hudsonian | Spruce-fir forest; springs riparian vegetation | At elevations above GC in the White and San Francisco Mtns |
| 3400-3700 | Subalpine | Subalpine fir (*Abies lasiocarpa*), bristlecone pine (*Pinus aristata*), low shrubs, springs riparian vegetation | At elevations above GC in the White and San Francisco Mtns |
| 3600-3850 | Alpine | Tundra shrubs, graminoides | At elevations above GC in the San Francisco Mtns |

Table 1. Merriam life zones and vegetation zonation across elevation in the GCE

basin. The Verde River basin breaches the geologic province boundary, draining the central southern portion of the GCE. The Virgin River drains southwestern Utah and northwestern Arizona, also breaching the geologic province boundary. In addition, the GCE supports thousands of caves and springs (Stevens and Meretsky 2008), and thousands of km of escarpment edges.

Embedded wholly within the southwestern Colorado Plateau, GC is naturally separated into an eastern basin, which receives the flows of the Paria and LCR drainages, and a more open western basin that connects western Grand Canyon to the Mojave and Sonoran deserts to the west and south (Fig. 3; Billingsley and Hampton 1999, Stevens and Huber 2004). The two GC basins are separated by the steep, narrow 35 km-long Muav Gorge, which creates a formidable cliff-dominated barrier to upstream and downstream dispersal of numerous southwestern plant, invertebrate, and vertebrate taxa (Miller et al. 1982; Phillips et al. 1987; Schmidt and Graf 1990; Stevens and Polhemus 2008).

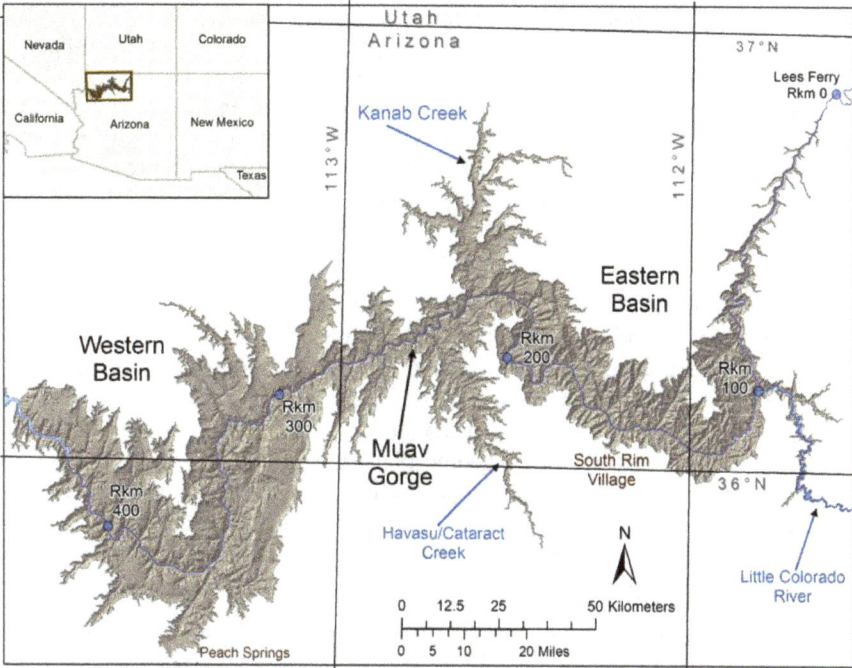

Fig. 3. Grand Canyon rim and major tributaries map, showing the eastern and western basins divided by the Muav Gorge, and distances along river from Lees Ferry (Rkm 0). Map redrawn from Billingsley and Hampton (1999), and prepared by J. D. Ledbetter.

## 3. Ecological gradients

### 3.1 Overview

LDCs create complex spatially and temporally intercorrelated environmental gradients, and ecological processes that often are stratified across elevation, strongly affecting the composition and structure of biotic assemblages. The major gradients in LDCs include: gravity, geology, regional climate, elevation, and aspect. Collectively, these affect, generate, or permit the solar radiation budget, micro- to synoptic climate, moisture availability, dip angle, pedogenesis, and natural disturbance regimes, all of which affect the development of canyon biotic assemblages over time. Here I briefly review the roles of these major physical gradients in the GCE.

### 3.2 Gravity

Gravity plays a profound role in species dispersal and ranges in LDCs. Passively dispersing taxa, including plants, non-flying invertebrates and vertebrates, and even some bird species more readily disperse and colonize downslope. Although the inner canyon corridor within 100 m of the river makes up less than 3 percent of the overall land area in the GC, that zone supports more than 760 plant species, more than 40% of the entire GC flora (Stevens and Ayers 2002, Busco et al. 2011). Downslope dispersal accounts for the in-canyon occurrence of many rim-dwelling species of grasses, forbs, shrubs (e.g., *Fallugia paradoxa, Parreyella filifolia*), and trees. These upper elevation "waif" or "Gilligan" species may reach the canyon-floor desert, far from their normal habitats and, while able to become established, they may not be able to reproduce or form permanent colonies. Such species are particularly likely to occur in aspect refugia: boreal colonists often occur in the desert on north-facing slopes. Numeorus insects (e.g, *Prionus heros*, various scarabaeids, *Pandora* moths), wild turkey (*Meleagris gallopavo*), and other rim fauna also readily disperse downslope, and have been regularly detected along the river (Brown et al. 1987; LaRue et al. 2001; Stevens unpublished data).

## 3.3 Geology and geomorphology

Geologic context, parent rock, structure, and tectonic history play dominant roles governing the development, functioning, and characteristics of LDC ecosystems. Geology and geologic processes control: 1) elevation, a primary determinant of local climate; 2) aspect, in turn influencing solar radiation budget, microclimate, productivity, pedogenesis, habitat availability, and species distributions; 3) geomorphology, including width-depth relationships of the canyon stream channel; and 4) rock color, which may differentially affect heat loading. The subject of microtopographic ecological gradient impacts on assemblages is generally poorly studied, but topographic differences of less than 10 m can affect local microclimate and species distributions, particularly near water sources where surface heating affects humidity.

Although much debated, drainage evolution of the Colorado River in the American Southwest has occurred over the past 40 million yr through conclusion of the Laramide orogeny and the onset of Basin and Range orogenic uplift and stream capture (Hunt 1956; Dickinson 2002; review in Blakey and Ranney 2008). On-going debate over development of the Colorado River basin is divided between advocacy for an old canyon (Oligocene-Miocene origination) as opposed to a young canyon (late Miocene-Pleistocene origination), and top-down drainage integration versus integration of variably independent sub-basins. Basin and Range crustal extension caused the uplift of the Sierra Nevada Range and numerous other north-trending ranges, which block the on-shore movement of moist Pacific airflow. This geologic deformation and development of the regional rain-shadow over the past 15 million ys changed the GCE from a low-lying, mesic, savannah-dominated landscape to its present-day uplifted, arid character. Also of great biogeographic impact was the tectonic/volcanic connection between South and North America 2.7 million years ago (Wallace 1876; Webb 2006). The ensuing land bridge permitted movement of biota between continents, resulting in the Great American Biotic Interchange, resulting in the movement and extinction of numerous species. The interchange is largely responsible for the contemporary co-occurrence of 150 genera of plants and animals in South America and the American Southwest. Following integration of the Colorado River, western Grand Canyon was dammed by numerous lava flows over the past 750,000 yr, forming large Pleistocene lakes and perhaps large outburst floods (Hamblin 1994; Crow et al. 2008). Biogeographic analyses of aquatic nepomorph Hemiptera (Stevens and Polhemus 2008) demonstrate significant upstream attenuation in species richness, suggesting that such flooding reversed or retarded upriver colonization in Pleistocene time. Interestingly, the ranges of aquatic Heteroptera taxa in that study also reveal the "biological shadow" of an earlier Paleogene east- and north-flowing Nevada river system into western Grand Canyon, a river unrelated to the present-day Colorado River.

Fluvial geomorphology has been thoroughly studied in GC (e.g., Howard and Dolan 1981; Schmidt and Graf 1990; Topping et al. 2005). Rapids form at the confluence of tributary canyons when rare debris flows deliver large boulders into the Colorado River, damming the mainstream (Webb et al. 1987). Rapids create recirculation zones (eddies) that, on descending flows, deposit fine sediments in characteristic locations both upstream and downstream of the debris fans. Thus, unlike alluvial rivers, the location of sandbars in the geomorphically constrained Colorado River are fixed within 13 geomorphic reaches (Table 2). More than 550 debris fan complexes generate a distinctive suite of fluvial microhabitats, each with discrete grain sizes, soils, stage elevation relationships, inundation frequencies, and stage-zoned riparian vegetation (Turner and Karpiscak 1980; Schmidt and Graf 1990; Melis et al. 1997). Analyses of flow regulation influences at a large suite of debris fan complexes revealed that flow regulation allowed extensive development of fluvial marshes, enhancing riverine plant species richness (Stevens et al. 1995; Waring et al. in press). Clearwater releases and flood control also greatly increased aquatic productivity within the river which, coupled with riparian vegetation expansion, has led to increased waterbird biodiversity, and peregrine falcon (*Falco peregrinus anatum*) population increases and the potential for formation of novel post-dam trophic cascades (Brown et al. 1992; Stevens et al. 1997a,b; Stevens et al. 2009).

Thus, flow regulation reduced flooding disturbance and increased productivity of this naturally flood-prone river ecosystem, changing the river ecosystem energetics from allochthonous to autochthonous sources (Carothers and Brown 1991), and increasing riverine species richness and trophic complexity, in accord with the predictions of both Connell's (1978) intermediate disturbance concept and Huston's (1979) dynamic equilibrium model (Stevens and Ayers 2002). But also of scientific and stewardship interest, flow regulation differentially influenced geomorphological influences on subaqueous versus subaerial biota, colonization, and production. Flow regulation has swamped geomorphic influences on biological organization in the aquatic domain of the river, reducing differences in benthic standing stock or species among various aquatic microhabitats (Stevens et al. 1997b). In contrast, flow regulation has enhanced variation in biotic development on the various geomorphic microhabitats in debris fans in the riparian domain (Stevens et al. 1995, 1997a). Differential responses of the river ecosystem to flow regulation on the aquatic versus the riparian domains vastly complicate environmental management of focal species and habitats, creating challenging administrative tradeoffs (Stevens et al. 2001; Lovich and Melis 2007).

### 3.4 Elevation

Elevation strongly affects local climate, productivity, species composition, and microsite ecology. Following the model of Von Humboldt and Bonpland (1805, Jackson 2009), C.H. Merriam was the first American naturalist to quantify the influences of elevation on biotic zonation across an Arizona transect from the floor of Grand Canyon to the top of the San Francisco Peaks and out into the Painted Desert (Merriam and Steineger 1890). Merriam attributed the discrete zonation of trees across elevation primarily to temperature and latitude; however, Holdridge (1947) and others subsequently recognized the importance of seasonal and total precipitation, evapotranspiration, and other factors controlling vegetation and biome development (Olson et al. 2001).

| Geomorphic Reach | Distance from Lees Ferry, AZ (km) | Dominant Bedrock Strata | Mean River Width (m) | Mean Floodplain Width (m) | Estimated Floodplain Area (ha) | % Potential Solar Energy |
|---|---|---|---|---|---|---|
| Glen Canyon Reach | -25 - 0 | Mesozoic sandstones | 85.3 | --- | 95 | 65.3 |
| Permian Reach | 0-17.7 | Permian-Pennsylvanian limestones, sandstone, shale | 70.0 | 20.5 | 73 | 78.0 |
| Supai Gorge | 17.7-36.2 | Pennsylvanian limestone, sandstone, shale | --- | 14.4 | 53 | 64.7 |
| Redwall Gorge | 36.2 - 62.8 | Mississippian limestone | 67.1 | 17.6 | 94 | 62.5 |
| Marble Canyon | 62.8-98.2 | Cambrian limestones, sandstone, shale | 106.7 | 23.9 | 169 | 66.9 |
| Furnace Flats | 98.2-123.9 | Late Precambrian shales, basalts, quartzite | 118.9 | 24.2 | 125 | 81.8 |
| Upper Granite Gorge | 123.9-189.5 | Early Precambrian schist and granite | 57.9 | 18.1 | 237 | 68.1 |
| The Aisles Reach | 189.5-202.7 | Cambrian sandstone, shale | --- | 14.8 | 39 | 73.2 |
| Middle Granite Gorge | 202.7-225.4 | Early Precambrian schist, granite | --- | 18.3 | 83 | 73.6 |
| Muav Gorge | 225.4-257.4 | Cambrian limestones | 54.9 | 15.3 | 98 | 54.8 |
| Lower Canyon Reach | 257.4-344.3 | Cambrian limestones, sandstone, shale | 94.5 | 17.1 | 297 | 70.3 |
| Lower Granite Gorge | 344.3 - 386.2 | Early Precambrian schist, gneiss,granite | 73.2 | --- | --- | 88.1 |
| Upper Lake Mead | 386.2-448.1 | Early Precambrian schist, gneiss,granite; Cambrian limestone, sandstone, shale | ca. 125 | --- | --- | --- |
| Total | 473.1 | All | 85.0 | 19.0 | ca. 1800 | 69.5 |

Table 2. Geomorphic reaches of the Colorado River in Grand Canyon.

Nonetheless, elevation remains an overwhelmingly important ecological state variable due to its strong negative relationship with air temperature and freeze-thaw cycle frequency, and its positive relationship to precipitation and relative humidity. The global adiabatic lapse rate is -6.49 °C/km. Analysis of paired daily minimum and maximum air temperature from 1941-2003 at Phantom Ranch (elevation 735 m) on the floor of GC with the South Rim (2100 m) produces a GC-specific lapse rate of -8.7 °C/km. The >1.3-fold steeper lapse rate in GC is likely a function not only of the dark red and black bedrock color of the inner canyon, but also to aspect. Steep, S-facing slopes in the GCE, particularly those with darker rock color, absorb and re-radiate more heat than do N-facing slopes, which often are shaded from direct sunlight, and are cooler and more humid than S-facing slopes across elevation. Overall, elevation strongly and broadly influences synoptic climate, while aspect exerts strong local control over microclimate and microsite potential evapotranspiration and therefore productivity.

### 3.5 Aspect and solar radiation

Solar radiation limitation is an important factor limiting ecosystem productivity of rivers and lakes (e.g., Yard et al. 2005; Karlsson et al. 2011). Depending on the latitude, depth, and cardinal orientation, LDCs also can be strongly limited by solar radiation, particularly east- or west-oriented canyons at higher latitudes (Fig. 4; Stevens et al. 1997b). During pre-dam time, the Colorado River's naturally high suspended load prevented sunlight from reaching the floor of the river (Topping et al. 2005); however, sediment retention in Lake Powell and resulting clearwater releases now allow illumination of the river floor in the upper reaches, enormously increasing benthic productivity. Yard et al (2005) measured mid-day solar radiation in the river at 25 km intervals from Glen Canyon Dam to Diamond Creek and described variation in the availability of photosynthetically active radiation (PAR) of the Colorado River aquatic domain. They reported that river surface PAR varied strongly in relation to cardinal orientation of the canyon and in relation to tributary-contributed suspended inorganic sediment load, ranging from little limitation in upstream, wide, and east-west oriented reaches to significantly reduced PAR in downstream (more turbid), narrow, north-south-oriented reaches. These results help explain the low pre-dam productivity of the pre-dam sediment-laden Colorado River, postdam stairstep decreases of benthic standing stock at the confluences of the Paria River and the LCR, and spikes in mainstream productivity at the mouths of N- or S-flowing tributary (Stevens et al. 1997b).

Yard et al. (2005) focused attention on interactions of physical solar radiation limitation on turbidity, canyon geomorphology, and aquatic productivity. To better understand potential solar radiation variation on the Colorado River floodplain in GC, I used a solar pathfinder (SPF; Solar Pathfinder 2008) to measure the percent of mean monthly potential solar radiation at riverside and at the 10-year return flood stage at each bend in the river, typically every 1-2 km between Glen Canyon Dam and Diamond Creek (Table 2). The general model for this analysis includes latitude (sun angle) and canyon configuration (width, depth, cardinal orientation), but does not account for cloud cover or atmospheric aerosol obstruction of solar radiation intensity (Fig. 4). The use of a SPF is more accurate than landscape modeling of solar reception on the river banks because the floodplain is often narrow, and minor cliff projections and large rocks strongly influence shading patterns.

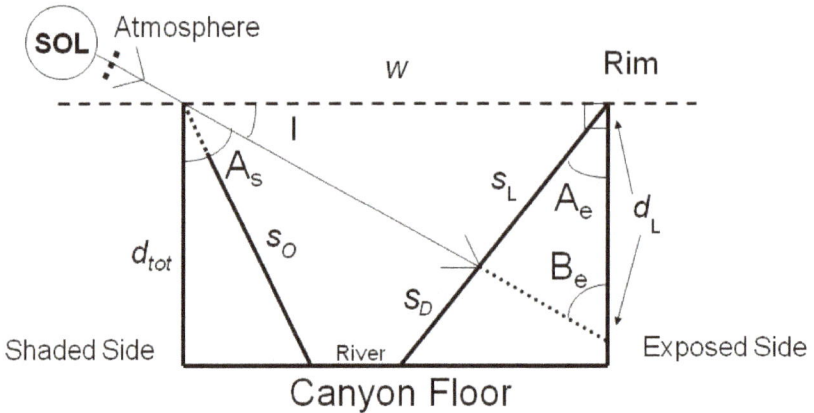

**Exposed Side:**  $s_L = [(w \sin I)/\sin(180° - (B_e + A_e))] \sin B_e)$

$s_D = (d_{tot}/\cos A_e) - s_L$

**Shaded Side:**   If I > $A_s$ then $s_o = d_{tot}/\cos A_s$ , else $s_o = 0$

Fig. 4. A conceptual schematic of solar insolation limitation in large deep canyons. $A_e$- slope angle on the sun-exposed side; $A_s$-slope angle on the shaded side; $B_e$-angle of incidence on the sun-exposed side; $d_L$-depth of canyon on sunlit side; $d_{tot}$-total depth on the shaded side; I-angle of solar incidence; $S_L$ - sunlit slope, $S_D$ -unlit dark slope; $S_O$-slope length on the shaded side; w-width of the canyon. The full solar budget of a canyon is the sum of solar radiation integrated across momentary to annual time scales. This model does not illustrate or take into account atmospheric interference, which reduces the canyon's total solar energy budget.

These data demonstrate that limitation of solar radiation is substantial on the Colorado River floodplain in GC, varying by the cardinal orientation of the canyon and both local and rim cliff structure (Fig. 5). The overall average percent of potential sunlight received by the Colorado River floodplain in GC is 69.5% of that available on the rims, varying from an average low of 54.8% with virtually no direct wintertime insolation in the steep, narrow, west-flowing Muav Gorge, to 75-88.1% in the relatively wide Permian, Furnace Flats, and Lower Granite Gorge reaches. The most extreme limitations of solar limitation occur on north-facing slopes in east-west flowing segments of the canyon in the deepest portions of the Canyon. Extremely steep, narrow, deep geomorphic reaches, such as the Muav Gorge, receive no direct insolation during winter months, regardless of aspect, allowing mesic rim grasses, such as galleta-grass (*Pleuraphis* spp) to grow even on S-facing slopes. In less steep but still narrow reaches (e.g., the Aisles) galleta-grass grows only on refugial N-facing slopes, and otherwise is rare to non-existent in the lower elevations of GC.

River corridor vegetation composition varies in relation to aspect and the solar radiation budget, with greatest differences on N- versus S-facing slopes. N-facing slopes, along the river, such as those at Rkm 63-65R, 193-198L, 228-232R, and 315-318L, typically support upland and boreal Great Basin Desert plant species whose ranges are normally 600 m higher

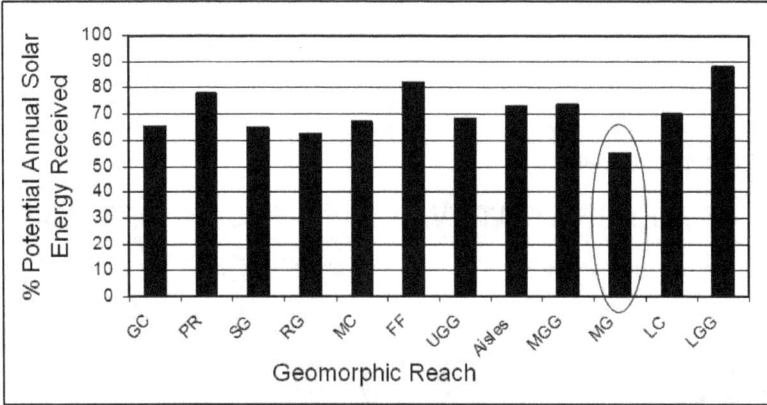

Fig. 5. Results of solar radiation budget measurements among 12 reaches on the floodplain of the Colorado River between Glen Canyon Dam and Diamond Creek in Grand Canyon. Reach abbreviations are listed in Table 2. The Muav Gorge (MG; circled) is the most deeply canyon-bound reach, receives the least solar radiation, and is a significant barrier to upriver and downriver range extensions.

in elevation, including netleaf hackberry (*Celtis laevigata* var. *reticulata),* galleta grass (*Pleuraphis* spp.), juniper (*Juniperus* spp.), buffalo berry (*Shepherdia rotundifolia*), and other perennial grasses, shrubs, and woodland taxa. In contrast and with little compositional overlap, S-facing slopes in those segments support Sonoran Desert vegetation, including several cacti taxa, brittlebush (*Encelia farinosa*), and upper riparian zone western honey mesquite (*Prosopis glandulosa* var. *torreyana*), but little desert grass cover. Analysis of the aspect of all stands of western honey mesquite and netleaf hackberry reveal strong, opposite affinity for S- and N-facing slopes, respectively. E-and W-facing slopes support a mix of N-facing slope boreal and S-facing slope desert taxa, and thus have roughly twice the species density as N- or S-facing slopes. Additional study is needed to distinguish plant compositional differences between E-facing slopes that receive early warming sunlight, from W-facing slopes, which receive hotter, late afternoon radiation. Aspect similarly influences small desert mammal composition, with N-facing slopes supporting a mixture of canyon deer mice (*Peromyscus crinitus*), woodrats (*Neotoma* spp.), and seed-feeding *Chaetodipus* spp. pocket mice, while S-facing slopes primarily support a lower diversity of weedy rodent species, dominated by western cactus deer mouse (*Peromyscus eremicus*) and less common white-tailed antelope squirrel (*Ammospermophilus leucurus*).

Elevation and aspect interactively affect solar radiation, which affects vegetation at at springs throughout the GCE. Cliff Spring near Cape Royal on the North Rim of Grand Canyon is a hanging garden (a contact spring on a cliff face; Fig. 6A). It is a high-elevation, E-facing site that receives direct insolation during winter, allowing it to thaw and warm quickly after cold winter nights. But due to the overhanging cliff, it is protected from direct insolation during the summer months. This configuration moderates the springs' summertime microclimate, allowing the highest elevation population of *Primula specuicola* wall plants to persist there. In comparison, Vaseys Paradise is a gushet spring along the Colorado River at Rkm 51R (Fig. 6B). Also E-facing, its depth in the canyon precludes direct insolation during the winter, but its relatively warm (16°C) water and rapid warming in the morning hours, coupled with early shade, allows the endangered Kanab ambersnail (*Oxyloma haydeni kanabensis*) to persist there, one of only three naturally occurring *Oxyloma* populations in Arizona (Meretsky and Stevens 2000). More detailed modeling of elevation-aspect relationships will be productive for predicting potential climate change impacts on springs and regional vegetation.

## SOLAR ENERGY AT TWO EAST-FACING SPRINGS

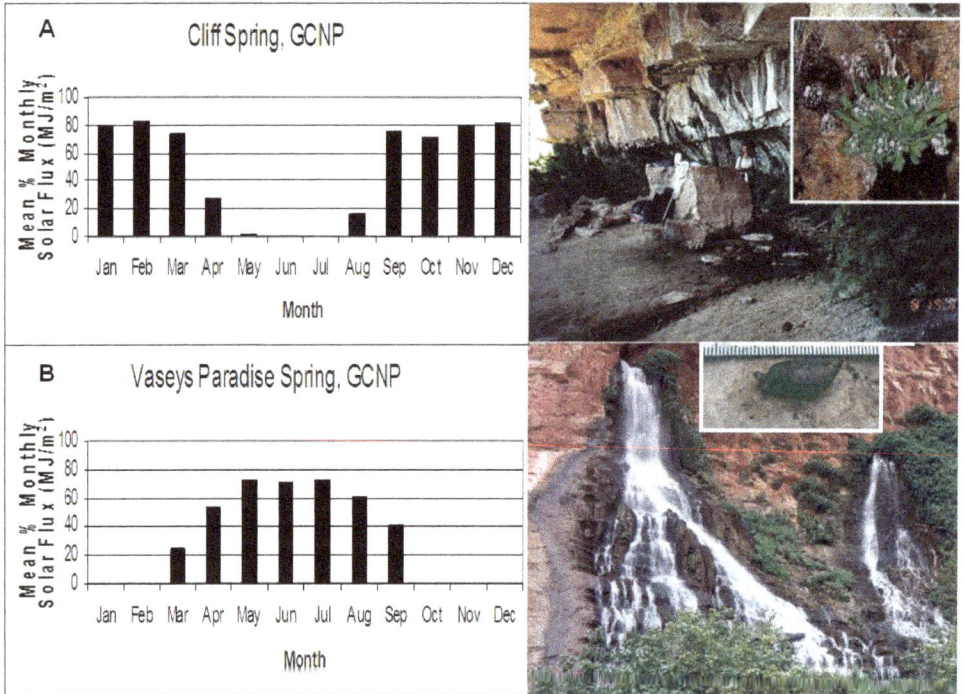

Fig. 6. Solar radiation budgets at two east-facing springs in GC: A – Cliff Spring is a hanging garden with direct insolation in winter, but is shaded by the overhanging cliff during summer, permitting persistence of a small population of *Primula specuicola* (inset); B – Vaseys Paradise is a gushet spring that received morning light throughout the year, but is shaded in early afternoon during the hot summer months, factors that allow endangered Kanab ambersnail (*Oxyloma haydeni kanabenss*; inset) to persist.

These observations, models, results, and studies demonstrate strong, pervasive, and controlling impacts of solar radiation limitation on LDC ecosystem ecology. The extent and patterns of solar radiation limitation created by cliffs in the GCE indicate that, depending on LDC width and depth, solar radiation limitations are most influential in deep E-W oriented canyons, impacts that increase with latitude and canyon depth. Solar limitation in LDCs also may exert impacts on canyon geomorphology. At the latitude of GC (35° N), N-facing slopes freeze in winter and may remain frozen for prolonged periods, while S-facing slopes generally receive direct solar radiation every day and thus undergo daily freeze-thaw cycles.

Can such aspect differences lead to faster erosion rates and cliff retreat of S-facing slopes as compared to N-facing slopes? This question has yet to be studied in detail, but may help account for the order-of-magnitude greater width of the canyon from river to rim of the S-facing North Rim of GC, as compared to that of the N-facing South Rim (Fig. 3).

## 4. LDC biogeography

### 4.1 Biodiversity

GC likely supports more than 10,000 species of (non-microbial) macrobiota, while the GCE may support 15,000 or more taxa (Table 3; estimated from Carothers and Aitchison 1976; Suttkus et al. 1978; Hoffmeister 1986; Harper et al. 1994; Busco et al. 2011; Stevens unpublished data). Plants and vertebrates have garnered most inventory attention, and recent and on-going advances in invertebrate biodiversity research are expanding understanding of the biogeographic and evolutionary significance of GC as an LDC.

| Taxon | Estimated Total No. Spp in GCE | Relative Species Richness | Overall Taxon Vagility | Biogeographic Effects | | | | References |
|---|---|---|---|---|---|---|---|---|
| | | | | Corridor | Barrier | Refuge | Null | |
| Plants | 2200 | High | Low | Strong | Strong | Weak | Strong | Phillips et al. 1987 |
| Mollusca | 59 | High | Low | Weak | Strong | Moderate | Weak | Spamer and Bogan 1994 |
| Odonata | 89 | High | High | Strong | Weak | Weak | Strong | Stevens and Bailowitz 2009 |
| Aquatic Hemiptera | 89 | High | Low | Strong | Moderate | Moderate | Moderate | Stevens and Polhemus 2008 |
| Butterflies | 140 | High | High | Strong | Moderate | Moderate | Strong | Garth 1950 |
| Tiger Beetles | 44 | Low | Low | Strong | Strong | Strong | Weak | Stevens and Huber 2004 |
| Mosquitoes | 18 | Low | Low? | Weak | Strong | Weak | Weak | Stevens et al. 2009 |
| Bees | 500+ | High | High | Strong | Strong | Strong | Strong | Stevens et al. 2007 |
| Other Arthropoda | 10000 | Moderate | Variable | Strong | Strong | Moderate | Moderate | Stevens, unpublished |
| Fish | 24 | Low | High | Strong | Weak | Weak | Moderate | Minckley 1991 |
| Herpeto-fauna | 60 | High | Moderate | Strong | Strong | Weak | Weak | Miller et al. 1982; Stevens 1990 |
| Birds | 340 | High | High | Strong | Weak | Weak | Strong | Brown et al. 1987; LaRue et al. 2001 |
| Mammals | 104 | High | Variable | Moderate | Strong | Weak | Strong | Hoffmeister 1987; Stevens 1990; Frey 2005 |

Table 3. Summary of selected GCE taxa, relative species richness, overall taxon vagility, biogeographical responses to GC as a LDC.

## 4.2 LDC landform effects

As a landform, GC influences species ranges and gene flow processes in four primary ways, each with complex subprocesses: 1) a partial or full corridor of low elevation riverine and desert habitats through the uplifted Colorado Plateau; 2) a partial or full barrier across the Plateau or in an upstream-downstream fashion; 3) a refuge, particularly for species requiring rare microhabitats, such as springs, caves, and rim edges; and 4) a null effect, not limiting gene flow across or within the landscape (Stevens and Polhemus 2008; Table 3). Below I elaborate on each of these types of landform influences using GCE biodiversity data and I discuss biogeographic anomalies in GC and the GCE.

*Corridor Effects:* Several biogeographic corridor functions exist in GC, including range, movement, and migration corridor effects.

*Full upriver range corridor:* The Colorado River corridor provides a swath of low elevation desert habitat and numerous Sonoran and Mohave Desert terrestrial biota extend their ranges upstream into the Colorado Plateau through GC, including cacti (e.g., *Echinocactus polycephalos*, various *Platyopuntia, Mammillaria*), buckwheats (e.g., *Eriogonum inflatum*), amphibians (e.g., *Bufo* spp.), reptiles (e.g., *Coleonyx variegatus, Sauromalus ater*), and numerous desert mammals (e.g., aquatic mammals, and *Peromyscus eremicus* cactus deer mouse, *Neotoma lepida* pack rat, *Chaetodipus intermedius* pocketmouse), and other taxa.

*Partial upriver range corridor:* Although also considered a partial barrier/filter effect, many species ranges extend part way up the Colorado River corridor in GC, with the Muav Gorge serving as a barrier to upstream dispersal of numerous desert plants (e.g., *Yucca whipplei* yucca, *Phoradendron californicum* mistletoe, *Larrea tridentata* creosote-bush, *Fouquieria splendens* ocotillo), invertebrates (e.g., *Brechmorhoga mendax* clubskimmer dragonfly, *Hetaerina americana* damselfy, *Abedus h. herberti* waterbug), reptiles (e.g., five *Crotalus* rattlesenake species, *Heloderma suspectum* Gila monster, *Leptotyphlops humilis* blind snake), and other taxa (Stevens and Polhemus 2008). Further upstream, the Redwall Gorge serves as a second range limit for common Sonoran Desert plant species (e.g., *Prosopis glandulosa* var. *torreyana* mesquite, *Acacia greggii* catclaw, *Encelia farinosa* brittlebush; Waring et al. in press).

*Full downriver range corridor:* the Colorado River also serves as a corridor for dispersal of Colorado Plateau taxa from Utah and the upper Colorado River basin downstream through GC into the lower Colorado River basin (e.g., *Corispermum americanum* nonnative goosefoot, *Salix exigua* coyote willow, and *Baccharis salicina* seep willow).

*Partial downriver range corridor:* Some Utah and upper Colorado River basin species ranges descend part way through GC, including: *Pariella filifolia* (Fabaceae), *Falugia paradoxa* Apache plume, *Symphoricarpos oreophilus* snowberry, *Quercus turbinella* scrub oak and *Q. gambellii* Gambel's oak, *Boehmeria cylindrica* at Rkm 55.5R Spring, *Carex specuicola, Oxyloma* ambersnails, and *Rana pipiens* northern leopard frog). The larger faults of Grand Canyon also provide access down from the rims, and are actively used by ungulates, terrestrial predators, humans, and passively wind-dispersing "aerial plankton".

*Annual migratory corridor:* Prior to impoundment the canyon served as a long-linear migratory corridor for several native fish (e.g., *Xyrauchen texanus* razorback sucker and *Ptychochelius lucius* pikeminnow; Minckley 1973, 1991) and still provides that function for numerous migratory waterbirds (Stevens et al. 1997a), and probably bats and monarch butterflies (Garth 1950). An autumn migration route along the east side of the East Kaibab Monocline brings high densities of raptors across Grand Canyon from the north, and those birds use rising thermal air currents to ascend out of Grand Canyon on the South Rim at Lipan and Yaqui Points (Hoffman et al. 2002; Smith et al. 2008). Another hawk flyway likely exists along the Grand Wash and Hurricane Cliffs, but has not been studied.

*Through-canyon movement corridor:* The Colorado River provides a movement corridor for numerous aquatic species, including: 8 native fish species (4 of which have been extirpated since 1963) and 19 non-native fish species, which transport at least 17 non-native fish parasites (Minckley 1991; Choudhury et al. 2004); summer breeding waterbirds (e.g., *Anas playtrhynchos* mallard, *Ardea herodias* great blue heron); and aquatic mammals (e.g., *Castor canadensis* beaver, *Ondatra zibitheca* muskrat, and the now likely extinct *Lontra canadensis sonora* river otter; Hoffmeister 1986).

*Short-term lateral migration/movement corridor:* Several taxa undergo temporal movements into or out of GC. Rabe et al. (1998) documented summertime daytime roosting of spotted bat (*Euderma maculatum*) along the river in central GC, with nocturnal forays of more than 38 km/night to North Rim meadows, likely to forage on abundant coniferous forest meadow moths and beetles. On a seasonal time scale, numerous taxa move into GC from the rims in autumn and winter, including rim- and montane-dwelling taxa as diverse as *Culiseta incidens* mosquitoes, American crow and common raven (*Corvus brachyrhynchos* and *C. corax*, respectively), desert mule deer (*Odocoileus hemionus*) and mountain lion (*Puma concolor kaibabensis*; Stevens et al. 2008a; Brown et al. 1987; Stevens unpublished observations). Thus, the biogeographic corridor effect operates in numerous, complex ways, affecting a great array of taxa in this LDC.

**Barrier/Filter Effects:** The evolutionary consequences of GC as a barrier are well known through studies of the divided distribution of Abert's tassel-eared squirrel (*Sciurus aberti aberti*) from Kaibab squirrels (*S. a. kaibabensis*; Lamb et al. 1997). These two taxa were divided by the Pleistocene-Holocene climate transition, which eliminated suitable habitat within GC, isolating *S.a. kaibabensis* on the North Rim. The color shift between these two taxa involves a minor genetic change, and black *S.a. aberti* individuals have been reported south of GC (e.g., Allred 1995). Numerous other, but less well known examples of taxa with ranges divided by GC exist, and cryptic speciation may be commonplace. Taxa occurring north, but not south of GC include the landsnail family Orcohclicidae with the large genus *Oreohelix* (Bequaert and Miller 1973), *Satyrium behrii* hairstreak butterfly (Garth 1950), *Plestiodon skiltonianus* skink and *Pituophis catenifer deserticola* Great Basin gopher snake (Miller et al. 1982), and *Thomomys bottae planirostris* and *T. talpoides kaibabensis* pocket gophers (Hoffmeister 1986). Examples of taxa occurring on the South Rim but not the North Rim include the landsnail family Helminthoglyptidae with the large genus *Sonorella* (Bequaert and Miller 1973), *Coenomorpha tullia furcae* ringlet butterfly (Garth 1950), *Plestiodon gilberti* in Peach Springs Wash (L.E. Stevens personal observation) and *P. multivirgatus* skinks, *Pituophis catenifer affinis* Sonoran gopher snake (Miller et al. 1982), the oddly disjunct population of *Tantilla hobartsmithi* lyre snake between Lees Ferry and the LCR (Brennon and Holycross 2006), (from NE to SW) *Thomomys bottae alexandrae, T.b. aureus, T.b. fulvus,* and *T.b. desertorum* pocket gophers, and extirpated *Panthera onca* jaguar (Hoffmeister 1986).

The Colorado River corridor itself presents a barrier to the downstream distribution of *Phrynosoma platyrhinos* horned lizard and other species from southern Utah, and similar upstream exclusion of common Sonoran and Mohave Desert plant species from Lake Mead, such as *Yucca brevifolia* Joshua tree and *Y. mohavensis* Mohave yucca, *Parkinsonia* spp., and *Psorothamnus spinosus* desert shrubs, which probably have been excluded by the lack of suitable low-gradient bajada habitat within the walls of GC.

*Refugia:* The two basins of GC generally support discrete endemic assemblages, with relatively few endemic species occurring in both. However, *Camissonia specuicola hesperia* Kaibab suncup and *Polistes kaibabensis* Kaibab paper wasp occur relatively widely through GC, and *Rosa stellata abyssa* wildrose occurs on the rims, bridging both basins.

*The Eastern Basin:* The eastern basin of GC supports several endemic plant species, including: *Agave phillipsiana* in Deer, Tapeats, and Phantom Creeks; *Agave utahensis* var. *kaibabensis* on calcareous or sandstone outcrops; *Euphorbia aaron-rossii* across elevation from river level in the Permian-Redwall Gorges up to 2,000 m at Cane Springs in House Rock Valley; *Argemone arizonica* in Bright Angel Creek (Arizona Rare Plants Committee 2001); *Scutellaria potosina* var. *kaibabensis* mint along the East Rim of the Kaibab uplift (Rhodes and Ayers 2010); *Cirsium rydbergii* in the Paria River gorge and downstream to Saddle Canyon; and *Silene rectiramea* in the shaded N-facing upper elevations of Red, Garden Creek, and Hermit Canyons on the south side. Endemic invertebrates in the eastern basin, include: an undescribed stonefly at Thunder River; 1-2 other new stoneflies in North Canyon; *Brechmorhoga pertinax* masked clubskimmer dragonfly; 3 *Nebria* ground beetle species; the recently described *Schinia immaculata* noctuid moth; *Cicindela hemorrhagica arizonae* tiger beetle; and *Eschatomoxys tanneri* cave darkling beetle, among others. Among the herpetofaunae, the endemic *Crotalus oregonus abyssus* Grand Canyon pink rattlesnake is virtually the only rattlesnake in the eastern basin of GC, detected thus far river to rim from Rkm 27.5 to 262 (however, a single *Crotalus viridis nuntius* Hopi western rattlesnake was photographed in Bright Angel Canyon). *C. o. abyssus* is fully replaced downstream from the Muav Gorge by at least 5 other crotalids, including: *C. atrox* western diamondback rattlesnake, *C. mitchelli* speckled rattlesnake, *C. mollosus* blacktailed rattlesnake, *C. oregonus lutosus* Great Basin rattlesnake, and *C. scutulatus* Mohave rattlesnake (Miller et al. 1982; Brennon and Holycross 2006). Lack of access through the cliff-bound Muav Gorge likely protects *C. o. abyssus* from upriver movement of these five western GC basin rattlesnake taxa into the eastern basin.

*The Western Basin:* Plant species endemic to the western GC basin include: *Camissonia exilis* suncup (Onagraceae) in NW Arizona and *Camissonia specuicola hesperia* in dry, gravelly washes and Colorado River sandbars; *Penstemon distans* beardtongue in Whitmore, Parashant, and Andrus Canyons; and *Astragalus lentiginosus* var. *trumbullensis* locoweed (Fabaceae) on Mt. Trumbull. Spence (2008) noted a similar pattern among endemic springs plants from Glen Canyon through Grand Canyon and into Zion Canyon. Among the endemic western basin invertebrates are the likely new waterbug *Belostoma* nr. *flumineum* in Vulcans Well Spring (289L), several cave invertebrates (see below). Among western basin herpetofaunae, Oláh-Hemmings et al. (2009) analyzed mitochondrial DNA, reporting that the highly isolated population of *Rana* near *yavapaiensis* lowland leopard frog in Surprise Canyon (Rkm 399R) is the result of a middle Pleistocene separation from the main *R. yavapaiensis* clade.

*Caves:* Several specific refugial habitats in GC support numerous restricted, rare, or endemic taxa. Caves are abundant in the 120-200 m-thick Redwall limestone, and evince ancient and some still-active groundwater channels. GC caves support: many of the Canyon's 22 known bat species and their parasites; the endemic Grand Canyon cave larcid pseudoscorpion *Archeolarca cavicola* – known only from Cave of the Domes in central GC); an undescribed genus of sphaeropsocid bark louse (Psocoptera) and *Loxosceles kaiba* recluse spider, the primary prey of which are the endemic cave darkling beetles *Eschatomoxys pholitor* and *E. tanneri*. Recent exploration of GC caves by J.J. Wynn (U.S. Geological Survey, Flagstaff) and K. Voyles (U.S. Bureau of Land Management, St. George) have revealed additional new invertebrate species, including another unique spider, two new millipede species, an undescribed rhaphidophorid cave cricket genus, and at least one new coleopteran species. Although GC caves are numerous, they appear to support highly individualistic assemblages, with relatively low similarity among caves, and with endemic taxa often restricted to one or a few closely-related caves.

*Springs:* Springs are well known as refugial habitats in arid regions (Stevens and Meretsky 2008). In the GCE springs habitat makes up <0.001 % of the overall landscape, but springs support numerous springs-specialist plant species: at least 9.5% of the regional flora are springs-specialist taxa. Several groups of these springs-specialist plants occur in the GCE: 1) facultative springs plants, such as cardinal monkeyflower (Scrophulariaceae: *Mimulus cardinalis*) also occur in protected habitats along perennial streams. 2) obligate springs species may either be widespread across the GCE, such as *Epipactis gigantea* heleborine orchid, *Primula specuicola* cave primrose, and *Cirsium rydbergii* thistle, or 3) species very narrowly restricted to one or a few springs (e.g., Asteraceae: *Flaveria mcdougallii* ragweed along the Colorado River between Rkm 218-285; *Ranunculus uncinatus* buttercup at North Rim springs). Some plant species may be relatively common nationally, but only have been detected in GCE at springs, such as *Persicaria amphibia* water knotweed at Rkm -14.4L and *Boehmeria cylindrica* false nettle at 55.5R.

GCE springs also support rare and some endemic invertebrates, and springs assemblages are often highly individualistic. Stevens and Polhemus (2008) reported that 53% of the 89 aquatic Hemiptera taxa in the GCE were restricted to 3 or fewer of 444 localities (primarily springs) at which aquatic Hemiptera were detected. For example: *Ochterus rotundus* (Ochteridae) only occurred at GC springs, far outside of its range in south-central Mexico (Polhemus and Polhemus 1976). *Belostoma* nr. *flumineum* (Belostomatidae: likely a new taxon based on its year-round reproductive behavior) occurs only at Vulcans Well Spring (Rkm 289L); *Abedus breviceps* (Belostomatidae) occurs only at Boucher Spring and Creek (Rkm 154.5L); and *Abedus herberti utahensis* occurs only at a springs complex in the lower Virgin River drainage. While high levels of endemism among low-vagility Nepomorpha waterbugs might not be surprising, Stevens and Bailowitz (2009) reported a similar, although weaker pattern, among the far more dispersive Odonata of the GCE: 15 (16.9%) of Odonata species were rare and restricted to 3 or fewer localities, and 4 (4.5%) species were detected at only a single locality. Like *Ochterus rotundus*, the masked clubskimmer dragonfly occurs along several springfed desert streams, far from its range in Central America. In addition, at least one endangered fish species, *Gila cypha* humpback chub reproduce successfully only in the outflow of springs in the lower LCR, at one subaqueous springs complex on the floor of the Colorado River at Rkm 48.3, and in the lower, springfed reaches of Kanab and Havasu Creeks.

*Plateau Refugia:* The plateau lands around GC support numerous endemic species, some of which only occur on the rims or at upper elevations in GC, including: five *Pediocactus* (Cactaceae) taxa, of which *P. bradeyi* principally occurs at the contact between the Mesozoic

Moenkopi Formation and the underlying Kaibab Limestone, *P. paradinei* along the east side of the Kaibab Plateau, and *P. peeblesianus* var. *fickeiseniae*, which occurs more broadly across northern and northwestern Arizona between 1310-1660 m elevation. House Rock Valley and the Paria Plateau area support endemic *Sclerocactus sileri* (Cactaceae), and the Little Colorado River Platform supports *Phacelia welshii* scorpionweed, and other xeric-adapted mid-elevation plants. *Sclerocactus parviflorus* var. *intermedius* pineapple cactus occurs on the Esplanade Platform on both sides of the Canyon in both basins. *Talinum validulum* (Portulacaceae) occurs on both sides of GC in the western basin. Between elevations of 2450 and 2800 m, the uplifted North Kaibab Plateau supports several microendemic taxa, including *Castilleja kaibabensis* paintbrush, *Lesquerella kaibabensis* (Brassicaceae), *Selaginella watsonii* spikemoss, and *Cicindela terricola kaibabensis* tiger beetle in subalpine meadows. Also, *Thelypodiopsis purpussii* mustard occurs on the Esplanade of the North Rim of western GC in the Toroweap area.

*Escarpment Edges:* Canyon rim edges are remarkably important refugial and ecotone habitats. Rim edges are abundant throughout the GCE, with its many steep, long escarpments. Rim edges have been poorly studied, but appear to be subject to greater climatic stresses than are other habitats, sustaining more severe temperatures from subsidence of cold air during winter and higher temperatures from rising hot air from desert canyons during the summer. In addition, such habitats are subject to strong, erratic winds during all seasons. As a consequence, many of the plant species that are restricted to rims are mat- or clump-forming or otherwise sturdy, low-growing shrubs. GC species wholly restricted to these harsh rim habitats include *Astragalus cremnophylax* var. *cremnophylax* sentry milk-vetch, an endangered legume that occurs in a few small patches primarily on the South Rim (Maschinski et al. 1997; Busco et al. 2011). Numerous other rare and restricted, but non-endemic plant species occur on canyon rims and in rim open areas, including: *Phyllodoce* near *empetriformis* heather on the South Rim; *Rosa stellata* var. *abyssus*; *Penstemon pseudoputus* and *P. rydbergii* beardtongues; and *Ostrya knowletonii* hornbeam, *Paronychia sessiflora* pink, *Phacelia filiformis* scorpionweed, *Pteryxia petraea* rock wing parsley on both rims. In addition, two butterfly species also are rim-edge specialists: the canyon ringlet (Nymphalidae: *Coenomorpha tullia furcae*) on the South Rim, and the indra swallowtail (Papilionidae: *Papilio indra kaibabensis*) on the North Rim and to Cameron, Arizona.

*Specific Drainages as Refugia:* Specific drainages within GC and the GCE are particularly rich in rare or endemic species. a) In the overall GCE, the Virgin River, the White Mountains of Arizona, and the Verde River all support high levels of endemic and rare invertebrates and fish. b) At least 6 regionally endemic species occur in the 10 km² North Canyon on the North Rim, including: *Cimicifuga arizonica* (Ranunculaceae - Arizona bugbane), *Speyeria atlantis schellbachi* fritillary butterfly, *Nebria* n.sp. ground beetle, two undescribed species of Plecoptera, the inocellid *Negha inflata* snakefly (the only occurrence of this family, genus, and species in Arizona), and an early-translocated population of Apache trout (Salmonidae: *Oncorhynchus gilae apache*). In addition, the forest floor there supports abundant *Formica rufa* complex (Formicidae) ants, which also may be unique and are ecologically dominant (G. Alpert, Global Ant Coordinator, personal communication). c) Deer Creek (Rkm 219 R) supports 1-2 endemic *Nebria* groundbeetles and the only known population of the large spirobolid millepede *Tylobolus utahensis* in Arizona (Shelley and Stevens 2004). d) Another canyon of note is Peach Springs Wash (Rkm 363L), which supports several herpetofauna not detected elsewhere in GC, including Gila monster, desert horned lizard, zebra-tailed lizard, western Gilbert's skink, and western diamondback rattlesnake (L. Stevens unpublished data). These taxa reflect the connection of Peach Springs Wash across the southwestern corner of the Colorado Plateau into the Basin and Range geologic province. e) At a finer spatial scale, Montezuma Well is a large travertine mound limnocrene springs pool in the Verde River basin on the edge of the Colorado Plateau. This springs complex supports the highest concentration of endemic species (8 taxa) of any point in North America to my knowledge (Blinn 2008).

*Null Effects:* The ranges of several widely-distributed animal taxa appear to be little affected by Grand Canyon as an LDC. Common raven (*Corvus corax*) are abundant across elevation in the summer months, sometimes retreating to the rims and extra-canyon habitats in winter. Desert bighorn rams (*Ovis canadensis nelsoni*) are wide-ranging herbivores, occasionally swimming the river, moving across elevation, and up onto the rims. Merriam and Steineger (1890) reported bighorn sheep on the top of Mt. Humphreys outside of Flagstaff, Arizona at an elevation of 3850 m, a habitat that no longer supports this species or the now-extinct native Merriam's elk (*Cervus canadensis merriami*). In contrast, bighorn ewes and young are generally restricted to much smaller home ranges within individual geologic strata in GC, where a matriarch guides the flock across access routes to known pastures and water sources, and where they respond negatively to the noise of sight-seeing helicopters (Stockwell et al. 1991; L.E. Stevens, personal observations). Numerous other flying organisms occur widely across elevation, both in and outside of GC, including: *Pantala hymenaea, Rhionaeschna multicolor,* and other large dragonflies (Stevens and Bailowitz 2009); *Aquaris remigis* water striders (Gerridae; Stevens and Polhemus 2008); *Culiseta incidens* mosquitoes (Culicidae; Stevens et al. 2008a); and numerous bird species. Large size and flight capability are not prerequisites for ubiquitous distribution in GC. *Litaneutria minor,* a diminutive ground-dwelling mantid, and side-blotched lizard (*Uta stansburiana*) are both commonly encountered across elevation on both sides of GC, as well as in extra-canyon habitats; however, microspeciation has not been investigated among non-volant, cursorial organisms.

For many species, too few detections have been made to understand landform impacts on their ranges in GC. For example, a total of three individuals of poison ivy (*Toxicodendron rydbergii*) exist, one each at three riverside springs (Vaseys Paradise, Lower Deer Creek Spring, and Mile 142L Spring). *Scolopendron viridis* giant centipedes have been detected twice in GC to my knowledge, once in the lower riparian zone at Rkm 49R on the north side of the Colorado River, and once at Grandview Point on the South Rim (Stevens unpublished data). A single specimen of the Amblypygi, *Paraphrynus* sp. was collected in GC at Rkm 219L. Additional inventory is needed to clarify the role of GC on these distributions, if any.

*Biogeographic Anomalies:* Several common southwestern desert taxa are conspicuously missing from the Colorado River corridor in GC, and their absence highlights otherwise difficult-to-discern LDC ecological processes. Among the missing taxa are: horned lizards (*Phrynosoma* spp.), kangaroo rats (*Dipodomys* spp.), and lagomorphs (i.e., *Sylvilagus* spp. and *Lepus* spp.). All of these taxa are common on the canyon rims and around Lakes Powell and Mead, and all but the kangaroo rats are found at Lees Ferry; however, none have been detected along the Colorado River in Grand Canyon except at the Lake Mead boundary. In the case of the lagomophs, rabbits occur on both rims, and down onto the Esplanade Platform (a high-elevation platform) on the North (but not the South) Rim, and I observed a single *Sylvilagus* on the Tonto Platform at Cottonwood Creek in November 2001, the only individual seen by me in more than 10,000 km of trekking through the inner Grand Canyon. Predation by wall-nesting and migrating raptors likely causes extinction probability to exceed colonization potential, thus preventing lagomorph populations from becoming established in the inner GC. In the cases of horned lizards and kangaroo rats, the steepness of the surrounding terrain likely limits habitat patch suitability, and despite relatively likely downriver/downslope colonization over millennia via rafting, extinction probability is apparently too high to permit establishment.

Another biogeographic anomaly is the much-reduced presence and ecological role of termites (Isoptera) in Grand Canyon. Termites are abundant desert southwestern decomposers and provide substantial food resources to a wide array of insectivores (Ueckert et al. 1976). Their relative scarcity and small colony sizes in Grand Canyon are surprising, and few species have been detected (Jones 1985). The generally steep terrain of the canyon means that fallen wood is unstable as habitat, and driftwood piles along the river were, in pre-dam time, too commonly wetted or moved by flooding to provide suitable, long-term termite habitat. Thus, like the missing vertebrate taxa, habitat size and stability probably limit termite colonization on the floor of GC. One consequence of the absence of termites is that driftwood piles in Grand Canyon contain logs that regularly exceed 1,000 yr in age (A. McCord, University of Arizona Tree Ring Laboratory, personal communication), providing a largely overlooked wealth of paleoecological information.

Lastly, many records of highly disjunct species distributions exist in GC. For example: a single, large, isolated stand of *Canotia holocantha* cruxifixion-thorn (Celastraceae) exists in a small canyon at Rkm 196L; a single specimen of *Xantusia vigilis* was collected in the middle reaches of Clear Creek (Rkm 135R; Miller et al. 1982); the damselfly *Coenagrion resolutum* occurs in a few ponds on the North Rim (Stevens and Bailowitz 2009); the termite *Incisitermes minor* (Isoptera: Kalotermitidae) was collected at Cardenas Creek (Rkm 114L) but otherwise is known only from the Pacific Coast (Jones 1985); the Surprise Canyon relict leopard frog occurs in a highly isolated population in western GC; and numerous erratic bird records exist in GC, including a scissortail flycatcher (*Tyrannus forficatus*), a painted bunting (*Passerina ciris*) and 2 records of magnificent frigatebird (*Fregata magnificens*; Brown et al. 1987; LaRue et al. 2001). A perplexing skeleton of a collared peccary (*Pecari tajacu*) was found in middle Spring Creek (Rkm 328R; Stevens unpublished data, skull housed at the National Park Service collection at Grand Canyon). Although this species is expanding its range northward from the southern deserts, it was not previously known to occur north of the Colorado River. Each of these records represents range extension of populations considerably external to GC, primarily but not exclusively from the south. Additional basic inventory is needed to better understand such apparently enigmatic range records.

## 4.3 LDC biogeographic hypotheses

Six questions illuminate the extent to which GC influences the distribution and evolution of biotic assemblages. Testing these questions required three critical analyses: a) a landscape analysis of land area by elevation of the GC within the larger GCE, which was accomplished using geographic information systems (GIS) analysis of 30 m digital terrain data; b) a compilation of species richness, elevation range, and rarity data for a broad array of organisms with differing dispersal capabilities within GC and the surrounding GCE, including plants, vertebrates, and selected groups of invertebrates, work that is on-going through the Museum of Northern Arizona in Flagstaff, Arizona; and c) biogeographic affinity information for the taxa under consideration. The latter data are now available for most vascular plants and vertebrates, but only for a few of the better-known arthropod goups.

**1) Do landform corridor, barrier or refuge effects differentially influence species richness over refuge and null effects?** If GC is a biogeographically significant landform, then corridor, barrier/filter, and refugial effects on species ranges should greatly predominate over null effects. Among the plants, as mentioned above, nearly 46% of the flora is found in the river corridor, demonstrating a pronounced corridor effect. Instances of refuge effects and restricted distributions of unique genomes are becoming more widely recognized (e.g., Stevens and Polhemus 2008; Bryson et al. 2010). While partial corridor effects are numerous, relatively few non-endemic plant taxa are restricted to one or the other rims, suggesting that barrier effects among plants are relatively weak (Table 3).

Among invertebrates, high resolution data on Odonata and aquatic Hemiptera revealed the following patterns. For the 58 Odonata taxa found in GC (Stevens and Bailowitz 2009):

Partial or full corridor (58.6% of species) >> Refuge (19.0%) >
Barrier/filter (13.8%) > Null (10.3%).

Assuming that null and corridor effects should prevail equally (expected values of 33.3% each), and that barrier and refuge effects might be half as important for this generally vagile taxon (expected values of 16.7% each), the hypothesis was supported ($X^2_{df=3}$ = 19.33, $P$ = 0.0002), with 89.7% of the species ranges affected by the LDC landform though corridor, barrier, or refuge effects. Somewhat similarly, among the 54 aquatic Heteroptera taxa in GC, the following pattern was detected:

Partial or full corridor (37.0%) > Barrier/filter (27.8%) >
Null (18.5%) > Refuge (16.7%).

Thus, a total of 81.5% of GC aquatic Hemiptera demonstrated range patterns related to LDC landform configuration. Differences in the strength of landform impacts between Odonata and aquatic Hemiptera may be related to vagility, with many Odonata being more capable of long-distance dispersal. Evidence from these studies also points to a more concentrated landform impact on both of these taxa in GC as compared to the general ecotonal impacts of the Mogollon Rim (Polhemus and Polhemus 1976).

Among the GC herpetofauna, the role of partial and full corridor distributions also is dominant, as mentioned above (Table 3). Several species are widespread and occur from river to rim throughout GC (e.g., the lizards *Uta stansburiana* and *Urosaurus ornatus*). Those and other lizard taxa are found throughout the river corridor, including: *Coleonyx variegatus* gecko, *Sceloporus magister* spiny lizard, *Sauromalus ater* chuckwalla; and *Aspidoscelis tigris* whiptail (Brennon and Holycross 2006). However, many lizard and snake species ranges extend partway up the river corridor, including *Heloderma suspectum* Gila monster to Rkm 325L, *Callisaurus draconoides* zebra-tailed lizard in sand dunes at the mouth of Diamond Creek (Rkm 363L, habitat from which they were recently extirpated; Stevens 2011), and *Xantusia vigilis* night lizard. We discount the report of *H. suspectum* from Clear Creek in the eastern basin of GC (Miller et al. 1982) – that observation was likely an immature *Sauromalus ater*. Rather few reptile species ranges extend partway down the Colorado River corridor, excepting Crotaphytidae: *Crotaphytus collaris*, which gives way to *C. bicinctores* in lower GC, and *Phrynosoma platyrhinos*, which only occurs along the Colorado River downstream to Lees Ferry, and does not occur within GC except partway down Peach Springs Wash. Similarly, *Gambelia wislizenii* occurs at Lees Ferry and in upper Diamond Creek, but not in inner corridor of GC. Another distinctive example of N-S barrier effects among the herpetofauna is the range of *Pituophis catenifer* gopher snake, with *P.c. affinis* exclusively south of the Colorado River and *P. c. deserticola* only found north of the river (Brennon and Holycross 2007).

At least 178 mammal taxa have existed in the GCE in historic times, with 128 species among 67 genera in 23 families in 7 orders (Durrant 1952, Hoffmeister 1986, Flinders et al. 2005). The GCE fauna is overwhelmingly dominated by Rodentia, with 94 taxa, followed by Carnivora with 28 taxa and Chiroptera with 24 taxa. At least 113 mammal taxa exist in or on the immediate rims of Grand Canyon, with 97 species among 59 genera in 22 families. Excluding humans, 10 mammal species are non-native. Hoffmeister (1986) lists 145 recent mammal species in Arizona, a number that has changed somewhat due to further collecting and improved taxonomy. Therefore, the GCE supports at least 88% of the state's fauna, and the GC supports two thirds of the State's fauna. Hoffmeister (1986) reported 55 widely-distributed species in the GCE, with at least 14 taxa restricted to upper elevations, five species restricted to the north of the Colorado River, and 10 species restricted to south of the river. Thus, neotropical influences on the GCE mammal fauna prevail over those of the nearctic region, a pattern similar to that of other taxa.

My review of the data on 104 mammal taxa in GC for which data are available indicates the following biogeographic pattern:

Null (34.6%) = Barrier (34.6%) > Corridor/Partial Corridor (18.3%) > Refuge (12.5%).

Thus, barrier, corridor, and refuge effects collectively dominate over null effects ($X^2_{df=1}$ = 9.846, $P < 0.002$), indicating that GC is a significant biogeographic feature for mammals. Although the landform impact on GC species ranges is overwhelmingly evident, the significance of GC as an IDC requires closer analysis of the distribution and evolution of endemic taxa (see question 6, below).

**2) Does biogeographic affinity influence assemblage composition in GC?** Greater-than-expected species richness of GC and the GCE in general is a partial result of topographic complexity, as well as the position of the GCE as a mixing zone during late Cenozoic time. The GC and GCE support several suites of taxa. a) A distinctive Madrean biota is comprised of Central American, Mexican, and pan-tropical plant and animal taxa whose ranges likely extended into the region during warm interglacial phases, and these taxa generally occupy lower elevations (e.g., Phillips et al. 1987; Stevens and Polhemus 2008; Stevens et al. 2008a,b). Peculiar to this group are the "Guatemalan taxa": isolated GCE populations with ranges otherwise occurring in Central and northern South America. For example, masked clubskimmer dragonfly and *Ochterus rotundus* waterbug are found at GC springs and also in Central America (Stevens and Polhemus 2008; Stevens and Bailowitz 2009). b) Boreal Rocky Mountain taxa extended their ranges southward and downslope during glacial phases and still occupy refugia at upper elevations and in mesic, north-facing habitats at lower elevations. Such taxa include the coniferous forest and Great Basin shrub plants, *Coenagrion* *resolutum* damselfly, various gerrid waterstriders (Hemiptera), and other taxa. c) In addition to species with exogenous biogeographic affiliations, a large suite of broadly distributed species exists that are more-or-less centered in their ranges (e.g., *Anax junius* and *Rhionaeschna multicolor* dragonflies; *Danaus plexippus* monarch and *Vanessa cardui* painted lady butterflies; and others). d) Also, in the case of Odonata, as many as 18 (20.2%) of the 89 GCE taxa have ranges that include or are centered on the Pacific Coast of North America, an intriguing pattern suggesting that some dragonfly ranges may predate the Basin and Range orogeny. The back-and-forth and elevational adjustment of assemblages during glacial advances and retreats between climate oscillations, and the stability of refugial microhabitats within GC contribute to the relatively high biodiversity of this LDC.

**3) Does elevation influence species richness in a fashion analogous to the effects of latitude?** Many of the GCE faunal taxa studied thus far show the well-known latitudinal diversity gradient, a pattern of declining species richness across latitude (reviewed in Lomolino et al. 2010). Exceptions to the pattern include species, such as conifers, Salicaceae and tenthridinid sawflies, and ichneumonid wasps that are derived from boreal regions. Species richness also generally declines across elevation in a fashion analogous to latitude; however, the "mid-domain effect" of elevation on species richness commonly results in a unimodal peak of species richness at intermediate elevations (Romdal et al. 2005; review in Lomolino et al. 2010). Our work on plants (Figs. 7A, B), and macroinvertebrates (Fig. 8) demonstrate that although the mid-elevation richness peak is distinctive (e.g., among plants – Fig. 7A; aquatic Hemiptera and Odonata - Stevens and Polhemus 2008 and Stevens and Bailowitz 2009, respectively; and landsnails and non-melittid bees – Stevens unpublished data), the effect is largely accounted for by the species-area relationship: the ratio of the $log_{10}$-transformed insect species richness to the $log_{10}$ area of the GCE within 100 m belts revealed a strong negative response across elevation (Figs. 7B, 8). The GCE flora and fauna both consist of broad mixtures of Maderan and Rocky Mountain taxa. The 1600 m elevation zone may approximate the division between those two assemblages, and the sensitivity of composition and vegetation structural responses to climate change may warrant more focused research on that zone.

**4) Does vagility influence species richness (are highly mobile taxa relatively more species rich than less mobile taxa)?** I used the literature and available information to conduct a qualitative analysis of landform effects for a wide array of taxa for which biogeographic data are available from within the GC and across the GCE (Table 3). This analysis indicates that taxa with low overall vagility show stronger evidence of barrier/filter effects than those with higher vagility (e.g., strongly dispersive taxa), and the data indicate that taxa with high vagility are relatively more species rich than taxa with low vagility. While somewhat intuitive, these patterns reinforce the view that GC is an isolated, relatively young geologic feature, one still undergoing colonization and assortative assemblage development by species colonizing from the surrounding region and affecting in-canyon refugia.

**5) Does species richness attenuate upstream through the Colorado River corridor?** Plant distribution data in the Colorado River corridor indicate some support for this pattern. The ranges of numerous conspicuous species extend part-way up into Grand Canyon, including: *Yucca whipplei* yucca, *Fouquieria splendens* ocotillo, *Prosopis glandulosa* mesquite, *Acacia greggii* catclaw, *Ferocactus cylindrica* barrelcactus, several *Cylindropuntia* cholla cacti species, *Canotia holocantha* crucifixion-thorn, and *Larrea tridentata* creosote-bush, to name a few.

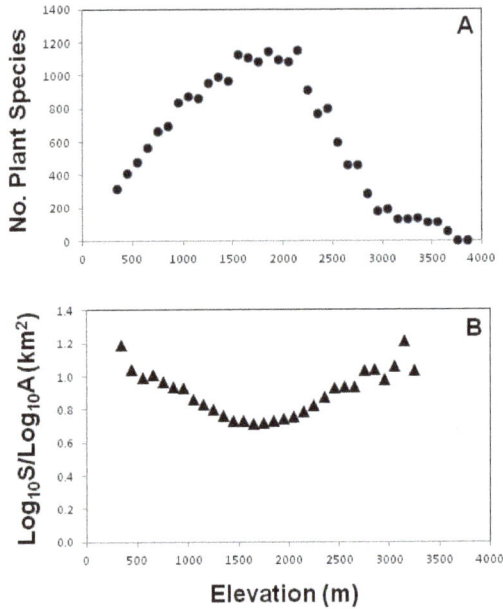

Fig. 7. Distribution of vascular plants among 100 m elevation belts in the Grand Canyon ecoregion. A – Raw plant species richness within 100 m elevation belts. B – $Log_{10}$ plant species richness/$log_{10}$ area ($km^2$) within an elevation belt, as a function of belt elevation. Modified with additional data from Stevens et al. (2007).

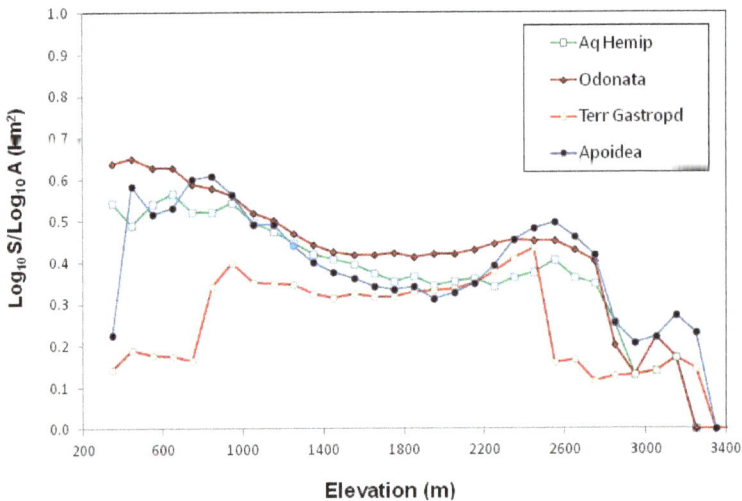

Fig. 8. $Log_{10}$ species richness/$log_{10}$ area ($km^2$) within an elevation belt, as a function of belt elevation for Grand Canyon ecoregion non-melittid bees, terrestrial gastropods, Odonata, and aquatic Hemiptera. Redrawn in part from Stevens and Polhemus (2008) and Stevens and Bailowitz (2009).

Among the insects, analysis of aquatic Hemiptera and Odonata ranges conclusively demonstrate upstream attenuation of species richness through the Colorado River corridor. Most Nepomorpha Hemiptera are found in the lower reaches of the Colorado River drainage, with only a few species of Corixidae, Notonectidae, one Gerridae, one Gelastocoridae, and two Veliidae common in the upper reaches of GC (Polhemus and Polhemus 1976; Stevens and Polhemus 2008). Even though they are generally far more vagile, Odonata diversity declines with distance upstream as well. Several provocative examples of exclusion exist: *Brechmorhoga pertinax* and *Hetaerina vulnerata* both replace widespread and common congeners in western basin of GC. These two species exist in the eastern basin of GC only along perennial springfed tributaries (Stevens and Bailowitz 2005, 2009). Whether exclusion of the more widespread Odonata congeners is the result of competitive superiority or other factors remains to be determined.

Among the herpetofauna, upstream species attenuation also is apparent, as described above (Fig. 9). Species richness on the rims is equivocal from the upper to the lower Canyon, with greater species richness on the North Rim likely attributable to the greater range of elevations there. However, both in the inner canyon and on the canyon floor, herpetofaunal species richness attenuates upstream markedly, with at least 7 taxa (24%) in the lower Canyon missing from the upper Canyon. Species-area influences may account for some of this attenuation, as the upper Canyon is narrower; however, habitat area is yet to be determined for most of these herpetofaunal taxa. Range restrictions are not clear for other terrestrial vertebrates.

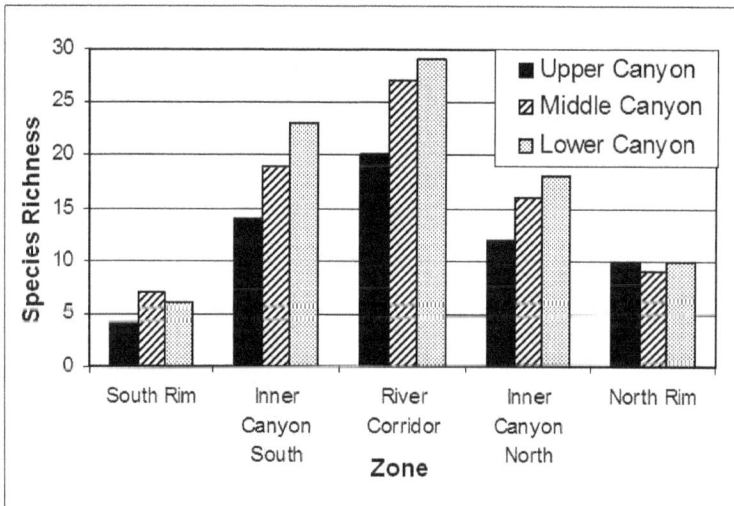

Fig. 9. Grand Canyon herpetofaunal species richness on the South Rim, the south side Inner Canyon, the Colorado River corridor, the north side Inner Canyon, and the North Rim, from the upper (eastern) to the lower (western) Grand Canyon.

**6) Are levels of endemism and rarity consistent with the geologic developmental history of GC?** Most endemic taxa in GC and the GCE are restricted to harsh, constant environments (e.g., high-conductivity limnocrene springs, south-facing desert slopes, rim edges, caves, alpine habitats). Based on aquatic Hemiptera data, levels of endemism and rarity previously were regarded as being low in GC, a phenomenon attributed to the youth of GC as a landform (Polhemus and Polhemus 1976), and supported by the greater frequency of subspecific or varietal-level endemism as compared to species-level endemism. Varietal level endemics in GC are numerous, including: plants, such as *Aletes m. macdougalii*, *Arabis g. gracilipes* (found elsewhere in the GCE), and an undescribed *Arctomecon californica* variety in western GC (Phillips et al. 1987; Brian 2000; Arizona Rare Plants Committee 2001).

Numerous endemic invertebrates exist in GC and the GCE (Table 4). Many butterfly taxa that occur at lower elevations in GC appear slightly different from other populations in the Southwest, and at least 4 endemic butterfly and skipper subspecies are known from GC: *Papilio indra kaibabensis*, *Speyeria atlantis schellbachi*, *Coenomorpha tullia furcae*, and *Agathymus alliae paiute* (Garth 1950; Stevens unpublished data). Three endemic tiger beetle subspecies are known from the GCE: *Cicindela hemorrhagica arizonae* in inner GC, *C. hirticollis coloradulae* in the LCR, and *C. terricola kaibabensis* in North Rim meadows (Stevens and Huber 2004). Other varietal-level endemics include: the likely new subspecies of Vulcans Well waterbug (*Belostoma* near *flumineum*; Stevens and Polhemus 2008), Grand Canyon rattlesnake (*Crotalus oregonus abyssus*), North Rim *Thomomys talpoides kaibabensis*, and Kaibab squirrel (Miller et al. 1982; Hoffmeister 1986).

| Taxon | No. Species | % Endemic | % Rare | References |
|-------|-------------|-----------|--------|------------|
| Hirundinea | ~10 | 20 | 20 | Stevens unpublished data |
| Mollusca | 59 | 1 | 3.4 | Baequaert & Miller, Spamer and Bogan 1994 |
| Plecoptera | 8 | 37.5 | 37.5 | Stevens unpublished data |
| Odonata | 89 | 1.1 | 5.6 | Stevens and Bailowitz 2009 |
| Orthoptera | 90 | 2.2 | ---- | Stevens unpublished data |
| Aq. Heteroptera | 89 | 10.1 | 25.6 | Stevens and Polhemus 2008 |
| Tiger Beetles | 44 | 11.9 | 9.5 | Stevens and Huber 2004 |
| Darkling Beetles | 143 | 0.7 | 0.7 | Stevens unpublished data |
| Butterflies and skippers | 140 | 4.3 | 7.1 | Garth 1950; Stevens and Bailowitz unpublished data |
| Trichoptera | 109 (AZ) | --- | --- | Blinn and Ruiter 2006, 2009 |
| Mosquitoes | 18 | 0 | 72.2 | Stevens et al. 2008 |
| Chironomidae | 38 | 10.5 | 10.5 | Sublette et al. 1998 |
| Total | 745 | 9.7 | 19.2 | All |

Table 4. Percent of endemism and rarity among selected GCE invertebrate taxa. Rarity was evaluated as the percent of localities at which a species was detected in relation to the total number of localities at which members of that taxon were detected.

While early focus on these taxa suggested GC was not an evolutionarily significant landform, more collecting and observation in recent decades has revealed many more endemic species-level taxa in GC, including several neotropical isolates (Table 4; Stevens and Huber 2004; Arizona Rare Plants Committee 2001; Stevens and Polhemus 2008; Spence 2008; Stevens and Bailowitz 2009; Utah Native Plant Society 2011). More than 150 endemic plant species exist in the GCE. Endemic plant species in GC include: plants such as *Agave phillipsiana*, *Camissonia specuicola* and *C. confertiflora*, *Argemone arizonica*, *Flaveria mcdougallii*, and *Silene rectiramea* (Phillips et al. 1987; Brian 2000; Arizona Rare Plants Committee 2001; Hodgson 2001); the above-mentioned cave endemic invertebrates; Kaibab monkey grasshopper - *Morsea kaibabensis*; at least four chironomid midge species (Sublette et al. 1998); Tapeats robber fly - *Efferia tapeats* (Scarbrough et al. in press); three unique *Nebria* ground beetles (D. Kavanaugh, California Academy of Sciences, personal communication); *Schinia immaculata* (Pogue 2004); Kaibab paper wasp - *Polistes kaibabensis* (Snelling 1974); and other species. While no species-level vertebrates are endemic to GC, 19.2% of 775 invertebrate species that have been studied in detail are rare and 9.7% of those species are endemic (Table 4). With at least 30 endemic faunal species in and on the periphery of GC, and with more than 200 varietal- and species-level endemic taxa in the surrounding GCE, GC is emerging as a far more important evolutionary landscape than previously recognized.

## 5. Discussion and conclusions

### 5.1 Comparison of LDCs with mountains and islands

As a LDC, GC clearly exerts a profound biogeographical influence on the ranges of biota in and around it, and increasing evidence points to the evolutionary importance of this landform. But how do LDC biogeographical processes compare with those of mountain range and island landscapes, landforms that are well known to affect species distribution and evolution (MacArthur and Wilson 1967; Brown 1971 - contested by Lawlor 1998; Lomolino et al. 2010)?

A comparison of the characteristics of these three types of landforms indicates that LDCs differ in biogeographic function from the other two types, but are somewhat more similar to mountain ranges than they are to islands or archipelagos (Table 5). The focal feature of LDCs are dendritic drainage networks, with directional, gravity-facilitated flow, sediment, nutrient, and species transport, usually in a downstream direction, but also upstream through Aeolian transport for some components. LDCs are more strongly characterized by connectivity and gravity-facilitated movement, with productivity, growing season length and species richness increasing towards the focal feature (the canyon floor). In contrast, the peaks and ridges that characterize mountain ranges are harsh, somewhat to extremely unproductive, difficult to access, with short growing seasons, and are generally inhospitable habitats. Although montane slopes may be used by wide-ranging terrestrial taxa and by migrating birds, mountain ranges appear to be considerably less likely to facilitate gene flow than are LDCs. Dispersal among islands is restrictive, often limiting an island's impact on gene flow to passive processes or to active habitat searching by highly vagile taxa. Rockfall and flooding are more dominant forms of natural disturbance in LDCs than in the other two landforms, and all three landforms provide various refugial habitats. The longevity of LDCs and islands is usually geologically shorter than that of mountain ranges, and the several to tens of millions of years that LDCs exist may not be sufficient to create as much genetic isolation.

| Characteristic or Process | LDCs | Mountain Ranges | Islands |
|---|---|---|---|
| Extent of solar energy limitation | Large | Moderate | Variable |
| Effect of elevation on local climate | Large | Large | Variable |
| Role of latitude | Large | Large | Large |
| Productivity | Increases with depth and proximity to water | Usually decreases at upper elevations | Variable |
| Role of gravity | Positive | Negative | Neutral-negative |
| Ecological connectivity | Dendritic | Linear | Interrupted |
| Species Richness | Accumulative, but with upstream attenuation (GC); area and latitude dependent | Upslope attenuation; area and latitude dependent | Area, proximity to other islands, and latitude dependent |
| Common natural disturbances | Flooding, rockfall | Flooding, rockfall, fire, storms, volcanic activity; | Flooding, storms, volcanic activity |
| Trophic structure | Aerial predators (raptors) common; trophic cascades likely | Exclusion of bears, social carnivores; aerial predators (raptors) common; trophic cascades likely | Area and proximity to other islands regulates predator composition,influencing potential for trophic cascades |
| Role of cardinal orientation and aspect | Large | Moderate | Low |
| Nutrient cycling | Moderate-very long spirals | Short-long spirals | Short-long spirals |
| Ecosystem energetics | Allochthonous, accumulating downslope | Autochthonous-shedding upslope | Autochthonous-variable |
| Common anthropogenic disturbances | Hunting, grazing, damming, pollution, recreation, introduction of non-native species, climate change | Hunting, grazing, recreation, development, introduction of non-native species, climate change | Hunting, grazing, recreation, development, introduction of non-native species, climate change |

Table 5. Comparison of biogeographic characteristics and roles of large, deep canyons in comparison with those of mountain ranges and islands.

Likely the largest difference between LDCs and mountain ranges is that the former are generally narrower than the latter, a difference that limits the extent of isolation on the rims and in refugia. The corridor functions of LDCs contribute to, and can facilitate, regional gene flow, and LDC aspect influences enhance in-canyon species retention and genetic diversity. In contrast, the corridor function of mountain ranges may be less significant, and vicariance effects stronger, filtering gene flow around the peidmont peripheries. The several *Ensatina* salamander taxa whose non-overlapping ranges encircle the Sierra Nevada Mountains in California is a well-known example (Pereira and Wake 2009). However, such processes also occur around LDCs. A ring clade of pocket gopher (*Thomomys bottae*) subspecies has been reported around GC (Hoffmeister 1986). Moving clockwise from Lees Ferry around GC are the non-overlapping ranges of *T.b. alexandrae*, *T.b. aureus*, *T.b. fulvus* and *T.b. desertorum* on the south side of GC, and *T.b. planirostris* and *T.b. fulvus* on the north side (Hoffmeister 1986). Whether or not such subspecific differentiation represents cryptic speciation or morphological noise (e.g., Rios and Álvarez-Casteñeda 2007) remains to be determined, but it does suggest a process similar to that of the *Ensatina* salamanders of California.

Overall, and depending on landform size and structure, LDCs can play a significant role in regional biogeography, affecting species dispersal and gene flow. LDCs play complex roles, broadly functioning as corridors, barriers, and refugia and affecting the majority of species in the landscape, and with potentially strong evolutionary consequences on regional diversity. Additional research on gene flow in and around LDCs is warranted to better understand biogeographic patterns and processes, and comparative studies are needed to compare species-area relationships among LDCs of different sizes, cardinal orientation, and latitude.

## 5.2 Landform development and biogeography

Badgley (2010) proposed that mammalian diversity is greatest in tectonically active landscapes, where ecotones are abundant, habitat diversity is greatest, and ecologically gradients are steepest. The biotic assemblages of the Mogollon Rim ecotone and, to some extent, GC support this hypothesis (e.g., Stevens and Polhemus 2008; Stevens and Bailowitz 2009), but not completely. In particular, her predictions 4 ("endemism...should reflect origination within the region rather than range reduction from larger areas") and 7 ("species originating in topographically complex regions should colonize adjacent lowlands more often than the reverse pattern") are not fully supported in GC, where relictual endemism prevails over adaptational radiation, and elevated species richness in refugia is more the result of colonization from surrounding biomes during favorable climate conditions. Nonetheless, the upriver attenuation of aquatic Heteroptera and Odonata taxa reported in our work suggests that species richness is related to proximity to geologic province boundaries and regional topographic diversity.

Landform evolution is evident not only in regional geology, but also through the distribution and genomes of present-day species. However, the role of past landscape change on contemporary biogeography is difficult to determine. Mitochondrial analysis of GC *Hyla arenicolor* treefrogs indicates a discrete episode of introgressive hybridization with *H. eximia* in the latest Miocene and recent or on-going hybridization with *H. wrightorum*

(Bryson et al. 2010). The timing of those introgressions may reflect initial opening of the Colorado River to the Gulf of Mexico, and post-Pleistocene faunal mixing, respectively. But many other examples of ancient landform change and contemporary biogeography are likely to exist. Is the range restriction of most GC nepomorph Hemiptera to westernmost GC the shadow of the ancient (Oligocene?) river drainage from what is now southern Nevada into Arizona and north into Utah (Stevens and Polhemus 2008)? Does the Pacific Coast affinity of at least 3, and perhaps as many as 18 dragonfly species reflect pre-Basin and Range orogenic connectivity? Why do more than twice as any neotropical/Mexican mammals reach their northern range limits at the Colorado River, as compared to boreal species (Hoffmeister 1986). Why are *Ochterus rotundus*, 8 other aquatic GCE Hemiptera, and masked clubskimmer dragonflies neotropical isolates, with ranges otherwise centered in central and southern Mexico? Further research, including distributional, autecological, and genetics analyses of these generally poorly known taxa, is needed to determine whether and at at what scale environmental changes influence present-day LDC biogeography in the GCE.

## 5.3 Conservation biogeography

Human impacts on the GCE have profoundly altered ecosystem structure, composition, and biogeography through three processes: habitat alteration, extirpation, and the introduction of non-native species. The Colorado River is one of the most regulated rivers in North America, with a dozen large dams and thousands of small impoundments throughout its catchment (Hirsch et al. 1990). Dams and irrigation systems have fundamentally altered flow, flood dynamics, sediment transport, water temperature and chemistry, and the distribution of riverine biota. The loss of 4 of the 8 native fish species from GC due to habitat changes and interruption of migration has been thoroughly described (Minckley 1991), but less recognized has been the impact of greatly diminished connectivity on plant colonization processes, and interruption of range among tiger beetles, herpetofauna, southwestern river otter (*Lontra canadensis Sonora*), and other river corridor biota (Stevens 2011). So to, the rims of GC have sustained significant human impacts from fire suppression and alteration of forest structure (Fulé et al. 2002), the loss of large predators through federal extermination programs (Rasmussen 1941), and the degradation of more than 90% of rim springs and natural ponds (Grand Canyon Wildlands Council, 2002).

GC is one of the world's great landscape parks, a vast wilderness, and a United Nations-designated World Heritage Site, so it comes as a surprise that rather many biota have been lost during its protection by the National Park Service (Table 6; Newmark 1995; Stevens 2011). At least 20 and perhaps as many as 29 native vertebrate taxa have been extirpated since 1919, including all large predators (i.e, *Canis lupus youngi*, *Ursus arctos*, *Panthera onca*, and the 1+ m-long native predatory Colorado pikeminnow - *Ptychocheilus lucius*) except mountain lion (*Puma concolor*) and black bear (*Ursus americanus*; Rasmussen 1941; Hoffmeister 1986). Loss of top carnivores has likely had large but poorly understood consequences on GC ecosystems (Ramussen 1941; e.g., Estes et al. 2011). The loss of other ecologically strongly interacting species, such as prairie dog (*Cynomys gunnisoni*) likewise has reduced habitat availability for numerous other species. In addition, the population status of many other species is unknown, particularly of insects and plants that have only rarely or singly been detected (Stevens 2011).

| Taxon | Extirpated Taxa | At-Risk Taxa or Insufficient Data |
|---|---|---|
| Plants | Yerba santa | |
| Plants | | Numerous cacti and other plants |
| Plants | | Sentry milk-vetch * |
| Turbellaria | | *Leucochloridium cyanocittae* |
| Gastropoda | | Kanab ambersnail * |
| Arthropoda | | Numerous rare, poorly known taxa |
| Fish | Colorado pikeminnow * | |
| Fish | Razorback sucker * | |
| Fish | Roundtail chub | |
| Fish | Bonytail chub | |
| Fish | | Humpback chub * |
| Herpetofauna | Northern leopard frog | |
| Herpetofauna | Relict leopard frog | |
| Herpetofauna | Zebra-tailed lizard | |
| Avifauna | Burrowing owl | |
| Avifauna | California condor ** | |
| Avifauna | Sage grouse | |
| Avifauna | Yellow-billed cuckoo | |
| Avifauna | Pileated woodpecker | |
| Avifauna | Southwestern willow flycatcher * | |
| Mammals | Prairie dog | |
| Mammals | | Muskrat |
| Mammals | River otter | |
| Mammals | | Hog-nosed skunk |
| Mammals | Black-footed ferret ** | |
| Mammals | Badger | |
| Mammals | Jaguar | |
| Mammals | Gray wolf | |
| Mammals | Miriam elk | |
| **Total** | **21 species** | **9 taxa** |

Table 6. Extirpated, endangered (*), extirpated-but-reintroduced (**), and at-risk taxa, including those for which insufficient information exists to evaluate conservation status.

The other important human impact on GC has been the introduction of numerous non-native species. At present, at least 194 non-native vascular plant species are known from GC National Park, 10.7% of the entire flora, and the same percent of non-native plants as exists in the United Kingdom (Stevens and Ayers 2002; Busco et al. 2011; Stevens unpublished data). As in Stohlgren et al. (1999), Stevens and Ayers (2002) reported that hotspots of native plant species richness, such as springs and riparian habitats, also support higher numbers of non-native species. They also documented numerous non-native fauna in GC. Among the non-native invertebrates: *Procambarus clarkii* crayfish and *Dreissena rostriformis bugensis* quagga mussel are encroaching from Lake Mead; *Potamopyrgus antipodarum* New Zealand mudsnail is widespread in the upper Colorado River in GC; *Diorhabda* spp tamarisk leaf beetles have recently become abundant; *Anatis lecontei* is a common, predatory ladybird beetle; *Pieris rapae* cabbage white butterflies recently have been detected in GC; the large moth *Noctua pronuba* has recently invaded upper GC; and hybrid Africanized honey bees (*Apis mellifera scutellata*) are ubiquitous across elevation in GC and throughout the Southwest. Among fish, 20 coldwater and warmwater species have been introduced, threatening the remaining 4 native species, and among herpetofauna, *Apalone spinifera* softshell turtles occur in Lake Mead up to Rkm 389. Non-native birds include: rock dove (*Columbia livia*), starling (*Sturnus vulgaris*), and house sparrow (*Passer domesticus*) in GC; and chukar (*Alectoris chukar*), wild turkey, and blue grouse (*Dendragapus obscurus*) are regularly encountered on the rims. Eurasian collared dove (*Streptopelia decaocto*) were first detected in Havasu Canyon in 1994 (L. Stevens, unpublished data), and are now ubiquitous south of the South Rim in Arizona. Nonnative mammals include house mouse (*Mus musculus*) and Norway rat (*Rattus norvegicus*), Rocky Mountain elk (*Cervus elaphus nelsoni*), and feral horse (*Equus caballus*) on the South Rim, and cattle x *Bison bison* hybrids on the North Rim. Feral burro (*Equus asinus*) were widespread in GC from 1900-1982, but have been removed from the National Park. However, burros still occur on the Hualapai Reservation and on lands to the west of GC. Feral housecats (*Felis catus*) occasionally occur at Lees Ferry and on the rims.

Improved understanding of the levels of habitat alteration and loss, the extent and on-going threats of extirpation, and the role and impacts of non-native species on the region's ecosystems are essential to protect the native species, natural resources, and biogeographic processes of GC and the GCE. Notable successes have been made in the restoration of native species and natural ecological processes in the GCE, including restoration of riparian habitats at Lees Ferry, and restoration of springs habitats in northwestern Arizona by Grand Canyon Wildlands Council, Inc (www.grandcanyonwildlands.org). Protection of endangered sentry milkvetch from trampling at the South Rim has enhanced population viability (Maschinski et al. 1997). Population reintroduction/restoration successes in the GCE include: the reintroduction of formerly endangered peregrine falcons (Brown et al. 1992) and California condors (*Gymnogyps californianus*; California Condor Recovery Team 2007); on-going attempts to reintroduce black-footed ferret (*Mustela nigripes*); and protection and translocation of endangered Kanab ambersnail into an off-river springs complex in middle Royal Arch Creek (Meretsky and Stevens 2000). A robust non-native plant control program has been implemented by the National Park Service: several common exotic plant species have been eliminated in the Park, and riparian habitat restoration is being attempted. Selective removal of non-native rainbow trout (*Oncorhynchus mykiss*) may have contributed to increased population size of endangered humpback chub (*Gila cypha*) near the LCR (Coggins and Yard 2011). These examples demonstrate that focused conservation actions can be effective for protection and restoration of native species and natural ecosystems in this internationally recognized LDC.

## 6. Summary

Large deep canyons (LDCs) are relatively common landforms on Earth, but their regional biogeographic roles and significance has received little scientific attention. Here I summarize information on ecological gradients, species richness, and ecosystem structure on the world's best known LDC, Grand Canyon (GC) of the Colorado River in the context of the surrounding GC ecoregion (GCE) on the southern Colorado Plateau. I first describe the extent and influences of major physical gradients in LDC biogeography affecting its biodiversity, including geomorphology, elevation, gravity, and climate. By virtue of its depth and narrowness, the inner canyon is naturally light-limited, receiving only 69.5% of ambient solar radiant energy. I then briefly review the ecology of the Colorado River ecosystem and the impacts of Glen Canyon Dam, reporting that nearly 50 yr of flow regulation has swamped geomorphic differences and limited assemblage composition in the aquatic domain, but had the opposite effect on the riparian domain. Next, I use regional biodiversity and range information on GC and GCE plants, invertebrates, and vertebrates to evaluate the biogeographic influences of GC on its biota. As a landform, GC influences species ranges and gene flow in four ways, as: 1) a partial or full range corridor of low elevation riverine and desert habitats through the uplifted Colorado Plateau; 2) a partial or full barrier across the Plateau; 3) a refuge, particularly in microhabitats like caves, springs, and escarpment rim edges; and 4) a null effect, not limiting gene flow across the landscape. Available data indicate that GC functions primarily as a corridor and barrier/filter, and also supports refugial functions, and the ranges of relatively few taxa are unaffected by GC as a landform. GC has greater species richness than expected because it is a mixing zone of: a) Maderan (Mexican and neotropical) taxa occupying lower elevations and south-facing slopes; b) boreal and upland taxa occupying higher elevations and north-facing slopes; and c) range-centered taxa occupying middle elevations. Aspect refugia likely acquire taxa during climate extremes and support those populations well into climate transitions. Strongly vagile (e.g., flying taxa like butterflies, dragonflies, birds, and bats) tend to be relatively more species rich than low-vagility taxa (e.g., non-volant taxa, such as land Mollusca and non-flying beetles). Endemism is not as low in the region as previously reported, with 9.7% endemism among 745 invertebrate species in 10 orders studied thus far. Factors limiting development of endemism include the relatively young age of the landform (5-17 million years old), climate changes, and damming of the river by volcanic eruptions during the past half million years. At least 20 and perhaps as many as 29 vertebrate taxa, including nearly all large, wide-ranging predators have been extirpated from GC and the GCE in the past century, and more than 200 non-native plant and animal taxa have been introduced into GC, substantially altering the trophic structure of GC ecosystems. As an LDC, GC exerts a profound effect on the biota within and around it, functioning differently and in a more complex fashion than do other kinds of landforms. Due in part to this complexity, the assemblages and ecological functions of GC are susceptible to numerous human alterations, even when the best conservation practices are adopted.

## 7. Acknowledgements

I thank the publisher and particularly Ms. Dragana Manestar,for the opportunity to present this information. I received much-appreciated support from the Annenburg/Explore fund to prepare this manuscript. I warmly thank the Museum of Northern Arizona, Dr. Breunig its Director, and the MNA staff for administrative and office support. It has been my deepest pleasure and honor to work with numerous collaborators and governmental agencies on these topics over the past 4 decades, as reflected in the bibliography. Insights and stimulating discussions with these collaborators have allowed me to greatly deepen my understanding of regional biogeography; however, any errors herein are mine, not theirs. I warmly thank Ms. Jeri Ledbetter for her extraordinary skill with map preparation, and I thank Peter W Price for his critical review of the manuscript.

# 8. References

Allred, W. S. 1995. Black-bellied form of an Abert squirrel (*Sciurus aberti aberti*) from the San Francisco Peaks Area, Arizona. The Southwestern Naturalist 40:420.

Arizona Rare Plant Committee. 2001. Arizona Rare Plant Field Guide: A Collaboration of Agencies and Organizations. U.S. Government Printing Office, Washington.

Badgley, C. 2010. Tectonics, topography, and mammalian diversity. Ecography 33: 220–231.

Bequaert, J.C., and W.B. Miller. 1973. The Mollusks of the Arid Southwest. The University of Arizona Press. Tucson.

Billingsley, G.H. and H.M. Hampton. 1999. Physoigraphic rim of the Grand Canyon, Arizona: a digital database. U.S. Geological Survey Open File Report 99-30, Washington, D.C.

Blakey, R. and W. Ranney. 2008. Ancient landscapes of the Colorado Plateau. Grand Canyon Association, Grand Canyon.

Blinn, D.W. 2008. The extreme environment, trophic structure, and ecosystem dynamics of a large, fishless desert spring: Montezuma Well, Arizona. Pp. 98-126 *in* Stevens, L.E. and V.J. Meretsky, editors. Aridland Springs in North America: Ecology and Conservation. University of Arizona Press, Tucson.

Brennan,T.C. and A.T. Holycross. 2007. A Field Guide to Amphibians and Reptiles in Arizona. Arizona Game and Fish Department, Phoenix.

Brian, N.J. 2000. A Field Guide to the Special Status Plants of Grand Canyon National Park. Grand Canyon National Park Science Center, Grand Canyon. Available on-line at: http://www.nps.gov/grca/naturescience/upload/plant_guide_1.pdf (accessed 20 Sept., 2011).

Brown, W.H. 1971. Mammals on mountaintops: nonequilibrium insular biogeography. American Naturalist 105:467-478.

Brown, B.T., S.W. Carothers, and R.R. Johnson. 1987. Grand Canyon Birds. University of Arizona Press, Tucson.

Brown, B.T., G.S. Mills, R.L. Glinski, and S.W. Hoffman. 1992. Density of nesting peregrine falcons in Grand Canyon National Park, Arizona. Southwestern Naturalist 37:188-193.

Bryson, R.W. Jr., A. Nieto-Montes de Oca, J.R. Jaeger, and B.R. Riddle. 2010. Elucidation of cryptic diversity in a widespread nearctic treefrog reveals episodes of mitochondrial gene capture as frogs diversified across a dynamic landscape. Evolution 64-8: 2315–2330

Busco, J., E. Douglas, and J. Kapp. 2011. Preliminary pollination study on sentry milk-vetch (*Astragalus cremnophylax* Barneby var. *cremnophylax*), Grand Canyon National Park;s only endangered plant species. The Plant Press 35:

Butman, B, D.C. Twichell, P.A. Rona, B.E. Tucholke, T.J. Middleton, and J.M. Robb. 2006, Sea floor topography and backscatter intensity of the Hudson Canyon region offshore of New York and New Jersey. U.S. Geological Survey Open-File Report 2004-1441, Version 2.0. http://pubs.usgs.gov/of/2004/1441/index.html.

California Condor Recovery Team. 2007. Review of the Second Five Years of the California Condor Reintroduction Program in the Southwest. California Condor Recovery

Team and U.S. Fish and Wildlife Service, California/Nevada Operations Office, Sacramento.

Canals, M., R. Danovaro, S. Heussner, V. Lykousis, P. Puig, F. Trincardi, A.M. Calafat, X. Durrieu de Madron, A. Palanques, and A. Sànchez-Vidal. 2009. Cascades in Mediterranean submarine grand canyons. Oceanography 22(1):26–43.

Carothers, S.W. and S.W. Aitchison, editors. 1976. An ecological survey of the riparian zone of the Colorado River between Lees Ferry and Grand Wash Cliffs. Colorado River Technical Report 10, Grand Canyon National Park, Grand Canyon.

Carothers, S.W. and B.T. Brown. 1991. The Colorado River through Grand Canyon: Natural History and Human Change. University of Arizona Press, Tucson.

Choudhury, A., T.L. Hoffnagle, and R.A. Cole. 2004. Parasites of native and non-native fishes of the Little Colorado River, Grand Canyon, Arizona. Journal of Parasitology 90:1042-1053.

Coggins, L.G., Jr., and Yard, M.D., 2011. An experiment to control nonnative fish in the Colorado River, Grand Canyon, Arizona: U.S. Geological Survey Fact Sheet 2011-3093.

Connell, J.H. 1978. Diversity in tropical rain forests and coral reefs. Science 199:1302–1310.

Crow, R., K.E. Karlstrom, W. McIntosh, L. Peters, and N. Dunbar. 2008. History of Quaternary volcanism and lava dams in western Grand Canyon based on LIDAR analysis, $^{40}$Ar/$^{39}$Ar dating, and field studies: Implications for flow stratigraphy, timing of volcanic events, and lava dams. Geosphere 4:183-206.

Dickinson, W.R. 2002. The Basin and Range Province as a composite extensional domain. International Geological Review 44:1–38.

Durrant, S.D. 1952. Mammals of Utah. University of Kansas Press, Lawrence.

Estes, J.A., J. Terborgh, J.S. Brashares, M.E. Power, J. Berger, W.J. Bond, S.R. Carpenter, T.E. Essington, R.D. Holt, J.B.C. Jackson, R.J. Marquis, L. Oksanen, T. Oksanen, R.T. Paine, E.K. Pikitch, W.J. Ripple, S.A. Sandin, M. Scheffer, T.W. Schoener, J.B. Shurin, A.R.E. Sinclair, M.E. Soulé, R.Virtanen, and D.A. Wardle. 2011. Trophic downgrading of planet Earth. Science 333:301-306.

Flinders, J.T., D.S. Rogers, J.L.Webber-Alston, and H.A. Barber. 2002. Mammals of the Grand Staircase-Escalante National Monument: a literature and museum survey. Monographs of the Western North American Naturalist 1:1-64.

Fulé, P.Z., W.W. Covington, M.M. Moore, T.A. Heinlein, and A.E.M. Waltz. 2002. Natural variability in forests of Grand Canyon, USA. Journal of Biogeography 29:31–47.

Garth, J.S. 1950. Butterflies of Grand Canyon National Park. Grand Canyon Natural History Association, Grand Canyon.

Grand Canyon Wildlands Council, Inc. 2002. A hydrological and biological inventory of springs, seeps and ponds of the Arizona Strip, final report. Arizona Water Protection Fund, Phoenix.

Gurnell, A.M. and G.E. Petts, editors. 1995. Changing River Channels. Wiley, Chichester.

Hamblin, W.K. 1994. Late Cenozoic lava dams in the western Grand Canyon. Geological Society of America Memoir 183:1-135.

Harper, K.T., L.L. St. Claire, K.H. Thorne, and W.M. Hess. 1994. Natural History of the Colorado Plateau and Great Basin. University Press of Colorado, Boulder.

Hirsch, R.M., J.E. Walker, J.C. Day, and R. Kollio. 1990. The influence of man on hydrologic systems. Pp. 329-359 *in* M.G. Wolman and H.C. Riggs, editors. Surface Water Hydrology. Geological Society of America Decade of North American Geology 0-1.

Hodgson, W.C. 2001. Taxonomic novelties in American Agave (Agavaceae). Novon 11:410-416.

Hoffman, S. W., J. P. Smith, and T. D. Meehan. 2002. Breeding grounds, winter ranges, and migratory routes of raptors in the Mountain West. Journal of Raptor Research 36:97–110.

Hoffmeister, D. 1986. Mammals of Arizona. University of Arizona Press, Tucson.

Holdridge, L.R. 1947. Determination of world plant formations from simple climatic data. Science 105:367-368.

Howard, A. and R. Dolan. 1981. Geomorphology of the Colorado River in the Grand Canyon. Journal of Geology 89:269-298.

Hunt, C.B. 1956. Cenozoic geology of the Colorado Plateau. U.S. Geological Survey Professional Paper 279, Washington DC.

Huston, M.A. 1979. A general hypothesis of species diversity. American Naturalist 113: 81–99.

Jackson, S.T., editor. 2009. Essay on the Geography of Plants: Alexander Von Humboldt and Aimé Bonpland (translated by S. Romanowski). University of Chicago Press, Chicago.

Jones, S.C. 1985. New termite records for the Grand Canyon. Southwestern Entomologist 10:137-138.

Karlsson, J., P. Brystöm, J. Ask, P. Ask, L. Persson, and M. Jansson. 2011. Light limitation of nutrient-poor lake ecosystems. Nature 460:506-509.

Lamb, T., T.R. Jones, and P.J. Wettstein. 1997. Evolutionary genetics and phylogeography of tassel-eared squirrels (*Sciurus aberti*). Journal of Mammalogy 78:117-133.

LaRue, C.T., L.L. Dickson, N.L. Brown, J.R. Spence, and L.E. Stevens. 2001. Recent bird records from the Grand Canyon region, 1974-2000. Western Birds 32:101–118.

Lawlor, T.E. 1998. Biogeography of Great Basin mammals: paradigm lost? Journal of Mammalogy 79:1111-1130.

Lomolino, M.V., D.R. Riddle, R.J. Whittaker, and J.H. Brown. 2010. Biogeography, 4th edition. Sinauer Associates, Sunderland.

Lovich, J.E. and T.S. Melis. 2007. The state of the Colorado River ecosystem in Grand Canyon: lessons from 10 years of adaptive ecosystem management. International Journal of River Basin Management. 5:1-15.

MacArthur, R.H. and E.O. Wilson. 1967. The Theory of Island Biogeography. Princeton University Press, Princeton.

Maschinski, J., R. Frye, and S. Rutman. 1997. Demography and population viability of an endangered plant species before and after protection from trampling. Conservation Biology 11:990-999.

Melis, T.S., R.H. Webb, and P.G. Griffiths. 1997. Debris flows in Grand Canyon National Park: Peak discharges, flow transformations, and hydrographs. Pp. 727-736 *in* Chen, C-l., editor. Debris-flow Hazards Mitigation: Mechanics, Prediction, and Assessment. American Society of Civil Engineers, New York.

Meretsky, V.J. and L.E. Stevens. 2000. Kanab ambersnail, an endangered succineid snail in southwestern USA. Tentacle 8:8-9.

Merriam, C. H. and L. Steineger. 1890. Results of a biological survey of the San Francisco Mountain region and the desert of the Little Colorado, Arizona. U.S. Department of Agriculture Division of Ornithology and Mammalia North American Fauna Report 3, Washington.

Miller, D.M., R.A. Young, T.W. Gatlin, and J.A. Richardson. 1982. Amphibians and Reptiles of the Grand Canyon. Grand Canyon Natural History Association Monograph 4, Grand Canyon.

Minckley, W.L. 1973. Fishes of Arizona. Arizona Game and Fish Department, Phoenix.

Minckley, W. C. 1991. Native fishes of the Grand Canyon region: An obituary? *In:* Marzolf, G. R., editor. Colorado River Ecology and Dam Management. National Academy Press, Washington, D.C.

Newmark, W.D. 1995. Extinction of mammal populations in western North American national parks. Conservation Biology 9:512-526.

Oláh-Hemmings, V., J.R. Jaeger, M.J. Sredl, M.A. Schlaepfer, R.D. Jennings, C.A. Drost, D.F. Bradford, and B.R. Riddle. 2009. Phylogeography of declining relict and lowland leopard frogs in the desert Southwest of North America. Journal of Zoology 280: 343–354.

Olson, D.M. et al. 2001. Terrestrial Ecoregions of the World: A New Map of Life on Earth. BioScience 51:933-938.

Pereira, R.J. and D.B Wake. 2009. Genetic leakage after adaptive and nonadaptive divergence in the *Ensatina eschscholtzii* ring species. Evolution 68:2288-2301.

Phillips, B.G., A.M. Phillips III, and M.A. Schmidt-Bernzott. 1987. Annotated checklist of vascular plants of Grand Canyon National Park. Grand Canyon Natural History Association Monograph No. 7, Grand Canyon.

Pogue, M.G. 2004. A new species of *Schinia* Hübner from riparian habitats in the Grand Canyon (Lepidoptera: Noctuidae: Heliothinae). Zootaxa 788:1–4.

Polhemus, J.T. and M.S. Polhemus. 1976. Aquatic and semi-aquatic Heteroptera of the Grand Canyon (Insecta: Hemiptera). Great Basin Naturalist 36:221-226.

Rabe, M.J., M.S. Siders, C.R. Miller, and T.K. Snow. 1998. Long foraging distance for a spotted bat (*Euderma maculatum*) in northern Arizona. Southwestern Naturalist 43:266-269.

Rasmussen, D. I. 1941. Biotic communities of the Kaibab Plateau, Arizona. Ecological Monographs 11:229–275.

Rhodes, S.L. and T.J. Ayers. 2010. Two new taxa of *Scutellaria* section *Resinosa* (Lamiaceae) from northern Arizona. Journal of the Botanical Research Institute of Texas 4:19- 26.

Rios, E. and S.T. Álvarez-Casteñeda. 2007. Environmental responses to altitudinal gradients and subspecific validity in pocket gophers (*Thomomys bottae*) in Baja California Sur, Mexico. Journal of Mammalogy 88:926-934.

Romdal, T.S., R.K. Colwell, and C. Rahbek. 2005. The influence of band sum area, domain extent and range sizes on the latitudinal mid-domain effect. Ecology 86:235-244.

Scarbrough, A.G., L.E. Stevens, and C.R. Nelson. A review of the *albibarbis* group of *Efferia* Coquillett from the Grand Canyon region, southwestern USA (Diptera: Asilidae), including description of a new species. Pan-Pacific Entomologist, *in press.*

Schmidt, J.C., and J.B. Graf. 1990. Aggradation and degradation of alluvial sand deposits, 1965-1986, Colorado River, Grand Canyon National Park, Arizona. United States Geological Survey Professional Paper 1493:1-74.

Sellers, W.D., R.H. Hill, and M. Sanderson-Rae. 1985. Arizona climate: the first hundred years. University of Arizona Press, Tucson.

Shelley, R.M. and L.E. Stevens. 2004. Discovery of the milliped, *Tylobolus utahensis* Chamberlin, in Arizona (Spirobolida: Spirobolidae). Western North American Naturalist 63:541-542. https://ojs.lib.byu.edu/wnan/index.php/wnan/article/view/456/322.

Smith, J.P., C.J. Farmer, S.W. Hoffman, G.S. Kaltenecker, K.Z. Woodruff, and P.F. Sherrington. 2008. Trends in autumn counts of migratory raptors in western North America. Pp. 217-252 *in* Bildstein, K.L., J.P. Smith, E.R. Inzunza, and R.R. Veit, Editors. State of North America's Birds of Prey. American Ornithologists' Union Series in Ornithology 3:1-462.

Snelling, R.R. 1974. Change in the status of some North American *Polistes* (Hymenoptera: Vespidae). Proceedings of the Entomological Society of Washington 76:476-479.

Solar Pathfinder. 2008. Instruction Manual for the Solar Pathfinder. Linden, TN. Available on-line at: http://www.solarpathfinder.com/pdf/pathfinder-manual.pdf.

Spence, J.R. 2008. Spring-supported vegetation along the Colorado River on the Colorado Plateau: floristics, vegetation structure, and environment. Pp. 185-210 *in* Stevens, L.E. and V.J. Meretsky, editors. Aridland Springs in North America: Ecology and Conservation. University of Arizona Press, Tucson.

Stevens, L.E. 2011. Extirpated and declining species of management concern from Glen Canyon Dam to Lake Mead, Arizona. Grand Canyon Wildlands Council, Inc., Flagstaff.

Stevens, L.E. and T.J. Ayers. 2002. The biodiversity and distribution of alien vascular plant and animals in the Grand Canyon region. Pp. 241-265 *in* Tellman, B., editor. Invasive Exotic Species in the Sonoran Region. University of Arizona Press, Tucson.

Stevens, L.E. and R.A. Bailowitz. 2005. Distribution of *Brechmorhoga* clubskimmers (Odonata: Libellulidae) in the Grand Canyon region. Western North American Naturalist 65: 170-174.

Stevens, L.E. and R.A. Bailowitz. 2009. Odonata biogeography in the Grand Canyon ecoregion, southwestern U.S.A. Annals of the Entomological Society of America 102:261-274.

Stevens, L.E. and R.L. Huber. 2004. Biogeography of tiger beetles (Cicindelidae) in the Grand Canyon Ecoregion, Arizona and Utah. Cicindela 35:41-64.

Stevens, L.E. and G.P. Nabhan. 2002. Biodiversity: plant and animal endemism, biotic associations, and unique habitat mosaics in living landscapes. Pp. 41-48 *in* Center for Sustainable Environments, Terralingua, and Grand Canyon Wildlands Council, editors. Safeguarding the Uniqueness of the Colorado Plateau: An Ecoregional Assessment of Biocultural Diversity. Center for Sustainable Environments, Northern Arizona University, Flagstaff.

Stevens, L.E. and V.J. Meretsky, editors. 2008. Aridland Springs in North America: Ecology and Conservation. University of Arizona Press, Tucson.

Stevens, L.E. and J.T. Polhemus. 2008. Biogeography of aquatic and semi-aquatic Heteroptera in the Grand Canyon ecoregion, southwestern USA. Monographs of the Western North American Naturalist 4:38-76.

Stevens, L.E., J.C. Schmidt, T.J. Ayers, and B.T. Brown. 1995. Flow regulation, geomorphology and Colorado River marsh development in the Grand Canyon, Arizona. Ecological Applications 5:1025-1039.

Stevens, L.E., K.A. Buck, B.T. Brown and N. Kline. 1997a. Dam and geomorphic influences on Colorado River waterbird distribution, Grand Canyon, Arizona. Regulated Rivers: Research & Management 13:151-169.

Stevens, L.E., J.P. Shannon and D.W. Blinn. 1997b. Benthic ecology of the Colorado River in Grand Canyon: dam and geomorphic influences. Regulated Rivers: Research & Management 13:129-149.

Stevens, L.E., T.J. Ayers, J.B. Bennett, K. Christensen, M.J.C. Kearsley, V.J. Meretsky, A.M. Phillips III, R.A. Parnell, J. Spence, M.K. Sogge, A.E. Springer, and D.L. Wegner. 2001. Planned flooding and Colorado River riparian trade-offs downstream from Glen Canyon Dam, Arizona. Ecological Applications 11:701-710.

Stevens, L.E., T.L. Griswold, O. Messinger, W.G. Abrahamson, II, and T.J. Ayers. 2007. Plant and pollinator diversity in northern Arizona. The Plant Press 31:5-7.

Stevens, L.E., F.B. Ramberg, and R.F. Darsie, Jr. 2008a. Biogeography of Culicidae (Diptera) in the Grand Canyon region, Arizona, USA. Pan-Pacific Entomologist 84:92-109.

Stevens, L.E., D.S. Turner, and V. Supplee. 2008b. Background: wildlife and flow relationships in the Verde River watershed. Pp. 51-70 in Haney, J.A., D.S. Turner, A.E. Springer, J.C. Stromberg, L.E. Stevens, P.A. Pearthree, and V. Supplee. Ecological Implications of Verde River Flows. The Nature Conservancy, Phoenix.

Stevens, L.E., B.T. Brown, and K. Rowell. 2009. Foraging ecology of peregrine falcons along the dam-regulated Colorado River, Grand Canyon, Arizona. The Southwestern Naturalist 54:284-299.

Stockwell, C.A., G.C. Bateman, and J. Berger. 1991. Conflicts in national parks: A case study of helicopters and bighorn sheep time budgets at the Grand Canyon. Biological Conservation 56:317-328.

Stohlgren, T.J., D. Binkley, G.W. Chong, M. Kalkhan, L.D. Schell, K.A. Bull, Y. Otsuki, G. Newman, M.A. Bashkin, and Y. Son. 1999. Exotic plant species invade hotspots of native plant diversity. Ecological Monographs 69:25–46.

Sublette, J.E., L.E. Stevens, J.P. Shannon. 1998. Chironomidae (Diptera) of the Colorado River in Grand Canyon, Arizona, U.S.A., I: Taxonomy and ecology. The Great Basin Naturalist: 58:97-146. https://ojs.lib.byu.edu/wnan/index.php/wnan/article/view/ 826/1666 (accessed 14 Dec 2011).

Suttkus, R.D., G. Clemmer, and C. Jones. 1978. Mammals of the Riparian Region of the Colorado River in the Grand Canyon Area of Arizona. Occasional Papers of Tulane University, Belle Chasse.

Topping, D.J., J.C. Schmidt, and L.E. Vierra Jr. 2003. Computation and analysis of the instantaneous-discharge record for the Colorado River at Lees Ferry, Arizona – May 8, 1921, through September 30, 2000. United States Geological Survey Professional Paper Series 1677.

Topping, D. J., Rubin, D. M., and Schmidt, J. C., 2005, Regulation of sand transport in the Colorado River by changes in the surface grain size of eddy sandbars over multi-year timescales. Sedimentology 52:1133-1153.

Turner, R.M. and M.M. Karpiscak. 1980. Recent vegetation changes along the Colorado River between Glen Canyon Dam and Lake Mead, Arizona. United States Geological Survey Professional Paper No. 1132.

Ueckert, D.N., M.C. Bodine, and B.M. Spears. 1976. Population density and biomass of the desert termite *Gnathamitermes tubiformans* (Isoptera: Termitidae) in a shortgrass prairie: relationship to temperature and moisture. Ecology 57:1273–1280.

Utah Native Plant Society. 2011. Utah Rare Plant Guide [Internet]. A.J. Frates, A.J., editor. Utah Native Plant Society, Salt Lake City. Available on-line from: http://www.utahrareplants.org (accessed 14 Dec 2011).

Wake, D. 1997. Incipient species formation in salamanders of the *Ensatina* complex. Proceedings of the National Academy of Science 94:7761-7767.

Wallace, A.R. 1876. The Geographical Distribution of Animals, with a Study of the Relations of Living and Extinct Faunas as Elucidating the Past Changes of the Earth's Surface. Harper & Brothers Publishers, New York.

Waring, G.L., L.E. Stevens, B.G. Ralston, and S. Archer. The changing ecology of western honey mesquite (Fabaceae: *Prosopis glandulosa* var. *torreyana*) in Grand Canyon, Arizona, USA. Journal of the Arizona-Nevada Academy of Sciences, submitted.

Webb, D.S. 2006. The Great American Biotic Interchange: patterns and processes. Annals of the Missouri Botanical Garden 93(2):245-257.

Webb, R.H., P.T. Pringle, and G.R. Rink. 1989. Debris flows from tributaries of the Colorado River, Grand Canyon National Park, Arizona. United States Geological Survey Professional Paper 1492.

Yard, M.D., G.E. Bennett, S.N. Mietz, L.G. Coggins Jr., L.E. Stevens, S. Hueftle, and D.W. Blinn. 2005. Influence of topographic complexity on solar insolation estimates for the Colorado River, Grand Canyon, AZ. Ecological Modeling 183:157-172.

# 10

# Contributions
# of Cladistic Biogeography
# to the Mexican Transition Zone

Isolda Luna-Vega[1,*] and Raúl Contreras-Medina[2]
*[1]Laboratorio de Biogeografía y Sistemática, Departamento de Biología Evolutiva,
Facultad de Ciencias, Universidad Nacional Autónoma de México (UNAM),
[2]Escuela de Ciencias, Universidad Autónoma "Benito Juárez" de Oaxaca (UABJO),
México*

## 1. Introduction

The Mexican Transition Zone (MTZ) was defined by Halffter (1987) as the complex area where Neotropical and Nearctic biotic elements overlap, including the southern United States, Mexico and adjacent areas of Central America. The biota of this area has received the attention of several naturalists since the mid 19th century (e.g. Sclater, 1858) due to its placement between the Nearctic and Neotropical regions and its "hybrid" character (Morrone, 2005 and references therein), associated with high biotic diversity and a high degree of endemism of plant and animal taxa. This area also has received attention in conservation and biodiversity plans at a worldwide level, because its central portion, the Mesoamerican hotspot, was considered one of the main biodiversity hotspots recognized in the world (Flores-Villela & Gerez, 1994; Luna-Vega and Contreras-Medina, 2010; Mittermeier et al., 1998; Myers et al., 2000).

Recently, authors have proposed hypotheses to explain disjunct biotic patterns observed in the MTZ from different approaches and methodologies, such as panbiogeography (Álvarez & Morrone, 2004; Contreras-Medina & Eliosa-León, 2001; Escalante et al., 2004; Morrone & Márquez, 2001; Morrone & Gutiérrez, 2005), parsimony analysis of endemicity (Aguilar-Aguilar et al., 2003; Espinosa et al., 2000; Luna-Vega et al., 2001; Morrone et al., 1999; Morrone & Escalante, 2002), and cladistic biogeography (Contreras-Medina et al., 2007; Espinosa et al., 2006; Flores-Villela & Goyenechea, 2001; Liebherr, 1991; Marshall & Liebherr, 2000; Rosen, 1978). The latest studies based on cladistic biogeography approaches represent the basis of this chapter.

Our goal here is to compare and discuss the different contributions made to the biogeography of the Mexican Transition Zone by applying different cladistic biogeographical methods. It is also to contrast the different areas of endemism proposed in these studies with the objective of obtaining consensus on areas of endemism to be considered in future research.

---

* Corresponding Author

## 2. Cladistic biogeography: Basis and methods

Cladistic biogeography is an approach to historical biogeography that searches for patterns of relationships among areas of endemism based on the phylogenetic relationships of the taxa inhabiting them (Crisci et al., 2000; Humphries & Parenti, 1999; Morrone & Crisci, 1995). This approach emerged as a combination of principles of phylogenetic systematics of Willi Hennig (1966) and the biogeographic ideas formulated by Léon Croizat (1958, 1964) represented in his book Panbiogeography (Contreras-Medina, 2006; Craw, 1988; Crisci et al., 2000; Espinosa & Llorente, 1993; Flores-Villela & Goyenechea, 2001). Cladistic biogeography is based on a close relationship among systematics and biogeography, considering that taxonomic cladograms converted into area cladograms can elucidate the fragmentation sequence of studied areas. With this procedure, it is possible to propose hypotheses on historical relationships among areas of endemism (Contreras-Medina, 2006; Morrone, 1997). Cladistic biogeographic analysis essentially looks for congruent patterns among area cladograms of different taxa (Humphries & Parenti, 1999; Page & Lydeard, 1994) showed in a general area cladogram, which represents the final result of this kind of analysis (Contreras-Medina, 2006; Morrone, 1997). Interpretation of cladistic biogeographical results usually focuses on vicariance rather than on dispersal events, because vicariance affects different groups of organisms simultaneously (Morrone & Crisci, 1995; Nelson & Platnick, 1981).

The construction of area cladograms is simple when each taxon is endemic to a single area and each area has only one taxon (Morrone, 1997). Incongruence can occur between two or more area cladograms due to three main reasons, known as redundant distributions, widespread taxa and missing areas (Morrone et al., 1996). Redundant distributions occur when an area appears more than once in an area cladogram, because two or more terminal taxa are distributed in this area; widespread taxa occur when any of the terminal taxa inhabits two or more of the areas analyzed; missing areas occur when there is not a terminal taxon distributed in one of the areas analyzed, therefore this area is not represented in the area cladograms (Morrone, 2009). One or more of the methods that have been proposed in cladistic biogeography are then applied to these area cladograms, and the final result is a general area cladogram (Morrone, 1997).

The first method developed in cladistic biogeography was the reduced consensus cladogram proposed by Rosen (1978), which was only used by him. Another method proposed was the Brooks parsimony analysis or BPA (Brooks, 1990; Kluge, 1988), which has been extensively used in studies developed in the MTZ (Contreras-Medina et al., 2007; Espinosa et al., 2006; Marshall & Liebherr, 2000). BPA has been criticized by some authors as a suboptimal method, because it uses both dispersal and vicariance explanations to fit taxa and areas to the same tree (e.g. Siddall & Perkins, 2003); however, other authors (e.g. Brooks et al., 2001; Van Veller & Brooks, 2001) have defended it as a valid method. Nelson & Ladiges (1996) noted that when nodes and areas are associated in order to be included in a data matrix, geographic paralogy may result because of duplication or overlap in the distribution of taxa related by paralogous nodes. To this end, another method was proposed, known as paralogy-free subtrees (Nelson & Ladiges, 1995). This method consists in the reduction of complex cladograms to paralogy-free-subtrees, that means that geographic data are associated only with the informative nodes, and areas duplicated or redundant in the descendants of each node do not exist (Morrone, 2009). Once that paralogy-free-subtrees are obtained, these are represented in a component or a three-item matrix and analyzed with a parsimony algorithm (Crisci et al., 2000). A parsimony analysis of these paralogy-free subtrees (PAPS) may thus be used to generate a more robust hypothesis, because geographic paralogy has been removed. Another frequent method that has been cited in biogeographic literature is component analysis (Nelson & Platnick, 1981), but unfortunately this method has not been applied to the biota of the MTZ.

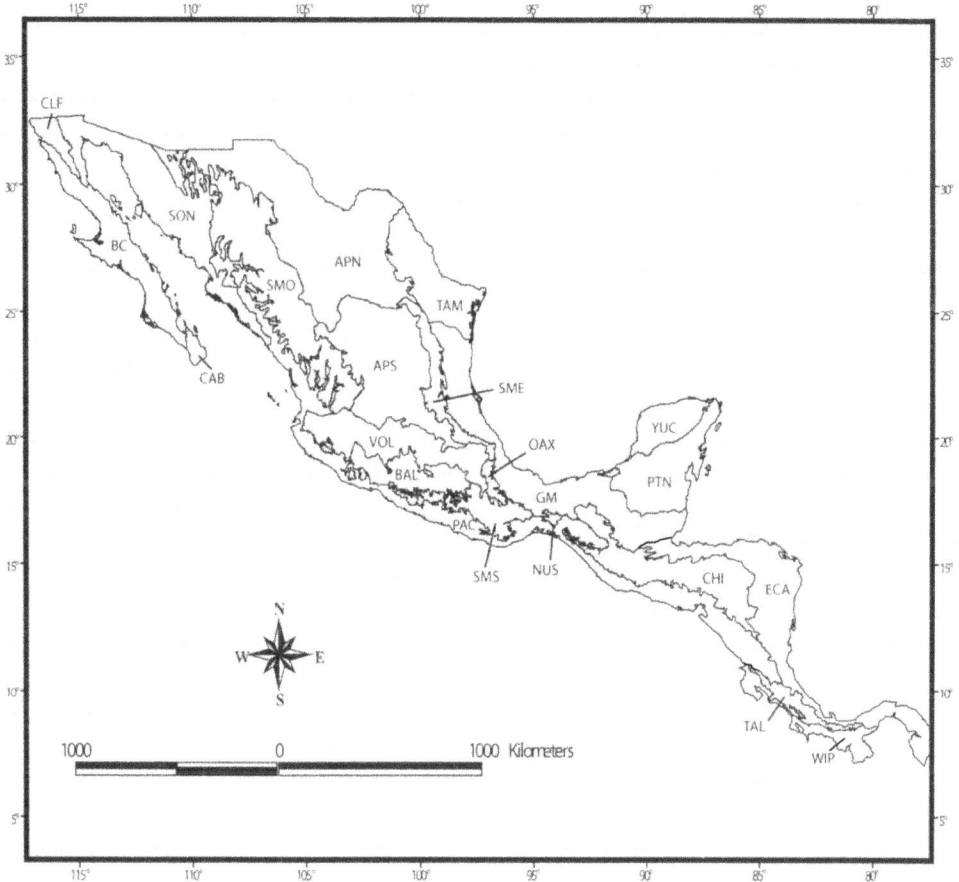

Fig. 1. The 19 biogeographic provinces of Mexico according to Arriaga et al. (1997). Abbreviations are: APN = Altiplano Norte, APS = Altiplano Sur, BAL = Depresión del Balsas, BC = Baja California, CLF = California, CAB = Del Cabo, CHI = Los Altos de Chiapas, GM = Golfo de México, NUS = Soconusco, OAX = Oaxaca, PAC = Costa del Pacífico, PTN = Petén, SME = Sierra Madre Oriental, SMO = Sierra Madre Occidental, SMS = Sierra Madre del Sur, SON = Sonorense, TAM = Tamaulipeca, VOL = Eje Volcánico, YUC = Yucatán. The provinces of Central America from Morrone (2001) are: CHI= Los Altos de Chiapas, ECA= Eastern Central America, WIP= Western Isthmus of Panamá, and TAL = Sierra de Talamanca.

## 3. Areas of endemism in the Mexican Transition Zone

Areas of endemism are defined by the overlapped distributions of endemic taxa (Contreras-Medina, 2006; Morrone, 1994). These kinds of areas represent the basic units in cladistic biogeography (Contreras-Medina et al., 2001); for this reason, the recognition and delimitation of areas of endemism represent a prerequisite to carry out cladistic biogeographic studies (Morrone, 1997).

Some proposals on regionalization of the Mexican Transition Zone have been suggested based on different criteria. One of the most supported schemes was that proposed by the Comisión Nacional para el Conocimiento y Uso de la Biodiversidad (CONABIO) (National Commission for the Knowledge and Use of Biodiversity, Mexico), which represents the effort of a set of several specialists on geology, vascular plants, reptiles and mammals working together to provide different proposals on regionalization of the Mexican territory with different points of view. Nineteen Mexican biogeographic provinces were recognized through this workshop (Arriaga et al., 1997). An important regionalization proposal for Central America also was made by Morrone (2001). This scheme was also followed in this study, and was also used to compare the areas of endemism found in the different revised studies on cladistic biogeography of the Mexican Transition Zone.

In the study of Espinosa et al. (2006), the parsimony analysis of endemicity (PAE) method (Rosen, 1988) was used in order to generate the 23 areas of endemism that were used later in the cladistic biogeography section of this paper. The areas of endemism were obtained by merging those areas supported by endemic (synapomorphic) species from PAE following the proposed provinces of Morrone (1994).

In many cases, the Mexican mountain ranges coincide with the areas of endemism (Fig. 2), so their recognition and delimitation represent necessary tools in future cladistic biogeographic studies. The origin and development of these mountain chains have strongly

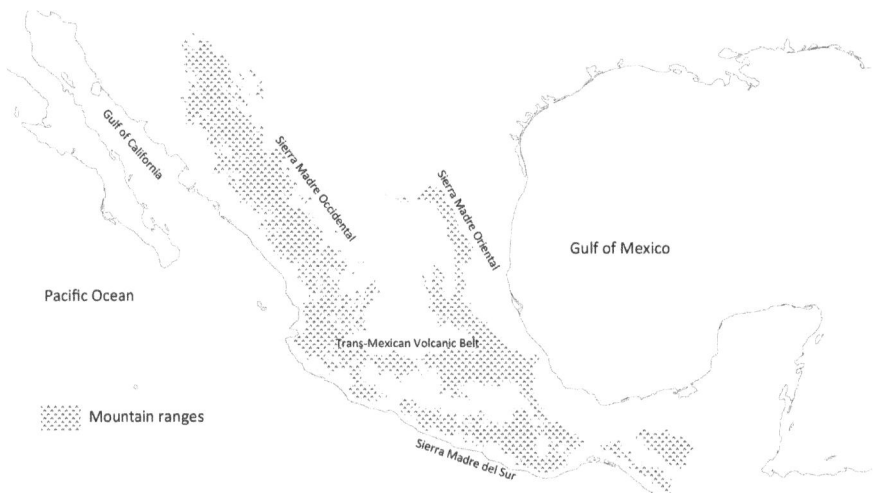

Fig. 2. The main Mexican mountain chains.

| Biogeographic provinces | Marshall & Liebherr (2000) | Espinosa et al. (2006) | Flores-Villela & Goyenechea (2001) | Contreras-Medina et al. (2007) |
|---|---|---|---|---|
| CLF | | | | Californian |
| BC | | | | Baja California |
| CAB | | El Cabo | | Baja California |
| SON | Sonoran Desert | Sonora | Sonoran Desert | Planicie Costera del Noroeste |
| SMO | Sierra Madre Occidental plus Central Plateau | | Sierra Madre Occidental | Sierra Madre Occidental |
| APN | Sierra Madre Occidental plus Central Plateau | | Chihuahuan Desert | Altiplano |
| APS | Sierra Madre Occidental plus Central Plateau | | Chihuahuan Desert | Altiplano |
| TAM | Sierra Madre Occidental plus Central Plateau | | Semiarid lands of Tamaulipas | Planicie Costera del Noreste |
| SME | Sierra Madre Oriental | | Sierra Madre Oriental | Sierra Madre Oriental |
| VOL | Sierra Transvolcánica | | Transvolcanic Axis of Central Mexico | Serranías Meridionales |
| BAL | | W and E Balsas | Balsas Depression | Balsas Basin |
| SMS | Sierra Madre del Sur | Tehuantepec | Sierra Madre del Sur | Serranías Meridionales |
| OAX | | Tehuacán-Cuicatlán Valley | Transvolcanic Axis of Central Mexico | Tehuacán-Cuicatlán Valley |
| GM | | | | Costa del Golfo de México |
| PAC | | Tuito, Papagayo, Armería-Coahuayana and Tehuantepec | Pacific lowlands of Mexico plus Balsas Depression | Costa Pacífica |
| NUS | | | | Soconusco |

| Biogeographic provinces | Marshall & Liebherr (2000) | Espinosa et al. (2006) | Flores-Villela & Goyenechea (2001) | Contreras-Medina et al. (2007) |
|---|---|---|---|---|
| YUC | Chiapan Guatemalan Highlands | | | Yucatán Península |
| PTN | Chiapan Guatemalan Highlands | | | Yucatán Península |
| CHI | Chiapan Guatemalan Highlands | | Highlands of Chiapas | Serranías Transístmicas |
| ECA | Chiapan Guatemalan Highlands | | Eastern Central American Atlantic lowlands | Eastern Central America |
| WIP | Talamancan Cordillera | | Western Central American Pacific lowlands | Costa Pacífica |
| TAL | Talamancan Cordillera | | Talamanca ridge | |

NOTES: When the same name of an area of endemism appears in two or more cells, it means that this area in this particular study is very large and includes two or more biogeographic provinces. Empty cells mean that the biogeographic province was not included or considered in our study.

Table 1. Comparison among the areas of endemism of the Mexican Transition Zone used in some of the studies cited in this chapter, in relation to the biogeographic provinces of Arriaga et al. (1997) and Morrone (2001).

Influenced the geographic distribution of the biota of the MTZ. The geological evolution of Mexico and Central America during the Cretaceous and Cenozoic has had a significant influence on the processes controlling the composition of their biota (Cevallos-Ferriz & González-Torres, 2005). The primary geological events shaping Mexico and Central America during these time intervals include the following: (1) the development of large magmatic provinces (Sierra Madre Occidental, Trans-Mexican Volcanic Belt, and Sierra Madre del Sur); (2) the fragmentation and displacement of continental segments including the opening of the Gulf of California (the separation of the Baja California Peninsula from the mainland) and the displacement of the Chortis Block to northern Central America; (3) the marine regression that outlined the current shape of Mexico; and (4) the formation and rise of an eastern orogenic belt known as the Sierra Madre Oriental (Cevallos-Ferriz & González-Torres, 2005; Ortega-Gutiérrez et al., 1994).

In Table 1 (above), two additional studies merit comment. Liebherr (1991) did not name areas of endemism, but designated them by letters, making it difficult to compare his areas with those of other authors. Rosen (1978) identified very small areas of endemism, only including small parts of the MTZ - these areas were mainly in Chiapas, eastern Central America, Petén, and Golfo de México.

## 4. Cladistic biogeography studies in the Mexican Transition Zone

Rosen (1978) was the first to apply a cladistic biogeography approach to the MTZ. In his biogeographic research, the author included the phylogenies of two fresh water fish of the genera *Heterandria* and *Xiphophorus*. Rosen (1978) recognized three coincidences of historical relationships, indicating the existence of a common ancestral distributional area that had been fragmented twice and that currently constitutes the basins of the Bravo River (in northern Mexico), the Pánuco River (in northeastern Mexico), and the Grijalva River (on the border between Mexico and Guatemala).

Another pioneering cladistic biogeographic study in the MTZ is that of Liebherr (1991). This author analyzed the history the mountain biota based on the phylogenies of two Coleoptera genera (*Elliptoleus* and *Calathus*). The complexity of the geologic history of southern Mexico is noted in the general area cladogram obtained, showing an ancient primary fragmentation that separated extra-tropical from tropical biota. Liebherr (1991) obtained a first clade that was divided in two subgroups, the first formed by the Sierra Madre Oriental and the second by the Sierra Madre Occidental and part of the Mexican Plateau. The tropical clade included the southern portions of the Sierra Madre Occidental, Sierra Madre Oriental and other Mexican mountain chains such the Sierra Madre del Sur and the Trans-Mexican Volcanic Belt.

Marshall & Liebherr (2000) published a cladistic biogeography study applying BPA and Assumption 0 methods based on 33 taxa (beetles, reptiles, fish and one plant genus) and using nine areas of endemism; they obtained four general area cladograms, one with BPA and three with Assumption 0. The congruence among these cladograms is shown in the relationships among the Arizona Mountains plus the Sierra Madre Occidental in one clade, and the Sierra Madre del Sur plus the Trans-Mexican Volcanic Belt related with the Chiapan-Guatemalan Highlands in another clade. Unfortunately, their approach included methodological confusion regarding the Assumption 0: the missing taxa option (or missing areas) in the matrix is treated by them as '0', indicating the lack of information regarding a taxon's occurrence. However, the Assumption 0 approach considers missing areas to be uninformative, as well as the other two assumptions (1 and 2) in a component analysis method (Morrone, 1998), so missing areas are located in all possible positions in the resolved area cladograms. Assumption 0 in BPA implies that areas inhabited by a widespread taxon are considered as sister areas (Morrone, 1997).

Flores-Villela & Goyenechea (2001) used 10 animal taxa in their biogeographical study (beetles, lizards, snakes and frogs) distributed in 13 areas of endemism; the method of reconciled trees implementing assumptions 0, 1, and 2 was applied in this study. The authors obtained six general area cladograms, but they did not achieve a consistent pattern of relationships among these cladograms. They concluded that each cladogram could only explain portions of the geographic history of the taxa studied, and that the results obtained were due to the geological complexity of the area under study.

Espinosa et al. (2006) generated hypotheses on the historical biogeography of tropical Mexico using the flowering plant genus *Bursera*. This study was based on the geographic distribution of 104 species in this genus analyzed with the BPA method. The consensus general area cladogram obtained was formed by two main clades: the first included different areas of the Greater Antilles, and the second 12 Mexican areas of endemism. In this second clade all the areas located on the Pacific slope are included in one group, and the basal branch of this clade is constituted by the unique area located on the Gulf of Mexico slope. This general area cladogram shows the proposed historical relationships among tropical areas of Mexico.

The study of Contreras-Medina et al. (2007) was based on 81 gymnosperm species of the genera *Ceratozamia*, *Dioon* and *Pinus*, applying BPA and paralogy-free subtrees methods. In the case of the Mexican provinces, the unique relationship of both general area cladograms was the Mexican Plateau plus the Sierra Madre Oriental related with the Sierra Madre Occidental; the lowland provinces were mostly related, as in the study of Espinosa et al. (2006) with *Bursera*. Contreras-Medina and collaborators considered that both peninsulas of Mexico (Baja California and Yucatán) had a different history in relation to the continental portion of the country.

## 5. Conclusions

The MTZ represents a challenge for cladistic biogeography, due (in part) to its location at the boundaries between the Nearctic and Neotropical biogeographic regions, allowing the mixture of different biotic elements (Escalante et al., 2007), as suggested by previous studies from panbiogeographic approach (Contreras-Medina & Eliosa-León, 2001; Escalante et al. 2004; Morrone & Márquez, 2001). Also from a tectonic perspective, this region has formed through the convergence of three tectonic plates (North American, Cocos, and Caribbean), contributing to its geological complexity (Ortega et al., 1994).

After discussing the differences derived from applying different methods to carry out cladistic biogeographic analyses, Morrone & Carpenter (1994) concluded that when the data are "clean" (including few widespread taxa, redundant distributions, and missing areas), the results obtained using different methods are consistent. Some MTZ vicariance patterns, particularly those influenced by widespread taxa, redundant distributions and missing areas, appear to be ambiguous or the result of biogeographic noise. It remains difficult to select appropriate cladistic biogeographic methods because all have advantages and disadvantages, and all of them have been criticized (Morrone, 2009). One suggestion is to apply several methods and compare their differences and similarities (i.e., Contreras-Medina et al., 2007).

In addition, differences among the size and number of areas of endemism used in the different studies carried out in the MTZ make the search for congruent patterns of area relationships difficult. Unfortunately, it is not possible to make generalizations and comparisons among some clades obtained from the different studies commented upon here because, for example, some of the areas included in a clade of one study were not included in another, or in some studies wider units of analysis (areas of endemism) were used than those used in other studies. In some cases, different studies can be complementary, as occurred in the case of Marshall and Liebherr (2000) and Espinosa et al. (2006): the former was based mainly on mountainous taxa, whereas the latter was based mainly on lowland taxa. Notwithstanding, incongruent patterns should not be taken as signs of dispersal; they may be the result of more than one vicariance pattern and/or to the reticulate biogeographic history in areas of endemism (Morrone, 2009).

A much needed analysis in the near future is consensus on the unification of areas of endemism in the MTZ. This will greatly strengthen cladistic biogeography interpretation, as well as the generation of new morphological and molecular phylogenies of the biota of the MTZ. Both activities represent a challenge, but it is the only way to advance in the comprehension of the complicated historical biogeography of this region. As Flores-Villela & Goyenechea (2001) noted in the title of their paper, we are still "seeking the lost pattern of the biota inhabiting the Mexican Transition Zone."

# 6. Acknowledgments

Dragana Manestar invited us to present this contribution. David Espinosa and Othón Alcántara assisted us with the figures and with useful comments on the manuscript. Lawrence E. Stevens made useful suggestions to this chapter. Finantial support was given by DGAPA-PAPIIT IN221711.

# 7. References

Álvarez, E. & Morrone, J.J. (2004). Propuesta de áreas para la conservación de aves de México, empleando herramientas panbiogeográficas e índices de complementariedad. *Interciencia* Vol. 29, pp. 112−120, ISSN 0378-1844

Aguilar-Aguilar, R.; Contreras-Medina, R. & Salgado-Maldonado, G. (2003). Parsimony Analysis of Endemicity (PAE) of Mexican hydrological basins based on helminth parasites of freshwater fishes. *Journal of Biogeography*, Vol. 30, pp. 1861−1872, ISSN 0305-0270

Arriaga, L.; Aguilar, C.; Espinosa, D. & Jiménez, R. (eds). (1997). Regionalización ecológica y biogeográfica de México. Worshop developed in the Comisión Nacional para el Conocimiento y Uso de la Biodiversidad (CONABIO), November 1997.

Brooks, D.R. (1990). Parsimony analysis in historical biogeography and coevolution: Methodological and theoretical update. *Systematic Zoology*, Vol. 39, pp 14−30, ISSN 0039-7989

Brooks, D.R.; Van Veller, M.G.P. & McLennan, D.A. (2001). How to do BPA, really. *Journal of Biogeography*, Vol. 28, pp. 345−358, ISSN 0305-0270

Cevallos-Ferriz, S.R.S. & González-Torres, E. A. (2005). Geological setting and phytobiodiversity in Mexico. In: *Studies on Mexican Paleontology*, F. Vega, T.G. Nyborg, M.C. Perrilliat, M. Montellano-Ballesteros, S. Cevallos-Ferriz & S. Quiroz-Barroso (Eds.). Springer-Verlag. Dordrecht, The Netherlands. pp. 1−18, ISBN 1-4020-3882-8

Contreras-Medina, R. (2006). Los métodos de análisis biogeográfico y su aplicación a la distribución de las gimnospermas mexicanas. *Interciencia* Vol. 31, No. 3, pp. 176−182, ISSN 0378-1844

Contreras-Medina, R. & Eliosa-León, H. (2001). Una visión panbiogeográfica preliminar de México, In: *Introducción a la Biogeografía en Latinoamérica: conceptos, teorías, métodos y aplicaciones*, J. Llorente & J.J. Morrone (Eds.), 197−211, Universidad Nacional Autónoma de México-CONABIO, México, D. F. ISBN 968-36-9463-2

Contreras-Medina, R.; Morrone, J.J. & Luna-Vega, I. (2001). Biogeographic methods identify gymnosperm biodiversity hotspots. *Naturwissenschaften* Vol. 88, No. 10, pp. 427 – 430, ISSN 0028-1042

Contreras-Medina, R.; Luna-Vega, I. & Morrone, J.J. (2007). Gymnosperms and cladistic biogeography of the Mexican Transition Zone. *Taxon* Vol. 56, No. 3, pp. 905 – 915, ISSN 0040-0262

Craw, R. (1988). Continuing the synthesis between panbiogeography, phylogenetic systematics and geology as illustrated by empirical studies on the biogeography of New Zealand and the Chatham Islands. *Systematic Zoology* Vol. 37, pp. 291 – 310, ISSN 1096-0031

Crisci, J. V. ; Katinas, L. & Posadas, P. (2000). *Introducción a la teoría y práctica de la biogeografía histórica*. Buenos Aires, Sociedad Argentina de Botánica, ISBN 987-97012-4-0 (English translation: 2003, *Historical biogeography : An introduction*, Cambridge, Mass., Harvard University Press).

Croizat, L. (1958). Panbiogeography. Published by the author. Caracas. 1731 p. ISBN 978-0854860340

Croizat, L. (1964). *Space, time, and form: The biological synthesis*. Published by the author. Caracas. ISBN 978-0854860364

Escalante, T.; Rodríguez, G. & Morrone, J.J. (2004). The diversification of the Nearctic mammals in the Mexican transition zone: A track analysis. *Biological Journal of the Linnean Society*, Vol. 83, pp. 327 – 339, ISSN 1095 8312

Escalante, T.; Rodríguez, G.; Cao, N.; Ebach, M.C. & Morrone, J.J. (2007). Cladistic biogeographic analysis suggests an early Caribbean diversification in Mexico. *Naturwissenschaften*, Vol. 94, pp. 561 – 565, ISSN 0378-1844

Espinosa, D.; Morrone, J.J. & Llorente, J. (2006). Historical biogeographical patterns of the species of *Bursera* (Burseraceae) and their taxonomic implications. *Journal of Biogeography*, Vol. 33, pp. 1945 – 1958, ISSN 0305-0270

Espinosa, D. & Llorente, J. (1993). *Fundamentos de biogeografías filogenéticas*. Universidad Nacional Autónoma de México -CONABIO, México, D. F. ISBN 968-36-2984-9

Espinosa, D.; Morrone, J.J.; Aguilar, C. & Llorente, J. (2000). Regionalización biogeográfica de México: provincias bióticas. In: *Biodiversidad, taxonomía y biogeografía de artrópodos de México: hacia una síntesis de su conocimiento*, J. Llorente, E. González & N. Papavero (Eds.), Volume 2, Universidad Nacional Autónoma de México-CONABIO, México, D. F., pp. 61 – 94. ISBN 978-9683648570

Espinosa, D.; Llorente, J. & Morrone, J.J. (2006). Historical biogeographical patterns of the species of *Bursera* (Burseraceae) and their taxonomic implications. *Journal of Biogeography* Vol. 33, pp. 1945 – 1958, ISSN 0305-0270

Flores-Villela, O. & Gerez, P. (1994) *Biodiversidad y conservación en México: vertebrados, vegetación y uso de suelo*. Second Edition. Universidad Nacional Autónoma de México, México, D. F. ISBN 968-36-3992-5

Flores-Villela, O. & Goyenechea, I. (2001). A comparison of hypotheses of historical area relationships for Mexico and Central America, or in search for the lost pattern. Pp. 171--181 In: Johnson, J. D., Webb, R. G. & Flores-Villela, O. (eds.), *Mesoamerican herpetology: Systematics, zoogeography and conservation*. University of Texas, El Paso, Texas, ISBN 79968-0533

Halffter, G. (1987). Biogeography of the montane entomofauna of Mexico and Central America. *Annual Review of Entomology*, Vol. 32, pp. 95−114, ISSN 0066-4170

Hennig, W. (1966). *Phylogenetic systematics*. University of Illinois Press, Urbana IL, 280 p., ISBN 978-025-2068-140

Humphries, C.J. & Parenti, L.R. (1999). *Cladistic biogeography*. Oxford University Press, New York, ISBN 019-854818-4

Kluge, A.G. (1988). Parsimony in vicariance biogeography: A quantitative method and a Greater Antillean example. *Systematic Zoology* Vol. 37, pp. 315−328, ISSN 0039-7989

Liebherr, J.K. (1991). A general area cladogram for montane Mexico based on distributions in the Platynine genera *Elliptoleus* and *Calathus* (Coleoptera: Carabidae). *Proceedings of the Entomological Society of Washington*, Vol. 93, No. 2, pp. 390−406, ISSN 0013-8797

Luna-Vega, I.; Morrone, J.J.; Alcántara, O. & Espinosa, D. (2001). Biogeographical affinities among Neotropical cloud forests. *Plant Systematics and Evolution* Vol. 228, pp. 229−239, ISSN 0378-2697

Luna-Vega, I. & Contreras-Medina, R. (2010). Plant biodiversity hotspots and biogeographic methods. In: *Biodiversity hotspots*, Rescigno, V. & S. Maletta (Eds.), 181−191, Nova-Science Publishers, New York. ISBN 978-1-60876-458-7

Marshall, C.J. & Liebherr, J.K. (2000). Cladistic biogeography of the Mexican transition zone. *Journal of Biogeography*, Vol. 27, pp. 203−216, ISSN 0305-0270

Morrone, J.J. (1994). On the identification of areas of endemism. *Systematic Biology* Vol. 43, pp. 438−441, ISSN 1063-5157

Morrone, J.J. (1997). Biogeografía cladística: conceptos básicos. *Arbor* Vol. 158, pp. 373−388, ISSN 0210-1963

Morrone, J.J. (2001). *Biogeografía de América Latina y el Caribe*. SEA y M & T Tesis, Zaragoza, Spain, ISBN 84-922495-4-4

Morrone, J. J. (2005). Cladistic biogeography: identity and place. *Journal of Biogeography* Vol. 32, pp. 1281−1284, ISSN 0305-0270

Morrone, J.J. (2009). *Evolutionary biogeography: An integrative approach with case studies*. Columbia University Press, New York, ISBN 978-0-231-14378-3

Morrone, J. J. & Crisci, J. V. (1995). Historical biogeography: Introduction to methods. *Annual Review of Ecology and Systematics* Vol. 26, pp. 373−401, ISSN 0066-4162

Morrone, J.J. & Márquez, J. (2001). Halffter's Mexican transition zone, beetle generalized tracks, and geographical homology. *Journal of Biogeography* Vol. 28, pp. 635−650, ISSN 0305-0270

Morrone, J.J. & Escalante, T. (2002). Parsimony analysis of endemicity (PAE) of Mexican terrestrial mammals at different area units: When size matters. *Journal of Biogeography* Vol. 29, pp. 1095−1104, ISSN 0305-0270

Morrone, J. J. & Gutiérrez, A. (2005). Do fleas (Insecta: Siphonaptera) parallel their mammal host diversification in the Mexican transition zone? *Journal of Biogeography* Vol. 32, pp. 1315−1325, ISSN 0305-0270

Morrone, J. J.; Espinosa, D. & Llorente, J. (1996). Manual de biogeografía histórica. Universidad Nacional Autónoma de México, México, D. F. ISBN 968-36-4842-8

Morrone, J. J.; Espinosa, D.; Aguilar, C. & Llorente, J. (1999). Preliminary classification of the Mexican biogeographic provinces: A parsimony analysis of endemicity based on plant, insect, and bird taxa. *Southwestern Naturalist* Vol. 44, pp. 508–515, ISSN 0038-4909

Morrone, J.J. & Carpenter, J.M. (1994). In search of a method for cladistic biogeography: An empirical comparison of component analysis, Brooks parsimony analysis, and three-area statements. *Cladistics*, Vol. 10, No. 2, (June 1994), pp. 99–153, ISSN 1096-0031

Myers, N.; Mittermeier, R.A.; Mittermeier, C.G.; da Fonseca, G.A.B. & Kent, J. (2000). Biodiversity hotspots for conservation priorities. *Nature* Vol. 403, pp. 853–858, ISSN 0028-0836

Nelson, G. & Ladiges, P.Y. (1995). TAX: MSDOS computer programs for systematics. Published by the authors, New York and Melbourne.

Nelson, G. & Ladiges, P.Y. (1996). Paralogy in cladistic biogeography and analysis of paralogy-free subtrees. *American Museum Novitates* Vol. 3167, pp. 1–58, ISSN 0003-0082

Nelson, G. & Platnick, N.I. (1981). *Systematics and biogeography: Cladistics and vicariance.* Columbia University Press, New York, ISBN 0-231-04574-3

Ortega-Gutiérrez, F.; Sedlock, R. L. & Speed, R. C. (1994). Phanerozoic tectonic evolution of Mexico. In: *Phanerozoic evolution of North American continent-ocean transitions*, R.C. Speed (Ed.), pp. 265–306, The Geological Society of America, Boulder, Colorado. ISBN 978-0813753058

Page, R.D.M. & Lydeard, C. (1994). Towards a cladistic biogeography of the Caribbean. *Cladistics*, Vol. 10, No. 1, (March 1994), pp. 21–41, ISSN 1096-0031

Rosen, D.E. (1978). Vicariant patterns and historical explanation in biogeography. *Systematic Zoology* Vol. 27, pp. 159–188, ISSN 0039-7989

Sclater, P.L. (1858). On the geographical distribution of the Class Aves. *Journal of the Linnean Society of London, Zoology*, Vol. 2, pp. 130–145, ISSN 0368 2935

Siddall, M.E. & Perkins, S.L. (2003). Brook parsimony analysis: a valiant failure. *Cladistics* Vol. 19, pp. 554–564, ISSN 1096-0031

Van Veller, M. G. P. & Brooks, D.R. (2001). When simplicity is not parsimonious: a priori and a posteriori methods in historical biogeography. *Journal of Biogeography* Vol. 28, pp. 345–358, ISSN 0305-0270

# 11

# Rare and Endemic Species
# in Conacu-Negreşti Valley, Dobrogea, Romania

Monica Axini
*"Monachus" Group of Scientific Research and Ecological Education,*
*Romania*

## 1. Introduction

Since antiquity, man has been concerned with biological diversity. Initially, concerns were related to knowledge of living things in order to use them as sources of food, clothing, to treat various diseases or simply knowledge of the environment. With the accumulation of knowledge, man was concerned about the classification of life forms – first, in the form of empirical systems and later, as scientific systems (Bavaru et al. 2007; Bleahu, 2004).

However, officially, the notion of biodiversity emerged only in 1986. Launched as a purely scientific term in National Forum on Biological Diversity, held in Washington that year and organized by the National Research Council (NRC), the term was formalized in 1988 by entomologist E. O. Wilson in the book *"Biodiversity"*, a paper published in the proceedings of the Forum (Bavaru et al. 2007).

Although the term is now widely used in biology, ecology and biogeography, with the signing and entry into force of the Convention on Biological Diversity in Rio de Janeiro in 1992, the term has become a widely used concept, and more or less understood by specialists in other fields (economics, ethnography, law, political science), as well as politicians and public (Skolka et al., 2005; Bavaru et al. 2007).

Until recently, the term was understood only in the sense of specific diversity (all species, subspecies, races, etc. existing in the world). But today, the concept includes other components, such as: genetic diversity, eco-diversity, anthropogenic diversity.

Genetic diversity includes all genetic resources, expressed or not, included in the genetic heritage of life forms on Earth.

Eco-diversity (eco-system diversity, diversity of habitats and ecological systems of life support) is diversity viewed as a whole with the living environment. It includes ecological systems, all species senior biological systems and hydro-geomorphological sistems. It is an important consideration because the species can not protected individually but only in a whole with their environment (Bavaru et al. 2007; Bleahu, 2004; Skolka et al., 2005).

The role and place of man in nature, human evolution, and interconnections between human society and nature, are issues that have gained increasing recognition and stimulated interest in. Human diversity, including linguistic diversity, ethno-cultural diversity, socio-economic diversity (Bleahu, 2004; Botnariuc, 1989; Skolka et al., 2005).

Biodiversity issues in all its aspects must be addressed in terms of protection, rather than conservation, which implies a transition from static to a dynamic management, in which man does not simply collect and store museum artifacts, but is actively and consciously involved in stewardship of all life forms (Bavaru et al. 2007).

During the evolution of life, from its appearance on Earth more than 3.5 billion years ago, specific diversity has changed continuously, and microbial, plant and animal species in now number (5-50 million). Over geological time there have been numerous changes in earth`s crust, cosmic events, alternating periods of warm with many glacial periods. At least five mass extinction events have differentially selected for or against taxa (Skolka et al., 2005). Humans have played and continue to play a major role in extinction, and have initiaded a sixth major extinction event. This is particulary alarming as the many species disappear before they are even scientifically recognized, and natural ecosystems are deteriorating rapidly.

Experts estimate that about 50.000 species worldwide are lost to extinction each year, many of which have not even described (Bavaru et al. 2007; Bleahu, 2004; Botnariuc, 1989; Skolka et al., 2005).

Ecosystems and habitats sustain multiple damages, including desertification, overexploitation of natural resources, the introduction of alien species, expansion of agro-ecosystems and intensive agricultural practices that are detrimental to the environment, the appearance and expansion of urban anthropogenic ecosystems, pollution and other impact. All these are examples of cases with serious consequences locally, regionally and globally (Bavaru et al. 2007; Bleahu, 2004; Botnariuc, 1989; Skolka et al., 2005).

Here I describe the geography, geology, and biogeography of the Dobrogea Region and specifically the Conacu-Negreşti Valley in eastern Romania. I describe the biota that occupy the valley, and describe in detail the ecology and biogeography of a dozen of its little-known rare or endemic plant species. This is a poorly known valley that harbors numerous rare and unique species, and deserves more recognition for its wealth of life and unique natural heritage.

## 2. Dobrogea region

Dobrogea is located on the northern Balkan Peninsula in southeastern Central Europe – (44°17´03,77´´N, 28°21´53,27´´E, Figures 1, 2). It occupies an area of approximately 23.142 km², of which 15.570 km² are located in Romania (making up 6.52% of the total area of Romania; Skolka et al., 2005; Dobrogea, 2011) and 7.572 km² in Bulgaria (Dobrici Region, 2011; Silistra Region, 2011). The Dobrogea Region is bordered by the lower Danube River to the southwest, west, northwest and north, the Danube Delta to the northeast, the Black Sea to the east and the Ludogorie Plateau to the southeast and south (Peahă, 1982). The most easterly point of Dobrogea lies at 29°41´24´´ east longitude, corresponding area of the Sulina, Romania.

North Dobrogea is located in the south –eastern of Romania and is composed of two subunits, each with distinctive physical-geographical, soil and climate features: continental Dobrogea (Dobrogea Plateau) and maritime Dobrogea (Peahă, 1982; Skolka et al., 2005).

Fig. 1. Map of Europe (from Google Earth) – showing the position of Dobrogea and Conacu-Negrești Valley in Europe

Fig. 2. Geographical position of the Conacu-Negrești Valley in the Dobrogea region (area circled in yellow) (image from Google Earth)

The Dobrogea Plateau is divided into three geographical units: Northern Dobrogea with an average altitudes of 200 m, Central Dobrogea, and Southern Dobrogea, the latter two units with average altitudes of 100 m.

North Dobrogea encompassed several distict geographic subunits: the Măcin Mountains, the Niculițel Plateau, the Tulcea Hills and the Babadag Plateau. The Măcin Mountains are the oldest mountain range in Europe, formed during the late Paleozoic Hercynian orogeny, with a maximum altitude is 476 m on Țuțuiatu Peak or Greci Peak (Axini M., 2009).

Central Dobrogea contains the Casimcea Plateau, the oldest topographic relief in Dobrogea, formed during the early Paleozoic Caledonian orogeny.

South Dobrogea includes the Oltina and Negru-Vodă Plateaus.

The maritime portion of Dobrogea contains the Danube Delta (a fluvial-marine plain undergoing continuous genesis), the Razelm-Sinoie Lagoon complex, and the supra-littoral zone of the Black Sea (Peahă, 1982; Skolka et al., 2005).

## 3. Conacu-Negreşti Valley

Conacu-Negreşti Valley is located in the center of Cobadin Plateau, subunit of Negru -Vodă Plateau (Iana, 1970), falls within the following geographical coordinates: the parallel 43°58'48,93'' north latitude and the meridian 28°10'05,12'' east longitude (Figures - 1-3). This location in south-eastern Romania explains, the continental climate characteristics of the regions, which influences all environmental characteristics (Axini, 2006, 2009, 2011b; Brezeanu, 1997; Coteţ, 1969).

The landscape has formed predominantly on Cretaceous and Sarmatian limestone, on Precambrian basement lithology and covered by a thick blanket of 40 m of Quaternary loess. The Preterozoic foundation is composed of crystalline schists and sedimentary superstructure, which are distinguished by two types of Paleozoic-Mesozoic and Neozoic formations.

Paleozoic Silurian formations are composed of clay schists with Devonian diaclases, consistion of thick marl clay, marl-limestone, etc. Mesozoic formations are composed of alternating calcareous and detritic deposits. Jurassic Period strata are composed of alternating limestone and diatomites. The Cretaceous period is represented by reef limestone and, marl-limestone, with sand, glaucenic sandstones, and microconglomerates in the middle strata, and marl clay at the base. Uppermost strata are dominate by debris facies with calcareous sandstones, microconglomerates, chalk, etc. These formations have been subjected to folding with foundation blocks revealing wavy structure.

In South Dobrogea, the Conacu-Negreşti Valley is located on a north-west to south-east axis. It is a *"canara"* (Iana, 1973), a term specific to Dobrogea, and meaning a valley generally short, narrow, with limestone slopes, high and steep walls with small caves, partly covered with SubMediterranean xerophyte meadows and scrub forest vegetation.

The Conacu-Negreşti Lake is of recent geological origin, formed by natural damming. Sixty years ago catastrophic flooding occurred after heavy rains, and water from Plopeni Lake located in the south and at an altitude higher than Conacu-Negresti Valley, overflowed into it. Alluvium was deposited by the flood, and the lake formed and is maintained by groundwater sources in the valley and precipitations (Basarabeanu, 1969; Gâştescu & Breier, 1969; Godeanu, 2002; Axini, 2011b).

Until 2003, the valley was known only from the geographical and geological field studies and research of Mrs. Dr. Sofia Iana, University of Bucharest, on the Negru -Vodă Plateau. In 2003, the "Monachus" Group of Scientific Research and Environmental Education, Constanta, Romania developed a study and research project on the biodiversity of the Conacu-Negresti Valley. The „Monachus" G.S.R.E.E. effort, became a permanent research program, undertaking new research and public education projects for environmental protection of this valley.

## 4. Biodiversity of Conacu-Negreşti Valley

Conacu-Negreşti Valley, part of South Dobrogea, is distinguished by spectacular landscape beauty and is characterized by rich and diverse assemblage, with many rare or endemic species specific to the Dobrogea Province. Its significance also is derived from its geological, geomorphological and paleontological characteristics.

The valley previously was dominated by one dense forest of pubescent oak (*Quercus pubescens*), the remnants of which can be seen on the lake bottom. The proof that the valley was dominated by an oak forest is the angle of inclination of both limestone walls of the valley and the presence of herbaceous and shrub plant and invertebrates species, which normally live together in a oak forest (Axini et al., 2010).

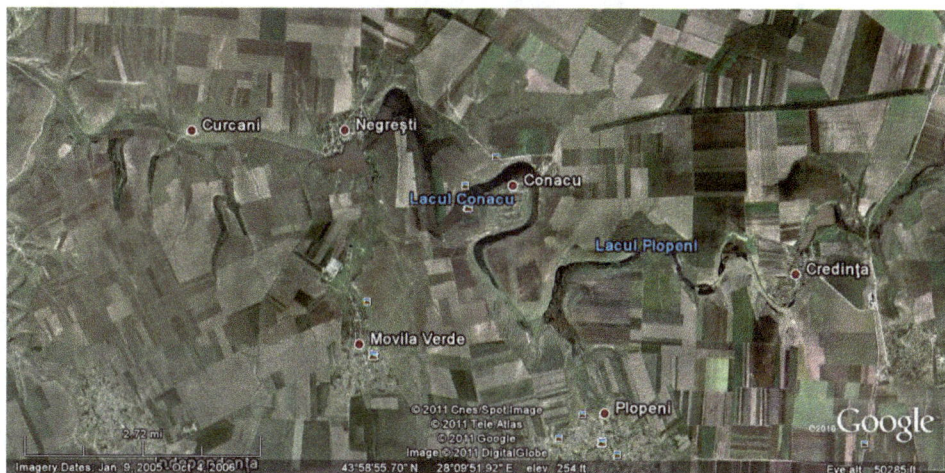

Fig. 3. Conacu-Negreşti Valley (from Google Earth)

Thus far, we have detected a total of 301 plant species, among 32 orders and 62 families (Andrei, 2003; Axini, 2009a, 2009b; Ciocâlan, 2000a, 2000b; Cristurean & Liţescu, 2002a, 2002b; Morariu & Todor, 1972; Prodan, 1935, 1936a, 1936b; Prodan & Buia, 1966; T. Săvulescu (ed), 1952-1976; Todor, 1968; Tutin et al. (ed), 1964-1980a, 1964-1980b, 1996).

In numerical order, the families represented the largest number of species are: *Compositae* - 42 taxa, *Graminaceae* - 30 taxa, *Labiatae* - 25 taxa, *Caryophyllaceae* - 14 taxa; *Rosaceae*, *Leguminosae* and *Cruciferae* families, each of them with - 13 taxa (Figure 4).

The dominant shrub species in the valley include: *Prunus spinosa* L. and *P. (Padus) mahaleb* (L.) Borkh. (Fam. Rosaceae Juss.), - both shrub species found only the large canyon (one of the two canyons in the south-west of the valley); *Crataegus monogyna* Jacq. (Fam. Rosaceae Juss.), found on the plateau, on walls with southeast and northeast aspect (all three in the central valley), and in the large canyon of the south-west of the valley; and *Cornus mas* L. (Fam. Cornaceae Link.), found only in the small canyon (other canyon in the south-west of the valley, which is lower compared to that mentioned above) (Axini, 2011b).

Invasive plant species also occur in the region, including: *Ailanthus altissima* (Mill.) Swingle (Fam. Simaroubaceae L. C. Rich.), *Elaeagnus angustifolia* L. (Fam. Elaeagnaceae R. Br.),

*Gleditshia triacanthos* L. (Fam. Cesalpiniaceae), *Hedera helix* L. (Fam. Araliaceae Vent.), *Robinia pseudacacia* L. (Fam. Leguminosae Juss.), and *Morus* sp. (Fam. Moraceae Lindl.). These species occur on the limestone walls, plateaus, grassy hills and in the canyons of the southwest part of the valley (Axini M., 2011b).

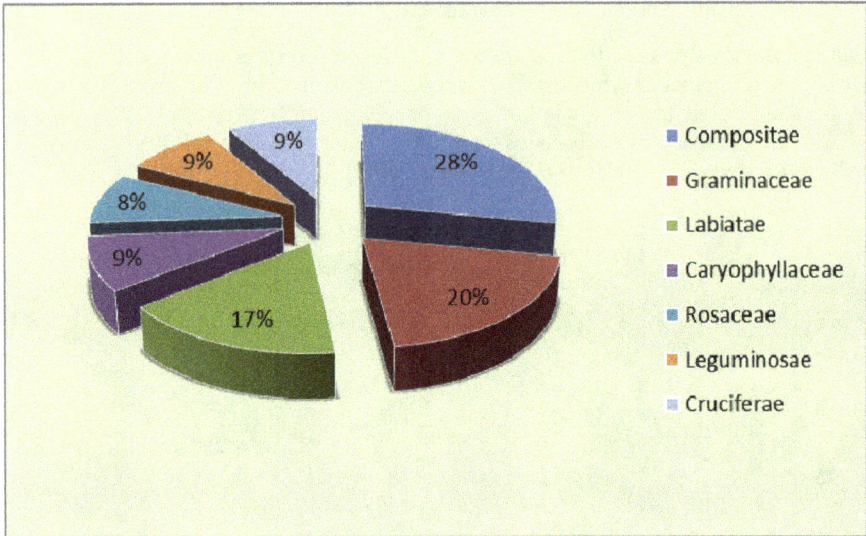

Fig. 4. The proportion of plants families with the highest number of species identified in the valley so far

Conacu-Negreşti Valley hosts many species of invertebrates, of which we so far have identified a total of 101 terrestrial and aquatic species, belonging to 17 orders and 57 families (Figure 5). Of these, insects dominate with 70 taxa, followed by gastropods with 22 taxa (Axini, 2006, 2009b, 2011a, 2011b; Axini & Skolka, 2010; Axini et al., 2010; Chiriac & Udrescu, 1965; Müller, 2002a, 2002b; Müller & Tomescu, 2002; Skolka, 2002, 2008). Of all groups of insects identified, the dominant order is Diptera, including nine families, followed by the Heteroptera with seven families, and Coleoptera with six families. The Lepidoptera and Himenoptera each are reprezented by four families (Figure 7; Skolka, 2002, 2008; Skolka et al., 2005). Among the Heteroptera, only the aquatic taxa have been identified thus far. Among those, *Hebrus pusillus* Fallen (Hebridae) is a rare species, known so far only from the Danube Delta (Axini, 2006, 2009b, 2011b; Axini & Skolka, 2010; Kiss, 2002).

Of the gastropods, identified thus far, the 22 species include, both terrestrial and aquatic taxa among, nine families in three orders (Figure 6; Axini, 2006, 2009b, 2011a; Axini & Skolka, 2010; Grossu, 1986, 1987; Müller, 2002a; A. Negrea, 2002). Of these Mollusca, two are endemic, seven are relicts, and two taxa are rare. One species (*Helix pomatia* Linnaeus - Helicidae) is of European importance and included on the Habitats Directive and Bern Convention lists.

Coleoptera are well represented in Dobrogea fauna, with both terrestrial and aquatic forms. In terms of species from Conacu-Negreşti Valley, they are little known to date with - 11 species belonging to 6 families. Among those species identified, 4 species are important: *Platambus maculatus* Linnaeus is of Euro-Central Asian afinity, *Hydroporus dobrogeanus* Ieniştea is endemic to Dobrogea, and *Cybister lateralimarginalis* De Geer is of Euro-Siberian afinity, (all three species are in the Dytiscidae), and *Cerambyx cerdo* L. (Cerambycidae) is an **endangered** and **rare** species, characteristic of wooded areas (Figure 8). This species is affiliaded with the oak forest in the valley (Cojocaru & Popescu, 2004; Niţu & Decu, 2002; Panin, 1957; Panin & N. Săvulescu, 1957).

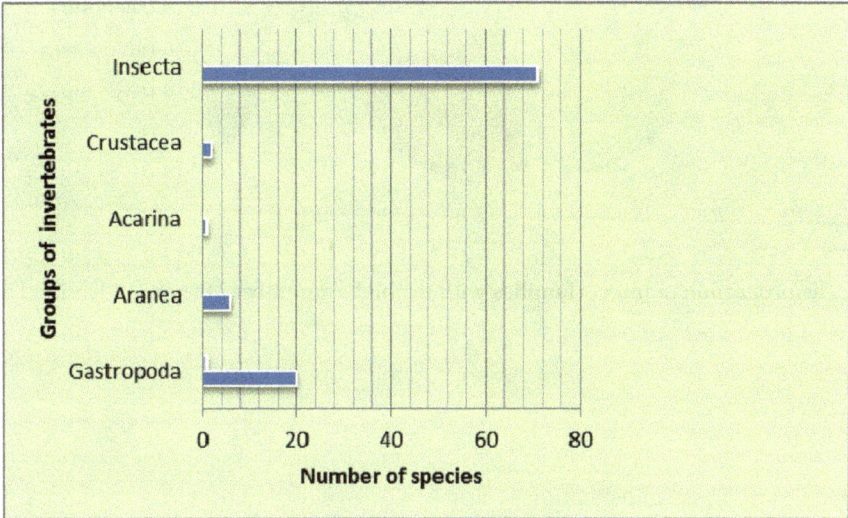

Fig. 5. Groups of invertebrates identified to date in the valley

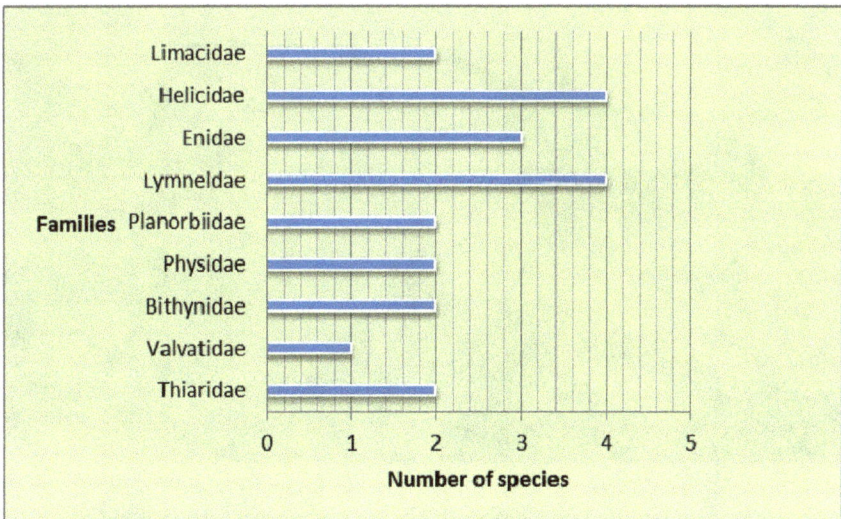

Fig. 6. Families of gastropods found in the valley

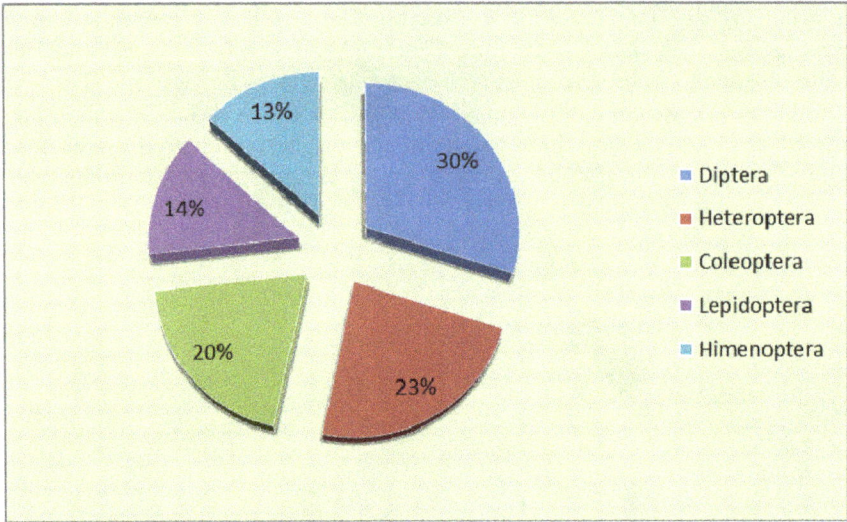

Fig. 7. The proportion of insects families with the highest number of species identified in the valley

Fig. 8. *Cerambix cerdo* identified on a grassy hill in the south-west corner of the valley (photo. M. Axini, 2009)

Lepidoptera are numerous species in Dobrogea, but are less studied in the Conacu-Negreşti Valley – 7 species belong to 4 families. One species larvae are adapted to aquatic environments - *Parapoynx stratiotata* L. (Fam. Crambidae). It was identified only in the Danube Delta (Căpuşe, 1968; Ruşti, 2002) and its presence in the lake indicated that the valley is a former arm of the Danube River. This, also is evident from field observations on the shape and orientation of the valley and canyons and geological data.

Until now, a total of 150 species of vertebrates have been detected, including: 13 species of fish, 7 species of amphibians, 12 species of reptiles, 94 species of birds, and 24 species of mammals) (Figure 9).

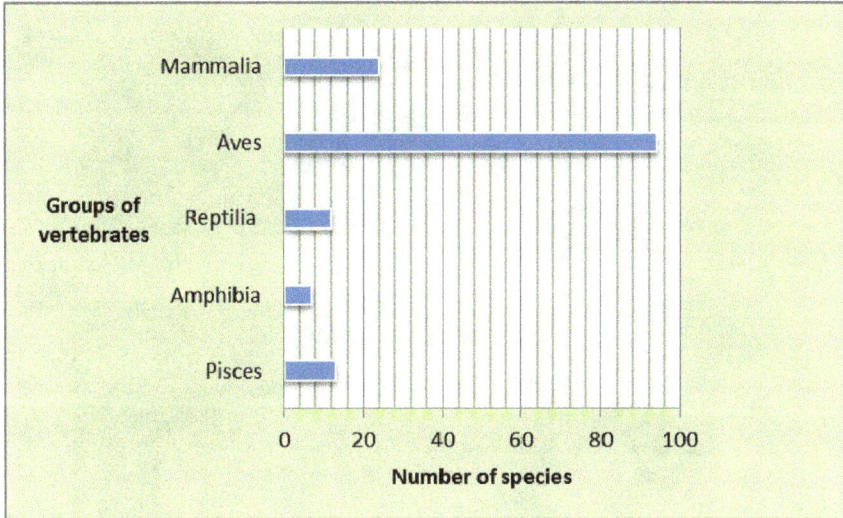

Fig. 9. Groups of vertebrates identified to date in the valley

Reptile populations are well represented in the Conacu-Negreşti Valley (Iftimie, 2001, 2005). Lizards include: *Lacerta (Podarcis) taurica* Pallas (Lacertidae) is abundant and occupies numerous habitats - on the walls of limestone cliffs in the day, on limistone tablelands, in the southwest canyons of the valley (Figure 10); *Lacerta trilineata dobrogica* Fuhn et Mertens (Lacertidae), a rare and localized - species characteristic of Dobrogea (Figure 11); *Lacerta (Podarcis) muralis maculiventris* (Lacertidae) is a **rare** and localized species (Figure 12). Turtles are represented by *Testudo graeca ibera* Linnaeus (Fam. Testudinidae Gray), a **vulnerable** species in Europe (Figure 13) and snakes are represented by *Dolicophis caspius* Gmel. (Colubridae Boulenger) , a **vulnerable** species in Romania, what inhabits limestone walls.

*Podarcis muralis* subsp. *maculiventris* was found only in wooded areas and with rocky walls in southern and northern Dobrogea, while in the Conacu-Negreşti Valley it inhabits only northeast facing limestone wall in the middle of the valley. *Lacerta (Podarcis) muralis* is a **rare** and localized species, in Dobrogea by subspecies *maculiventris*.

*Testudo graeca ibera* is a Dobrogean turtle, monument of nature, is abundant on plateaus and also in the canyons. The same cannot be said about *Emys orbicularis*, the water turtle, which is declining due to anthropogenic factors.

Fig. 10. *Lacerta taurica* from the limistone plateau in central valley (photo. M. Axini, 2009)

Fig. 11. *Lacerta trilineata dobrogica* (photo. M. Skolka, 2009)

The number of birds identified thus far in the valley include 32 families. Larks are most abundant but are more or less sedentary, migrating only in very cold winters. However, lark populations from Conacu-Negresti Valley are in numerical decline due to the anthropogenic activities.

Among the bird species of conservation concern identified in the valley thus far are: *Ardeola ralloides* Scopoli, *Ixobrychus minutus* Linnaeus (Ardeidae), *Ciconia ciconia* Linnaeus

(Ciconidae), *Burhinus oedicnemus* Linnaeus (Burhinidae), *Sterna hirundo* Linnaeus (Sternidae), *Melanocorypha calandra* Linnaeus (Alaudida), *Lanius minor* Gmelin (Laniidae), *Dendrocopos syriacus balcanicus* Hemprich et Bibr. (Picidae) (Bănică, 2006; Birds Directive; Flocea, 2004; Munteanu, 2005).

Fig. 12. *Lacerta (Podarcis) muralis maculiventris,* from the limestone wall of the central valley, with north-eastly aspect (photo. M. Skolka, 2009)

Fig. 13. *Testudo graeca ibera* from the limestone plateau in the central valley (photo. M. Axini, 2009)

Among the mammal species identified in the valley is the ground squirrel, *Citellus citellus* Pallas (Sciuridae), which is an endemic species to Europe and which inhabits steppe grasslands. Unfortunately, the species is in decline due to expansion of agricultural activities. In the valley, it occurred until recently as a large population, but currently it is in decline due to its popularity (recreation area) that led to the increasing number of tourists and fishermen. It is a species of conservation interest and require more rigurous protection (Habitats Directive, Annex 2, 4 and Natura 2000). The same applies to the mole *Talpa europaea* (Botnariu & Tatole, 2005; Habitats Directive; Iordache et al., 2004; Murariu, 2005). The ground squirrel occupies fragmented habitat by the Carpathian Mountains in: Pannonian Basin and the southeastern part (southern Romania, Bulgaria, Moldova and Ukraine). It is a critically endangered species due to the pronounced population declines and requires more rigorous protection (Habitats Directive, Natura 2000).

Among carnivorous mammals, foxes - *Vulpes* sp. - are more common and manages to survive despite human impacts on their habitats. They burrow in sandy canyon walls. In contrast, wolves, *Canis lupus* Linnaeus, are in decline throughout southern Dobrogea due to exntense deforestation. In the past, Cobadin had extensive forests of oak, and a large population of wolves. Records of *Canis aureus* exist, which entered the study area from Bulgaria.

## 4.1 Rare and endemic plants of Conacu-Negreşti Valley

A total of 12 rare and endemic plant species have been detected in the valley (Table 1; Dihoru & Dihoru, 1994; Dihoru & Negrean, 2009; Făgăraş et al., 2010; Olteanu et al., 1994).

| Orders | Families | Species |
|---|---|---|
| Ranunculales | Ranunculaceae | *Adonis volgensis* Steven ex DC. |
| Urticales | Urticaceae. | *Parietaria serbica* Panč |
| Centrospermae | Caryophyllaceae | *Dianthus pseudoarmeria* M. Bieb. |
| | | *Minuartia bilykiana* Klokov in Kotov |
| | | *Silene exaltata* Friv. |
| Ligustrales | Oleaceae | *Jasminum fruticans* L. |
| Tubiflores | Labiatae | *Satureja coerulea* Janka in Velen |
| Sinandrales | Compositae | *Achillea clypeolata* Sibth. et Sm |
| | | *Centaurea napulifera* Roch. |
| | | *Scolymus hispanicus* L. |
| Liliales | Liliaceae | *Ornithogalum oreoides* Zahar. |
| Graminales | Graminaceae | *Koeleria lobata* (M. Bieb.) Roem. & Schult. |

Table 1. Rare and endemic plant species identified from the Conacu-Negreşti Valley

### 4.1.1 *Adonis volgensis* Steven ex DC. (Ranunculaceae)

*Adonis volgensis* Steven ex DC. (syn. *Adonis vernalis* L. subsp. *volgensis* (Steven ex DC.) Korsh; *Anemone cicutaria* Gandoger; *Adonanthe volgensis* (Steven ex DC.) Chrtek & Slavíková) (Figure 14) is a perennial plant, up to 45 cm high. Its leaves have toothed lobes; and reproduces sexually, with direct or/and indirect pollination by insects. Its fruits and seeds are dispersed by ants. This species blooms in April-May (Szabó, 1973).

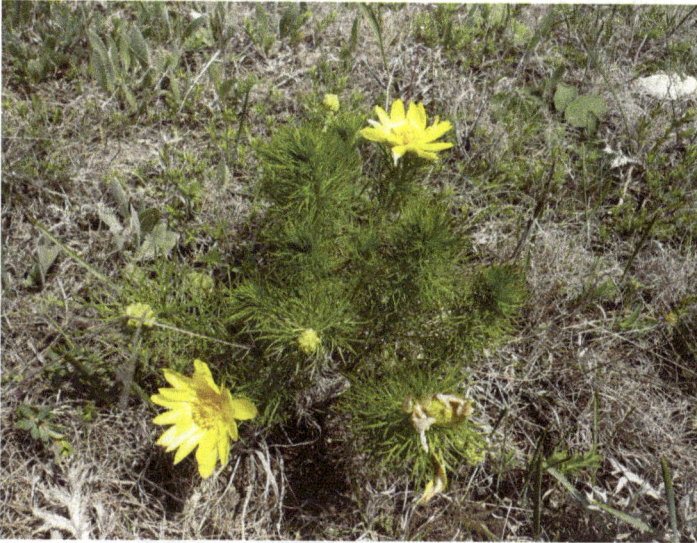

Fig. 14. *Adonis volgensis* (photo. M. Axini, 2009)

It is a heliophyte and xerophyitici species, widespread in the plains and hills, on dry, sunny soils that are neutral in pH and low in mineral nitrogen soils, in steppe grasslands (Szabó, 1973).

In Romania, it was occurs in: Transylvania, Moldavia and Dobrogea. In Dobrogea it was previously known only from the northern region, between Jurilofca and Gagebischi Liman, where it was identified by G. Grinţescu, Romanian botanist, under the name *Adonis villosa-vernalis* (copy found in Herbarium of Romanian Institut of Biology of Romanian Academy in Bucharest, BUCA), and it was found Tulcea (identified by the same botanist as the *A. walziana*). In the southern region, it occurs near resorts close to the Conacu-Negreşti Valley(Basarabi, Fântâniţa-Murfatlar Reserve and Cobadin) (Dihoru et al., 1965; Grecescu, 1898; Zahariadi, 1964).

In Conacu-Negreşti Valley, *Adonis volgensis* was detected only on a limestone plateau (43°59'19,11" N, 28°19'42,60"), approximately in the center of the valley. On the plateau edges, in the east, southwest, southeast, east, agro-ecosystems may threaten this species (Figure 15). There, it co-occurss with: *Asperula tenella* Heuff. ex Degen (Rubiaceae), *Adonis flammea* Jacq. (Ranunculaceae), etc..

*Adonis volgensis* is continental element, with geographical range in south-eastern Europe, from southeastern Hungary to the Central Ural Mountains, northeast Anatolia and Central Asia (Tutin et al., 1964-1980, 1996).

The scientific and practical importance of *Adonis volgensis* are high, because it is the medicinal and toxic plant, and it listed as vulnerable (VU), in Romania at the southwestern limit of its distribution.

Unfortunately, *Adonis volgensis* is threatened by reduction and degradation of its habitats by grubbing and overgrazing, expansion of agricultural area and exccessive collections. These impacts haves, lead to population reduction. To these impacts can be added natural factors related to the biology of the species, with low seeds production, failure of vegetative propagation, and occasional attack by microscopic fungi.

Fig. 15. Limestone plateau that supports *Testudo graeca ibera* and *Adonis volgensis* (Google Earth)

### 4.1.2 *Parietaria serbica* Panč (Urticaceae)

*Parietaria lusitanica* L. subsp. *serbica* (Pančić) P.W. Ball (syn. *P. serbica* Pančić, *P. chersonensis* Grec.; Figure 16) is an annual plant (T), 5-30 cm tall, that reproduces sexually and has indirect anemophilous pollination and fruit and seeds dissemination in its habitats. It blooms from May – September (Dihoru & Negrean, 2009).

*Parietaria serbica* lives on rocks and stones, and is shade-loving, often occurring in cracks in rocks. It is a calciphile, that grows in hilly and sub-mountainous areas. It does not typically co-occur with other species in the low quality, rocky soils.

In Romania, *Parietaria serbica* is widespread, especially in the south-west and southeast. In south-western Romania it has been detected in the Danube and Sohodol Gorges, since 1870, in various forms: *P. lusitatica*, *P. lusitatica* var. *chersonensis*, *P. chersonensis*, *P. serbica* (Păun et al., 1970; Păun & Popescu, 1978). It is known in Dobrogea from the central (Gura Dobrogei; Cheia) and southern regions (Canaraua Fetii Forest/Forest Reserve; Esechioi Forest Reserve; among rocks of Independența; in the Snakes Valley of Hagieni Forest; and among rocks, on banks and near sulphurous waters from Mangalia, near Tatlageac Lake) (Andrei & Popescu, 1966; Horeanu, 1973, 1976 a, 1976 b; Mihai et al., 1964; Ciocârlan, 2001; Dihoru & Negrean, 2009).

In the Conacu-Negrești Valley, *Parietaria serbica* was identified in a small canyon (43⁰59'34,68" N, 28⁰10'54,25" E), at the entrance to Conacu Village - in the central eastern part of the valley, with north-western aspect (Figure 17) and a limestone wall with north-eastly

Fig. 16. *Parietaria serbica* from a small canyon in the eastern valley (photo. M. Axini, 2009)

Fig. 17. A small canyon, located at the entrance of Conacu Village, that supports *Parietaria serbica* (Google Earth, 2011)

aspect – near the central - west valley (43⁰58'55,85'' N, 28⁰09'38,10'' E) (Figure 18). The existence of cattle, sheep and goat herds may endanger the future existence of this species (Figure 17). There, it co-occurs with: *Asplenium ruta-muraria* L. (Polypodiaceae), *Minuartia bilykiana* Klokov in Kotov (Caryophyllaceae), *Sanguisorba minor* subsp. *balearica* (**Bourg. ex Nyman) Muñoz Garm. & C. Navarro** (Rosaceae) and other species.

*Parietaria serbica* is a Daco-Moesian-Dobrogean element, with **a range** in the northern part of Balkan Peninsula and Romania. **This subspecies** is endemic to Europe (Dihoru & Negrean, 2009). It is a threatened species (EN) (IUCN, 2010), of considerable scientific importance of the species due to its extreme rarity and small range in Europe. The species has practical importance as an herb with the same therapeutic properties as *Parietaria officinalis*.

Fig. 18. A limestone wall in the central valley (Google Earth, 2011)

*Parietaria serbica* is a short, annual species, that grows in small populations in rocks caverns, with low power propagation It is a plant that needs to rebuild their herd seasonally. The specimens of Romania is at the northern limit of the area. All this shows the degree of endangerment of the species.

Conservation measures include: prevent quarrying in its habitats, its conservation within protected areas, conservation of seeds in gene banks and growning the species in botanical gardens. In Romania is present in protected areas the Porțile de Fier Natural Park and Sohodol Gorges in southwestern Romania, and in Hagieni Forest, Canaraua Fetii, Gura Dobrogei, Cheia Natural Reserves in the territory of Dobrogea.

### 4.1.3 *Dianthus pseudoarmeria* M. Bieb. (Caryophyllaceae)

*Dianthus pseudoarmeria* M. Bieb. is a annual-biennual plant (T-TH), short hairy, and without vegetative shoots. It has dense array of pink flowers, with sexual reproduction through direct entomophilous pollination and well as anemophily, and through zoochory („eating" seeds by animals and their dissemination through faeces material). It blooms in June-July

(Dihoru & Negrean, 2009). *Dianthus pseudoarmeria* is a heliophyte, termophyte, calciphili species (Dihoru & Doniţă, 1970; Dihoru & Negrean, 2009).

In Romania, it has been detected in Dobrogea and the Hanu Conachi (Galati). In Dobrogea, it has been detected: on Măcin Mountain, to Greci, to Cerna, on Pricopan Hill; to Babadag-Caugagia; the Dolojman Peninsula and on Saele Litoral Dunes (last, located in the Danube Delta Biosphere Reserve); the Histria Ancient City; on Capidava Hill; to Hârşova; to Medgidia; to Basarabi; in the Hagieni Forest, in southern of Cotul Văii, in Mare Valley and in Topolog Valley (Cristurean & Ţeculescu, 1970; Prodan, 1935).

In Conacu-Negreşti Valley, *D. pseudoarmeria* was identified on a south - facing limestone wall in the north-western portion of the valley ($44^000^021,29^{\circ}$ N, $28^008^051,34^{\circ}$ E; Figure 19). It also was reported on a grassy hill at the bottom of the lake, in the south-west ($43^058^038,28^{\circ}$ N, $28^010^015,43^{\circ}$ E; Figure 20) and in the large canyon in the south-west of the valley (.($43^058^003,98^{\circ}$ N, $28^010^030,27^{\circ}$ E; Figure 20).

In these settings, it co-occurs with: *Achillea setacea* W. et K., *A. coarctata* Poir and *Carduus thoermeri* Weinm. (Compositae), and other species.

Fig. 19. *D. pseudoarmeria* location on the limestone wall in the north-westest portion of the valley (image from Google Earth)

It is a Pontic-Tauric-Balkan element, with an range centered in south-east Europe (Dihoru & Negrean, 2009). It is a low-risk species (LR) (IUCN, 2011).

It is an important ornamental plant and is grown in flower gardens.

Currently, *D. pseudoarmeria* is listed within the territory of protected areas, including: Hanu Conachi, Hagieni, Fântâniţa-Murfatlar, Măcin Mountain National Park, Danube Delta Biosphere Reserve. This species deserves protection due to habitat loss, and because it is parasitized by microbial fungi. *"Ex situ"* cultivation could be achieved in botanical gardens.

Fig. 20. *D. pseudoarmeria* locations on the limestone hill and the canyons of the south-west valley (Google Earth, 2011)

Fig. 21. *Minuartia bilykiana*, on a limestone wall (photo. M. Axini, 2009)

### 4.1.4 *Minuartia bilykiana* Klokov in Kotov (Caryophyllaceae Family Juss.)

*Minuartia bilykiana* Klokov in Kotov (syn. *M. tenuifolia*) (Figure 21) is a low-growing, glandulous, annual plant, less than 10 cm tall. It is patent-branched and reproduces sexually through entomophily, and with anemochory and zoochory dispersal. The stamens mature before the stigmatas. *M. bilykiana* presents hermaphroditic flowers, with female reproductive organs mature before the male reproductive organs. It blooms from May-July (Dihoru & Negrean, 2009).

It is a plant of hills, a heliophytic, thermophyte, who living on dry soils that are low in nitrogen, and in xerophytic meadows and on rocks (Dihoru & Negrean, 2009; Zahariadi, 1965).

In Romania, it has been documented thus far only in Dobrogea northeast of the town of Baia, Agighiol and Dolojman Peninsula, between 1966 and 2000, by Romanian botanist Gavril Negrean. In central and southern Dobrogea, it was identified, between 1961-2004, by the same botanist in north of the town of Topalu; on Allah-Bair Mountain; in the Seid-Orman Valley, near the towns of Târguşor and Palazu Mic; to Basarabi, Fântâniţa; to Techirghiol; in Hagieni Forest, Şerpilor Valley; to Mangalia, and the south-southeast of Cotul Văii, Mare Valley, „la Ic". In 1930 and 1933, Romanian botanist E. I. Nyárády identified this species on Allah-Bair Mountain and in Chirişlic, as *M. villosa*. Romanian botanist C. Zahariadi botanized at two resorts in southern Bassarabia (now Moldova), in 1929 and 1933, identifying it as *M. tenuifolia*. In 1965, he synonomized *M. tenuifolia* with *Minuartia bilykiana*.

In Conacu-Negreşti Valley, *M. bilykiana* was identified on a southeast – facing limestone wall in the central valley (43°59´39,02¨ N, 28°10´38,98¨) (Figure 22). It also has been identified in the small canyon (43°59´34,68¨ N, 28°10´54,25¨ E), located in the entrance of Conacu Village (Figure 17). It co-occurs with: *Satureja coerulea* Janka (Labiatae), *Ornithogalum oreoides* Zahar. (Liliaceae), and other species – on the limestone wall, and with: *Asplenium ruta-muraria* L. (Polypodiaceae), *Sanguisorba minor* subsp. *balearica* **(Bourg. ex Nyman) Muñoz Garm. & C. Navarro** (Rosaceae), *Parietaria serbica* Panč. (Urticaceae), and other species – in the small canyon.

It is a Scythian element and is European endemite, with a range limited by the northern Black Sea (Dihoru & Negrean, 2009). Although it is a low-risk (LR) species (IUCN, 2011), its scientific importance is quite high, because it is a rare and ephemeral species, requiring future taxonomic research.

It requires protection because of reduced habitat availability and because it is parasitized by species of microbial fungi. As a conservation measure, *"ex situ"* cultivation in botanical gardens would help protect this species.

### 4.1.5 *Silene exaltata* Friv. (Caryophyllaceae)

*Silene exaltata* Friv. (syn. *Otites exaltata* (Friv.) Holub.) is a biennual to perennial plant species up to 2 m tall, with sexual reproduction and with entomophilous pollination and with anemochory dispersal. *S. exaltata* blooms from May to July (Dihoru & Negrean, 2009) and it is a heliophytic, thermophyte that lives on dry, neutral soils, on limestone slopes (Dihoru & Negrean, 2009).

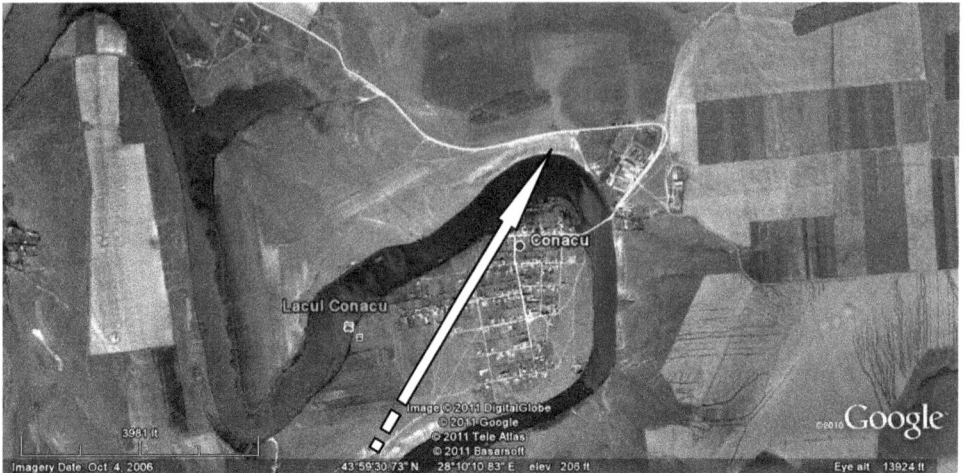

Fig. 22. Location of *Minuartia bilykiana* in the study area on the southeast –facing limestone wall of the central valley (Google Earth, 2011)

In Romania, *S. exaltata* been identified thus far only in Dobrogea. It was reported between 1978 and 1983 by the Romanian botanist Gavril Negrean in the northern Dobrogea: on the Sărăturile Coastal Dunes from Sfântu Gheorghe, east of Enisala Ancient City, in the southeastern Heraclea Ancient City, north of „Caramanchioi" and on Dolojman Peninsula. In southern Dobrogea, it was detected in Agigea, Basarabi and in the Hagieni Forest (in "Cascaia"). Currently, *S. exaltata* is protected within Danube Delta Biosphere Reserve, Fânțânița-Murfatlar Forest, Maritime Dunes Reserve from Agigea and Hagieni Forest.

In Conacu-Negrești Valley, *S. exaltata* was identified on the limestone plateau (43⁰59˙19,11˙ N, 28⁰19˙42,60˙˙), in the centre of the valley where agriculture on the southern and eastern sides may threaten this species (Figure 15). It also has been identified in the small canyon in the south-west of the valley (43⁰58˙18,04˙˙ N, 28⁰10˙11,11˙˙ E) (Figures 20, 23), a canyon with expanded population of *Ailanthus altissima* (Mill.) Swingle - "cenușerul" (Simaroubaceae), an invasive species that is native to eastern Asia. This endangers the existence of *S. exaltata* and other native plant species existing in the small canyon. It co-occurs with: *Satureja coerulea* Janka (Labiatae), *Ornithogalum oreoides* Zahar. (Liliaceae), and other species – on the plateau, and with: *Dianthus armeria* (Caryophyllaceae), *Campanula bononiensis* L. (Campanulaceae), *Asperula tenella* Heuff. ex Degen (Rubiaceae), and other species – in the small canyon.

It is an East Balkan element, endemic to Europe, with a range in south-east Europe. In Wrigley`s opinion, this species replaces *S. chersonensis* from north of the Danube River, in Romania (Dihoru & Negrean, 2009).

It is an threatened species (EN) (IUCN, 2011) due to its reduced habitat area and attack by parasitic microbial fungi.

The scientific importance of this species is quite large due to its taxonomic relation to the *Otites* group in Romania and in relation to its corology/arealography as a European species.

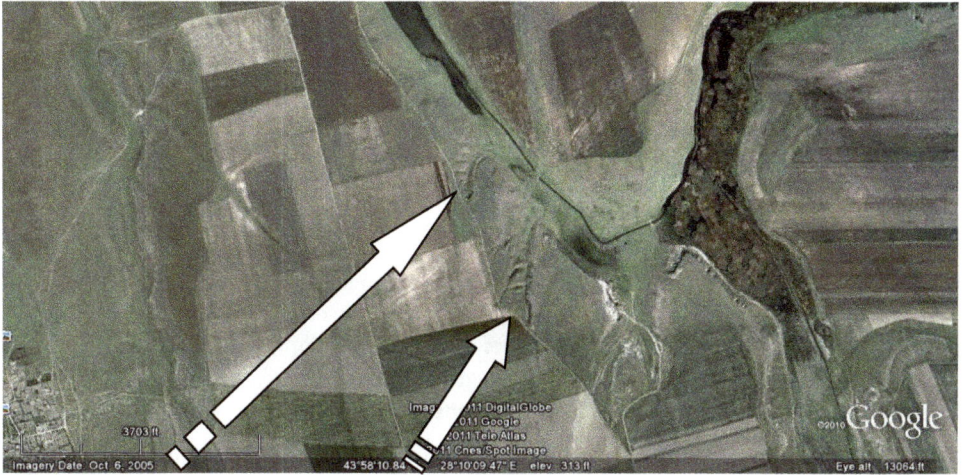

Fig. 23. The small canyon and the large canyon in the south-west of the valley, where *S. exaltata* has been detected (Google Earth, 2011)

### 4.1.6 *Jasminum fruticans* L. (Oleaceae Family)

*Jasminum fruticans* L. is a shrub with evergreen leaves (M), up to 3 m in height, with 4-angular branches; alternate leaves with three-leaflets and yellow flowers. The plant is pollinated by insects and dispersed by zoochory. It blooms from May to July (Dihoru & Negrean, 2009). *J. fruticans* is a heliophytic, thermophyte species that occurs on dry, pH-neutral and low nitrogen soils, unassuming plant to the substrate, but sensitive to frost (Tarnavschi & Diaconescu, 1965, cited in Dihoru & Negrean, 2009). In Hagieni Forest Natural Reserve (Dobrogea), this species occurs with shrubs on steep slopes, including *Carpinus orientali*s („cărpiniţă"), *Paliurus spina-christi* („păliur"), *Crataegus monogyna*, *Achillea clypeolata, Asparagus verticillatus, Asphodeline lutea, Ononis pusilla* and *Salvia ringens* (Cristurean & Ţeculescu, 1970).

In Romania, this species is widespread only in central and southern Dobrogea. To date, it has been found near the Cernavodă, Medgidia, Basarabi, Fântâna Mare, Gura Dobrogei towns, south of Mircea Vodă Station, north of Şipote, north of Mangalia, south-southeast of Cotul Văii, 2 Mai Village, in Seid-Orman and Stârghina Forests. Also, it found on the territory of nature reserves: Cheia Jurrasic Reefs Nature Reserve, the Fântâniţa-Murfatlar Reserve, Dumbrăveni Forest, the „Canaraua Fetii" and „Esechioi" Forest Reserves, and the Hagieni Forest (Andrei & Popescu, 1966; Borza, 1944; Cristurean & Ţeculescu, 1970; Dihoru et al., 1965; Dihoru & Negrean, 2009; Horeanu, 1976a; Morariu, 1961, as cited in Dihoru & Negrean, 2009; Parincu et al., 1998).

In Conacu-Negreşt Valley, it was identified only on the southeast facing limestone, in the central valley (43⁰59'39,02' N, 28⁰10'38,98") (Figure 22). There, it co-occurs with: *Crataegus monogyna* Jacq. (Rosaceae), *Prunus spinosa* L. (Rosaceae), *Achilea coarctata* and *Centaurea napulifera* Roch. (Compositae), *Satureja coerulea* Janka (Labiatae), *Ornithogalum oreoides* Zahar. (Liliaceae), and other species. The existence of the species in this valley is endangered due to several invasive species including: *Gleditshia triacanthos* L. Leguminosae), *Ailanthus altissima* (Mill.) Swingle (Simaroubaceae), *Elaeagnus angustifolia* L. (Elaeagnaceae).

*J. fruticans* is a Mediterranean element, with a range represented by the South Europe (Dihoru & Negrean, 2009) and is a vulnerable species (VU) ( IUCN, 2011).

The plant has a great scientific importance due to its highty isolated range, which indicate its great age (Boşcaiu, 1976, cited in Dihoru & Negrean, 2009).

It also is threatened by its small populations, lack of suitable habitat, the impacts of parasitic microbial fungi. These factors indicate that specific conservation measures should be undertaken, including both *"in situ"* and *"ex situ"* propagation options.

### 4.1.7 *Satureja coerulea* Janka in Velen (Labiatae Family Juss.)

*Satureja coerulea* Janka in Velen (Figure 24) is a sub-shrub (Ch) up to 25 cm tall, with deep roots; hairy stems and linear, ciliate leaves what are 1,5-2 mm wide. This species has pink fowers 7-10 mm wide, is self-pollinated and its fruits and seeds are disseminated by zoochory. *S. coerulea* blooms from July to September (Dihoru & Negrean, 2009).

Fig. 24. *Satureja coerulea* identified on the limestone plateau located in the central valley (photo. M. Axini, 2009)

*S. coerulea* is a heliophytic, thermophytic, calciphile and is ultra-xerophyle species who occurs on dry, neutral soils, on hills, in arid rocks meadows, and on rocks.

In Romania, this species is widespread only in Dobrogea. To date, it has been reported from the Măcin Mountains, the Babadag Plateau, the Urlichioi, Limanu, Adamclisi and Sevendic Valleys, on the southern shore of Tatlageac Lake, to Basarabi, Negru Vodă, Cheia, Casian, Gura Dobrogei, Palazu Mic, Castelu, Medgidia, Mangalia and Adamclisi. It is found within several protected areas: the Măcin Mountains National Park, the Cheia Jurassic Reefs, the Fântânița-Murfatlar Forest, the Hagieni Forest and the Dumbrăveni Forest (Borza, 1944; Burduja & Horeanu, 1976, as cited in Dihoru & Negrean, 2009; Dihoru & Doniță, 1970; Horeanu, 1976b; Morariu, 1970, as cited in Dihoru & Negrean., 2009; Răvăruț, 1961, as cited in Dihoru & Negrean, 2009; Mihăilescu–Firea et al., 1965, as cited in Dihoru & Negrean, 2009; Parincu et al., 1998; Țopa & Marin, 1968, 1970, 1973; Țopa & Marin, 1981, as cited in Dihoru & Negrean, 2009; Zahariadi, 1965).

In Conacu-Negrești Valley, S. coerulea was identified on the southeast facing limestone wall (43°59'39,02" N, 28°10'38,98") (Figure 22) and on the limestone plateau (43°59'19,11" N, 28°19'42,60") (Figure 15), both located in the central valley. Here, it co-occurs with: Crataegus monogyna Jacq. (Rosaceae), Prunus spinosa L. (Rosaceae), Jasminum fruticans L. (Oleaceae), Achilea coarctata and Centaurea napulifera Roch. (Compositae), Minuartia bilykiana Klokov in Kotov (Caryophyllaceae), Ornithogalum oreoides Zahar. (Liliaceae), and other species. In this valley, the species is endangered by non-native species: Gleditshia triacanthos L. (Leguminosae), Ailanthus altissima (Mill.) Swingle (Simaroubaceae), Elaeagnus angustifolia L. (Elaeagnaceae).

S. coerulea is an Scythian-Thracian-Anatolian element, with a range encompassing southeastern Europe (Dihoru & Negrean, 2009) and is a vulnerable species (VU) (IUCN, 2011).

The plant has practical importance as a decorative species and of importance in bee-keeping for its melliferous nectar production.

Although, it is conserved within protected areas, as mentioned above, S. coerulea requires aditional conservation measures, both „in situ" and „ex situ", including cultivation in botanical gardens and population restauration. Such actions are warranted because it tipically forms small populations and its rock surface habitats has been reduced.

### 4.1.8 *Achillea clypeolata* Sbth. et Sm. (Compositae)

*Achillea clypeolata* Sibth. et Sm (syn. *A. alexandri-borzae* Prodan) (Figure 25) is a perennial plant (H) with leaves in one plane, larges incisions that reach the median rib; inflorescences with peduncles 2 mm long, yellow flowers; sexual vegetative reproduction, with many stems growing from a single root; entomophilous and anemophilous pollination, and dispersal of fruits and seeds in its habitats. It blooms from June to July. It preserved well by vegetative reproduction by rhizomes. By yellow flowers and compact inflorescences, the species can be confused with *A. coarctata* Poiret and with *A. thracica* Velen. This latter species is a European endemite, a Dobrogean-Thracian element with potential for existence in the valley. *A. clypeolata* hybridizes with *A. neilreichii*, *A. setacea*, and *A. pannonica* (the first species identified in the valley), demostrating the vigurous nature of this species (Dihoru & Negrean, 2009; Prodan, 1931, 1939).

*A. clypeolata* is a heliophytic, thermophyte what occupies dry, neutral soils, in arid meadows (Cristurean & Ionescu-Țeculescu, 1970; Dihoru & Doniță, 1970).

Fig. 25. *Achillea clypeolata* identified on the limestome plateau located in the central valley (photo. M. Axini, 2009)

*A. clypeolata* is a Balcanic element and a European endemit to Albania, Bulgaria, Greece, Serbia, Muntenegru, Romania, and Turkey (Richardson, 1976, as cited in Dihoru & Negrean, 2009). It is listed as a critically endangered species (CR) ( IUCN, 2011). In Romania, this species is widespread only in Dobrogea. To date, it has been identified from Babadag, Jurilovca, Ceamurlia de Sus, Mangalia and the Coroanei Valley, Cotul Văii (Mare Valley, in the south) and Allah-Bair Hill, although subsequent searches have failed to detect it there. It is present in the Hagieni Forest Reserve (Ciocârlan & Costea, 1997; Mititelu et al., 1968, as cited in Dihoru & Negrean, 2009; Prodan, 1931; Prodan & Nyárády, 1964, as cited in Dihoru & Negrean, 2009).

In Conacu-Negrești Valley, it was found only on the limestone plateau located in the central valley (43°59'19,11" N, 28°19'42,60") (Figure 15). There, it co-occurs with: *Achillea setacea* W. et K. and *Echinops ritro* L. ssp. *ruthenicus* (M. Bieb.) Nyman (Compositae), *Salvia nutans* L. (Labiateae), *Allium saxatile* MBieb. (syn. *Allium globosum* MBieb ex DC) (Liliaceae). A agricultural practice to the west, southwest, southeast and east sides, endanger the future existence of this species.

The plant is economically important, as a decorative plant, and has been identified as important to the plant genetic fund. In addition, it has scientific importance as it reaches the north-eastern limit of its limited world range. Because of the reduction of Dobrogea steppes habitats by plowing and because it is parasitized by the microbial fungi, this species requires additional conservation protection.

### 4.1.9 *Centaurea napulifera* Rochel (Compositae)

*Centaurea napulifera* Rochel (Figure 26) is a short, perennial plant (H), only10-15 cm tall, what spreads by rhizomes and sometimes runners. It has thickened roots, leaves with "felt" (tiny) white-silver hairs, on its lobe-shaped or lyre-shaped, basal leaves arranged in a rosette. *C. napulifera* undergoes vegetative reproduction via its long, thin runners. It reproduces sexually, with both anemophilous and entomophilous (by ants) pollination. This species blooms in April-May. Two subspecies of *C. napulifera* are recognized in Romania in Flora Europaea: *C. napulifera napulifera* and *C. napulifera thirkei* (Schults Bip.), and two forms: *tuberosa* (Vis.) Gugler and *albiflora* Prodan (Ciocârlan, 2000; Dihoru et al., 1965, as cited in Dihoru & Negrean, 2009; Prodan, 1930).

Fig. 26. *Centaurea napulifera* from the limestome plateau located in the central valley (photo. M. Axini, 2009)

It is Dobrogean-Balkan element and is reported to be endemic species in Europe (Bulgaria, Greece, Serbia, Muntenegru, Romania, Moldova, Turkey) (Dostál, 1976, as cited in Dihoru & Negrean, 2009). In Romania, this species is widespread only in Dobrogea. To date, it has been detected in the Consul Mountain and on Denis-Tepe Hill, from the Casimcea Plateau and Caugagiei Valley, at the base of Slovan-Bair Hill, from Casimcea, Seidorman, Tatlageac, Sevendic Valleys, and from Basarabi, Mangalia, Adamclisi, Medgidia, Valul lui Traian, and 23 August Village. It is found within the following protected areas: the Cheia and Fântânița-Murfatlar Reserves, the Dumbrăveni and Hagieni Forests, Valul lui Traian, Allah-Bair Hill, Consul Hill, and Măcin Mountain National Park (Andrei & Popescu, 1966; Dihoru, 1965, as cited in Dihoru & Negrean, 2009; Dihoru & Doniță, 1970; Horeanu, 1975; 1976a, 1976b; Morariu, 1970, as cited in Dihoru & Negrean, 2009; Prodan, 1939; T. Săvulescu, 1953; Zahariadi & Negrean, 1969).

*C. napulifera* is a heliophytic, thermophytic, calciphile, a xerophytic species that lives on dry, neutral –pH soils. Species grows in green lands with rare trees, on dry meadows, and by bushes (Dihoru & Doniță, 1970). In the Conacu-Negrești Valley, it was found only on the limestone plateau located in the central valley (43°59'19,11" N, 28°19'42,60") (Figure 15). There, it co-occurs with: *Achillea setacea* W. et K. and *Echinops ritro* L. ssp. *ruthenicus* (M. Bieb.) Nyman (Compositae), *Salvia nutans* L. (Labiateae), *Allium saxatile* MBieb. (syn. *Allium globosum* MBieb ex DC) (Liliaceae).

*C. napulifera* is a vulnerable species (VU) (IUCN, 2001) due to agricultural alteration of its habitat in the west, southwest, southeast and east sides of the study area.

This species has considerable scientific importance because its unusual life history characteristics particularly is vegetative reproduction and short life cycle. It also has practical importance as a decorative plant and for its nectar and pollen production during spring.

The plant is endangered in part due to its short stature, which does not allow for widescale dispersal of fruit (usually only a few cm), and it may be attacked by microbial fungi. To these threats are added the impacts of intensive livestock grazing and other agricultural alternation of its natural grassland habitats.

Consequently, protective measures are needed, including research, resolution of its subspecific taxonomic, and propagation in protected settings, including botanical gardens.

### 4.1.10 *Scolymus hispanicus* L. (Asteraceae)

*Scolymus hispanicus* (Figure 27) is a thorny and hairy biennual to perennial species (TH-H). It was divided basal-leaves, basal leaves and petiolate, sessile leaves - on the stem and yellow flowers. *S. hispanicus* reproduces sexually, with anemophilous and entomophilous pollination, and dissemination of fruit and seeds through anemochory and gravity. It blooms from June-August (Nyárády, 1965, as cited in Dihoru & Negrean, 2009; Walters, 1976, as cited in Dihoru & Negrean, 2009).

Fig. 27. *Scolymus hispanicus* from the south end of the lake (photo. M. Axini, 2009)

It is a heliophytic thermophyte that lives on dry, nitrogen rich soils, on cliffs, and in ruderal settings (Dihoru & Doniţă, 1970).

In Romania, this species has been detected thus far only in Dobrogea and occasionally, localy near Sibiu and Giurgiu (where it was reported as adventitious). In Dobrogea, it was detected in the coastal zone (Sulina, north of Periboina to Portiţa, Chituc and Saele Litoral Dunes –Danube Delta; Midia Peninsula; "3 Papuci" Beach, south of Constanţa; Agigea; Eforie Nord; Eforie Sud; Tuzla; Mangalia; 2 Mai and Vama Veche Villages; Tatlageac and Techirghiol Lakes), of north (Măcin) and of south of Dobrogea (Sevendic Valley; Limanul Valley, Mare Valley –south-southeast of Cotul Văii, Hagieni Forest) (Borza, 1944; Dihoru & Negrean, 1976, 2009; Ionescu-Ţeculescu & Cristurean, 1967; Morariu, 1963, 1965; Prodan, 1939).

In the Conacu-Negreşti Valley, it has been detected on the southeast facing limestome wall in the north-west valley (44⁰00'21,29˝ N, 28⁰08'51,34˝ E) (Figure 19) and on the eastern shore of Conacu-Negreşti Lake, between Conacu Village and terminal side of the lake (43⁰59'07,38˝ N, 28⁰11'02,90˝ E) (Figure 28).

It is a Mediterranean element, with a range from southern Europe up to northern France (Albania, Azores, Baleares Islands, Bulgaria, Corsica, Crete, France, Greece, Spain, Italy, Serbia, Muntenegru, Portugal, Romania, Ukraine, Sardinia, Sicily, Turkey) (Walters, 1976, as cited in Dihoru & Negrean, 2009).

It is a plant with practical importance, as a medicinal herb, it is used as a diuretic and a treatment for eczema and liver failure; it also is used as a ornamental and a food plant.

It is regarded as a **vulnerable species** (VU) (IUCN, 2011). The species is characterized by small populations and therefore is threatened by loss of habitats. For these reasons, this species needs protection, both „*in situ*" (currently, there is within Danube Delta Biosphere Reserve) and „*ex situ*" (through cultivation in botanical gardens).

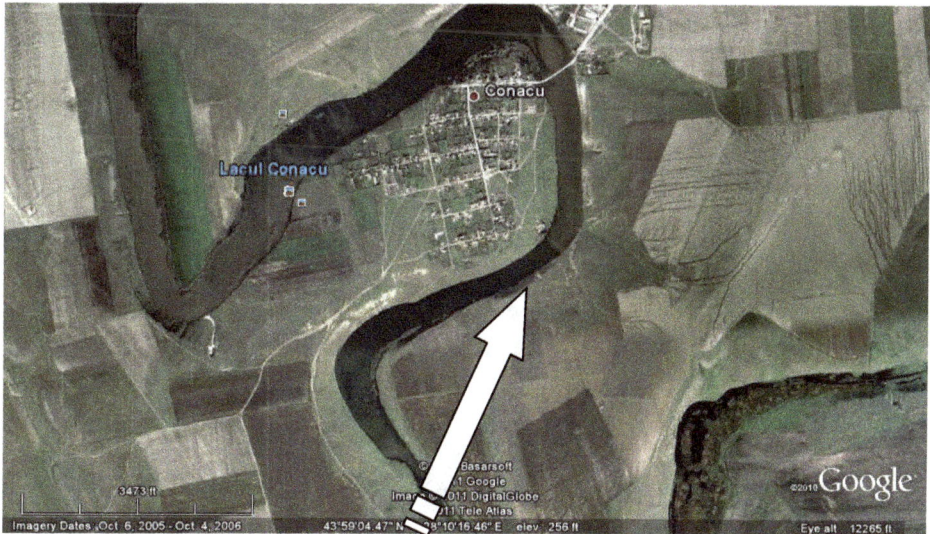

Fig. 28. Location of one site at which *Scolymus hispanicus* was detected (Google Earth, 2011)

## 4.1.11 *Ornithogalum oreoides* Zahar. (Liliaceae)

*Ornithogalum oreoides* (Figure 29) is a rhizomatous plant (G), with bulbs with free scales, canaliculate leaves that are smooth edgeds. It is polinated by insects and its fruit and seeds are disseminated by gravity; however, it also undergoes vegetative reproduction. *Ornithogalum oreoides* blooms in -May (Zahariadi, 1966; Zahariadi, 1980, as cited in Dihoru & Negrean, 2009).

It is a heliophytic, calciphile, and is a subxerophytice species (Dihoru & Doniță, 1970).

In Romania, this species has been reported thus far from Dobrogea and in refugia in other parts of Romania. In Dobrogea, it was found to Caraorman litoral sands from the Danube Delta, on the Babadag Plateau, and in „Canarale" from Hârşova, Gura Dobrogei, the Cheia Reserve, on Allah-Bair Hill, in the „Canaraua Fetei" and in the „Esechioi"Forest Reserves, and elsewhere (Andrei & Popescu, 1966; Ciocârl, 1994; Cîrţu, 1979; Dihoru & Doniță, 1970; Horeanu, 1976a; Lungeanu, 1972).

*O. oreoides* was detected in the Conacu-Negreşti Valley only on the limestone plateau located in the central valley (44°00`21,29` N, 28°08`51,34`` E) (Figure 15).

This is Scythian element, and is endemic in Europe. This range extends across the Northwestern Black Sea in Bulgaria, Romania, Moldova, Ukraine (Dihoru & Negrean, 2009).

It is classified as a vulnerable species (VU) (IUCN, 2011). Although an important ornamental and nectar producing species, it is a rare and is highly restricted to specific microsites, and may be attacked by fungi. Consequently, it deerves protected measures, both "in situ" (currently, it is in the territory of Cheia Jurassic Reefs Reserve, „Canaralele" from Hârşova Port, Gura Dobrogei Cave, and the Danube Delta Biosphere Reserve) and "ex situ" (through cultivation in botanical gardens and through restocking).

Fig. 29. *Ornithogalum oreoides* on the limestone plateau located in the central valley (photo. M. Axini, 2009)

## 4.1.12 *Koeleria lobata* (M. Bieb.) Roem. & Schult. (Poaceae)

*Koeleria lobata* (syn. *K. brevis* Steven; *K. degenii* Domin) is a perennial brush-shaped plant (H), sometimes forming small clusters. It has bulbous root base and many stems emerge from a single root. Along with its vegetative reproduction capacity, it is amemophilous and its seeds are dispersed by gravity and by „seed eating" animals. It blooms from May-July (Dihoru & Negrean, 2009).

*K. lobata* is a heliophytic, thermophytic, calciphile, an ultraxerophytic species that grows on dry, pH-neutral soils (Andrei et al., 1965; Cristurean & Ionescu-Ţeculescu, 1970; Ionescu-Ţeculescu & Cristurean, 1967; Dihoru & Doniţă, 1970).

In Romania, this species has been reported only from Dobrogea: in the Lupilor Litoral Dunes, the Letea Forest, Razim Lake, the Dolojman Peninsula from Danube Delta, Carvan Hill, the Babadag Plateau, the Măcin Mountains, Gura Dobrogei, Cheia Reserve, Valul lui Traian, Fântâniţa-Murfatlar, the Şerpilor Valley from Hagieni Reserve, and in south-southeast of Mare Valley from Cotul Văii (Andrei et al., 1965; Andrei & Popescu, 1966; Dihoru & Doniţă, 1970; Dihoru et al., 1965; Ionescu-Ţeculescu, 1967; Horeanu C., 1973, 1975, 1976a; Sârbu et al., 2000).

In the Conacu-Negreşti Valley, it was detected on a limestone wall (alt. 89 m) from the bottom of the lake in the south -central valley, on south-eastern exposure (43⁰58'53,66" N, 28⁰10'09,29" E) (Figure 30). Its presence is confirmed in the large canyon, on the limestone wall in the northwest valley and on the limestone plateau above the canyon, where co-occurs with *Parietaria*.

Fig. 30. Location of the limestone wall that supports *Koeleria lobata* (Google Earth, 2011)

*K. lobata* is a Continental element, with a range extending from Turkey to Southeast Russia in Turkey, Bulgaria, Romania, Ukraine, Crimeia, eastern Russia (Dihoru & Negrean, 2009).

This spesies is a vulnerable species (VU) (IUCN, 2011) due to loss of its habitats, and it is of scientific interes because of its vegetation reproduction. It also is a pioneer species, stabilizing rocky lands.

Protective measure are necessary, including - "in situ" through preservation in reserves, such as the Hagieni Forest, Fântânița-Murfatlar, the Babadag Forest, the Uspenia Monastery, and Măcin Mountains National Park. It also may require "ex situ" protection through cultivation in botanic gardens.

## 5. Conclusions

Data presented in this work were compiled from field and laboratory studies in 2003–2010. This research is part of a program developed by the "Monachus" Group for Research and Environmental Education in Constanța, Romania in partnership with the Faculty of Natural Sciences, "Ovidius" University in Constanța, Romania, aimed at identifying the biodiversity of the Conacu-Negresti Valley, the biology and ecology of which was not recognized until 2003.

Our field and laboratory studies lead us to conclude that the Conacu-Negreşti Valley is characterized by a rich and diverse flora and fauna, with numerous rare and endemic species for Romania and Dobrogea. Many of these species require improved management, protection, and preservation status. At present, the valley does not have designated conservation status, and a part of the lake is leased and managed as a fish farm. Human impacts on different aspects may contribute to future declines and even disappearance of some rare species. Therefore, urgent protective measures should be taken that will lead to improved biodiversity conservation and landscape protection in this valley.

Due to its unique nature, which result from its geological and paleontological past, the valley hosts many species important to science and human well-being, and some of which have not yet been described. Such efforts will be the focus of future work in this research program.

## 6. Acknowledgment

As director of the reseach and environment education program on Conacu-Negreşti Valley, I want to thank my colleagues on the research team: Associate Professor Ph.D. Marius Skolka, Ph.D. Gavril Negrean, Professor PhD. Rodica Bercu, and Ph.D. Gabriel Bănică.

## 7. References

Andrei, M. (2003). *Romania Flora. Ferns Determinate*, Sigma Primex Publisher, ISBN 973-8068-62-2, Bucharest (in Romanian)

Andrei, M., Dihoru, G. & Popescu, A. (1965). Two new plants for the flora of Romania Socialist Republic, In: *Acta Bot. Horti Bucurest*, (1965-1966) (in Romanian)

Andrei, M. & Popescu, A. (1966). Contribution to the study of flora and vegetation of „Gura Dobrogii" Nature Reserse, In: *Ocrot. Nat.* 10(2), 163-176 (in Romanian)

Axini, M. (2006). Lake and Conacu - Negreşti Valley – Study made to their proposal for entry in the list of protected areas in Romania. Thesis, Faculty of Agricultural Sciences and Natural Sciences, "Ovidius" University, Constanța, Romania (in Romanian).118 pag.

Axini, M. (2009). The ferns of Măcin Mountains. Scientific importance, protection and conservation, In: *Journal of Environment Protection and Ecology* (JEPE), Vol. 10, No. 1, (April 2009), pp. 121-130, ISBN 1311-5065

Axini, M. (2009). Conacu-Negreşti Valley. The ecological reevaluation. Dissertation, Faculty of Natural Sciences and Agricultural Sciences, „Ovidius" University, Constanţa, Romania (in Romanian)

Axini, M. 2011). Gastropods`s diversity in Conacu-Negreşti Valley, *Ovidius University Annals - Biology-Ecology Series*, ISBN 1453-1267 (in press)

Axini, M. (2011). Biodiversity and anthropogenic impact in Dobrogea –case study: Conacu – Negreşti Valley, *Ovidius University Annals - Biology-Ecology Series*, ISBN 1453-1267 (in press)

Axini, M. & Skolka, M. (2010). Invertebrates fauna in Conacu-Negreşti Valley (southern Constantza County), *Scientific Annals of "Al. I. Cuza" Univ.from Iaşi, New Series, Section I, Animal Biology*, ISBN 1224-581x (in press)

Axini, M., Skolka, M., Bănică, G. & Negrean G. (2010). Conacu-Negreşti Valley-biodiversity, threats, protective measure, *Scientific Annals of "Al. I. Cuza" Univ. from Iaşi, New Series, Section I, Animal Biology*, ISBN 1224-581x (in press)

Bavaru, A., Godeanu, S., P., Butnaru, G. & Bogdan, A. (2007). *Biodiversity and the environment protection*, Romanian Academy Publisher, Buchares, ISBN 978-973-27-1569-7 (in Romanian);

Basarabeanu, N. (1969). The role of rain water on actual relief modeling of Dobrogea. *Geographical Studies of Dobrogea, Proceedings of the First Symposium on the Geography of Dobrogea*, Constanţa, October 1968 (in Romanian)

Bănică, G. (2006). Study on the avifaunistical diversity in the coastal area between Cape Midia (Romania) and Cape Kaliakra (Bulgaria), *Vol. with the works of the Conference in Constanta Comparative studies on the biodiversity of coastal habitats, human impact and possibilities of conservation and restoration of habitats of European importance between Cape Midia and Cape Kaliakra*, ISBN 978-973-644-840-9, Mamaia, 26-28 September 2008

Birds Directive 79/409 EEC Council Directive on the conservation of wild birds, adopted on 2 April 1979, Available from: http://ec.europa.eu/environment/nature/legislation/birdsdirective/index_en.htm

Botnariuc, N. & Tatole, V. (Ed(s).) (2005) *Cartea Roşie a Vertebratelor din România*, Romanian Academy/„Grigore Antipa" National Natural History Museum, ISBN 973-0-03943-7, Bucharest (in Romanian)

Brezeanu, D. G. (1997). *The monograph of Cobadin Village*, Cobadin Village Hall, Constanţa County (in Romanian)

Bleahu, M., D. (2004). *Noe`s Arks in XXI century – Protected areas and nature protection*, National Publisher, ISBN 973-659-084-4, Bucharest (in Romanian)

Borza, A. (1944). Materials for Mangalia florula, In: *Bul. Grăd. Bot. Cluj* (Timişoara), Vol. 26, No. 1-2, (1944), pp. 1-47 (in Romanian)

Botnariuc, N. (1989). *Genetic fund and problems of its protection*, Ştiinţifică şi Enciclopedică Publisher, ISBN 973-29-0100-4, Bucharest (in Romanian)

Căpuşe I., 1968. *Fauna of Romania-Lepidoptera Tineidae*, 10, Academy Press, Bucharest (in Romanian)

Ciocârlan, V. (1994). *Flora of Danube Delta*, Ceres Publisher, Bucuresti (in Romanian)

Ciocârlan, V. (2000). *Illustraded Flora of Romania. Pteridophyta et Spermatophyta* (second edition), Ceres Publisher, Bucharest

CIOCÂRLAN, V. (2000). *Illustrated flora of Romania. Pteridophyta and Spermatophyta,*"Ceres" Publisher House, Bucharest *(in Romanian)*

Ciocârlan, V. (2001). Contributions to the knowledge of the flora of Romania, In: *Bul. Grăd. Bot. Iași*, Vol. 10, pp 55-57

Ciocârlan, V. & Costea, M. (1997). Flora of Alah Bair Hill Botanical Reserve (Constanța County), In: *Acta Horti Bot. București*, (1995-1996), pp. 97-104

Chiriac, E. & Udrescu, M. (1965). *Naturalist guide in the freshwater world*, Scientific Publishing House, Bucharest (in Romanian)

Cîrțu, D. (1979). Geobotany and agro-productive study of grasslands between Jiu-Desnățui-Craiova and Danube, Gorj County. Summary thesis, Timișoara (in Romanian).

Cristurean, I. & Ionescu-Țeculescu V. ( 1970). Plant associations in „Hagieni Forest" nature Reserve, In: *Acta Bot. Horti Burest.*, (1968), pp. 245-279

Cristurean, I. & Lițescu S. (2002). Phylum Pteridophyta, In: *The diversity of the living world. Illustrated Identification Manual of the Flora and Fauna of Romania*, S. P. Godeanu (ed.), pp. 182-184, Bucura Mond Publishing, ISBN 973-98248-5-4, Bucharest (in Romanian)

Cristurean, I. & Lițescu S. (2002). Phylum Magnoliophyta (Angiospermae). Flowering plants, In: *The diversity of the living world. Illustrated Identification Manual of the Flora and Fauna of Romania*, S. P. Godeanu (ed.), pp. 185-207, Bucura Mond Publishing, ISBN 973-98248-5-4, Bucharest (in Romanian)

Cojocaru, I. & Popescu, I., E. (2004). La diversitédes coléoptères aquatiques (Insecta, Coleoptera) du Marais de Văcărești (Bucarest), In: *Scientific Annals of "Al. I. Cuza" Univ.from Iași, New Series, Section I, Animal Biology*, Tom L, Issue 2004, pp. 77-83, ISBN 1224-885x

Coteț, P. (1969). South Dobrogea - Genesis and Evolution. *Geographical Studies of Dobrogea, Proceedings of the First Symposium on the Geography of Dobrogea*, Constanța, October 1968 (in Romanian)

Dihoru, G. & Dihoru, Al. (1994). Rear, endangered and endemic plants of Romanian Flora – Red list, Bucharest, *Acta Botanica Horti Bucurestiensis, Botanical Garden Papers*, Bucharest, Ser. 1993-1994, 173-197 (1994) (in Romanian)

Dihoru, G. & Doniță, N. (1970). *Flora and vegetation of Babadag Plateau*, Rep. Soc. Rom. Academy Publisher, Bucharest, 438 pag.

Dihoru, G. & Negrean, G. (1976). Flora of the Danube Delta. *Peuce (Bot.)*, Vol. 5, pp. 217-251

Dihoru, G. & Negrean, G. (2009). *Red Book of Vascular Plants in Romania*, Romanian Academy Publishing House, ISBN 978-973-27-1705-9, Bucharest (in Romanian)

Dihoru, G., Țucra, I. & Bavaru, A. (1965). Flora and vegetation of „Fîntînița"Reserve from Dobrogea. *Ocrot. Nat.*, Vol. 9, No. 2, pp. 167-184

Dobrici Region, 20011, Available from: http://ro.wikipedia.org/wiki/Regiunea_Dobrici (in Romanian)

Dobruja, 2011, Available from: http://en.wikipedia.org/wiki/Dobrudzha (in Romanian)

Făgăraș, M., Anastasiu, P. & Negrean, G. (2010). Rare and threatened plants in the Black Sea coastal area between Cape Midia (Romania) and Cape Kaliakra (Bulgaria). *Bot. Serbica*, Vol. 34, No. 1, pp 37-43

Flocea, F. (2004). Rare, vulnerable and protected birds from the Repedea-Bârnova area- Iași County. *Scientific Annals of "Al. I. Cuza" Univ.from Iași, New Series, Section I, Animal Biology*, Tom I, Issue 2004, pp. 311-318

Gâștescu, P. & Breier, A. (1969). The lakes of Dobrogea. *Geographical Studies of Dobrogea, Proceedings of the First Symposium on the Geography of Dobrogea*, Constanța, October 1969 (in Romanian)

Godeanu, S. (2002). Continental waters, overview, In: The diversity of the living world. Illustrated Identification Manual of the Flora and Fauna of Romania, S. P. Godeanu (ed.), pp. 1-24, Bucura Mond Publishing, ISBN 973-98248-5-4, Bucharest (in Romanian)

Grecescu, D. (1898). *Romanian Flora Conspect*, Dreptatea Typography, Bucharest (in Romanian)

Habitats Directive. Council Directive 92/43 EEC of 21 May 1992 on the conservation of natural habitats and wild flora and fauna, adopted on 12 May 1992, Available from: http://ec.europa.eu/environment/nature/legislation/habitatsdirective/index_en.htm

Horeanu, C. (1973). Floral aspects from Cheia (Dobrogea), *Ocrot. Nat.*, Vol. 17, No. 1, pp. 83-88 (in Romanian)

Horeanu, C. 1975. The study of *flora and vegetation of Casimcea Plateau*, Rezumatul tezei de doctorat, Univ. Iași, Fac. Biol. –Geogr.

Horeanu, C. (1976). Flora of „Cheia" natural Reserve (Constanta County). *Ocrotirea Naturii Dobrogene*, pp. 142-157 (litogr.), Cluj-Napoca

Horeanu, C. (1976). Proposals for the establishment of new reserves in Casimcea Plateau. *Ocrotirea Naturii Dobrogene*, pp. 159-168 (litogr.), Cluj-Napoca

Iana, S. (1970). South West Dobrogea (The Oltina, Negru Vodă Plateau); Physico-geographical study, with overview of biogeography. Thesis summaries, Faculty of Geology-Geography, University of Bucharest, Multiplication Center of the University of Bucharest, pp. 42, 4 maps, 4 legends (in Romanian).

Iana Sofia, 1973 – Fauna of Dobrogea Canarals (extras), *Scientific Studies and Research (Serie Biology)*, Inst. Pedag. of Bacău, Min. Educ. și Cerc. (in Romanian)

Iftimie, Al. (2001). Red list of amphibians and reptiles from Romania. *Protection of nature and the environment*, 44-45, 39-49

Iftimie, AL. (2005). Reptiles, In: *Red Book of Vertebrates in Romania*, N. Botnariuc, V. Tatole. (ed(s).), 173-194, Curtea Veche Trading S.R.L. Publishing House, ISBN 973-0-03943-7, Bucharest, (in Romanian)

Ionescu-Țeculescu, Venera & Cristurean, I. (1967). Floral research in „Hagieni Forest" Natural Reserve. *Ocrot. Nat.*, Vol. 11, No. 1, pp. 25-36 (in Romanian)

Iordache, I., Oprea, A., Zamfirescu, Ș., Ion, Ctin., Ion, E. (2004). The conservation of the terrestrial vertebrates from the protected areas and natural reserves of Moldavia. *Scientific Annals of "Al. I. Cuza" Univ. from Iași, New Series, Section I, Animal Biology*, Tom I, Issue 2009, 279-292, ISBN 1224-581x

IUCN 2011. IUCN Red List of Threatened Species. Version 2011.1. <www.iucnredlist.org>. Downloaded on 26 August 2011.

Kiss, Șt. (2002). Ordo Heteroptera. Water bugs, In: *The diversity of the living world. Illustrated Identification Manual of the Flora and Fauna of Romania*, S. P. Godeanu (ed.), pp524-530, Bucura Mond Publishing, ISBN 973-98248-5-4, Bucharest (in Romanian)

Los, W. (ed.) (2011). *Fauna Europaea*, Available from: http://www.faunaeur.org

Lungeanu, I. (1972). Contributions to the cariologic study of the genus Ornithogalum. *Acta Bot. Horti Bucurest.* (1970-1971), pp. 147-151 (in Romanian)

Mihai, G., Vițalariu, G., Chifu, T. (1964). Contributions of study of Dobrogea flora, *Stud. Cercet. Biol. - Bot.*, Vol. 16, No. (6), pp. 471-476 (in Romanian)

Morariu, I. (1963). Contributions at study of Black Sea coast and Dobrogea, *Lucr. Şti.*, Vol. 6, pp. 55-88 (in Romanian)

Morariu, I. (1965). Some aspects from Black Sea coast flora, *Stud. Cercet. Biol.-Bot.*, Vol. 17, No. 4-5, pp.503-509 (in Romanian)

Morariu, I., Todor, I. (1972). *Sistematique Botany*, The Revised Second Edition, Didactic and Pedagogic Publisher, Bucharest (in Romanian)

Müller G., I.(2002). Phylum Arthropoda. Arthropods, In: *The diversity of the living world. Illustrated Identification Manual of the Flora and Fauna of Romania*, S. P. Godeanu (ed.), pp. 372-374, Bucura Mond Publishing, ISBN 973-98248-5-4, Bucharest (in Romanian)

Müller, G., I. & Tomescu, M. (2002). Ordo Isopoda. Izopods, In: *The diversity of the living world. Illustrated Identification Manual of the Flora and Fauna of Romania*, S. P. Godeanu (ed.), pp. 472-474, Bucura Mond Publishing, ISBN 973-98248-5-4, Bucharest (in Romanian)

Munteanu, D. (2005). Birds, In: *Red Book of Vertebrates in Romania*, N. Botnariuc, V. Tatole. (ed(s).), 85-166, Curtea Veche Trading S.R.L. Publishing House, ISBN 973-0-03943-7, Bucharest (in Romanian)

Murariu, D. (2005). Mammals, In: *Red Book of Vertebrates in Romania*, N. Botnariuc, V. Tatole. (eds.), Curtea Veche Trading S.R.L. Publishing House, ISBN 973-0-03943-7 Bucharest, 11-80 (in Romanian)

Nițu, E., Decu, V. (2002). Ordo Coleoptera. Beetles, In: *The diversity of the living world. Illustrated Identification Manual of the Flora and Fauna of Romania*, S. P. Godeanu (ed.), pp. 554-571, Bucura Mond Publishing, ISBN 973-98248-5-4, Bucharest (in Romanian)

Olteanu, M, Popescu, A, Roman, N, Dihoru, G, Sanda, V & Mihăilescu, S. (1994). Romanian Red List of the higher plants. *Studies, Summaries, Documents of Ecology*, Bucharest, Vol. 1, pp. 1-52 (in Romanian)

Panin, S. (1957). *Fauna of Romania* – Coleoptera Scarabeidae II, Vol. 10, No. 4, Academy Press, Bucharest (in Romanian)

Panin, S., Săvulescu ,N. (1957). *Fauna of Romania* – Coleoptera Cerambycidae, Vol. 10, No. 5, Academy Press, Bucharest (in Romanian)

Parincu, M., Mititelu, D. & Aniței, L. 1998). Vascular flora from Dumbrăveni Forest Botanical Reserve (Constanta County). *Bul. Grăd. Bot. Iași*, Vol. 6, No. 2, pp. 353-358 (in Romanian)

Păun, M., Cîrțu, D. & Popescu, G. (1970). *Schedae ad „Floram Olteniae Exsiccatam" a Horto Bot. Univ. Craiov. Edit.* Cent. VII (in Romanian)

Păun, M. & Popescu, G. (1970). Aspects from vegetation of Sohodol Valley, Gorj County. *Stud. Cercet.*, 93-97, Comit. Cult. Educ. Soc. Jud. Gorj.

Peahă, Mircea (Ed.) (1982). *General Geographic Atlas*, „Didactic" și Pedagogică" Publisher, Bucharest (in Romanian)

Prodan, I. (1930). *Centaureele României*. Inst. Arte Grafice „Ardealul', Cluj (in Romanian)

Prodan, I. (1931). Achileele României. *Bul. Acad. Inalte Studii Agron. Cluj Memorii*, Vol. 2, pp. 1-28, „Cartea Românească" Tipography, Cluj.

Prodan, I. (1939). *Flora for determination and description of plants growing in Romania* (first, second, third edition), Tipografia Cartea Românească", Cluj. (in Romanian)

Proda, I., (1935, 1936). *Conspectus of Dobrogea Flora, Bul. Acad. Agron.* (1934). I. Vol. 5, No. 1, pp.175-342; (1935-1936). II.Vol. 6, pp. 206-259, Cluj

Prodan, I., (1939). *Conspectus of Dobrogea Flora,* III. *Bul. Fac. Agron.* (1938) Vol. 7, No. 16-102. Cluj (in Romanian)

Prodan, I., (1935). *Conspectus of Dobrogea Flora,* First part, Tipografia Națională Publ. House. Cluj-Napoca, 170 pp and 62 fig. (in Romanian)

Prodan I., (1936). *Conspectus of Dobrogea Flora.* Second part, Tipografia Nationalǎ Publ. House. Cluj-Napoca (in Romanian)

Proda, I., (1936). *Conspectus of Dobrogea Flora.* Third part, Tipografia Națională Publ. House. Cluj-Napoca (in Romanian)

Prodan, I. & Buia, Al. (1966). *The Romania Illustrate Small Flora,* Fourth edition, Agro-Silver Publisher, Bucharest (in Romanian)

Ruști, D. (2002). Ordo Lepidoptera. Butterflies, moths, In: *The diversity of the living world. Illustrated Identification Manual of the Flora and Fauna of Romania,* S. P. Godeanu (ed.), pp. 572-575, Bucura Mond Publishing, ISBN 973-98248-5-4, Bucaherst (in Romanian)

Sǎvulescu, T (ed) (1952-1976) *Flora of Popular Republic of Romania – Flora of Socialis Republic of Romania,* 1-13, Publ. House of Rep. Soc. Academy. Bucharest (in Romanian)

Sârbu, I., Ștefan, N., Oprea, A., Zamfirescu, O., (2000). Flora and vegetation of Grindul Lupilor Natural Reserve (Danube Delta Biosphere Reserve), *Bul. Grǎd. Bot. Iași,* Vol. 9, pp. 91-124

Silistra Region, 2011, Available from: http://ro.wikipedia.org/wiki/Regiunea_Silistra (in Romanian)

Skolka, M., 2002. General Entomology, „Ovidius" University Press, Constanța, 239 pp. (in Romanian).

Skolka M. (2002). Class Insecta. Insects, In: *The diversity of the living world. Illustrated Identification Manual of the Flora and Fauna of Romania,* S. P. Godeanu (ed.), pp. 475-480, Bucura Mond Publishing, ISBN 973-98248-5-4, Bucharest (in Romanian).

Skolka, M., Fǎgǎraș, M. & Paraschiv, G. M. (2005). The Dobrogea biodiversity, Work performed under grant CNCSIS 880/2004, Ovidius University Press, 296 pp., ISBN 973-614-232-9 (in Romanian).

Skolka, M. (2008). Invertebrate diversity in the western part of Black Sea coast: Cape Midia-Cape Kaliakra Zone, *Vol. with the works of the Conference in Constanța* Comparative studies on the biodiversity of coastal habitats, human impact and possibilities of conservation and restoration of habitats of European importance between Cape Midia and Cape Kaliaka, ISBN 978-973-644-840-9, Mamaia, September 2008

Szabó, A, (1973). Corologia si incadrarea fitocenotica a populatiilor de conservare Adonis volgensis in Romania. *Notulae Bot. Horti Agron. Clujensis,* Vol. 7, pp. 39-47 (in Romanian).

Todor, I. (1968). *Little Atlas of Plants from R.S.R. Flora,* Didactic and Pedagogic Publisher, ISBN Bucharest (in Romanian)

Țopa, E., Marin, E. (1968, 1970, 1973). Schedae ad floram Moldaviae et Dobrogeae exsiccatam a Horto Bot. Univ. „Al. I. Cuza" Iasssiensis edit., Cent. I., 24 pag., Cent. II, 24 pag. Cent. III, 26 pag. (in Romanian).

Tutin, T., G., Burges, N., A., Chater, A., O., Edmonson, J., R., Heywood, V., G., Moore, D., M., Valentine, D., H., Walters, S., M. & Webb, D., A. (1996) (Ed(s).s.). *Flora Europaea*, 1, 2nd ed., Cambridge University Press, Cambridge

Tutin, T., G, Heywood, V., H, Burges, N., A, Moore, D. ,M, Valentine, D.,.H, Walters, S. M & Webb, D., A. (Ed(s).) (1964-1980). *Flora Europaea*. vols. 1-5. Cambridge University Press, Cambridge

Tutin, T., G., Heywood, V., G., Burges, N., A., Valentine, D., H., Walters, S., M. & Webb, D., A. (1964-1980) (Ed(s).). *Flora Europaea*, 1-5, Cambridge University Press, Cambridge

Zahariadi, C. (1964). Taxonomy of some phanerogams from R.P.R. flora. *Stud. Cercet. Biol.-Bot.*, Vol. 16, No. 3, pp. 205-220 (in Romanian).

Zahariadi, C. (1965). Fîntîniţa Natural Reserve. *Stud. Cercet. Biol.-Bot.*, Vol. 17, No. 4-5, pp. 497-502 (in Romanian).

Zahariadi, C. & Negrean, G. (1969). Liliaceae nesemnalate sau dubioase in Romania, *Stud. Cercet. Biol.-Bot.*, Vol. 21, No. 6, pp. 403-408 (in Romanian).

# 12

# Aquatic Crustaceans in the Driest Desert on Earth: Reports from the Loa River, Atacama Desert, Antofagasta Region, Chile

Patricio De los Ríos-Escalante[1] and Alfonso Mardones Lazcano[2]
*[1]Universidad Católica de Temuco, Facultad de Recursos Naturales,*
*Escuela de Ciencias Ambientales,*
*[2]Universidad Católica de Temuco, Facultad de Recursos Naturales,*
*Escuela de Acuicultura,*
*Chile*

## 1. Introduction

Nortern Chile includes the Atacama Desert, which is characterized by scarce, shallow, saline shallow endorheic lakes, small intermittent streams, and a few rivers (Niemeyer & Cereceda, 1984). One of the main rivers of this zone is the Loa River which is 440 km long and is the longest river in Chile. The river basin occupies 33,570 km². Situated in the Antofagasta region of Chile, it originates in the Andes Mountains, close to Bolivia, and receives flow from four tributaries rivers: the Salado, San Pedro, Toconce, and San Salvador Rivers. The Loa basin contains two reservoirs, the Conchi and Sloman (Pumarino, 1978; Niemeyer & Cereceda, 1984; Gutierrez et al., 1998). Studies of the native Loa aquatic fauna to date have only described the presence of the freshwater shrimp, *Cryphiops caementarius* (Molina, 1782) (Jara et al., 2006), amphipods such as *Hyalella fossamanchini* and *H. kochi* (González, 2003), and the native silverside *Basilichthys* near *semotilus*; Dyer 2000a,b; Ruiz & Marchant 2004; Vila et al., 2006). Introduced fish taxa in the Loa River and its tributaries include *Oncorhynchus mykiss* and *Salmo trutta* (Pumarino, 1978; Wetzlar, 1979; Iriarte et al., 2005). Overall, there is little detailed information about aquatic species, their distribution, population status, and associations in the Loa River.

The Loa River and its basin are subjected to strong human influences due to water use for mining, domestic needs, and agriculture (Niemeyer & Cereceda, 1984; Gutiérrez et al., 1998). The biological resources of the basin also are known to be under pressure from fishing, particularly in the case of the shrimp, *C. caementarius* (Meruane et al., 2006a,b), and rainbow and brown trout populations (Pumarino, 1978; personal observations). Nevertheless, it is difficult to determine the status of the aquatic fauna in the Loa basin due to a lack of study stemming largely from the inaccessibility of water bodies in northern Chile (Chong, 1988). Our study aims to rectify this lack of knowledge by determining the faunal species inhabiting the Loa River, characterizing species associations along altitudinal and spatial gradients within the river, and testing for regulating factors potentially influencing aquatic community composition.

## 2. Material and methods

The Loa River originates in the Andes Mountains close to the boundary with Bolivia, and flows first from north to south, proceeds in a westerly direction, changes back to a northerly direction, and finally flows west to its confluence with the Pacific Ocean (Fig. 1). We compiled information from two two data sets, one collected during field studies in January 2008 that included nine sites along the Loa River (De los Ríos et al., 2010; Table I), and the other from field studies in 1980 from lower reaches of the Loa River between the outlet to the middle zone. The latter dataset included *C. caementarius* habitat (Alfaro et al., 1980; Table II). Altitude, in meters above sea level, was recorded for all sites. We tested for relationships between species richness and two physical variables, salinity (using a YSI-30 sensor) and altitude, using correlation coefficients (Rho-Spearman), calculated in the software SPSS v.12.

Crustacean community structure was explored using a co-occurrencenull model analysis, which tests whether species co-occur less frequently than expected by chance (Gotelli, 2000). A checkerboard score ("C-score") was calculated based upon an absence/ presence matrix, representing a quantitative index of co-occurrence. We used this method for analysis and evalutated results in relation to those of other studies (Tondoh, 2007; De los Ríos et al., 2008; De los Ríos-Escalante, 2011). A community may be structured by competition when the C-score is significantly larger than that expected by chance (Gotelli, 2000, 2001). In order to determine whether a particular score is statistically significant, a set of randomizations of the species occurrence data are performed and a null distribution for the coexistence index is created. Gotelli & Entsminser (1997) Tiho & Johens (2006), and Tondoh (2007), suggested the following three statistical models for creating randomized communities, with the species placed in rows and the sites in columns:

1. Fixed-Fixed Model. In this model, the row and column sums of the matrix are preserved. Thus, each random community contains the same number of species as the original community (fixed column) and each species occurs with the same frequency as in the original community (fixed row).
2. Fixed-Equiprobable Model. In this algorithm only the row sums are fixed and the columns are treated as equiprobable. This null model considers all the samples (column) as equally available for all species.
3. Fixed-Proportion Model. In this algorithm species occurrence totals are maintained as in the original community, and the probability that a species occurs at a site (column) is proportional to the column total for that sample.

All three of these models exhibit fairly reasonable combinations of Type I and Type II error rates, although model 3 has a high Type I error rate (false positives) using the C-score index, with differences in underlying assumptions and behaviour (Gotelli, 2000). The fixed-fixed model is suggested to be most appropriate for island species lists, in which species-area effects are expected, while the fixed-equiprobable model would be most appropriate for standardized samples in a homogeneous environment (Gotelli, 2000). The fixed-proportional algorithm represents an intermediate model, which might be most appropriate in our system due to habitat connectivity and heterogeneity, as well as differences in depth and width along the river. Differing results among models can provide insights into community structure. The null model analysis was using the software ECOSIM, version 7.0 (Gotelli & Etsminger, 1997).

Aquatic Crustaceans in the Driest Desert on Earth: Reports from the Loa River, Atacama Desert, Antofagasta Region, Chile

247

| | Santa Bárbara | Conchi | Salado | Chiuchiu | Chacance | Salvador | Iberia | Sloman | Quillagua |
|---|---|---|---|---|---|---|---|---|---|
| Geographical location (South latitude / West longitude) | 21° 58.7' 68° 36.7' | 22° 00.5' 68°36.7' | 20° 20.4' 68° 39.2' | 20° 20.4' 68° 39.2' | 22° 23.8' 69° 31.6' | 22° 23.8' 69° 31.6' | 21° 55.2' 69° 33.6' | 21° 51.2' 69° 30.9' | 21° 39.5' 69° 32.2' |
| Altitude (m a.s.l) | 3304 | 3272 | 2784 | 2768 | 1328 | 1328 | 1118 | 1085 | 866 |
| Salinity (g/l) | 1.8 | 1.5 | 4.2 | 1.5 | 8.3 | 4.8 | 10.4 | 11.5 | 12.0 |
| Cladocera | | | | | | | | | |
| Ceriodaphnia dubia (Richard, 1894) | | x | | | | | | | |
| Daphnia pulex (De Geer, 1877) | | x | | | | | | | |
| Chydorus sphaericus (O.F. Müller, 1785) | | | x | | | | | | |
| Copepoda | | | | | | | | | |
| Eucyclops serrulatus (Fisher, 1851) | | x | x | | | | | | |
| Unidentified cyclopoida | | | x | | | | | | |
| Tigriopus sp. | | | | | | | | | |
| Ostracoda | | | | | | | | | |
| Heterocypris panningi (Brehm, 1934) | | | x | | x | | x | x | |
| Cubacandona spp. (Broodbakker, 1983) | | | | | | | x | | |
| Amphipoda | | | | | | | | | |
| Hyalella fossamanchini (Cavalieri, 1959) | x | x | x | | | x | | | |
| H. kochi (González & Watling, 2001) | x | x | x | | | | | | |

Table 1. Geographical location, altitude, conductivity, salinity, and species reported for the studied sites in the Loa River during the 2008 sampling period.

## 3. Results and discussion

Our results revealed the presence of a small number of crustacean species in the Loa River (Table 1). . In the Quillagua River, no crustacean species were found, but the introduced fish species *Gambussia affinis* was abundant, whereas in Sloman and Chacance River, only the ostracod *Heterocypris panningi* was detected. By contrast, the Salado River harbored unidentified cyclopoid copepods, the cladoceran *Chydorus sphaericus*, *H. panningi*, and the amphipods *H. fossamanchini* and *H. kochi*. The Conchi Reservoir supported similarly high species richness, with *H. fossamanchini*, *H. kochi*, *Eucyclops serrulatus*, *Ceriodaphnia dubia*, and *Daphnia pulex* (Table I). Among data collected in 1980 in the lower Loa River basin, *H. panningi* was reported at all sites, and unidentified Harpacticoida were reported at all sites except in the Sloman reservoir, and *E. serrulatus* was found in El Borax, La Poroma and Angostura (Table 2).

| | Chacance | San Lorenzo | El Borax | Sloman | La Poroma | Angostura | Desembocadura |
|---|---|---|---|---|---|---|---|
| Geographical location | 22° 23.8′ | 22° 12.0′ | 21° 06.0′ | 21° 51.2′ | 21° 39.0′ | 21° 36.0′ | 21° 27.0′ |
| (South latitude / West longitude) | 69° 31.6′ | 69° 24.0′ | 69° 30.0′ | 69° 30.9′ | 69° 30.0′ | 69° 36.0′ | 71° 00.0′ |
| Altitude (m a.s.l) | 1328 | 1180 | 1100 | 1085 | 800 | 700 | 1.5 |
| Salinity (g/l) | 4.99 | 5.64 | 6.33 | 6.33 | 7.46 | 7.61 | 8.72 |
| Cladocera | | | | | | | |
| | | | | | | | |
| Cladocera | | | | | | | |
| | | | | | | | |
| *Ceriodaphnia dubia* (Richard, 1894) | | | | | | | |
| *Daphnia pulex* (De Geer, 1877) | | | | | | | |
| *Chydorus sphaericus* (O.F. Müller, 1785) | | | | | | | |
| Copepoda | | | | | | | |
| | | | | | | | |
| *Eucyclops serrulatus* (Fisher, 1851) | | | x | | x | x | |
| Unidentified cyclopoida | | | | | | | |
| | | | | | | | |
| Unidentified harpacticoida sp. | x | x | x | x | x | x | x |
| | | | | | | | |
| Ostracoda | | | | | | | |
| | | | | | | | |
| *Heterocypris panningi* (Brehm, 1934) | x | x | x | | x | x | x |
| *Cubacandona* spp. (Broodbakker, 1983) | | | | | | | |
| Amphipoda | | | | | | | |
| *H. fossamanchini* (Cavalieri, 1959) | | | | | | | |
| *H. kochi* (González & Watling, 2001) | | | | | | | |

Table 2. Geographical location, altitude, conductivity, salinity, and species reported for the studied sites in the Loa River during the 1980 sampling period.

Among data collected in 2008, Spearman rho correlation values indicated no relation between species number and salinity ($r$ = -0.39; $P$ = 0.149) however, a significant and positive relationship was detected between species richness and altitude ($r$ = 0.61; $P$ = 0.041) (Table 2). In contrast, among data collected in 1980, the Spearman rho correlation values indicated no relation between species number and salinity ($r$ = -0.391; $P$ = 0.149) or between diversity and altitude ($r$ = 0,189; $P$ > 0,05) (Table 3)).

|  | Salinity | | Altitude | |
|---|---|---|---|---|
| Number of species | R = -0.391; P = 0.149 | | R = 0.610; P = 0.041 | |
|  | Results of null model analysis | | | |
| Simulation | Observed index | Mean | Standard effect size | P |
| Fixed-Fixed | 1.250 | 1.132 | 1.066 | 0.155 |
| Fixed- Proportional | 1.250 | 1.105 | 0.443 | 0.372 |
| Fixed- Equiprobable | 1.250 | 1.873 | -2.301 | 0.982 |

Table 3. Results of correlation and null-model analyses. Correlation coefficients are provided between species richness and conductivity, salinity, and altitude, respectively, along the Loa River (sampling period 2008). The null-model analysis (see text) suggests that crustacean community structure is random.

The results of the null model analysis among all simulations within 2008 data revealed that crustacean community composition appears to be random (Table 4). However, small sample size can mask other underlying patterns (De los Ríos-Escalante, 2011). A different situation was observed within the 1980 data, where the fixed-proportional denoted the presence of regulatory factors, and the fixed-equiprobable analysis indicated a weak presence of regulator factor presence (Table 4).

Unfortunately chemical and other physical parameters were not measured in situ; however, human influences on the river may be potentially regulatory factors. Such influences have been documented on central Chilean rivers (Figueroa et al., 2003). The upper reaches of the Loa River are subject to lower levels of anthropogenic impact, given that human population size is small in relation to that in the lower reaches, which are subject to urbanization and and mining activities (Alvarez, 1999; Melcher, 2004).

The present literature on Chilean rivers only contains descriptions of invertebrate species associations in south-central rivers, and most studies attempt to use such information for bio-indication of water quality (Figueroa et al., 2003, 2007). These studies generally report differences in macroinvertebrate assemblages (mainly insects and crustaceans) in relation to water pollution along the river courses (Figueroa et al., 2003, 2006). These studies indicate that the fluvial aquatic fauna of Chilean is regulated by deterministic factors, namely water quality as influenced by the level of human alteration (Figueroa et al., 2003, 2006). However, Figueroa et al., (2003, 2006) only described the riverine biota to the family level, and this lack of species-level data precludes more precise statistical treatments of community structure (Gotelli & Graves, 1996; Gotelli & Ellison, 2000; Jaksic, 2001). Our study included only crustaceans, but emphasized species-level taxonomy, whereas Figueroa et al., (2003, 2006) studied all benthic invertebrates including aquatic insects, which are diverse compared to crustaceans. Future studies of riverine community structure should expand upon our studies in order to include aquatic insects at the species level. Aquatic insects are likely to ecologically interact with crustaceans in fluvial benthic communities, in turn affecting abundances and patterns of co-occurrence among both taxonomic groups (Parra et al, 2001).

Biogeographically, the presence of the amphipods *H. fossamanchini* and *H. kochi* in the Loa River has been previously described by Gonzalez (2003) and Jara et al. (2006). However, those descriptions did not specify details about the localities at which the species were found. Thus, our study presentsnew information to the knowledge of crustacean distribution. The absence of specimens of the northern river shrimp, *C. caementarius*, indicates that this species is threatened by excessive fishing as human food (Jara et al., 2006), and probably also by the presence of exotic fishes, such as *Gambusia affinis¸ Oncorhynchus mykiss* and *Salmo trutta*. All of those fish species are active predators on native aquatic invertebrates (Leyse et al., 2005).

Regarding the presence of ostracods in Chile, there are currently only records of *H. panningi* and *Cucacandona* spp., which also have been reported for other South American inland waters (Martens & Behen, 1994). Spatial differences in community composition were detected, with the presence of ostracods related to the absence of amphipods in the lower reaches of the Loa River, reaches that exhibit relatively high salinity and conductivity (Table 3). In contrast, the high-altitude zones of the Loa, as well as the Salvador River, have relatively lower salinity, with species associations between the amphipods *H. fossamanchini* and *H. kochi* (Table 1). The presence of microcrustaceans, specifically copepods and cladocerans in Conchi Reservoir, corroborates their distribution as reported in the literature (Araya & Zúñiga, 1985; Reid, 1985, Ruiz & Bahamonde, 1989). Although distributional data denote a segregation of species into low and high altitude reaches (Table 1), with significantly higher species richness at higher altitudes, the null model results suggest that crustacean species associations are random (Table 3). These results are seemingly in disagreement; however, the low total species richness and small number of study sites may preclude the detection of a true underlying pattern (De los Ríos-Escalante, 2011).

Nevertheless, the negative relationship between species richness with salinity and altitude (Table 4), is explained by predation by *C. caementarius* on microcrustaceans (Alfaro et al., 1980; López et al., 1986). This may represent a top-down trophic cascade that affects community structure. This scenario would be similar to descriptions of Chilean (Soto et al., 1994; De los Ríos, 2003) and Argentinean (Reissig et al., 2004) Patagonian lakes. Considering the present results, two main forces likely regulate aquatic community structure: salinity variation and predator presence. Such a combination of factors have been reported in shallow Andean mountain saline and sub-saline lakes (Hurlbert et al., 1986; De los Ríos-Escalante, 2011).

Our results contribute to understanding of the community ecology of inland water invertebrates in northern Chilean streams. Unfortunately, these ecosystems have been understudied to date due primarily to problems of accessibility. Further work on the crustacean and other aquatic faunae of northern Chilean rivers is clearly needed.

|  | Salinity | | Altitude | |
|---|---|---|---|---|
| Number of species | R = -0.031; P > 0,05 | | R = 0,189; P > 0,05 | |
| | Results of null model analysis | | | |
| Simulation | Observed index | Mean | Standard effect size | P |
| Fixed-Fixed | 2.833 | 2.764 | 0.272 | 0.346 |
| Fixed- Proportional | 2.833 | 1.506 | 1.715 | 0.039 |
| Fixed- Equiprobable | 2.833 | 1.665 | 1.523 | 0.057 |

Table 4. Results of correlation and null-model analyses. Correlation coefficients are provided between species richness and conductivity, salinity, and altitude, respectively, along the Loa River (sampling period 1980). The null-model analysis (see text) suggests that crustacean community structure is random.

## 4. Acknowledgements

The present study was supported by a grant to PR from the Research Direction of the Catholic University of Temuco (Funding for Development of Limnology). Also, we gratefully express our appreciation to Eliana Ibáñez, Elizabeth Escalante, Luis Escalante, and Elisa Pistán for their helpful logistical assistance during the field work.

## 5. References

Alfaro, D., Bueno G., Mardones A., Neira A., Segovia E. & Venegas E. (1980). Contribución al conocimiento de Cryphiops caementarius (molina, 1782) en el Río Loa, Antofagasta. Seminario para optar al título de Ingeniero (E) en Acuicultura. Universidad de Chile. Instituto de Investigaciones Oceanológicas. 58 pp.

Alvarez, A. (1999). Geo Biografía. Impresos Universitaria S.A., Santiago de Chile. 176 p.

Araya, J.M. & Zúñiga, L.R. (1985). Manual taxonómico del zooplancton lacustre de Chile. Boletín Limnológico, Universidad Austral de Chile, 8: 1-169 p.

Chong, G., (1988). The Cenozoic saline deposit of the Chilean Andes between 18°00 and 27°00 south latitude. Lecture Notes on Earth Sciences, 17: 137-151.

De los Ríos, P., (2003). Efectos de las disponibilidades de recursos energéticos, estructurales y de protección sobre la distribución y abundancia de crustáceos zooplanktónicos lacustres chilenos: 1-163. Doctoral Thesis, Austral University of Chile, Science Faculty.

De los Ríos, P., (2008). A null model for explain crustacean zooplankton species associations in central and southern Patagonian inland waters. Anales del Instituto de la Patagonia, 36: 25-33.

De los Ríos, P., Rivera N., & Galindo, M. (2008). The use of null models to explain crustacean zooplankton associations in shallow water bodies of the Magellan region, Chile. Crustaceana, 81: 1219-1228.

De los Ríos P., (2008). A null model for explain crustacean zooplankton species associations in central and southern Patagonian inland waters. *Anales del Instituto de la Patagonia* 36: 25-33.

De los Ríos-Escalante P., (2010). Crustacean zooplankton communities in Chilean inland waters. *Crustaceana Monographs* 12: 1-109 p.

De los Ríos-Escalante P., (2011). A null model to study community structure of microcrustacean assemblages in northern Chilean shallow lakes. *Crustaceana,* 84: 513-521.

Dyer, B, (2000a). Systematic review and biogeography of the freshwater fishes of Chile. *Estudios Oceanologicos* 19: 77-98.

Dyer, B, (2000b). Revisión sistemática de los pejerreyes de Chile (Teleostei, Atheriniformes). *Estudios Oceanologicos* 19: 99-127.

Figueroa, R., Valdovinos C., Araya E. & Parra, O. (2003). Macroinvertebrados bentónicos como indicadores de calidad de agua de ríos del sur de Chile. *Revista Chilena Historia Natural,* 76: 275-285

Figueroa, R, Palma A., Ruiz V. & Niell, X., (2007). Análisis comparativo de índices bióticos utilizados en la evaluación de la calidad de las aguas en un río mediterráneo de Chile: río Chillán, VIII Región. *Revista Chilena Historia Natural,* 80: 225-242.

González, E.R., (2003). Los anfípodos de agua dulce del género *Hyalella* Smith, 1874 en Chile (Crustacea: Amphipoda). *Revista Chilena Historia Natural,* 76: 623-637.

Gotelli, N.J., (2000). Null models of species co-occurrence patterns. *Ecology,* 81: 2606-2621.

Gotelli, N.J., (2001). Research frontiers in null model analysis. *Global Ecology and Biogeography,* 10: 337-343.

Gotelli, N.J. & G.L. Entsminger. (1997). EcoSim: Null models software for ecology. Version 7. Acquired Intelligence Inc. & Kesey-Bear. Jericho, VT 05465. http://garyentsminger.com/ecosim.htm.

Gotelli, N.J. & Graves, G.R. (1996). Null models in Ecology. Smithsonian Institution Press, Washington, USA, 357 p.

Gotelli, N.J. & Ellison, A.M. (2000). A primer of Ecological Statistics. Sinauer Associated Inc., Publishers, Sunderland, Massachussetts, U.S.A. 510 p.

Gutiérrez, J.R., López-Cortes F., & Marquet, P.A. (1998). Vegetation in an altitudinal gradient along the Rio Loa in the Atcama desert of northern Chile. *Journal of Arid Environments,* 40: 383-399

Hurlbert, S.H., Loayza, W., & Moreno, T., (1986). Fish-flamingo-plankton interactions in the Peruvian Andes. *Limnolology & Oceanography,* 31: 457-468.

Iriarte, A., Lobos G.A., & Jaksic, F. M., (2005). Invasive vertebrate species in Chile and their control and monitoring by governmental agencies. *Revista Chilena de Historia Natural,* 78: 143-151.

Jaksic, F., 2001. Ecología de Comunidades. Ediciones Pontificia Universidad Católica de Chile, Santiago de Chile.

Jara, C.G., Rudolph E.H., & González, E.R., (2006). Estado de conocimiento de los malacostráceos dulceacuícolas de Chile. *Gayana (Concepción),* 70: 40-49.

Leyse, K., S.P. Lawler & Strange, T., (2005). Effects of an alien fish *Gambusia affinis*, on an endemic fairy shrimp, *Linderella occidentalis*: implications for conservation of diversity in fishless waters. *Biological Conservation*, 118: 57-65

López, M., Segovia, E., & Alfaro, D., (1986). Microalgas: su importancia como recurso alimentario del Camarón de Río del Norte de Chile, *Cryphiops caementarius* (Molina, 1782). *Medio Ambiente*. 78: 39-47.

Martens, K.& Behen, F. (1994). A check list of the recent non marine ostracods (Crustacea, Ostracoda) from the inland waters of South America and adjacent islands. Ministerie des Affaires Culturelles. Travaux Scientifiques du Musee National D'Histoire Naturelle de Luxembourg, 1-84 p.

Melcher, G. (2004). El norte de Chile, su gente, desiertos y volcanes. Editorial Universitaria, Santiago de Chile. 149 p.

Meruane, J., Rivera J., Morales M., Morales C., Galleguillos C., & Hosokawa, H. (2006a). Juvenile production of the freshwater prawn *Cryphiops caementarius* (Decapoda: Palaemonidae) under laboratory conditions in Coquimbo, Chile. *Gayana (Concepción)*, 70: 228-236.

Meruane, J., Rivera J., Morales M., Morales C., Galleguillos C., Rivera M.A. & Hosokawa, H. (2006b). Experiencias y resultados de investigaciones sobre el camarón de río del norte *Cryphiops caementarius* (Molina 1782) (Decapoda: Palaemonidae): historia natural y cultivo. *Gayana (Concepción)*, 70: 280-292.

Niemeyer, H., & Cereceda, P. (1984). Hidrografía. Colección Geografía de Chile, VIII. Military Geographical Institute (Chilean Army), Santiago de Chile. 320 p.

Parra, O, N. Della Croce & Valdovinos, C. (2001). Elementos de limnología teórica y aplicada. Microart's Edizioni, Italia. 303 pp.

Pumarino, H., (1978). El Loa de ayer y hoy. Editorial Universitaria, Santiago de Chile. 203 p.

Reid, J., (1985). Chave de identificao e lista de referencias bibliográficas para as espécies continentais sudamericanas de vida libre da orden Cyclopoida (Crustacea, Copepoda). *Boletim Zoologico da Univerdidade do Sao Paulo*, 9: 17-143.

Reissig, M., B. Modenutti, Balseiro E. & Queimaliños, C., (2004). The role of the predaceous copepod *Parabroteas sarsi* in the pelagic food web of a large deep Andean lake. *Hydrobiologia*, 524: 67–77.

Soto D., Campos H., Steffen, W., Parra O., & Zúñiga, L. , (1994). The Torres del Paine lake district (Chilean Patagonia): a case of potentially N-limited lakes and ponds. *Archiv für Hydrobiologie*, 99: 181-197.

Tiho, S. & Johens, J., (2007). Co-occurrence of earthworms in urban surroundings: a null models of community structure. *European Journal of Soil Biology*, 43: 84-90.

Tondoh, J.E., (2006). Seasonal changes in earthworm diversity and community structure in central Côte d'Ivoire. *European Journal of Soil Biology*, 42: 334-340.

Ruiz, V. & Marchant, M., (2004). Ictiofauna de aguas continentales chilenas. Dirección de Docencia, Universidad de Concepción. 356 pp.

Silva, A., Franco, L. , & Iturra, N., (1985). Antecedentes sobre la reproducción y alimentación de la trucha arco iris *Salmo gairdneri* del Embalse Conchi, Antofagasta, Chile. *Biologia Pesquera*, 14: 32-39.

Vila, I., Pardo, R. , Dyer, B. , & Habit, E. , (2006). Peces limnicos: diversidad, origen y estado de conservación. In: Vila I, A Veloso, R Schlatter & C Ramírez (Eds). Macrófitas y vertebrados de los sistemas límnicos de Chile. Editorial Universitaria, Santiago de Chile, 73-102 p.

Wetzlar, H. (1979). Beiträge zur biologie und bewirschaftung von Forellen (Salmo gairdneri und S. trutta) in Chile. PhD Thesis, Universität Freiburg, 1-264 p.

# Part 4

## Evolutionary Biogeography of Macrotaxa

# 13

# Biogeography of Dragonflies and Damselflies: Highly Mobile Predators

Melissa Sánchez-Herrera and Jessica L. Ware
*Department of Biology, Rutgers The State University of New Jersey, Newark Campus,*
*USA*

## 1. Introduction

Dragonflies (Anisoptera) damselflies (Zygoptera) and Anisozygoptera comprise the three suborders of Odonata ("toothed ones"), often referred to as odonates. The Odonata are invaluable models for studies in ecology, behavior, evolutionary biology and biogeography and, along with mayflies (Ephemeroptera), make up the Palaeoptera, the basal-most group of winged insects. The Palaeoptera are thought to have diverged during the Jurassic (Grimaldi and Engel, 2005; Thomas et al., 2011), and as the basal-most pterygote group, odonates provide glimpses into the entomological past. Furthermore, few other insect groups possess as strong a fossil record as the Odonata and its precursors, the Protodonata, with numerous crown and stem group fossils from deposits worldwide.

Their conspicuous behavior, striking colors and relatively small number of species (compared to other insect orders) has encouraged odonatological study. Odonates are important predators during both their larval and adult stages. They are often the top predators in freshwater ecosystems, such as rivers and lakes. One of their most remarkable traits, however, is their reproductive behavior, which takes place in a tandem position with the male and female engaging in a "copulatory wheel" (Fig. 1).

Fig. 1. Odonata copulatory wheel (modified from Eva Paulson illustration Aug, 2010).

During the last 5 decades, our understanding about the ecology and evolution of Odonata has increased dramatically (e.g., Cordoba-Aguilar, 2008). A fair odonate fossil record coupled with recent advances in molecular techniques, have inspired several biogeographical studies of Odonata. The aim in this chapter is to review current understanding of odonate biogeography, to add new insights about the evolutionary history of this order, and to evaluate the contribution of odonatology to our overall understanding of biogeographical patterns. Furthermore, we discuss how this information may be used to develop predictions about relationships between current environmental alterations, such as climate change and deforestation, may affect the ranges and dispersal of Odonata. We also frame Odonata biogeography in the context of developing better mechanisms for the conservation of these important insects.

## 2. What are dragonflies and damselflies? Real hunters

The Odonata are one of the most ancient groups of extant insects. Fossils of the stem group order Protoodonata (stem fossil group, containing no extant representatives), recognizable progenitors of modern day dragonflies, date from middle Carboniferous Serpukhovian sediments formed almost 325 million years ago (Brauckman & Zessin, 1989). One such protoodonate, *Meganeura*, had a wingspan of over 30 cm (Brongniart, 1885; see Fig. 2 of *Typus permianus*, another Meganeuridae), and extremely dense wing venation. Protoodonate wing shapes suggest that they may have been capable of fast flight and although they were likely as voracious as present day dragonflies (Corbet, 1999), they may have been less agile due to their large size.

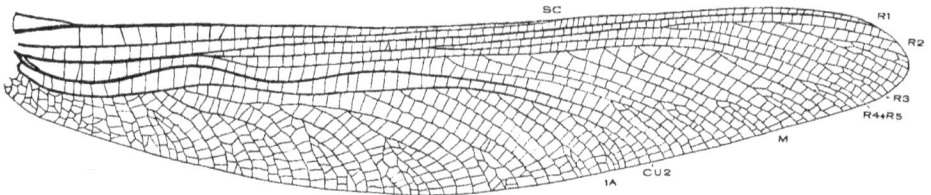

Fig. 2. The Protoodonate representative: *Typus permianus* forewing (modified from Carpenter 1931).

The first crown group odonate fossils (crown group fossil = extant representatives exist) date from the lower Permian period (ca. 250 million years ago); these fossils are not the huge Carboniferous monsters of Protoodonata, but rather the Protoanisoptera and early proto-zygopterans (Clarke, 1973; Carpenter, 1992; Wootton, 1981) were similar in size to modern dragonflies. Although all modern odonates have aquatic or semi-terrestrial juvenile stages (Watson, 1981), there is little fossil evidence to support that early odonates had aquatic larvae (Pritchard, 1993; Wootton, 1981). Larvae fossils are unknown before the Mesozoic, but by the Middle Triassic there is evidence the characteristic prehensile labium, ubiquitous in modern-day larvae, and of differences among larval habits in some zygopterans (coenagrionids) and anisozygoptererans. The former suggest that larvae possibly became aquatic during the Lower Permian (Corbet, 1999).

Odonata are relatively generalized insects; as members of the "Hemimetabola" they do not have a pupal stage between the larvae and the adult stage. Their larvae are confined to fresh or brackish waters and they develop rudimentary wing covers when they are about half grown. They show an incredible diversity of forms depending on the characteristics of the aquatic niche occupied, whether lentic or lotic.

## 2.1 Ecology and behavior

Dragonflies are voracious predators in larval and adult life stages, feeding exclusively on living prey. Larvae detect prey visually and with mechanoreceptors (Fig. 3), primarily as sit-and-wait predators. This is a successful strategy, in part due to a particularly distinctive odonate larval characteristic: prehensile mouthparts (labium) that can be extended to capture prey (Fig. 4 A). Several larger odonate taxa are considered top predators in the food chain of their freshwater ecosystems. As adults, odonates usually eat small flying insects, which they are able to detect using their globe-like eyes. Their spiny legs are used as a basket to net prey and move it forward during flight to their strong mandibles (Fig. 4 A). Most Odonata species feed during flight, which is not an easy task despite their being exceptional flyers (Corbet, 1999).

A.                                                 B.

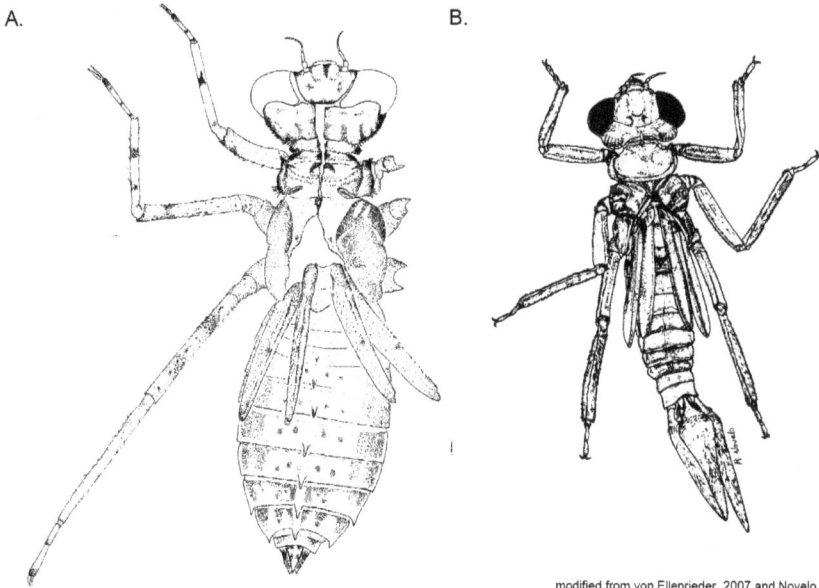

modified from von Ellenrieder, 2007 and Novelo, 1995.

Fig. 3. Odonata larvae. A. Dragonfly *Macrothemis hageni*. B. Damselfly *Amphypteryx longicaudata*.

Dragonflies and damselflies have long slender abdomens, short antennae, huge spherical eyes, (so large that they can sometimes make up the bulk of the head). Larval Zygoptera have caudal gills and swim by paddling with their legs, whereas larval Anisoptera have largely internal gills and move by jetting water from their abdomens. Adult Odonata have long wings with a conspicuous nodus and a pterostigma (Fig. 4 B). The latter is weighted and helps stabilize the wing during flight (Norberg, 1972). The wing apparently bends and flexes rather widely around the nodus during flight.

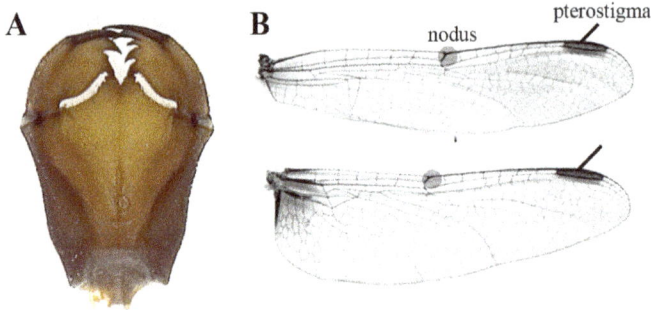

Fig. 4. A. *Neopetalia punctata* prenhesile labium. B. *Pachydiplax longipennis* wings showing the conspicuous nodus and pterostigma. Images: Jessica Ware

Unique among all Ptergyota is the odonate method of copulation, which involves indirect fertilization. Male dragonflies and damselflies have secondary genitalia at the base of their abdomens. Sperm is produced in the testes and released from the abdomen tip and then placed in the secondary copulatory organs on the underside of the second segment of the abdomen prior to copulation. During copulation, the female receives sperm from the male's *vesica spermalis*, a secondary penile structure at the base of the male abdomen, into her *bursa copulatrix*, or sperm storage organ. Females can mate multiple times, storing sperm in their body for later use. In turn, the male secondary organ can remove or displace the sperm of previous matings using the *penes* to increase their chances for paternity. Sperm competition in odonates has made them a well-studied taxonomic group in the field of sexual selection and reproductive behavior.

Fig. 5. *Erythemis vesiculosa* eating a Satyrinae butterfly. Image: Dr. Godfrey Bourne.

## 2.2 Species diversity and biogeography

Recently, a monogeneric third suborder was recognized with two extant species from Japan and the eastern Himalayas (*Epiophlebia*, Selys 1889; ). Anisozygopterans have some features recalling Zygoptera, such as petiolate wings, and some of Anisoptera, such as robust abdomens. Recent phylogenetic studies have reconstructed this suborder within the Anisoptera, and thus (Anisoptera + Anisozygoptera) have been called Epiprocta (e.g. Bechly, 1996, Lohmann, 1996; Bybee et al., 2008). Currently taxonomy suggests that there are eleven families in Anisoptera: Aeshnidae, Austropetaliidae, Gomphidae, Petaluridae, Cordulegastridae, Neopetaliidae, Chlorogomphidae, GSI (sensu Ware et al., 2007), Corduliidae, Macromiidae and Libellulidae, with Epiophlebiidae from Anisozygoptera considered by some to be an twelfth family of dragonflies. Although both Zygoptera and Anisoptera have roughly 3000 species, Zygoptera are divided into 21 families: Amphipterygidae, Calopterygidae, Chlorocyphidae, Coenagrionidae, Dicteriadidae, Euphaeidae, Hemiphlebiidae, Isostictidae, Lestidae, Lestoidedidae, Megapodagrionidae, Perilestidae, Philogangidae, Platycnemidae, Platystictidae, Polythoridae, Protoneuridae, Pseudolestidae, Pseudostigmatidae, Synlestidae, and Thaumatoneuridae.

Present-day distribution of Odonata reflects millions of years of geographic isolation and dispersal, coupled with adaptation over 300 million years of climate variation (Fig. 6). This has contributed to considerable speciation and endemism (Samways, 1992, 2006), particularly in the tropics, although speciation has been elevated among several Holarctic

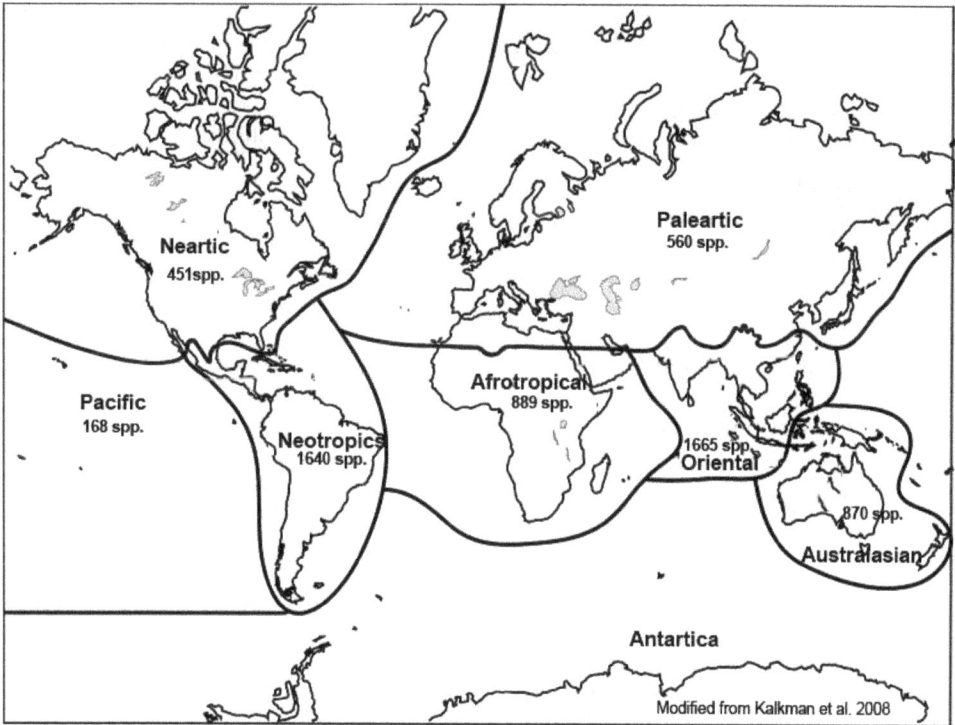

Fig. 6. Current odonate species diversity distribution.

taxa (e.g., Brown et al., 2000). In a warming global climate, current odonate biogeography will undoubtedly change. The habitat requirements of high elevation taxa may leave some montane specialists without suitable habitat (Samways, 1992, Stevens and Bailowitz, 2009), while tropical, warm-adapted taxa may expand their ranges to higher latitudes. Increasing isolation of populations in moist tropical environments due to deforestation, increasing temperatures, and flow regulation, with arid and unsuitable habitat in between (Samways, 2006), may lead to species loss.

## 2.3 Dispersal in Odonata, flight behavior and migration

Many Odonata are highly mobile and have varying levels of dispersal capabilities (e.g. Kormandy, 1961). *Anax*, the common green darner, *Pantala*, the wandering glider, and several other odonates, in fact, are capable of migrating very long distances. The north-south migration of *Anax junius* occurs for the most part in North America (e.g., Russell et al., 1998), with migrants moving up to 2800 km south (Wilkelski et al., 2006; May, 2008). Gene flow occurs among migrant and resident 'subpopulations' resulting in a panmictic population (Freeland et al., 2003; Matthews et al., 2007). *Pantala*, a highly migratory dragonfly, is even more cosmopolitan, with individuals found on all continents except Antarctica, although they not equally common on all continents (McLachlan, 1896; Wakana, 1959; Reichholf, 1973; Rowe, 1987; Russell et al., 1998; Corbet ,1999 ; Srygley, 2003; Feng et al., 2006; Buden, 2010). *Pantala* uses passive dispersal for example, to cross the Indian Ocean (e.g. Anderson et al., 2010;), while maintaining local island populations, such as those on Easter Island (Samways and Osbourne, 1998). The other 25-50 putative migratory Odonata (Kormondy, 1961) include *Sympetrum corruptum, Erythrodiplax umbrata*, several species of *Tramea*, and *Libellula quadrimaculata* (Artiss, 2004). Most odonates are thought to migrate by taking advantage of wind currents, and most migratory taxa are in the superfamily Libelluloidea.

Anisoptera are typically much stronger fliers than are damselflies. Heiser and Schmitt (2010) suggested that the relative dispersal capabilities of dragonflies and damselflies differentially influenced biogeographical patterns among Palaearctic taxa. An analysis of biogeographical patterns for that region showed that Anisoptera biogeographical patterns reflected historical vicariance and dispersal events, while relatively poorly dispersing Zygoptera showed distributions that seemed to reflect more the effect of climate (Heiser and Schmitt, 2010). However, dispersal capability alone does not determine observed biogeographical patterns. For example, *Hemicordulia*, appears to be capable of dispersing but may be a poor competitor (Dijkstra, 2007a, b), and its distribution may be more a reflection of simple vicariance.

## 3. Anisoptera phylogeny

Anisoptera are unequivocally monophyletic (e.g., Rehn 2004; Carle et al., 2008; Bybee et al., 2008). Frustratingly, interfamilal relationships among taxa have remained in conflict due to disagreements in recent phylogenetic hypotheses. In particular disagreement, yet of great interest, is the placement of Gomphidae, which has been recovered as sister to the Libelluloidea (e.g., Bybee et al., 2008, one analysis) and as sister to the Petaluroidea (e.g., Letsch, 2007). Gomphidae and Libelluloidea share exophytic oviposition behavior (i.e., eggs laid outside of plant tissue) and both have reduced or vestigal ovipositors, their egg laying aparati. Whether this is a synapomorphy or due to convergence is of phylogenetic interest.

Molecular phylogeographical analyses in Anisoptera are few, but several non-molecular studies directly or tangentially discuss anisopteran biogeography (e.g., Carle, 1995; Leiftinck, 1977). Further complicating matters is the fact that there are relatively few studies that have incorporated fossil and extant taxa in such analyses (with exceptions, see Bybee et al., 2008). With as geographically vast and geologically ancient a group as dragonflies, biogeographical analyses failing to incorporate fossils may not reveal much of their true history and the potential past impacts of ancient vicariant events.

### 3.1 Biogeography of the dragonfly superfamily Libelluloidea and Australian endemism

Ware et al. (2008) used molecular data to analyze the biogeography of an anisopteran taxon, the libelluloid *Syncordulia*. Their analysis suggested that the biogeography of the endemic South African taxon *Syncordulia* was a result of a southwestern Cape origin, approximately 60 million years ago. *Syncordulia* is a member of the family *Synthemistidae s.s.* (called "GSI" by Ware et al., 2007; previously considered to be members of the Corduliidae *s.l.*). The other taxa in this family are mostly Australasian endemics. The region studied has a high level of dragonfly endemism. Within the GSI, only *Gomphomacromia* (South American), *Idionyx* and *Macromidia* (IndoMalayan), and *Oxygastra* (Europe) are found outside of Australia and New Zealand, although other New World taxa, such as *Lauromacromia*, or *Neocordulia*, and African taxa such as *Neophya*, whose phylogenetic position have yet to be determined may ultimately be determined to be members of the Synthemistidae. The fossil record, however, suggests a more widespread distribution for several present-day Australasian endemics. The Mesozoic fossil taxon *Cretaneophya*, for example, thought to be sister to the extant West African *Neophya*, has been found in fossil deposits in Southeastern England (Jarzembowski & Nel, 1996). Similarly, the Argentinian fossil *Palaeophya argentina* is a putative member of the Cordulephyidae, a taxonomic group whose extant representatives are restricted to Australia (Petrulevicius & Nel, 2009).

Libelluloidea has been estimated to diverge during the Jurassic (e.g., Thomas et al., 2011) or Early Cretaceous (Jarzembowski & Nel, 1996; Fleck et al., 2008). Although at that time the continents were still in close proximity, the break-up of the supercontinent of Pangaea created the southwest Indian Ocean rift, splitting South America + Africa from East Gondwanaland and moving India away from Antarctica, and the North Atlantic- Caribbean rift, which separated Laurasia from South America and Africa (Dietz & Holden, 1970). If ancestral Libelluloidea (Fig. 7 A, B)were present on all landmasses, the subsequent isolation resulted in geographical vicariance that may have influenced divergence. Our unpublished estimate for the divergence of non-cordulegastrid taxa was 132 Mya, based on molecular data and estimated using a BEAST Bayesian analysis. This age estimate is similar to that of Carle (1995), who suggested that the radiation of non-cordulegastrid Libelluloidea began 'at least 140 million years ago'. During the early Cretaceous, Gondwanaland began to break apart more fully (e.g., Veevers, 2004), creating geographical barriers to dispersal and the isolation of populations. Vicariant events such as those have been suggested to drive the rate of speciation (e.g., Nelson, 1969; Rosen, 1975, 1978; Platnick & Nelson, 1978; Nelson & Rosen, 1980; Nelson & Platnick, 1981; Wiley, 1981). Tectonics may have resulted in increased uplift and increased inland water habitat (e.g., Hallam, 1993), and the occurrence of additional water sources may have encouraged Odonata dispersal.

By far the most extensive biogeographical study of Afrotropical libelluloid taxa was undertaken by Dijkstra (2007), who evaluated current distributions of odonates in tropical Africa and retrodicted past biogeographical patterns. African taxa are far less species rich than their Neotropical or Asian congeners; for example, African Aeshnidae, for example include 39 species among 5 genera, while there are 127 species among 15 genera in the Neotropics, and 138 species among 18 genera in the Orient (Dijkstra, 2007). Although species richness is low in several African regions, there are high levels of endemism (e.g., Clausnitzer & Dijkstra, 2005; Dijkstra, 2007).

Fig. 7. Several Odonate taxa involved in the biogeographical studies. A. Georgia River Crusier (*Macromia georgina*) B. Painted Skimmer (*Libellula semifaciata*) C. Sparkling Jewelwing (*Calopteryx dimidiata*) D. Slender Bluelet (*Enallagma traviatum*) E. Rubyspot (*Hetaerina occisa*) F. Megapodagrionidae (*Teinopodagrion macropus*) and G. Citrine Forktail (*Ischnura hastata*). Images A,B,C,D copyright from Dan Irizarry and Images E, F and G copyright from Adolfo Cordero.

## 4. Biogeography of Zygoptera

These slender and often rather small odonates are still taxonomically unresolved. Rehn (2003) proposed phylogenetic hypotheses that supported zygopteran monophyly based on morphological characters: nine morphological synapomorphies support Zygoptera. However several molecular Odonata phylogenetic analyses have failed to recover damselflies as a monophyletic group (Hasegawa & Kasuya, 2006, Saux et al., 2003). Recently, Bybee and colleagues (2008) used both molecular and morphological data and supported the monophyletic status of this taxon, suggesting that molecular data alone fails to recover a monophyletic Zygoptera due in part to limited taxon sampling. Nevertheless, internal familial relationships within the suborder remaining tangled, with families such as Megapodagrionidae, Perilestidae, Amphypteridae, Coenagrionidae, and Protoneuridae examples of putatively paraphyletic groups that will likely need to be reclassified (Bybee et al. 2008).

A limited number of studies have explored the effects of key biogeographical events on individual damselfly taxa (De Marmels, 2001; Dumont et al., 2005; Groeneveld et al., 2007; Polhemus, 1997; Turgeon et al., 2005). Today, patterns of damselfly distributions coincide with climatological zones: as temperature increases near the equator so too does the diversity of Zygoptera increase (Kalkman et al, 2008). Tropical regions hold the greatest number of species, and it has been suggested that this high diversity can be explained by aquatic habitat abundance in tropical forest (Orr, 2006). Moreover, tropical mountains provide a diversity of niches and regional refugia (Kalkman et al., 2008). The limited seasonality of tropical habitats increases the opportunities for specialist life-styles, thereby supporting the high diversity of tropical odonates and other taxa.

Within the damselflies the most successful family is indisputably the Coenagrionidae, in part due likely to their capacity for colonization. This family has been recovered as paraphyletic in recent systematic work (Bybee et al., 2008; O'Grady & May, 2003). Within this family, several genera (*Enallagma, Ischnura, Melagrion*) show a broad range of biogeographical patterns (Polhemus, 1997; Brown et al., 2000; Turgeon et al., 2002). *Enallagma* damselflies (Fig. 7 D) are present on all continents except Australia and Antarctica (Bridges, 1997). Their distribution shows two centers of diversification: North America and sub-Saharan Africa, with scattered species around the Asian and Palaearctic regions (Brown et al., 2000). This genus is one of the most species rich in North America, with 38 described species (Westfall & May, 1996). Recent molecular phylogenetic reconstructions of the Nearctic members of this genus suggest a radiation that relied on two recent progenitor lineages, the "*E. hageni*" and "*E. carunculatum*" clades (Brown et al., 2000). Turgeon et al. (2005) explored the diversification history of *Enallagma* across the Holarctic region using previous molecular phylogenetic work (Brown et al., 2000; Turgeon et al., 2002) and AFLP's as population genetic markers among species. There they suggested that the recent radiation of *Enallagma* was due to strong climate variation during Quaternary epoch.

The fork-tail damselflies (*Ischnura*, Fig.7 G) are the smallest members of the Coenagrionidae but have colonized most continents. Some species show female color polymorphism (e.g., I. *e.g., ramburii*; Cordero, 1990b, 1992; Fincke, 1987, 2004; Hinnekint, 1987; Johnson, 1964, 1966, 1975; Robertson, 1985; Robinson & Allgeyer, 1996; Sirot et al., 2003) and sperm competition (Cooper et al., 1996; Cordero, 1990a; Cordero & Miller, 1992; Miller, 1987; Waage, 1984). Recently, *Ischnura hastata*, a widespread species, was found to exhibit parthenogenesis in populations only in the Azores islands (Cordero et al., 2005), making this genus a good model for more extensive biogeographical analyses. Chippindale et al. (1999) explored phylogenetic relationships among North American *Ischnura* species, reporting a recent diversification along a latitudinal gradient. Realpe (2010) described two new species present in high altitudes of the Andes Cordillera in South America, and unpublished molecular data of those species suggest evidence of a recent radiation across elevation in the Neotropics (Realpe & Sanchez-Herrera, *pers. comm.*). The great dispersal ability of some fork-tail damselflies may have contributed to multiple rapid radiation events through their evolutionary history.

*Megalagrion* is one of the most species rich genera in the Pacific Region (Donelly, 1990). This genus contains 23 described species found on all the main Hawaiian Islands (Polhemus & Asquith, 1996; Daigle, 1996; Polhemus, 1997). Endemicity and species richness appear related to island age (Jordan et al., 2003). Molecular data on these species reveals two basic diversification patterns across the islands (Jordan et al., 2003). Some species dispersed in tandem as new islands were created, and over time those founding populations have lead to endemic species and assemblages on each island. Still other taxa show an adaptive burst as a single representative of a lineage colonized new islands (Jordan et al., 2003).

The family Megapodagrionidae (Fig. 7 F) appears to be paraphyletic (Bybee et al., 2008). De Marmels (2001) morphologically revised the Neotropical genus complex *Megapodagrion* s.str. *(Megapodagrion, Allopodagrion and Teionopodaprion)* which are distributed throughout South America. He suggested that high speciation rates in South American tropical forests were due to orogenic development of the southeastern Brazilian mountains and the Andes in Oligocene/Miocene times. Furthermore, he reported high degree of specialization within genera for particular Neotropical forests. Finally, he suggested a closer morphological relationship of this complex to taxa in the Malayan and Austral-Papuan region, based on the penile morphology (De Marmels, 2001).

The family Pseudostigmatidae is strikingly large but is restricted to Central and South America lowland montane forests (Fincke, 1992). The largest extant odonate is the helicopter damselfly *Megaloprepus coerulatus* with a wingspan approximately of 19 cm and an abdomen length of 10 cm. Recently, Groeneveld et al. (2007) addressed the evolution of gigantism among members of this family and an Eastern Africa endemic species, *Coryphagrion grandis*. The latter species had been placed in the family Megapodagrionidae; however, their habitat preferences, morphology, and behavior suggest they either lie in a monogeneric family or fall in the Pseudostigmatidae (Clausnitzer & Lindeboom, 2002; Rehn, 2003). Using molecular data from representative species of Pseudostigmatidae and *Coryphagrion,* Groeneveld et al. (2007?) suggest that gigantism evolved only once through the evolutionary history of this taxa. Their results support that gigantism in the endemic African genus was a reflection of phylogenetic history, and that this genus was a Gondwanaland relict (Groeneveld et al., 2007).

The family Calopterygidae is a monophyletic family within the Zygoptera (Bybee et al., 2008, Dumont et al., 2005). It is distributed worldwide, except for the Australasian region. All of the members of this family share remarkably similar habitat requirements (running waters) and morphology; however, males show a variety of mating displays (Buchholtz, 1995; Heymer, 1972). Dumont et al. (2005) evaluated phylogenetic relationships among species, but due to sparse Neotropical taxon sampling intrafamilial relationships remain unclear. Nevertheless, recent molecular phylogenetics among the Calopterygidae have clarified some biogeographical patterns (Dumont et al., 2005; Mullen & Andres, 2007). Fossil evidence and molecular dating techniques indicated that this family arose approximately 175 Ma and underwent rapid diversification approximately 150 Ma during the Cretaceous period (Dumont et al., 2005). Several Gondwanaland disjunctions exist among these taxa, such as the relictual distributions of *Irydictyon* and *Noguchiphaea*. Dumont et al. (2005) also reported that temperate taxa, such as *Calcopteryx* (Fig. 7 C), were affected by Pleistocene glaciation. Recently, Mullen and Andres (2007) used molecular systematic and phylogenetic methods on *Calcopteryx* and suggesting that this taxon has been present since the Miocene age; furthermore, they suggest that reproductive displays may be a result of reinforcement or ecological character displacement dating from when isolated populations came into secondary contact. Phylogenetic relationships remain unclear in the Neotropical genus *Hetaerina (Fig. 7 E)*, despite the fact that unlike other confamilial genera male genitalia are strongly divergent, suggesting mechanical isolation (Garrison, 1990).

Forest damselflies in the family Platystictidae are restricted to Central and northern South America, and to tropical Southeast Asia. Morphological and molecular data indicate they are a monophyletic taxon (Bechly, 1996; Rehn, 2003; van Tol et al., 2009). These forest dwellers have poor flying capacity, suggesting low dispersal capability, which is reflected in the small distributional ranges of most species (van Tol et al., 2009). Molecular analysis recovered Neotropical genera as the basal-most clades within the family. However, the morphological analyses suggest that *Sinosticta ogati* from southeastern China is instead the sister taxon to all other members of the family. Consequently, the ancestor of the family may have evolved in the Palaearctic and the Oriental regions (van Tol et al., 2009). This type of distribution has been recognized among other organisms, such as the Neotropical plant genus *Trigonobalanus* (van der Hammen & Cleef, 1983), and it is known as a "tropical amphitranspacific distribution" (van Steenis, 1962). This pattern is ascribed to dispersal from Africa to the northern hemisphere during the Late Cretaceous, with subsequent extinction in Africa due to Neogene desertification (Raven & Axelrod, 1974). van Tol et al. (2009) hypothesized an origin in eastern Africa, suggesting that the ancestor of this family evolved in eastern Gondwana, with subsequent dispersal into South America, Asia, and New Guinea. Although the family includes Neotropical taxa, additional Neotropical sampling is needed to test this hypothesis.

Finally, the distribution of members of the Afrotropical family Platycnemidae has been suggested to be the result of insular island biogeography. Dijkstra et al. (2007) described *Platycnemis pembipes*, a new species from the Pemba Island of Tanzania. He examined the morphology of all members of this genus, concluding that the new species *was* more related to Malagasy taxa than to Guineo-Congolian species, which have affinity to tropical Asia. The distribution of this new species suggests a remarkable colonization event probably due to wind dispersal across the Mozambique channel (Dijkstra et al., 2007).

Overall, many biogeographical mechanisms that have been proposed using damselflies as model organisms, but more thorough sampling and a greater variety of ecological experimentation is needed to further advance understanding of damselfly evolution and biogeography.

## 5. Conclusion

The biogeography of Odonata is a rich area of study that needs further attention. As one of the basal-most taxa in Insecta (Grimaldi and Engel, 2005), our understanding of the origin of flying insects will be greatly improved by additional study, particularly through research that includes thorough analyses of stem and crown group taxa. Future work should explore the biogeography of lesser-studied zygopteran groups from South America, and expand understanding of species rich groups like the Libelluloidea and Gomphidae. Dragonflies and damselflies have been heralded as model indicators for climate change, due in part to their great dispersal capabilities, and earlier emergence has been documented in our warming climate (e.g., Hassell et al., 2007). Range expansion of tropical taxa is predicted into higher latitudes. Although some Odonata ranges fluctuate with environmental changes, northward range expansions have been reported over the last 40 years among several European taxa (e.g., Hickling et al., 2005). The future biogeographical distribution of Odonata undoubtedly will be influenced directly and indirectly by anthropogenically-altered climate.

## 6. Acknowledgments

J.W. and M. S. H. acknowledge internal funding from Rutgers University in Newark. Moreover we thank Dr. Godfrey Bourne, Dan Irizarry, Dr. Adolfo Cordero and Eva Paulson for sharing the amazing pictures we used in our figures. Finally, we acknowledge Dr. Larry Stevens for the review edits that enrich our chapter.

## 7. References

Anderson, R.C. (2009). Do dragonflies migrate across the western Indian Ocean? *Journal of Tropical Ecology*, 25, 347–348.

*Artiss*, T. (2004). Phylogeography of a facultatively migratory dragonfly, *Libellula quadrimaculata* (*Odonata*: Anisoptera). *Hydrobiologia* 515, 225–234.

Bechly, G. (1996). Morphologische Untersuchungen am Flügelgeäder der rezenten Libellen und deren Stammgruppenvertreter (Insecta; Pterygota; Odonata). unter besonderer Berücksichtigung der Phylogenetischen Systematik und des Grundplanes der Odonata. *Petalura, special vol. 2*, 402 pp.

Brauckmen, C. & Zessin, W. (1989). Neue Meganeuridae aus dem Namurium von Hagen-Vorhalle (BRD). und die Phylogenie der Meganisoptera (Insecta, Odonata). *Deutsche entomologische Zeitschrift* 36, 177–215, pl 3–8.

Brongniart, C. (1885). Les insectes fossiles des terrains primaires. Coup d'oeil rapide sur la faune entomologique des terrains paléozoïques. *Bulletin of the Societe Amis Sci. nat. Rouen*, 1885, 50–68.

Brown, J. M., M. A. McPeek & May, M. L. (2000). A phylogenetic perspective on habitat shifts and diversity in the North American *Enallagma* damselflies. *Systematic Biology*, 49, 697–712.

Buden, D. W. (2010). *Pantala flavescens* (Insecta: Odonata) rides west winds into sal. Pacific Science, 64, 141–143.

Bybee, S.M., Ogden, H.T., Branham, M.A. & Whiting, M.F. (2008). Molecules, morphology and fossils: a comprehensive approach to odonate phylogeny and the evolution of the odonate wing, *Cladistics*, 23, 1–38.

Carle, F.L. (1995). Evolution, taxonomy, and biogeography of ancient Gondwanian libelluloides, with comments on anisopteroid evolution and phylogenetic systematics (Anisoptera: Libelluloidea). *Odonatologica*, 24, 383–506.

Carpenter, F.M. (1992). Volume 3: Superclass Insecta. In R.C. Moore, R.L. Kaesler, E. Brosius, J. Kiem and J. Priesener, (eds.), *Treatise on Invertebrate Paleontology, Part R, Arthropoda 4*. Boulder, Colorado: Geological Society of America; and Lawrence, Kansas: University of Kansas Press. 665 p.

Chippindale, P.T. et al., (1999). Phylogenetic Relationships of North American Damselflies of the Genus *Ischnura* (Odonata: Zygoptera: Coenagrionidae). Based on Sequences of Three Mitochondrial Genes. *Molecular Phylogenetics and Evolution* 11(1), 110–121.

Corbet, P. S. (1999). Dragonflies: Behavior and ecology of Odonata. Cornell University Press. Ithaca, New York.

Cooper, G., Miller, P. L., and Holland, P. W. H. 1996. Moleculargenetic analysis of sperm competition in the damselfly *Ischnura elegans* (Vander Linden). *Proc. R. Soc. London B* 263: 1343–1349.

Cordero, A. 1990a. The adaptive significance of the prolongued copulations of the damselfly, *Ischnura graellsii* (Odonata: Coenagrionidae). *Anim. Behav.* 40: 43–48.

Cordero, A. 1990b. The inheritance of female polymorphism in the damselfly *Ischnura graellsii* (Rambur) (Odonata:Coenagrionidae).*Heredity* 64: 341–346.

Cordero, A. 1992. Density-dependent mating success and colour polymorphism in females of the damselfly, *Ischnura graellsii* (Odonata: Coenagrionidae). *J. Anim. Ecol.* 61: 769–780.

Cordero, A., and Miller, P. L. 1992. Sperm transfer, displacementand precedence in *Ischnura graellsii* (Odonata: Coenagrionidae). *Behav. Ecol. Sociobiol.* 30: 261–267.

Cordero Rivera, A.; M.O. Lorenzo Carballa; C. Utzeri & V. Vieira. 2005. Parthenogenetic *Ischnura hastata* (Say), widespread in the Azores (Zygoptera: Coenagrionidae). Odonatologica, 34: 1-9.

Córdoba–Aguilar, A. (2008). Introduction. *In*: A. Córdoba–Aguilar (editor), *Dragonflies and Damselflies: Study Models in Ecological and Evolutionary Research*. Oxford University Press, Oxford, Pp. 1–3.

De Marmels J. (2001). Revision of *Megapodagrion* Selys, 1886 (Insecta, Odonata: Megapodagrionidae). (Diss. Doctor sci. nat.), Math.–naturwiss. Fak. Univ. Zurich. Zurich (Suiza). Pp. 218.

Dietz, R. S. & Holden, J. C. (1970). Reconstruction of Pangaea: breakup and dispersion of continents, Permian to present. *Jour. Geophys. Res.*, 75 (26), 4939–4956.

Dijkstra, K.-D.B. (2007). Demise and rise: the biogeography and taxonomy of the Odonata of tropical Africa. *In:* Dijkstra, K.-D.B. (Editor). *Demise and rise: the biogeography and taxonomy of the Odonata of tropical Africa*. PhD Thesis, Leiden University. 143–187.

Dijkstra, K.-D.B. (2007b). Gone with the wind: westward dispersal across the Indian Ocean and island speciation in *Hemicordulia* dragonflies (Odonata: Corduliidae). *Zootaxa* 1438, 27–48.

Feng, H.; K. Wu, Y. Ni, D. Cheng, & Guo, Y. (2006). Nocturnal migration of dragonflies over the Bohai Sea in northern China. *Ecol. Entomol.* 31, 511–520.

Fincke, O. M. 1987. Female monogamy in the damselfly *Ischnura verticalis* Say (Odonata: Coenagrionidae). *Odonatologica* 16: 129–143.

Fincke, O.M. 2004. Polymorphic signals from harassed females and the males that learn them support a novel frequency-dependent model. Animal Behavior 67:833-845

Fraser, F. C. (1943). A note on the 1941 immigration of *Sympetrum fonscolombii* (Selys). (Odon.). *J. Soc. Br. Entomol.*, 2, 133–136.

Fraser, F. C. (1945). Migration of Odonata. *Entomol. Monthly Mag.*, 81,73–74.

Freeland, J. R., M. May, R. Lodge, & Conrad, K. F. (2003). Genetic diversity and widespread haplotypes in a migratory dragonfly, the common green darner *Anax junius*. *Ecological Entomology*, 28, 413–421.

Garrison, R. W. (1990) A synopsis of the genus *Hetaerina* with descriptions of four new species (Odonata: Calopterygidae). *Transactions of the American Entomological Society*, 116(l): 175-259

Grimaldi, D. & Engel, M.S. (2005). Evolution of the Insects. Cambridge University Press.

Heiser, M. & Schmitt, T. (2010). Do different dispersal capacities influence the biogeography of the western Palearctic dragonflies (Odonata)? *Biological Journal of the Linnean Society*, 99, 177–195.

Jarzembowski, E.A. & Nel, A. (1996). New fossil dragonflies from the Lower Cretaceous of SE England and the phylogeny of the superfamily Libelluloidea (Insecta: Odonata). *Cretaceous Research 17*, 67–85.

Johnson, C. 1964. The inheritance of female dimorphism in in the damselfly *Ischnura damula*. *Genetics* 49: 513–519.

Johnson, C. 1966. Genetics of female dimorphism in *Ischnura demorsa*. *Heredity* 21: 453–459.

Johnson, C. 1975. Polymorphism and natural selection in ischnuran damselflies. *Evol. Theory* 1: 81–90.

Kormondy, E. J. (1961). Territoriality and dispersal in dragonflies (Odonata). *J. N.Y. Entomol. Soc.*, 69, 42–52.

Lieftinck, M. A. (1962). Insects of Micronesia, Odonata. *Insects Micronesia 5*, 1–95.

Lieftinck, M.A. (1977). New and little known Corduliidae (Odonata: Anisoptera). from the Indo–Pacific region. *Oriental Insects*, 11(2), 157–159.

May, M. L., & Matthews, J. H. (2008). Migration in Odonata: a case study of *Anax junius*, pp. 63–77. *In* A. Cordoba–Aguilar, (ed.), *Dragonflies and Damselflies. Model Organisms for Ecological and Evolutionary Research*. Oxford University Press, Oxford.

Matthews, J.H., S. Boles, C. Parmesan & Juenger, T. (2007). Isolation and characterization of nuclear microsatellite loci for the common green darner dragonfly *Anax junius* (Odonata: Aeshnidae). to constrain patterns of phenotypic and spatial diversity. *Molecular Ecology Notes*.

McLachlan, R. 1896. Oceanic migration of a nearly cosmopolitan dragonfly (*Pantala flavescens*, F.). *Entomol Mon. Mag. 7*, 254.

Nelson, G., & Platnick, N. I. (1981). Systematics and Biogeography: Cladistics and Vicariance. Columbia University Press, New York, 567 p.

Norberg, R. A. (1972). The pterostigma of insect wings as an inertial regulator of wing pitch. *J. Comp. Physiol.* 81, 9–22.

O'Grady, E.W. & May, M.L. (2003). A phylogenetic reassessment of the subfamilies of Coenagrionidae(Odonata: Zygoptera). *Journal of Natural History* 37, 2807–2834.

Peck, S. B. (1992). The dragonflies and damselflies of the Galapagos Islands, Ecuador (Insecta: Odonata). *Psyche (Camb.).* 99, 309–322.

Petrulevičius, J. F. & Nel, A. (2009). First Cordulephyidae dragonfly in America: A new genus and species from the Paleogene of Argentina (Insecta: Odonata). *Comptes Rendus Palevol*, 8 (4), 385–388.

Pilgrim, E. (2007).Systematics of the sympetrien dragonflies with emphasis on the phylogeny, taxonomy, and historical biogeography of the genus Sympetrum (Odonata: Libellulidae). Utah State University, 154 pp.

Pritchard, G., McKee, M. H., Pike, E. M., Scrimgeour, G. J. & Zloty, J. (1993). Did the first insects live in water or air? *Biol. J. Linn. Soc.* 49, 31–44.

Raven, P.H., & Axelrod, D.I. (1974). Angiosperm biogeography and past continental movements. *Ann. Mo. Bot. Gard.* 61, 539–673.

Realpe, E. (2010). Two new Andean species of the genus *Ischnura* Charpentier from Colombia, with a key to the regional species (Zygoptera: Coenagrionidae). *Odonatologica* 39(2), 121–131.

Reichholf, J. (1973). A migration of *Pantala flavescens* (Fabricius, 1798). along the shore of Santa Catarina, Brazil (Anisoptera: Libellulidae). *Odonatologica* (Utr.). 2, 121–124.

Rosen, D. E. (1975). A vicariance model of Caribbean biogeography. *Syst. Zool.* 24, 431–464.

Rosen, D. E. (1978). Vicariant patterns and historical explanation in bio- geography. *Systematic Zoology*, 27, 159–188.

Rowe, R. J. (1987). The dragonflies of New Zealand. Auckland University Press, Auckland, New Zealand.

Russell, R. W.; M. L. May, K. L. Soltesz, & Fitzpatrick, J. W. (1998). Massive swarm migrations of dragonflies (Odonata) in eastern North America. *Am. Midl. Nat.* 140,325–342.

Samways, M.J. (1992). Dragonfly conservation in South Africa: a biogeographical perspective. *Odonatologica*, 21, 165–180.

Samways, M. J., & Osborn, R. (1998). Divergence in a transoceanic circumtropical dragonfly on a remote island. *J. Biogeogr.*, 25, 935–946.

Samways, M.J. (2006). National Red List of South African dragonflies (Odonata). *Odonatologica* 35, 341–368.

de Selys Longchamps, E. de, (1889). *Palaeophlebia*. Nouvelle leagion de Calopterygides. Suivi de la description d'une nouvelle gomphine du Japon: *Tachopteryx pryeri*. *Annales de la Societeal Entomologique de Belgique* 33, 153–159.

Stevens, L. E. & Bailowitz, R. A. 2009. Odonata Biogeography in the Grand Canyon Ecoregion, Southwestern USA. *Annals of the Entomological Society of America* 102 (2), 261-274.

Srygley, R. B. (2003). Wind drift compensation in migrating dragonflies *Pantala* (Odonata: Libellulidae). *J. Insect Behav.* 16, 217–232.

Thomas, J.A., Trueman, J.W.H., Rambaut A. & Welch, J.J. (in press). Relaxed Phylogenetics and the Palaeoptera Problem: Resolving Deep Ancestral Splits in the Insect Phylogeny.

Turgeon,J., Stoks,R., & Thum, R.A., et al. (2005). Simultaneous Quaternary Radiations of Three Damselfly Clades across the Holarctic. *The American Naturalist* 165(4), 78–107.

van der Hammen, T and A. M. Cleef (1983) *Trigonobalanus* and the Tropical Amphi-Pacific Element in the North Andean Forest. *Journal of Biogeography.* 10(5), 437-440.

van Steenis, C.G.G.J., 1962. The land-bridge theory in botany, with particular reference to tropical plants. – Blumea 11: 235-542.

van Tol, J., Reijnen, B. T. & Thomassen, H. A. (2009). Phylogeny and biogeography of the Platystictidae (Odonata). pp. 3–70 *In*: van Tol, J. *Phylogeny and biogeography of the Platystictidae (Odonata)*. PhD. thesis, University of Leiden. x + 294pp.

Wakana, I. 1959. On the swarm and migratory flight of *Pantala flavescens*, an observation in Kawagoe area. *Tombo* 1, 26 –30 [in Japanese with English summary].

Ware, J.L., May, M.L., & Kjer, K.M. (2007). Phylogeny of the higher Libelluloidea (Anisoptera: Odonata): An exploration of the most speciose superfamily of dragonflies, *Molecular Phylogenetics and Evolution*, 45, 289–310.

Ware, J. L.; Simaika, J. P., Samways, & M. (2009). Biogeography and divergence estimation of the relic Cape dragonfly genus *Syncordulia*: global significance and implications for conservation. *Zootaxa*, 2216, 22–36

Wiley, E. O. (1981). Phylogenetics: the theory and practice of phylogenetic systematics. John Wiley and Sons, New York.

Wikelski,M.,D., Moskowitz, J.S., Adelman, J., Cochran, D.S. Wilcove & May, M. L. (2006). Simple rules guide dragonfly migration. *Biology Letters*, 1–5.

Wootton, R.J. (1981). Palaeozoic insects. *Annual Review of Entomology*, 26,319–344.

Watson, J. A. L. (1983). A truly terrestrial dragonfly larva from Australia (Odonata: Corduliidae). *Journal of the Australian Entomological Society*, 21, 309–11.

# 14

# Biogeography of Flowering Plants: A Case Study in Mignonettes (Resedaceae) and Sedges (*Carex*, Cyperaceae)

Santiago Martín-Bravo[1,*] and Marcial Escudero[1,2]
*[1]Department of Molecular Biology and Biochemical Engineering,*
*Pablo de Olavide University,*
*[2]The Morton Arboretum, Lisle,*
*[1]Spain*
*[2]USA*

## 1. Introduction

Biogeography is a multidisciplinary science that studies the past and present geographic distribution of organisms and the causes behind it. The combination of historical events and evolutionary processes has usually an outstanding role when explaining the shape of a species range. As already noted by Darwin more than 150 years ago, patterns of species distribution may often be seen as clear footsteps of their evolution and diversification (Darwin, 1859). It is now also well known that geological events (i.e. continental drift, orogeny or island formation) and climatic oscillations occurred during the recent geological history of the Earth, like the cooling and aridification that took place during the Pliocene (5.3 – 2.5 million years ago, m.a) and the Pleistocene glaciations (1.8 – 0.01 m.a), prompted great range shifts. These geological and/or climatic changes caused, in some cases, the extinction of species; in many others, they provided conditions of reproductive isolation and/or genetic divergence between populations and, eventually, produced speciation, the engine of biodiversity.

The development of molecular techniques to study biodiversity from the end of the XX[th] century has implied a great methodological revolution in the field of systematics, evolutionary biology and biogeography. They constitute valuable and powerful tools that allow tackling multiple biogeographic and evolutionary hypotheses, as well as to progress towards a natural classification of biodiversity that reflects its evolutionary history. In particular, these methodological advances have boosted the study of the principles and historical processes behind the geographical distribution of genetic lineages at low taxonomic levels, in the recently arisen discipline termed phylogeography (Avise et al., 1987; Avise, 2000). Phylogeography is a multidisciplinary science that integrates methods and concepts from population genetics ("microevolution") and systematics ("macroevolution") (Avise, 2000).

---

* Corresponding Author

Biogeographic studies using molecular approaches have helped to elucidate the origin of striking bipolar or intercontinental range disjunctions and to estimate the divergence time between allopatric populations or taxa (e.g. Dick et al., 2007; Donoghue, 2011; Givnish & Renner, 2004; Milne, 2006; Mummenhoff & Franzke, 2007; Popp et al., 2011; de Queiroz, 2005; Renner, 2005; Shaw et al., 2003; Wen & Ickert-Bond, 2009), the causes of disparate species or lineage richness in different territories (e.g. Ricklefs et al., 2006; Svenning et al., 2008; Valente et al., 2010, 2011), or the reconstruction of ancestral ranges (e.g. Drummond, 2008; Fernández-Mazuecos & Vargas, 2011; Mansion et al., 2009; Salvo et al., 2010). In this chapter, we briefly review the main methodological approaches that have enabled the rise of biogeographic and phylogeographic studies in plants based on molecular data during the last decade. Specifically, as a case study, we review researches that have successfully applied molecular methods to address biogeographic questions in two plant groups, a dicot (family Resedaceae) and a monocot (genus *Carex* L., Cyperaceae).

## 2. Methods in molecular biogeographic and phylogeographic studies

### 2.1 Molecular markers

One of the critical points in molecular biogeographic and phylogeographic studies of plants is the availability of DNA regions that provide an adequate level of reliable molecular variability at the studied taxonomic level (Schaal et al., 1998). Bio- and phylogeographic markers should preferentially be ordered (DNA sequences), rather than unordered (e.g. AFLP, ISSR, RAPD), thus containing a record of their own histories and providing information about genealogical relationships between alleles (e.g. Lowe et al., 2004; Schaal & Leverich, 2001; Schaal & Olsen, 2000).

Sequences from organellar genomes (mitochondrial and plastid DNA) have played a key role in phylogeographic studies. While mitochondrial DNA has been extensively used in animals, it has been scarcely used in plants (but see Sinclair et al., 1998; Tomaru et al., 1998) as this genome is usually not sufficiently variable in plants and is commonly submitted to intramolecular recombination (Palmer, 1992). Many plant phylogeographic studies have relied on the plastid genome, since is more variable than mitochondrial DNA, recombination processes are not frequent, and it is usually uniparentally inherited (maternally in most angiosperms; Harris & Ingram, 1991). It has also a reduced effective population size in comparison to the nuclear genome due to its haploid nature. This feature results in an increased effect of genetic drift reflected in a greater genetic differentiation of fragmented populations. Thus, plastid regions may longer retain phylogeographic signals of past migrations, range fragmentation and dispersal events (e.g. Hudson & Coyne, 2002; Kadereit et al., 2005; Newton et al. 1999; Petit et al., 2005; Rendell & Ennos, 2003; Schaal et al., 1998).

At the first stages of phylogeography, the most widespread technique for detecting molecular variation within plastid genome used to be restricted fragment length polymorphism (RFLPs, PCR-RFLPs; Hampe et al., 2003; Mason-Gamer et al., 1995; Wagner et al., 1987; review in Soltis et al., 1992). More recently, to avoid homoplasy problems, direct sequencing of non-coding regions of plastid genome has become prevalent (e.g. Hung et al., 2005; Jakob & Blattner., 2006; Koch et al., 2006). Nonetheless, due to its biparental inheritance, genealogical patterns inferred from nuclear markers are probably more representative of the true evolutionary history and gene flow patterns than those derived from organellar genomes (Harpending et al., 1998; Lowe et al., 2004). Nuclear ribosomal internal transcribed spacer (ITS) sequences have been widely used in phylogenetic and biogeographic studies. Among the characteristics that explain the success of this marker are the almost universal primers and a high number of copies in the genome, which greatly eases PCR amplification. In addition, ITS usually provides an appropriate level of variability at the generic and infrageneric level (reviews in Alvárez & Wendel, 2003; Calonje et al., 2009; Nieto Feliner & Rosselló, 2007). However, the use of this marker has been also criticized due to its multicopy nature and concerted evolution processes, which frequently results in high levels of homoplasy (Álvarez & Wendel, 2003; Bailey et al., 2003). Single or low-copy nuclear genes have been also used for phylogeographic studies, although its experimental tuning is usually complex (Caicedo & Schaal., 2004; Olsen, 2002; Olsen & Schaal, 1999; reviews in Hare, 2001; Pleines et al., 2009). In addition, in this case, issues derived from the dynamics of nuclear genome, like recombination and loci homology, may complicate the analysis of its molecular variation (Hare, 2001; Schaal et al., 1998; Schierup & Hein, 2000; Zhang & Hewitt, 2003). Fingerprinting techniques, like microsatellites (SSRs; review in Ouborg et al., 1999) and AFLPs (Rubio de Casas et al., 2006; Tremetsberger et al., 2006; review in Meudt & Clarke, 2007), and, to a lesser extent, RAPDs and ISSRs (e.g. Clausing et al., 2000; Hess et al., 2000), are currently also frequently used for phylogeographic studies, due to the generally high levels of variability retrieved (review in Pleines et al., 2009). SSRs have the advantage of being a codominant marker of known genetic origin, while AFLPs, ISSRs and RAPDs are anonymous and dominant markers (e.g. Lowe et al., 2004; Mueller & Wolfenbarger, 1999). In these cases, inferences about evolutionary history of populations are not interpreted from genealogical relationships of the markers, but from patterns of genetic diversity and differentiation of populations or groups of populations. Nevertheless, studies on the evolutionary history of species or populations based on genealogical relationships retrieved from unordered markers such as AFLPs have become more common in recent years (Beardsley et al., 2003; Tremetsberger et al., 2006; Pearse & Hipp, 2009). Finally, the latest advances in sequencing techniques, such as restriction-site associated DNA (RAD) and "genotyping-by-sequencing" (GBS) markers, are already being used in biogeographic and phylogeographic studies (Baird et al., 2008; Elshire et al., 2011; review in Davey et al., 2011).

## 2.2 Reconstruction of genealogical relationships using DNA sequences

The analysis and interpretation of genealogical relationships of alleles is one of the main issues of biogeography and phylogeography. The first approaches for disentangling genealogies of alleles were based on bifurcate phylogenetic reconstruction methods using parsimony (as implemented in PAUP (Swofford, 2003) or TNT (Goloboff et al., 2008)), maximum likelihood (as implemented in PAUP (Swofford, 2003) or PAML (Yang, 1997)) and/or Bayesian (as implemented in MrBayes; Ronquist & Huelsenbeck, 2003) approaches (review in Page & Holmes, 2004). The two latter apply evolutionary models of nucleotide substitution (see Posada, 2008). In the last years, reconstruction of genealogical relationship of alleles for phylogeographic studies is usually performed from the basic conceptual framework of coalescent theory (reviews in Ewens, 1990; Fu & Li, 1999; Hudson, 1990; Tavaré, 1984). Accordingly, the coalescent theory has been also implemented in phylogenetic methods (Drummond & Rambaut, 2007; BEAST software). This theory predicts the effects of different processes (genetic drift, mutation, selection) on the evolution of alleles. At low taxonomic levels, the frequent intervention of biological processes like reticulation and persistence of ancestral alleles may not always be accurately represented with standard, bifurcate phylogenetic trees. Therefore, allele genealogies are more appropriately represented with haplotype networks (Huson & Bryant, 2006; McBreen & Lockhart, 2006; Posada & Crandall, 2001). Haplotypes are inferred from nucleotide polymorphisms in DNA sequences. In a species, gene alleles (either sampled, unsampled or extinct) derive from a common ancestral allele in which all coalesce (Schaal & Leverich, 2001). Genealogical relationships of haplotypes may be interpreted together with the patterns of congruence between haplotype frequency and their geographical distribution. This may allow to infer historical processes in the evolutionary history of species and/or populations, like range expansion or fragmentation, geographical isolation, gene flow, genetic bottlenecks, or incomplete lineage sorting of ancestral polymorphisms (Schaal et al., 1998; Schaal & Olsen, 2000).

Reconstruction of genealogical networks, in parallel to that found for phylogenetic inference (review in Page & Holmes, 2004), may be based in distance or character methods (reviews in Huson & Bryant, 2006; McBreen & Lockhart, 2006; Morrison, 2005). Among the first stand split decomposition (Bandelt & Dress, 1992) and neighbor-net (Bryant & Moulton, 2004). On the other hand, character methods include joining networks (Median-joining network, Bandelt et al., 1999; median network, Bandelt et al., 2000) and, especially, the widely used statistical parsimony (Templeton et al., 1992), which represents each change between two haplotypes as a mutational step (Clement et al., 2000; TCS software). Networks may be also reconstructed from phylogenetic trees, such as with consensus networks (Holland et al., 2004, 2006) or super-networks (Huson et al., 2004).

Recently, species trees methods based on coalescent theory have been developed which may be useful to deal with phylogenetic incongruences between different genes or genomes, for example when processes of incomplete lineage sorting or hybridization are involved (Blair & Murphy, 2011; Degnan & Rosenberg, 2009; Liu et al., 2009; Zachos, 2009). Moreover, "isolation-with-migration" methods simultaneously model the differentiation of the species / populations and the hybridization/gene flow rates between them (Becquet & Przeworski, 2007, 2009; Hey & Nielsen, 2004, 2007; Wilkinson-Herbots, 2008). This area is currently at the fore of phylogenetics and holds considerable promise for the methodological development of the analysis and interpretation of genealogical relationships of alleles.

## 2.3 Reconstruction of genealogical relationships and genetic structure using fingerprinting data

When analyzing fingerprinting data for the reconstruction of genealogical relationships, phylogenies based on pairwise genetic distance among individuals, populations or species are widely used. Nei & Li's (1979) and Jaccard's coefficients are by far the most commonly used to calculate pairwise genetic distances between individuals from AFLP, RAPD or ISSR data (1/0 matrices), accounting only for allele presence matches (1). In contrast, the simple matching coefficient considers both presence (1) and absence (0) matches, which is considered less accurate (see Weising et al., 2005). Other genetic distances or coefficients widely used for AFLP, RAPD or ISSR (often to calculate genetic distances between populations) are pairwise $F_{st}$ values and Nei's distances (see Lynch & Milligan, 1994; Nei, 1978). For SSR data, DA genetic distance based on allele frequencies (Nei et al., 1983) is one of the most widely used. The election of the measure of genetic distance depends on the organism level (individual, population or species), the analyzed marker (AFLP, RAPD, ISSR or SSR) and the particular study goals. Pairwise genetic distances are depicted by dendrograms, which may be reconstructed with different methods: UPGMA (unweighted pair-group method using arithmetic averages; Sneath & Sokal, 1973; Sokal & Michener, 1958), minimum evolution (ME, Edwards & Cavalli-Sforza, 1963; Kidd & Sgaramella-Zonta, 1971; Rzhetsky & Nei, 1993) and neighbor joining (NJ; Saitou & Nei, 1987). At present, NJ and ME, which do not assume constant evolutionary rate along all branches, are the most widely used methods. Genetic relationships among individuals are also frequently represented with the principal coordinate analysis (PCO) for 1/0 matrices and the principal component analysis (PCA) for allele frequency matrices. As already noted, inferences about evolutionary history of populations are not usually interpreted from genealogical relationships of haplotypes, but from patterns of genetic diversity or population differentiation. Some of the most widely used genetic diversity indices are percentage of polymorphic loci (P), allelic richness (A), effective number of alleles ($A_e$), Shannon index and Nei's index (Lynch & Milligan, 1994; Nei, 1973). For estimation of genetic differentiation, F Statistics ($F_{st}$) and related measures have been widely used (Wright, 1951; see Weising et al., 2005); however, at present, AMOVA (Excoffier et al., 1992) is considered a more accurate approach as it does not assume Hardy-Weinberg equilibrium. In the last years, Bayesian analyses for disentangling population genetic structure have been developed, as implemented in STRUCTURE (Pritchard et al., 2000) or in BAPS (Corander et al., 2003) softwares.

## 2.4 Ancestral range reconstruction

Extinction, dispersal and vicariance are important historical factors to explain a current taxon range. The combination of them has frequently caused disjunct ranges, including remarkable patterns such as bipolar, intercontinental or trans-oceanic disjunctions. The study and interpretation of disjunctions have been one of the most fascinating aspects of plant biogeography and phylogeography (e.g. Givnish & Renner, 2004; Milne, 2006; Raven, 1972; Thorne, 1972; Wood, 1972; Zhengyi, 1983). One of the most widely used methodological approach to elucidate the origin of disjunctions and patterns of colonization and dispersal has been the estimation of divergence times (see 2.5) and ancestral range reconstruction. The latter method maps extant taxa distributions on molecular phylogenies. In the last years, new algorithms have been developed to improve the reconstruction of the evolutionary history of non-molecular characters using molecular phylogenies. Three different approaches have been implemented for ancestral character mapping and biogeographical inference in phylogenetic reconstruction: (1) parsimony (DIVA software: Ronquist, 1997; Mesquite software: Maddison & Maddison, 2010; S-DIVA software: Yu et al., 2010), (2) maximum likelihood (Schluter et al., 1997; Lagrange software: Ree et al., 2005, Ree & Smith, 2008; Mesquite software: Maddison & Maddison, 2010), and (3) Bayesian (SIMMAP software: Bollback, 2006; Bayes-DIVA software: Nylander et al., 2008, Sanmartín et al., 2008). Some of these tools (e.g. DIVA, Lagrange) were specifically developed for the reconstruction of geographical areas, while others were designed for the study of morphological character evolution (e.g. Mesquite, SIMMAP). However, mapping of ancestral distributions and its implementation in software intended for morphological character mapping have been criticized, because species distributions are not expected to follow the same models as morphological characters (Ree et al., 2005). Methods that assume dispersalist or center-of-origin mechanisms have been criticized by some authors who argue in favour of vicariance (Humphries & Parenti, 1999). Specific methods for ancestral range reconstruction and morphological character reconstruction were compared by Clark et al. (2008). They concluded that the methods which consider branch lengths (dispersal-extinction-cladogenesis and stochastic character mapping) usually yield the most plausible biogeographic hypotheses (see also Buerki et al., 2011; Ree & Sanmartín, 2009). Very recently, a new approach based on speciation, extinction and dispersal rates has been developed (geographic state speciation and extinction model, GeoSSE; Goldberg et al., 2011). It combines features of the constant-rates birth-death model with a three state Markov model. This software allows the codification of three different areas, of which two are contained in the third, a widespread one. The output parameterizes speciation, extinction and dispersal rates among the two distinct areas (Goldberg et al., 2011).

## 2.5 Estimation of divergence times

The development of methods for molecular dating is currently one of the most active fields of research within plant biogeography and phylogeography. The basis of molecular dating is that a direct relationship exists between the degree of molecular divergence between two taxa and the time elapsed since their divergence from a common ancestor. Therefore, it provides a temporal context that allows relating important processes during the evolutionary history of a taxon (speciation, extinction, radiation) with palaeogeologic/climatic events. Estimation of divergence times were, at the first stages of molecular systematics, based in proteins (Zuckerkandl & Pauling, 1962, 1965) that were progressively replaced by DNA sequences.

The molecular clock hypothesis, which assumes a relatively constant rate of molecular divergence through time, has been one of the most debated concepts of molecular biology (e.g. Hillis et al., 1996; Sanderson, 1998; Sanderson & Doyle, 2001). Many researchers soon rejected that DNA mutation rates were constant (Simpson, 1964; Mayr, 1965). Statistical approaches were subsequently developed to evaluate the constancy of the evolutionary rate among different lineages, like the relative ratio test (Sarich & Wilson, 1967) or the Langley & Fitch (1974) test. Nowadays it is well known that DNA evolves heterogeneously at different organization levels, from the nucleotide positions in a codon, to different genes or regions, genomes, and organisms (Britten, 1986; Bromham & Penny, 2003; Graur & Li, 2000; Li, 1997; Wolfe et al. 1987). Several factors may influence nucleotide mutation rate as much as or even more than chronological time, like the organism's generation time (Laird et al., 1969; Li et al., 1987; Ohta, 1995) or the metabolic rate (Martin & Palumbi, 1993).

If molecular clock hypothesis is assumed, estimation of divergence time may be calculated with the method of mean path length (Bremer & Gustafsson, 1997; Britton et al., 2002, PATH software) that applies a linear regression (Graur & Li, 2000; Nei, 1987). Constant mutation rate may be also optimized with maximum likelihood (Felsenstein, 1981; Langley & Fitch, 1974). Rejection of the constant rate hypothesis may lead to pruning tree branches that do not fit the hypothesis (linearized trees, Li & Tanimura, 1987), or applying "local" molecular clocks, which assume a relatively constant rate between related lineages (Hasegawa et al., 1989; Rambaut, 2001, maximum likelihood approach in RHINO software; Rambaut & Bromham, 1998, maximum likelihood approach in Qdate software; Yang, 1997, maximum likelihood approach in PAML/BASEML software; Yoder & Yang, 2000). More recently, methodological advances have enabled the application of variable mutation rates, as well as to integrate palaeontological information, which allows a more realistic and accurate estimation of divergence times (reviews in Magallón, 2004; Rutschmann, 2006; Sanderson et al., 2004). Some of the most widespread methods currently used implement mutation rate "smoothing", like the non-parametric rate smoothing (Sanderson, 1997) and the semiparametric method of penalized likelihood (Sanderson, 2002, 2003, r8s software), based on the maximum likelihod criterium. Bayesian inference approaches to molecular dating have been also developed (Aris-Brosou & Yang, 2002, PhyBayes software; Drummond & Rambaut, 2007, BEAST software; Huelsenbeck et al., 2000; Kishino et al., 2001; Thorne et al., 1998; Thorne & Kishino, 2002, Multidivtime software). The use of parametric methods for estimating divergence times is also becoming widespread, as they are believed to be more reliable than non-parametric or semiparametric approaches (Drummond & Rambaut, 2007, BEAST software; Thorne & Kishino, 2002, Multidivtime software).

When dealing with fairly recent events in the evolutionary history of organisms, molecular dating needs to be based on rapidly evolving markers such as DNA fingerprinting, due to the usual lack of nucleotide variability found among DNA sequences. Accordingly, a method for the estimation of absolute times of diversification using an AFLP clock approach has been developed (Kropf et al., 2009). They found that the degree of AFLP divergence between mountain phylogroups in different alpine species was significantly correlated with their time of divergence (as inferred from palaeoclimatic/palynological data), indicating constant AFLP divergence rates. Nevertheless, this method was criticized by Ehrich et al. (2009), because the relationship between genetic distance and time in Kropf et al.'s (2009) data was not always linear, and also due to the potential bias introduced by intrapopulation genetic diversity in the suggested genetic distance ($D_{72}$; Nei, 1972).

Molecular dating analyses usually result in a chronogram, this is, a phylogenetic tree that explicitly represents chronological time with branch lengths. It is generally assumed that estimates of divergence time must be considered cautiously (e.g. Hillis et al., 1996). One of the main problems in molecular dating is calibration, which is the inclusion of independent (non-molecular) chronological information within the phylogeny, in order to transform relative to absolute time (Heads, 2005; Magallón, 2004; Reisz & Müller, 2004; Rutschmann et al., 2007). Calibration should be based preferentially in fossils, or, failing that, on palaeogeological events (i.e. continental vicariance, ages of oceanic islands or mountain ranges). However, fossils only provide a lineage's minimum age (Benton & Ayala, 2003; Magallón, 2004). Incongruences between age estimates inferred from molecular data and fossil record have been often found (Pulquério & Nichols, 2007; Sanderson et al., 2004; Steiper & Young, 2008). Likewise, resulting age estimates may display significant differences according to sampling, methods, calibration points, or DNA region analysed (e.g. Linder et al., 2005; Magallón, 2004; Magallón & Sanderson, 2005; Sanderson et al., 2004; Sanderson & Doyle, 2001).

## 3. A case study in Resedaceae and *Carex* L. (Cyperaceae)

### 3.1 Study groups

#### 3.1.1 Resedaceae

The Resedaceae is a small family included in order Brassicales and composed of six genera (*Caylusea* A. St. Hil, *Ochradenus* Del., *Oligomeris* Cambess., *Randonia* Coss., *Reseda* L. and

*Sesamoides* All.; Fig. 1) and ca. 85 species mainly distributed in temperate areas of the Old World (Fig. 2A), with the main centre of diversity around the Mediterranean Basin. Most species grow in sunny and arid habitats, like steppes, deserts and dry slopes, and generally prefer basic soils. In five of the six genera of the family, some (in *Caylusea* and *Reseda*) or all of their species (in *Ochradenus, Oligomeris* and *Randonia*) live in desert and subdesert regions. There are also some ruderal species in genus *Reseda* that occur in waste grounds and disturbed places, and a few are confined to mountainous areas. Four genera are mostly composed of annual or perennial herbs (*Caylusea, Oligomeris, Reseda* and *Sesamoides*), while the remaining two are formed of shrubs (*Ochradenus, Randonia*).

Fig. 1. Representatives of family Resedaceae: A) *Reseda barrelieri* Müll. Arg.; B) *Oligomeris linifolia* (Vahl) J.F. Macbride; C) *Sesamoides purpurascens* (L.) G. López; D) *Caylusea hexagyna* (Forssk.) M.L. Green; E) *Reseda glauca* L.; F) *Randonia africana* Coss.; G) *Ochradenus baccatus* Del.

One of the three species of *Caylusea* occurs in desert regions of the Old World, from Cape Verde archipelago across N Africa to SW Asia, while the remaining two are found in the mountains of NE Tropical Africa (Abdallah & de Wit, 1978; Taylor, 1958; Fig. 2B). *Ochradenus* (9 spp.; Abdallah & de Wit, 1978; Miller, 1984) is distributed in desert regions from Central-North Africa to SW Asia (Fig. 2C). *Oligomeris* (3 spp.; Abdallah & de Wit, 1978) displays interesting range disjunctions, with two species endemic to SW Africa and another widespread from the Canary Islands to SW Asia through N Africa, which also includes disjunct populations in SW North America (Fig. 2D). *Randonia* is a monotypic genus (Abdallah & de Wit, 1978; Miller, 1984) confined to gypsum soils of Central and Western Sahara Desert (Fig. 2E). *Reseda*, with ca. 65 species (Abdallah & de Wit, 1978; Müller

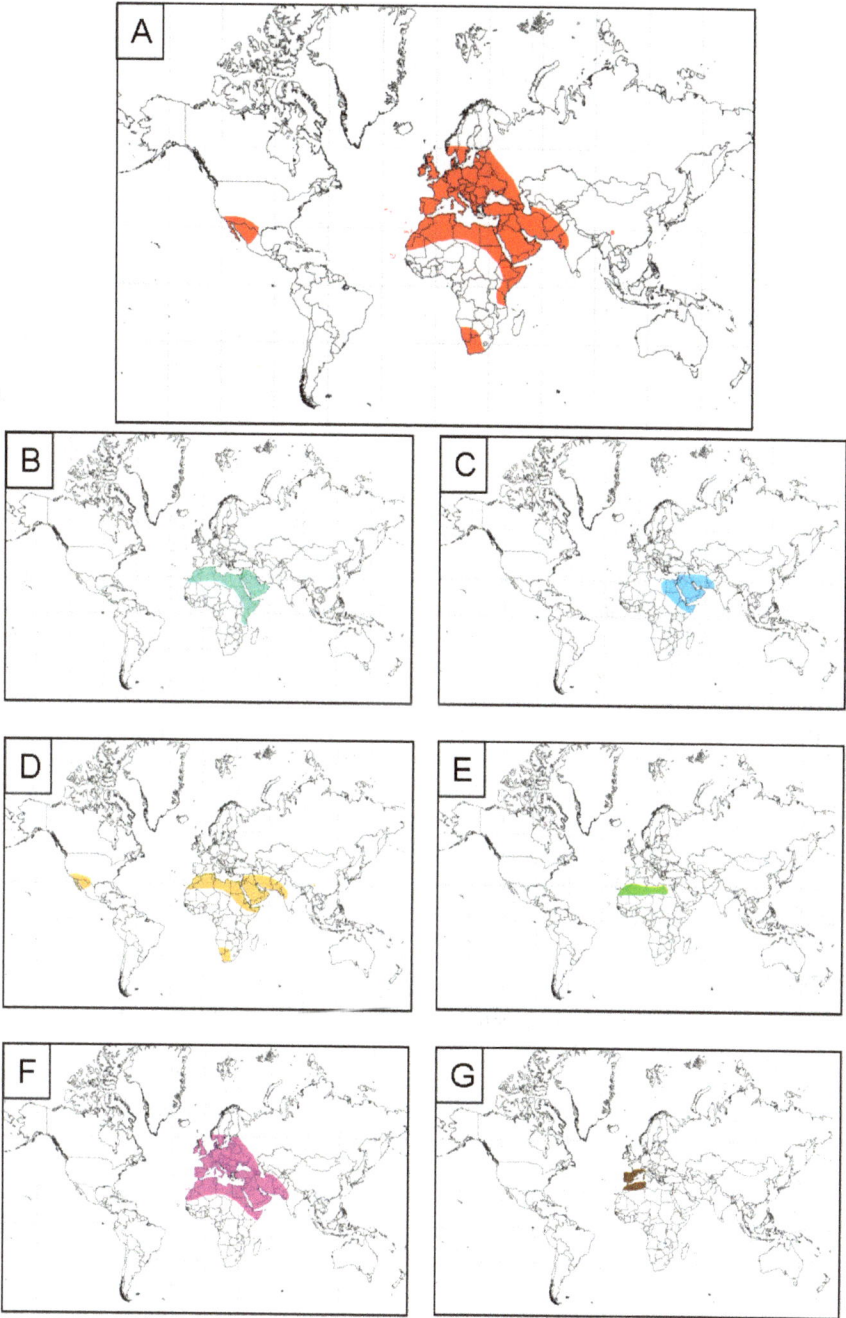

Fig. 2. Approximate distribution of A) family Resedaceae, and its genera: B) *Caylusea*; C) *Ochradenus*; D) *Oligomeris*; E) *Randonia*; F) *Reseda*; G) *Sesamoides*.

Argoviensis, 1868), is by far the largest genus of the family, and its distribution is clearly centered on the Mediterranean Basin (Fig. 2F), with two regions of high species richness, one in the Western and the other in the Eastern Mediterranean and SW Asia. A few of its species has spread as introduced weeds in many temperate regions of the world. Finally, *Sesamoides* (1-6 spp; Abdallah & de Wit, 1978; López González, 1993; Müller Argoviensis, 1868) is endemic to the Western Mediterranean region (Fig. 2G).

### 3.1.2 *Carex*

*Carex* L. (Cyperaceae) is distributed worldwide, with a diversification centre in temperate regions of the Northern Hemisphere. With ca. 2000 species, this genus probably ranks among the four most diversified of angiosperms, and is by far the largest in the temperate regions of the Northern Hemisphere (only the genera *Astragalus, Rosa* and *Euphorbia* may contain more species than *Carex*; Judd et al., 2007). Within genus *Carex*, we have studied two species groups. On one hand, there are six species (*C. canescens* L., *C. macloviana* D´Urv. and *C. maritima* Gunn. from subgenus *Vignea* (P. Beauv. Ex T. Lestib.) Peterm., *C. arctogena* Harry Sm. and *C. microglochin* Wahlenb. from subgenus *Psyllophora* (Degl.) Peterm., and *C. magellanica* Lam. from subgenus *Carex*; Fig. 3 A-F) which display a bipolar distribution (Fig. 4). This means that their populations occur in the high latitudes of both the Northern and Southern Hemispheres. There are only 30 known plant species with such striking pattern, which may be seen as the largest possible disjunction. Accordingly, 20% of bipolar plant species belong to genus *Carex*. On the other hand, *Carex* section *Spirostachyae* (Drejer) L.H. Bailey (subgenus *Carex*) is composed of ca. 40 species (Fig. 3 G-L) mainly distributed in the Mediterranean Basin, Europe and Eastern tropical Africa (Fig. 5). Following the molecular phylogeny, two subsections were recognised, namely *Spirostachyae* (11 spp.) and *Elatae* (Kük.) Luceño and M. Escudero (29 spp.; Fig. 4). Both have a centre of diversification in the Mediterranean region (Fig. 5 A,B, respectively), while subsection *Elatae* displays an additional centre of species diversity in the mountainous region of Eastern tropical Africa (Fig. 5B). In addition, sect. *Spirostachyae* includes some widely disjunct species growing in South America, South Africa, Australia, and the oceanic archipelagos of Macaronesia, Tristan da Cunha, Bioko Island and Mascarene Islands (Fig. 5A).

### 3.2 Methods

Molecular phylogenies were reconstructed for each of the studied groups based on nuclear and plastid DNA regions, with maximum parsimony and Bayesian inference criteria. In Resedaceae, a first phylogeny based on nuclear ribosomal DNA (nrDNA) ITS and plastid *trnL-F* sequences tested the monophyly of the family and established its main lineages (Martín-Bravo et al., 2007). Species diversity and endemism, together with the distribution of the main lineages and their phylogenetic relationships, were analysed to obtain biogeographic insights in Resedaceae (Martín-Bravo et al., 2007). The basic phylogenetic framework subsequently enabled the development of further studies focusing on different lineages which displayed interesting evolutionary or biogeographic features (*Oligomeris*, Martín-Bravo et al., 2009; *Reseda* section *Glaucoreseda* DC., Martín-Bravo et al., 2010). For both groups, phylogeographic analyses were performed, including the reconstruction of haplotype networks with statistical parsimony, using nrDNA ITS and cpDNA *trnL-F* and *rps16* sequences. Penalized likelihood analyses were performed to estimate divergence

Fig. 3. Representatives of the studied *Carex* species. Upper box, bipolar sedges:
A) *C. canescens*; B) *C. macloviana*; C) *C. arctogena*; D) *C. magellanica*; E) *C. maritima*;
F) *C. microglochin*. Lower box, *Carex* sect. *Spyrostachyae*: G) *C. borbonica*; H) *C. lainzii*;
I) *C. punctata*; J) *C. greenwayi*; K) *C. extensa*; L) *C. helodes*.

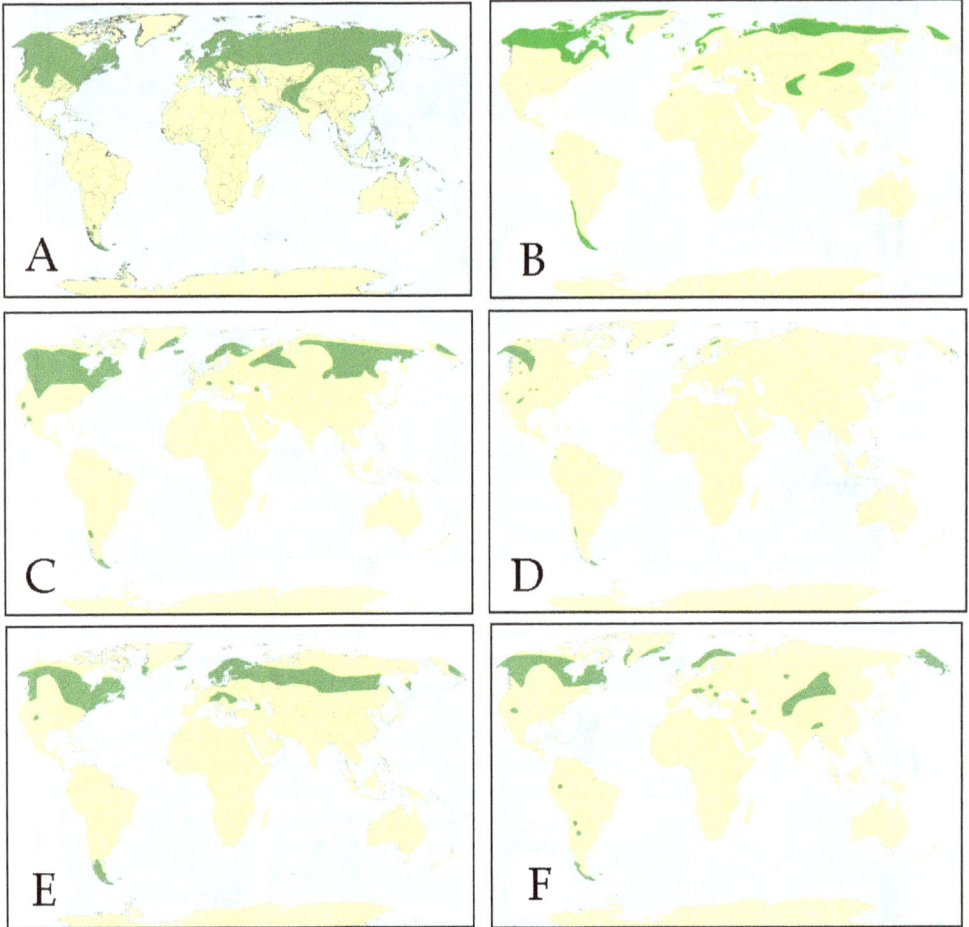

Fig. 4. Approximate distribution of bipolar *Carex* species: A) *C. canescens*; B) *C. maritima*;
C) *C. arctogena*; D) *C. macloviana*; E) *C. magellanica*; F) *C. microglochin*.

times, based on ITS and cpDNA sequences (*rbcL, matK, trnL-F*). Additionally, in *Oligomeris*,
an independent molecular clock approach was performed, namely a test of vicariance based
on nucleotide substitution rates. Finally, in *Reseda* sect. *Glaucoreseda*, a selected set of ITS
sequences was cloned to investigate the origin of intra-individual polymorphisms.

In bipolar *Carex* species, molecular phylogenies based on nuclear ribosomal DNA (ITS) and
plastid (*rps16*) sequences tested the monophyly of five of the six bipolar *Carex* species;
haplotype network reconstructions based on the statistical parsimony method and using
cpDNA (*rps16*) sequences were used to analyse the genetic-geographic structure within
them (Escudero et al., 2010a). In *Carex* sect. *Spirostachyae*, molecular phylogenies based on
nuclear ribosomal DNA ITS and plastid 5′*trnK* intron sequences tested the monophyly of
the section and established its main lineages (Escudero et al., 2008a; Escudero & Luceño,
2009). Subsequently, several studies addressing biodiversity and biogeographic questions

for sect. *Spirostachyae* as a whole were performed, using ancestral area reconstruction (dispersal-extiction-cladogenesis and stochastic methods using maximum likelihood and Bayesian inference approaches, respectively) and estimation of times of diversification (penalized likelihood and parametric uncorrelated log-normal methods using maximum likelihood and Bayesian inference approaches, respectively; Escudero et al., 2009, 2010b). More specifically, different lineages which displayed interesting phylogeographic patterns were studied with AFLP and SSR fingerprinting data, in addition to DNA sequences (*C. helodes* Link, Escudero et al., 2008b; *C. extensa* Good. and allies, Escudero et al., 2010c). For both groups, phylogeographic analyses were performed, including the reconstruction of haplotype networks with statistical parsimony using cpDNA sequences (*rps16* and 5′*trnK* intron for *C. helodes* and *C.* gr. *extensa*, respectively). Both penalized likelihood and parametric uncorrelated log-normal analyses were performed to estimate divergence times in *C.* gr. *extensa*, based on ITS and cpDNA sequences (5′*trnK* intron).

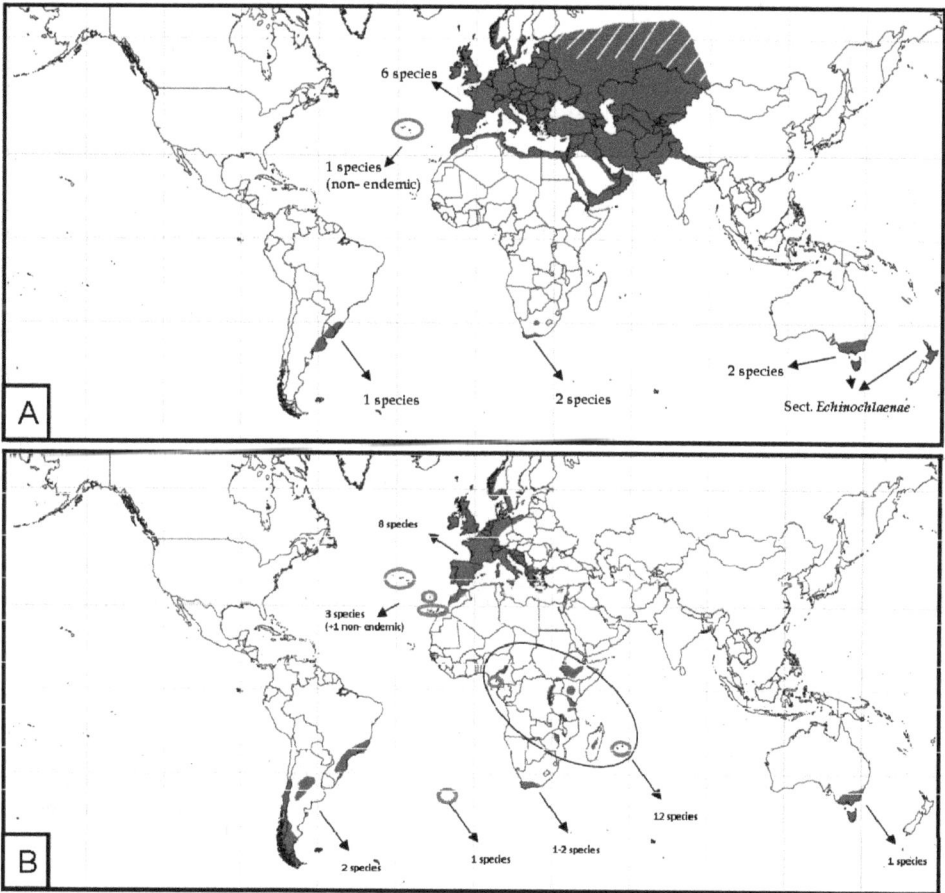

Fig. 5. Approximate distribution of *Carex* sect. *Spyrostachyae* and species richness in each of the main distribution areas (see details in Escudero et al., 2009): A) Subsect. *Spyrostachyae*; B) Subsect. *Elatae*.

## 3.3 Main findings and discussion

### 3.3.1 Patterns of distribution and centres of diversification

In Resedaceae, the Iberian Peninsula and NW Africa harbour one endemic (*Reseda* sect. *Glaucoreseda*) and two subendemic (genus *Sesamoides* and *Reseda* sect. *Leucoreseda* DC.) lineages. On the other hand, sect. *Reseda* appears to have diversified mainly in the Middle East. Section *Phyteuma* Lange is mostly composed of regional endemics restricted to either the Western or the Eastern Mediterranean regions, including such East-West disjunction between closely related species (*R. media* Lag. – *R. orientalis* (Müll. Arg.) Boiss; Martín-Bravo et al., in prep.). In sect. *Phyteuma*, the inferred speciation rate in the Eastern Mediterranean could be significantly higher than in the Western; likewise, dispersal from the Eastern to the Western Mediterranean appears to be predominant than the contrary (Martín-Bravo et al., in prep.). This biogeographical E-W Mediterranean pattern, repeatedly found at different taxonomic levels within Resedaceae (genera, sections and species) could be related to two major, disjunct centres of diversification at both sides of the Mediterranean. This also has been reported for other plant groups at different taxonomic levels (Plumbaginaceae subfamily Staticoideae, Lledó et al., 2005; *Cuminum*, Davis & Hedge, 1971; *Hedera*, Valcárcel & al., 2003; *Buxus balearica*, Rosselló et al., 2007; *Carex extensa*, Escudero et al., 2010c; *Erophaca baetica*, Casimiro-Soriguer et al., 2010; *Microcnemum coralloides*, Kadereit & Yaprak, 2008). Accordingly, both areas are considered melting pots of plant diversity and endemism (Médail & Quézel, 1997, 1999) within the Mediterranean Basin hotspot (Myers et al., 2000), and traditionally have been seen as critical for the diversification of Mediterranean flora. In contrast, non-Mediterranean Europe displays a low number of Resedaceae taxa, of which most are ruderal and widespread *Reseda* species (Martín-Bravo et al., 2007). This distribution pattern in Resedaceae leads us to propose a general scenario of glacial refugia and diversification in the Western and Eastern regions of the Mediterranean (Hewitt, 1996, 2001; Médail & Diadema, 2009; Taberlet et al., 1998). Species with greater colonization and dispersal ability could have postglacially recolonized Europe from those refugia. In areas of high diversification like the Iberian Peninsula and NW Africa (including the Canary Islands) geographical isolation could be regarded as an important driver of speciation (Martín-Bravo et al., 2010), probably as a result of the climatic, edaphic and topographic heterogeneity of these regions (Cowling & al., 1992, 1996; Martín-Bravo et al., 2007). Northeastern tropical Africa and the southern Arabian Peninsula feature an extraordinary endemism degree for Resedaceae (ca. 80%; Martín-Bravo et al., 2007). Two of the three *Caylusea* species and all *Reseda* species (nine) are endemic there. In addition, this region is clearly the center of diversity of genus *Ochradenus*, harbouring eight of its nine species, of which six are endemic. Active speciation processes have been related with the aridification of this area during the Pleistocene (Cane & Molnar, 2001; Chiarugi, 1933; Demenocal, 1995; Quézel, 1978), as well as with its topographic and geologic complexity. Genus *Reseda* appears to have diversified preferentially in the horn of Africa (six endemics), whereas *Ochradenus* displays a greater number of endemic species (four) in the southern Arabian Peninsula (Martín-Bravo et al., 2007).

As already pointed out, *Carex* is by far the largest angiosperm genera in the temperate and cold regions of the Northern Hemisphere. Interestingly, it displays an inverse latitudinal gradient of species richness (Hillebrand, 2004; Kaufman & Willig, 1998). This feature suggests that historical cold periods could have promoted *Carex* diversification (Escudero et al., in prep.). Bipolar *Carex* species are an interesting example of this distribution pattern in cold regions of the Northern Hemisphere. In contrast, *Carex* sect. *Spirostachyae* is a very peculiar group within the genus *Carex*, as it preferentially grows in warmer regions, and accordingly, is mainly distributed in the Mediterranean Basin and Eastern tropical Africa. Both subsections within *Spirostachyae* have a center of diversification in the Mediterranean region (with six of 11 species in subsect. *Spirostachyae*, and eight of 29 in subsect. *Elatae*; Fig. 5), while subsection *Elatae* displays an additional centre of species diversity in the mountainous region of Eastern tropical Africa (10 of 29 species; Fig. 5B). In addition, sect. *Spirostachyae* includes some widely disjunct species growing in remote regions in the world (Fig. 5A). On one hand, subsect. *Elatae* has disjunct species in South America (2 spp.), South Africa (1-2 spp.), Australia (1 sp.), and the oceanic archipelagos of Macaronesia (3 spp.), Tristan da Cunha (1 sp.), and Mascarene Islands (2 spp.). On the other hand, subsect. *Spirostachyae* includes disjunct species in South America (1 sp.), South Africa (2 spp.), Australia (2 spp.), and the archipelagos of Macaronesia (3 endemics and one species also distributed in the continental Old World). Interestingly, the section *Echinochlaenae* T. Holm (ca. 30 spp.; subendemic of the Northern Island of New Zealand, with ca. 2 spp. growing in Australia) forms a monophyletic group together with the Australian species of subsect. *Spirostachyae* (personal communication from Dr. Marcia Waterway, McGill University, Montreal, Quebec, Canada).

### 3.3.2 Range disjunctions: Long-distance dispersal or vicariance?

Our study groups show interesting patterns of disjunction at very different geographic scales (Figs. 2, 4, 5). Phylogenetic and phylogeographic analyses, molecular dating of lineage divergence, palaeogeological/climatic data, and ancestral range reconstruction (see methods in 3.2) were variously used to try to answer one of the prevailing questions in plant biogeography: the explanation of the causes of such disjunctions.

#### 3.3.2.1 Old – New World disjunctions

In Resedaceae, one species in genus *Oligomeris* (*O. linifolia*) is widespread in desert and arid areas of the Old World (N Africa – SW Asia), and also includes disjunct populations in the New World (SW North America), constituting the most remarkable disjunction within the family (Fig. 2). Two examples of Old - New World disjunctions between closely related species may be found in *Carex* sect. *Spirostachyae*: *C. extensa* (subsect. *Spirostachyae*) and *C. punctata* Gaudin (subsect. *Elatae*) grow in the Mediterranean Basin, Europe and SW Asia, while their sister species *C. vixdentata* (Kük.) G.A. Wheeler and *C. fuscula* D'Urv - *C. catharinensis* Boeck., respectively, occur in Central and Southern South America (Fig. 5).

This pattern of trans-oceanic disjunction, rarely found at the species level in plants, has been traditionally explained by vicariance hypotheses dating back at least to the Miocene (ca. 20 m.a; Axelrod, 1975; Stebbins & Day, 1967; Tiffney, 1985). More recently, molecular dating has explained this pattern of disjunction at the family and genus level, favouring long-distance dispersal in some cases (i.e. *Erodium*, Fiz et al., 2010; *Thamnosma*, Thiv et al., 2011) and vicariance in others (reviewed in Wen & Ickert-Bond, 2009). In our *Oligomeris* study, relatively recent estimates of divergence time for *O. linifolia* (Upper Pleistocene), the low level of genetic differentiation between the disjunct populations, and the distribution of the rest of the family, suggests a long-distance dispersal event from the Old World, probably occurred during the Quaternary, to account for this disjunction (Martín-Bravo et al., 2009). Our hypothesis is congruent with the results obtained for other species from arid regions that display a similar disjunction (*Senecio mohavensis*, Coleman et al., 2003; *Plantago ovata*, Meyers & Liston, 2008). These studies also have estimated Pleistocene as the temporal framework and the same direction of the dispersal event (from the Old to the New World). However, both species have epizoochoric dispersal syndromes (Coleman et al., 2003; Meyers & Liston, 2008), while *Oligomeris* lacks apparent specific mechanisms for long-distance dispersal. Results from our *Spirostachyae* study are mostly congruent to those from *Oligomeris*. Firstly, estimation of divergence times in sect. *Spirostachyae* also discards a vicariance pattern and supports trans-hemisphere, Old to New World, long-distance dispersal. In addition, as in *Oligomeris*, the *Carex* species studied apparently lack specific mechanisms for long-distance dispersal. Nevertheless, estimation of the split between Old and New World species are not as recent as in *Oligomeris*, dating back to Pliocene times (Escudero et al., 2009, 2010b).

### 3.3.2.2 Northern – Southern Hemisphere disjunctions

In Resedaceae, two *Oligomeris* species (*O. dipetala*, *O. dregeana*) are endemic to SW Africa, in contrast to the mostly Mediterranean distribution of the Resedaceae (Fig. 2). Many species show a disjunct distribution between arid regions of Southern and Northern Africa (Goldblatt, 1978; Thulin, 1994; de Winter, 1971). Both dispersalist and vicariance hypotheses have been invoked to explain this pattern of disjunction (i.e. Beier et al., 2004; Thiv et al., 2011; Thulin, 1994). The latter has been based on the presence of an arid corridor through Eastern Africa which intermittently connected Northern – Southern Africa during the Pliocene – Pleistocene (Jürgens, 1997; Verdcourt, 1969; van Zinderen Bakker, 1978), although the exact age of this corridor is still disputed (Thiv et al., 2011). These palaeoclimatic data, together with our Lower Pleistocene time estimates for the origin of genus *Oligomeris* (Martín-Bravo et al., 2009) does not allow us to clarify the process involved in the origin of the SW African endemics.

As stated above, the genus *Carex* includes six of the 30 known plant species which display a bipolar disjunction (Moore & Chater, 1971). This has traditionally been one of the most intensively studied patterns of disjunct distributions, and both long-distance dispersal and vicariance hypotheses have been proposed (Moore & Chater, 1971, Vollan et al., 2006). The later were based on the presence of trans-tropical land bridges through mountain ranges during the Mesozoic age (250 - 65 m.a; du Rietz, 1940). Phylogenetic relationships and the low level of genetic differentiation among Northern-Southern populations of *Carex* species suggest long-distance dispersal as the most plausible cause of the bipolar disjunction for the five sampled species (all except for *C. arctogena*, not included in Escudero et al., 2010a). In addition, haplotype genealogical relationships point to a southward direction of dispersal in three species (*C. macloviana*, *C. magellanica* and *C. canescens*; Escudero et al., 2010a). Nevertheless, the timing of these bipolar disjunctions and the alternative dispersal hypotheses (i.e. direct long-distance dispersal or mountain hopping; Escudero et al., 2010a) involved in their origin remain to be investigated. A very recent study of the bipolar disjunction displayed by genus *Empetrum* (Popp et al., 2011; see also Donoghue, 2011) also explained it with a North to South long-distance dispersal colonization. In addition, they dated the event to Pleistocene times and postulated that direct dispersal rather than mountain hopping was at the origin of the disjunction. On the other hand, *Carex* sect.

*Spirostachyae* shows an interesting pattern of Northern – Southern Hemisphere disjunctions. In addition to the above cited species in South America (see Old – New World disjunctions), there are two species in South Africa (*C. ecklonii* Nees and *C. burchelliana* Boeck.) whose origin, according to ancestral range reconstruction, may be placed in the Northern Hemisphere Old World (ancestors of *C. gr. extensa* and *C. gr. distans*, respectively; Escudero et al., 2009). Estimations of times of diversification discard a vicariance process, and indicate that the most plausible hypothesis to explain these disjunctions is also North to South long-distance dispersal in Late Miocene - Pliocene times (Escudero et al., 2009, 2010b).

### 3.3.2.3 Colonization of continental and oceanic archipelagos

Despite the Resedaceae and *Carex* sect. *Spirostachyae* apparently lack specific mechanisms for long-distance dispersal, they have successfully colonized various continental and oceanic archipelagos. *Oligomeris linifolia*, whose trans-oceanic disjunction has already been commented above, grows in many islands or archipelagos throughout its large range (Fig. 2B), including most islands off the Californian coast, the Eastern Canary Islands and islands in the Persian Gulf (Martín-Bravo et al., 2009). Many of these islands are of oceanic origin and their indigenous floras a consequence of dispersal from mainland. Moreover, some of them are situated a considerable distance from the mainland, such as the Canary Islands (ca. 100 km), the Channel Islands (20-100 km), and Guadalupe Island (260 km). Therefore, *O. linifolia* seems to have great dispersal and colonization ability, despite its unassisted dispersal syndrome. Other organisms with apparent low dispersal ability display a similar pattern of oceanic dispersal (review in de Queiroz, 2005).

At least four independent long-distance dispersal events could have been involved in the colonization of the oceanic Canary Islands by the four species of the family growing there (the widespread *Oligomeris linifolia* and *Reseda luteola* L., and the endemic *R. crystallina* Webb & Berthel. and *R. scoparia* Willd.), which are placed in distinct clades of the phylogeny (Martín-Bravo et al., 2007). The origin of the Canarian endemic *R. crystallina* could have taken place following dispersal from NW Africa, as inferred from its close phylogenetic relationship with the NW African *R. lutea* subsp. *neglecta* (Müll. Arg.) Abdallah & de Wit (Martín-Bravo et al., 2007). *Carex* sect. *Spirostachyae* is represented in Macaronesian archipelagos by three endemics from subsect. *Elatae*, *C. perraudieriana* Gay in Canary Islands, *C. lowei* Bech. in Madeira and *C. hochstetteriana* Gay in Azores, as well as by the presence of two widespread species, *C. extensa* in subsect. *Spirostachyae* and *C. punctata* in subsect. *Elatae*. Estimation of species diversification times in sect. *Spirostachyae* support a fairly old colonization of Macaronesia by the endemic species, probably predating Pliocene, and recent colonization of the widespread species (Escudero et al., 2009, 2010b). Phylogenetic relationships indicate multiple colonization events although it is difficult to know the exact number of colonizations (at least three; Escudero et al., 2009). In Macaronesia, a single dispersal event and subsequent colonization has usually been reported for taxa lacking specific dispersal syndromes, whereas recurrent colonization has rarely been reported for such plants (Vargas, 2007). In contrast, multiple colonizations have been frequently proposed for taxa with specific dispersal mechanisms (endozoocory: *Hedera, Ilex, Juniperus, Olea*; hidrocory: *Euphorbia, Lavatera*; reviewed in Vargas, 2007, but see *Cistus*, Guzmán & Vargas, 2009a).

Likewise, two different allopatric speciation processes could have taken place in the colonization of Socotra archipelago by Resedaceae, represented by two phylogenetically distinct endemics, *Reseda viridis* Balf. f. and *Ochradenus socotranus* A.G. Mill. (Martín-Bravo et al., 2007). Socotra is of continental origin, but has remained isolated from continental masses for a long geologic term (18 - 15 m.a; van Damme, 2009; Fleitmann et al., 2004). Its age, together with the estimated Miocene origin of Resedaceae (10 - 16 m.a; Martín-Bravo et al., 2009, 2010), and the relatively low genetic differentiation between the endemics and their closest relatives, point again to independent long-distance dispersal events to account for the colonization of Socotra.

The colonization of the Atlantic archipelago of Tristan da Cunha archipelago (*C. thouarsii* Carmichael), and of the Mascarene Islands in the Indian Ocean (*C. boryana* Schkuhr and *C. borbonica* Lam.), seems to be recent (Escudero et al., 2009, 2010b). Specifically, low levels of genetic differentiation with their most closely related continental species, lead to infer relatively recent long-distance dispersal events from Western South America (*C. fuscula – C. catharinensis*) and Eastern tropical Africa (tropical African group of subsect. *Elatae*) to explain the colonization of Tristan da Cunha Archipelago Island and Mascarene islands, respectively (Escudero et al., 2009, 2010b). Plant colonization of Tristan da Cunha Archipelago has been scarcely study, and as far as we know, no similar pattern of colonization (dispersal from Western South America) has been previously described for angiosperms. Nevertheless, long-distance dispersal from Africa to Tristan da Cunha archipelago has been already proposed (Anderson et al., 2001; Richardson et al., 2003). On the other hand, our data support a single colonization of the Mascarene archipelago from Eastern tropical Africa or Madagascar. Uncertainty about the presence of sect. *Spirostachyae* in Madagascar does not allow us to clarify between both possibilities.

### 3.3.2.4 Ibero-North African disjunctions across the strait of Gibraltar

The Iberian Peninsula and NW Africa have close biogeographic affinities (Médail & Quézel, 1997). The Ibero-North African floristic element is well represented in the flora of the Iberian Peninsula, especially in its southern part (Blanca et al., 1999). Thus, southern Iberia and northern Morocco share about 75% of 3,500 species (Valdés, 1991), with more than 500 endemics (Quézel, 1978). The two regions were connected from the end of the Miocene to the upper Pliocene due to the partial desiccation of the Mediterranean Sea following an increase in aridity (Messinian salinity crisis, 5.9-5.3 million years ago; Hsü et al., 1977; Duggen et al., 2003; Rouchy & Caruso, 2006), allowing contacts between the floras of the two continents.

*Reseda battandieri* Pit. is a Moroccan endemic species within the otherwise Iberian section *Glaucoreseda*. Pleistocene divergence times estimated for sect. *Glaucoreseda* and *R. battandieri* (Martín-Bravo et al., 2010) are later than the Mio-Pliocene opening of the Strait of Gibraltar (ca. 5.3 m.a; Krijgsman, 2002). This fact, together with genealogical relationships of haplotypes, suggests dispersal from the Iberian Peninsula across the Strait of Gibraltar during the Pleistocene, to account for the colonization of NW Africa by sect. *Glaucoreseda* and the allopatric differentiation of *R. battandieri* (Martín-Bravo et al., 2010).

*Carex helodes* (sect. *Spirostachyae*) is an endemic species from SW Iberian Peninsula and Northern Morocco. Nuclear and plastid sequences, AFLPs and cytogenetic counts demonstrate that European and African populations constitute two independent lineages, although with low genetic divergence between them (Escudero et al., 2008b). Genealogical relationships of haplotypes and the pattern of genetic diversity suggest dispersal from the Iberian Peninsula into NW Africa, probably later than the Mio-Pliocene opening of the Strait of Gibraltar (ca. 5.3 m.a; Krijgsman, 2002), and subsequent differentiation of Moroccan populations (Escudero et al., 2008b).

These results from *Reseda* and *Carex* underline the importance of the Strait of Gibraltar as a barrier to plant gene flow, although its specific role greatly depends on the study group. In most cases, a certain degree of differentiation at the specific or population level is observed, whereas continuity in the genetic structure has been reported in a few cases (review in Rodríguez-Sánchez et al., 2008). Overall, results from studies of other plant groups point to a dispersal rather than a vicariance pattern to explain Ibero-North African disjunctions (e.g. Guzmán & Vargas, 2009b; Jiménez-Mejías et al., 2011; Rodríguez-Sánchez et al., 2008; Rubio de Casas et al., 2006).

### 3.3.2.5 Ecological vicariance

Four of the five species in *Reseda* sect. *Glaucoreseda* (*R. complicata* Bory, *R. glauca* L., *R. gredensis* (Cutanda & Willk.) Müll. Arg. and *R. virgata* Boiss. & Reuter) are endemics with an allopatric distribution in the high mountain ranges and plateaus of the Iberian Peninsula. We studied the possible correlation between Quaternary glaciations, historical range dynamics and speciation processes (Martín-Bravo et al., 2010). Molecular dating points to a late Pleistocene diversification of sect. *Glaucoreseda*, which together with the current distribution of endemics and their cytogenetic features, suggest an ecological vicariance hypothesis, characterised by the interglacial range fragmentation of an ancestral species and subsequent allopatric speciation (Martín-Bravo et al., 2010). Ecological specialization in different kind of substrates could have acted in concert with geographical isolation to promote morphological differentiation and subsequent speciation (Dixon et al., 2007; Martín-Bravo et al., 2010) in the different Iberian species of sect. *Glaucoreseda*. The clear morphological differences between sect. *Glaucoreseda* species (Abdallah & de Wit, 1978; Martín-Bravo, 2009; Valdés Bermejo, 1993), coupled with the low level of molecular divergence among them, lead us to suggest that phenotypic has been faster than genotypic differentiation in sect. *Glaucoreseda*, as suggested for other Resedaceae genera (Martín-Bravo et al., 2007). In summary, our results lend support to the important role of range shifts induced by Quaternary climatic oscillations in the diversification of European mountain plant groups.

Within *Carex* sect. *Spirostachyae*, subsect. *Elatae* shows an interesting pattern of species distribution with two disjunct centres of diversification, one in the Mediterranean and the second in the mountains of Eastern tropical Africa (Cronk, 1992; Escudero et al., 2009). Tropical African species of subsect. *Elatae* might have been abundant in subtropical and tropical African woodlands until the Late Pliocene, when these habitats were extensive (Cronk, 1992). Subsequently, these habitats were dramatically reduced during the Pleistocene aridification of Africa (Cronk, 1992), probably entailing extinction of many populations of these species. The Miocene origin estimated for this group (Escudero et al., 2009, 2010b) supports an ecological vicariance hypothesis in which the general cooling and aridification of Africa and the formation of Sahara Desert during the Late Pliocene-Pleistocene could have interrupted a previously continuous range. Subsequently, allopatric speciation processes were probably responsible for the diversification of subsect. *Elatae* (Escudero et al., 2009).

## 4. Conclusions

Our biogeographic studies in Resedaceae and *Carex* provide examples of how the development of molecular biology and bioinformatics from the end of the XXth century has contributed to a revolution in the field of evolutionary biology, and, specifically, in biogeography. These advances have enabled us to gain insights into many long-standing questions in plant biogeography, like the location of ancestral areas and centers of diversification, the processes behind remarkable range disjunctions (vicariance vs. dispersal), and the important role of geography in plant evolution and biodiversity. With respect to range disjunctions, our results point to the outstanding importance of long-distance dispersal events rather than vicariance to explain different patterns of plant disjunction at diverse geographic scales, independently of the presence of specific mechanisms for long-distance dispersal. Remarkably, at least 23 long-distance dispersal connections (15 in *Carex* and 8 in Resedaceae) were inferred in our study groups (Fig. 6). We provide several documented examples of allopatric speciation, which confirm geographical isolation as one of the most important drivers of plant evolution. Geographical, coupled with reproductive isolation among populations, may lead to genetic divergence and, eventually, to speciation when gene flow is precluded. Ecological vicariance induced by range shift dynamics caused by climatic changes during Pliocene and Pleistocene also promoted speciation events and deeply shaped the current genetic structure of many species.

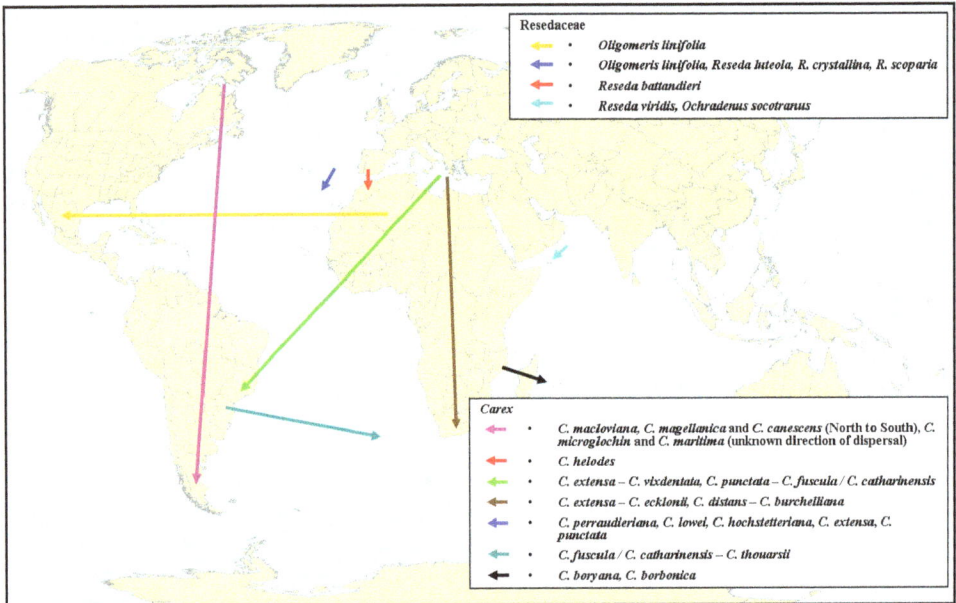

Fig. 7. Documented examples of long-distance dispersal connections inferred in our study groups (Resedaceae and *Carex*).

## 5. Acknowledgments

We thank Dr. M. Luceño, J.M. André, A. Al-Sirhan and P. Aurousseau for granting permission to publish some of the photographs included in this chapter, T. Villaverde for comments on the manuscript, and E. Maguilla for helping with the distribution map of bipolar sedges.

## 6. References

Abdallah, M.S. & de Wit, H.C.D. (1978). The Resedaceae: a taxonomical revision of the family (final instalment). *Mededelingen Landbouwhoogeschool Wageningen*, vol. 78, pp. 1-416

Álvarez, I. & Wendel, J.F. (2003). Ribosomal ITS sequences and plant phylogenetic inference. *Molecular Phylogenetics and Evolution*, vol. 29, pp. 417-434

Anderson, C.L., Rova, J.H.E. & Andersson, L. (2001). Molecular phylogeny of the tribe Anthospermeae (Rubiaceae): Systematic and biogeographic implications. *Australian Systematic Botany*, vol. 14, pp. 231-244

Aris-Brosou, S. & Yang, Z. (2002). Effects of models of rate evolution on estimation of divergence dates with special reference to the metazoan 18S ribosomal RNA phylogeny. *Systematic Biology*, vol. 51, pp. 703-714

Avise, J.C., Arnold, J., Ball, R.M., Jr, Bermingham, E., Lamb, T., Neigel, J.E., Reeb, C.A. & Saunders, N.C. (1987). Intraspecific phylogeography: the mitochondrial DNA bridge between population genetics and systematics. *Annual Review of Ecology and Systematics*, vol. 18, pp. 489–522

Avise, J.C. (2000). *Phylogeography: the history and formation of species*, Harvard University Press, Cambridge, Massachusetts, USA

Axelrod, D.I. (1975). Evolution and biogeography of the Madrean-Tethyan sclerophyll vegetation. *Annals of the Missouri Botanical Garden*, vol. 62, pp. 280-334

Baird, N.A., Etter, P.D., Atwood, T.S., Currey, M.C., Shiver, A.L., Lewis, Z.A., Selker, E.U., Cresko, W.A. & Johnson, E.A. (2008). Rapid SNP discovery and genetic mapping using sequenced RAD markers. *PLoS ONE*, vol. 3, art. no. e3376

Bailey, C.D., Carr, T.G., Harris, S.A. & Hughes, C.E. (2003). Characterization of angiosperm nrDNA polymorphism, paralogy, and pseudogenes. *Molecular Phylogenetics and Evolution* , vol. 29, pp. 435-455

Bandelt, H. & Dress, A.W.M. (1992). Split decomposition: A new and useful approach to phylogenetic analysis of distance data. *Molecular Phylogenetics and Evolution*, vol. 1, pp. 242-252

Bandelt, H., Forster, P. & Röhl, A. (1999). Median-joining networks for inferring intraspecific phylogenies. *Molecular Biology and Evolution*, vol. 16, pp. 37-48

Bandelt, H., Macaulay, V. & Richards, M. (2000). Median networks: Speedy construction and greedy reduction, one simulation, and two case studies from human mtDNA. *Molecular Phylogenetics and Evolution* , vol. 16, pp. 8-28

Beardsley, P.M., Yen, A. & Olmstead, R.G. (2003). AFLP phylogeny of *Mimulus* section *Erythranthe* and the evolution of hummingbird pollination. *Evolution*, vol. 57, pp. 1397-1410

Becquet, C. & Przeworski, M. (2007). A new approach to estimate parameters of speciation models with application to apes. *Genome Research*, vol. 17, pp. 1505-1519

Becquet, C. & Przeworski, M. (2009). Learning about modes of speciation by computational approaches. *Evolution*, vol. 63, pp. 2547-2562

Beier, B.A., Nylander, J.A.A. Chase, M.W. & Thulin, M. (2004). Phylogenetic relationships and biogeography of the desert plant genus *Fagonia* (Zygophyllaceae), inferred by parsimony and Bayesian model averaging. *Molecular Phylogenetics and Evolution*, vol. 33, pp. 91-108

Benton, M.J. & Ayala, F.J. (2003). Dating the tree of life. *Science*, vol. 300, pp. 1698-1700

Blair, C. & Murphy, R.W. (2011). Recent trends in molecular phylogenetic analysis: Where to next? *Journal of Heredity*, vol. 102, pp. 130-138

Blanca, G., Cabezudo, B., Hernández-Bermejo, E., Herrera, C.M., Molero Mesa. J., Muñoz, J. & Valdés, B. (1999). *Libro Rojo de la flora silvestre amenazada de Andalucía, vol. 1, Especies en peligro de extinción*, Consejería de Medio Ambiente, Junta de Andalucía, Sevilla, Spain.

Bollback, J.P. (2006). SIMMAP: Stochastic character mapping of discrete traits on phylogenies. *BMC Bioinformatics*, vol. 7, art. no. 88

Bremer, K. & Gustafsson, M.H.G. (1997). East Gondwana ancestry of the sunflower alliance of families. *Proceedings of the National Academy of Sciences of the United States of America*, vol. 94, pp. 9188-9190

Britten, R.J. (1986). Rates of DNA sequence evolution differ between taxonomic groups. *Science*, vol. 231, pp. 1393-1398

Britton, T., Oxelman, B., Vinnersten, A. & Bremer, K. (2002). Phylogenetic dating with confidence intervals using mean path lengths. *Molecular Phylogenetics and Evolution*, vol. 24, pp. 58-65

Bromham, L. & Penny, D. (2003). The modern molecular clock. *Nature Reviews Genetics*, vol. 4, pp. 216-224

Bryant, D. & Moulton, V. (2004). Neighbor-Net: An agglomerative method for the construction of phylogenetic networks. *Molecular Biology and Evolution*, vol. 21, pp. 255-265

Buerki, S., Forest, F., Alvarez, N., Nylander, J.A.A., Arrigo, N. & Sanmartín, I. (2011). An evaluation of new parsimony-based versus parametric inference methods in biogeography: A case study using the globally distributed plant family Sapindaceae. *Journal of Biogeography*, vol. 38, pp. 531-550

Caicedo, A.L. & Schaal, B. (2004). Population structure and phylogeography of *Solanum pimpinellifolium* inferred from a nuclear gene. *Molecular Ecology*, vol. 13, pp. 1871-1882

Calonje, M., Martín-Bravo, S., Dobes, C., Gong, W., Jordon-Thaden, I., Kiefer, C., Kiefer, M., Paule, J., Schmickl, R. & Koch, M.A. (2009). Non-coding nuclear DNA markers in phylogenetic reconstructions. *Plant Systematics and Evolution*, vol. 282, pp. 257-280

Cane, M.A. & Molnar, P. (2001). Closing of the Indonesian seaway as a precursor to east African aridification around 3 million years ago. *Nature*, vol. 411, pp. 157-162

Casimiro-Soriguer, R., Talavera, M., Balao, F., Terrab, A., Herrera, J. & Talavera, S. (2010). Phylogeny and genetic structure of *Erophaca* (Leguminosae), an East-West Mediterranean disjunct genus from the Tertiary. *Molecular Phylogenetics and Evolution*, vol. 56, pp. 441-450

Chiarugi, A. (1933). Paleoxilologia della Somalia Italiana. *Giornale Botanico Italiano*, vol. 40, pp. 306-307

Clark, J.R., Ree, R.H., Alfaro, M.E., King, M.G., Wagner, W.L. & Roalson, E.H. (2008). A comparative study in ancestral range reconstruction methods: retracing the uncertain histories of insular lineages. *Systematic Biology*, vol. 57, pp. 693-707

Clausing, G., Vicker, K. & Kadereit, J.W. (2000). Historical biogeography in a linear system: Genetic variation of Sea Rocket (*Cakile maritima*) and Sea Holly (*Eryngium maritimum*) along European coasts. *Molecular Ecology*, vol. 9, pp. 1823-1833

Clement, M., Posada, D. & Crandall, K.A. (2000). TCS: A computer program to estimate gene genealogies. *Molecular Ecology*, vol. 9, pp. 1657-1659

Coleman, M., Liston, A., Kadereit, J.W. & Abbott, R.J. (2003). Repeat intercontinental dispersal and Pleistocene speciation in disjunct Mediterranean and desert *Senecio* (Asteraceae). *American Journal of Botany*, vol. 90, pp. 1446-1454

Corander, J., Waldmann, P. & Sillanpää, M.J. (2003). Bayesian analysis of genetic differentiation between populations. *Genetics*, vol. 163, pp. 367-374

Cowling, R.M. & Holmes, P.M. (1992). Endemism and speciation in a lowland flora from the Cape Floristic Region. *Biological Journal of the Linnean Society*, vol. 47, pp. 367-383

Cowling, R.M., Rundel, P.W., Lamont, B.B., Arroyo, M.K. & Arianoutsou, M. (1996). Plant diversity in Mediterranean-climate regions. *Trends in Ecology and Evolution*, vol. 11, pp. 362-366

Cronk, Q.C.B. (1992). Relict floras of Atlantic islands: Patterns assessed. *Biological Journal of the Linnean Society*, vol. 46, pp. 91-103

van Damme, K. (2009). Socotra archipelago, In: *Encyclopedia of islands*, Gillespie, R.G. & Clague, D.A., (Eds.), pp. 846-851, University of California Press, Berkeley, California, USA

Darwin, C.R. 1859. *On the origin of the species*. John Murray, London, UK

Davis, P.H. & Hedge, I.C. (1971). Floristic links between NW Africa and SW Asia. *Annalen Naturhistorisches Museum Wien*, vol. 75, pp. 43-57

Davey, J.W., Hohenlohe, P.A., Etter, P.D., Boone, J.Q., Catchen, J.M. & Blaxter, M.L. (2011). Genome-wide genetic marker discovery and genotyping using next-generation sequencing. *Nature Reviews Genetics*, vol. 12, pp. 499-510

Degnan, J.H. & Rosenberg, N.A. (2009). Gene tree discordance, phylogenetic inference and the multispecies coalescent. *Trends in Ecology and Evolution*, vol. 24, pp. 332-340

Demenocal, P.B. (1995). Plio-Pleistocene African climate. *Science*, vol. 270, pp. 53-59

Dick, C.W., Bermingham, E., Lemes, M.R. & Gribel, R. (2007). Extreme long-distance dispersal of the lowland tropical rainforest tree *Ceiba pentandra* L. (Malvaceae) in Africa and the Neotropics. *Molecular Ecology*, vol. 16, pp. 3039-3049

Dixon, C.J., Schönswetter, P. & Schneeweiss, G.M., 2007. Traces of ancient range shifts in a mountain plant group (*Androsace halleri* complex, Primulaceae). *Molecular Ecology*, vol. 16, pp. 3890-3901

Donoghue, M.J. (2011). Bipolar biogeography. *Proceedings of the National Academy of Sciences of the United States of America*, vol. 108, pp. 6341-6342

Drummond, A.J. & Rambaut, A. (2007). BEAST: Bayesian evolutionary analysis by sampling trees. *BMC Evolutionary Biology*, vol. 7, art. no. 214

Drummond, C.S. (2008). Diversification of *Lupinus* (Leguminosae) in the Western New World: Derived evolution of perennial life history and colonization of montane habitats. *Molecular Phylogenetics and Evolution*, vol. 48, pp. 408-421

Duggen, S., Hoernle, K., van de Bogaard, P., Rüpke, L. & Phipps Morgan, J.M. (2003). Deep roots on the Messinian salinity crisis. *Nature*, vol. 422, pp. 602-606

Edwards, A.W.F. & Cavalli-Sforza, L.L. (1963). The reconstrution of evolution. *Annals of Human Genetics*, vol. 27, pp. 105-106

Elshire, R.J., Glaubitz, J.C., Sun, Q., Poland, J.A., Kawamoto, K., Buckler, E.S. & Mitchell, S.E. (2011). A robust, simple genotyping-by-sequencing (GBS) approach for high diversity species. *PLoS ONE*, vol. 6, art. no. e19379

Ehrich, D., Eidesen, P.B., Alsos, I.G. & Brochmann, C. (2009). An AFLP clock for absolute dating of shallow-time evolutionary history - Too good to be true?. *Molecular Ecology*, vol. 18, pp. 4526-4532

Escudero, M., Valcárcel, V., Vargas, P. & Luceño, M. (2008a). Evolution in *Carex* L. sect. *Spirostachyae* (Cyperaceae): A molecular and cytogenetic approach. *Organisms Diversity and Evolution*, vol. 7, pp. 271-291

Escudero, M., Vargas, P., Valcárcel, V. & Luceño, M. (2008b). Strait of Gibraltar: An effective gene-flow barrier for wind-pollinated *Carex helodes* (Cyperaceae) as revealed by DNA sequences, AFLP, and cytogenetic variation. *American Journal of Botany*, vol. 95, pp. 745-755

Escudero, M. & Luceño, M. (2009). Systematics and evolution of *Carex* sects. *Spirostachyae* and *Elatae* (Cyperaceae). *Plant Systematics and Evolution*, vol. 279, pp. 163-189

Escudero, M., Vargas, P., Valcárcel, V. & Luceño, M. (2010a) Bipolar disjunctions in *Carex*: Long-distance dispersal, vicariance, or parallel evolution? *Flora: Morphology, Distribution, Functional Ecology of Plants*, vol. 205, pp. 118-127

Escudero, M., Hipp, A.L. & Luceño, M. (2010b). Karyotype stability and predictors of chromosome number variation in sedges: a study in *Carex* section *Spirostachyae*. *Molecular Phylogenetics and Evolution*, vol. 57, pp. 353-363

Escudero, M., Vargas, P., Arens, P., Ouborg, J. & Luceño, M. (2010c). The East-West-North colonization history of the Mediterranean and Europe by the coastal plant *Carex extensa* (Cyperaceae). *Molecular Ecolology*, vol. 19, pp. 352-370

Ewens, W.J. (1990). Population genetics theory - the past and the future, In: *Mathematical and statistical developments of evolutionary theory*, Lessard, S. (Ed.), pp. 177-227, Kluwer Academic Publishers, New York, USA

Excoffier, L., Smouse, P.E. & Quattro, J.M. (1992). Analysis of molecular variance inferred from metric distances among DNA haplotypes: Application to human mitochondrial DNA restriction data. *Genetics*, vol. 131, pp. 479-491

Felsenstein, J. (1981). Evolutionary trees from DNA sequences: A maximum likelihood approach. *Journal of Molecular Evolution*, vol. 17, pp. 368-376

Fernández-Mazuecos, M. & Vargas, P. (2011). Historical isolation *versus* recent long-distance connections between Europe and Africa in bifid toadflaxes (*Linaria* sect. *Versicolores*). *PLoS ONE*, vol. 6, art. no. e22234

Fiz-Palacios, O., Vargas, P., Vila, R., Papadopulos, A.S.T. & Aldasoro, J.J. (2010). The uneven phylogeny and biogeography of *Erodium* (Geraniaceae): radiations in the Mediterranean and recent recurrent intercontinental colonization. *Annals of Botany*, vol. 106, pp. 871-884

Fleitmann, D., Matter, A., Burns, S.J., Al-Subbary, A. & Al-Aowah, M.A. (2004). Geology and Quaternary climate history of Socotra. *Fauna Arabia*, vol. 20, pp. 27-43

Fu, Y.-X. & Li, W.-H. (1999). Coalescing into the 21st century: An overview and prospects of coalescent theory. *Theoretical Population Biology*, vol. 56, pp. 1-10

Givnish, T.J. & Renner. S.S. (2004). Tropical intercontinental disjunctions: Gondwana breakup, immigration from the boreotropics, and transoceanic dispersal. *International Journal of Plant Sciences*, vol. 165 (Supplement 4), pp. S1-S6

Goldberg, E.E., Lancaster, L.T. & Ree, R.H. (2011). Phylogenetic inference of reciprocal effects between geographic range evolution and diversification. *Systematic Biology*, vol. 60, pp. 451-465

Goldblatt, P. (1978). An analysis of the flora of southern Africa: its characteristics, relationships, and origins. *Annals of the Missouri Botanical Garden*, vol. 65, pp. 369-436

Goloboff, P.A., Farris, J.S. & Nixon, K.C. (2008). TNT, a free program for phylogenetic analysis. *Cladistics*, vol. 24, pp. 774-786

Graur, D. & Li, W.-H. (2000). *Fundamentals of molecular evolution* (2nd ed.), Sinauer Associates, Sunderland, Massachussets, USA

Guzmán, B. & Vargas, P. (2009a). Historical biogeography and character evolution of Cistaceae (Malvales) based on analysis of plastid *rbcL* and *trnL-trnF* sequences. *Organisms, diversity and evolution*, vol. 9, pp. 83-99

Guzmán, B. & Vargas, P. (2009b). Long-distance colonization of the Western Mediterranean by *Cistus ladanifer* (Cistaceae) despite the absence of special dispersal mechanisms. *Journal of Biogeography*, vol. 36, pp. 954-968

Hampe, A., Arroyo, J., Jordano, P. & Petit, R.J. (2003). Rangewide phylogeography of a bird-dispersed Eurasian shrub: contrasting Mediterranean and temperate glacial refugia. *Molecular Ecology*, vol. 12, pp. 3415-3426

Hare, M.P. (2001). Prospects for nuclear gene phylogeography. *Trends in Ecology and Evolution*, vol. 16, pp. 700-706

Harris, S.A. & Ingram, R. (1991). Chloroplast DNA and biosystematics: The effects of intraspecific diversity and plastid transmission. *Taxon*, vol. 40, pp. 393-412

Hasegawa, M., Kishino, H. & Yano, T. (1989). Estimation of branching dates among primates by molecular clocks of nuclear DNA which slowed down in Hominoidea. *Journal of Human Evolution*, vol. 18, 461-476

Harpending, H.C., Batzer, M., Gurven, M., Jorde, L.B., Rogers, A.R. & Sherry, S.T. (1998). Genetic traces of ancient demography. *Proceedings of the National Academy of Sciences of the United States of America*, vol. 95, pp. 1961-1967

Heads, M. (2005). Dating nodes on molecular phylogenies: A critique of molecular biogeography. *Cladistics*, vol. 21, pp. 62-78

Hess, J., Kadereit, J.W. & Vargas, P. (2000). The colonization history of *Olea europaea* L. in Macaronesia based on internal transcribed spacer 1 (ITS-1) sequences, randomly amplified polymorphic DNAs (RAPD), and intersimple sequence repeats (ISSR). *Molecular Ecology*, vol. 9, pp. 857-868

Hey, J. & Nielsen, R. (2004). Multilocus methods for estimating population sizes, migration rates and divergence time, with applications to the divergence of *Drosophila pseudoobscura* and *D. persimilis. Genetics*, vol. 167, pp. 747-60

Hey, J. & Nielsen, R. (2007). Integration within the Felsenstein equation for improved Markov chain Monte Carlo methods in population genetics. *Proceedings of the National Academy of Sciences of the United States of America*, vol. 104, pp. 2785-2790

Hewitt, G.M. (1996). Some genetic consequences of ice ages, and their role in divergence and speciation. *Biological Journal of the Linnean Society*, vol. 58, pp. 247-276

Hewitt, G.M. (2001). Speciation, hybrid zones and phylogeography - or seeing genes in space and time. *Molecular Ecology*, vol. 10, 537-549

Hileman, L.C., Vasey, M.C. & Parker, V.T. (2001). Phylogeny and biogeography of the Arbutoideae (Ericaceae): Implications for the Madrean-Tethyan Hypothesis. *Systematic Botany*, vol. 26, pp. 131-143

Hillebrand, H. (2004). On the generality of the latitudinal diversity gradient. *The American Naturalist*, vol. 163, pp. 192-211

Hillis, D.M., Mable, B.K. & Moritz, C. (1996). Applications of molecular systematics, In: *Molecular systematics* (2nd ed.), Hillis, D.M., Moritz, C. & Mable, B.K. (Eds.), pp. 515-543, Sinauer Associates, Sunderland, Massachusetts, USA

Holland, B.R., Huber, K.T., Moulton, V. & Lockhart, P.J. (2004). Using consensus networks to visualize contradictory evidence for species phylogeny. *Molecular Biology and Evolution*, vol. 21, pp. 1459-1461

Holland, B.R., Jermiin, L.S. & Moulton, V. (2006). Improved consensus network techniques for genome-scale phylogeny. *Molecular Biology and Evolution*, vol. 23, pp. 848-855

Hsü, K.J., Montardet, P., Bernoulli, D., Citá, M.B., Erickson, A., Garrison, R.E., Kidd, R.B., Mèlierés, F., Müller, C. & Wright, R. (1977). History of the Mediterranean salinity crisis. *Nature*, vol. 267, pp. 399-403

Hudson, R. (1990). Gene genealogies and the coalescent process, In: *Oxford surveys in evolutionary biology*, Futuyma, D. & Antonovics, J. (Eds), vol. 7, pp. 1-44, Oxford University Press, Oxford, UK

Hudson, R.R. & Coyne, J.A. (2002). Mathematical consequences of the genealogical species concept. *Evolution*, vol. 56, pp. 1557-1565

Huelsenbeck, J.P., Larget, B. & Swofford, D. (2000). A compound Poisson process for relaxing the molecular clock. *Genetics*, vol. 154, pp. 1879-1892

Hung, K., Hsu, T., Schaal, B.A. & Chiang, T. (2005). Loss of genetic diversity and erroneous phylogeographical inferences in *Lithocarpus konishii* (Fagaceae) of Taiwan caused by the Chi-Chi Earthquake: Implications for conservation. *Annals of the Missouri Botanical Garden*, vol. 92, pp. 52-65

Huson, D.H., Dezulian, T., Klöpper, T. & Steel, M.A. (2004). Phylogenetic Super-networks from partial trees. *IEEE/ACM Transactions on Computational Biology and Bioinformatics*, vol. 1, pp. 151-158

Huson, D.H. & Bryant, D. (2006). Application of phylogenetic networks in evolutionary studies. *Molecular Biology and Evolution*, vol. 23, pp. 254-267

Jakob, S.S. & Blattner, F.R. (2006). A chloroplast genealogy of *Hordeum* (Poaceae), long-term persisting haplotypes, incomplete lineage sorting, regional extinction, and the consequences for phylogenetic inference. *Molecular Biology and Evolution*, vol. 23, pp. 1602-1612

Jiménez-Mejías, P., Escudero, M., Guerra-Cárdenas, S., Lye, K.A. & Luceño, M. (2011). Taxonomical delimitation and drivers of speciation in the Ibero-North African *Carex* sect. *Phacocystis* river-shore group (Cyperaceae). *American Journal of Botany*, vol. 98, pp. 1855-1867

Judd, W.S., Campbell C.S., Kellogg, E.A., Stevens, P.F. & Donoghue, M.J. (2007). *Plant Systematics: A Phylogenetic Approach* (3rd Edition), Sinauer Assocciates, Sunderland, Massachussets, USA

Jürgens, N. (1997). Floristic biodiversity and history of African arid regions. *Biodiversity and Conservation*, vol. 6, pp. 495-514

Kadereit, J.W., Arafeh, R., Somogyi, G. & Westberg, E. (2005). Terrestrial growth and marine dispersal? Comparative phylogeography of five coastal plant species at a European scale. *Taxon*, vol. 54, pp. 861-876

Kadereit, G. & Yaprak, A.E. (2008). *Microcnemum coralloides* (Chenopodiaceae-Salicornioideae): An example of intraspecific East-West disjunctions in the Mediterranean region. *Anales del Real Jardín Botánico de Madrid*, vol. 65, pp. 415-426

Kaufman, D.M. & Willig, M.R. (1998). Latitudinal patterns of mammalian species richness in the New World: The effects of sampling method and faunal group. *Journal of Biogeography*, vol. 25, pp. 795-805

Kidd, K.K. & Sgaramella-Zonta, L.A. (1971). Phylogenetic analyses: concepts and methods. *American Journal of Human Genetics*, vol. 23, pp. 235-252

Kishino, H., Thorne, J.L. & Bruno, W.J. (2001). Performance of a divergence time estimation method under a probabilistic model of rate evolution. *Molecular Biology and Evolution*, vol. 18, pp. 352-361

Koch, M., Kiefer, C., Ehrich, D., Vogel, J., Brochmann, C. & Mummenhoff, K. (2006). Three times out of Asia minor: the phylogeography of *Arabis alpina* (Brassicaceae). *Molecular Ecology*, vol. 15, pp. 825-839

Krijgsman, W. (2002). The Mediterranean, Mare Nostrum of Earth Sciences. *Earth and Planetary Science Letters*, vol. 205, pp. 1-12

Kropf, M., Comes, H.P. & Kadereit, J.W. (2009). An AFLP clock for the absolute dating of shallow-time evolutionary history based on the intraspecific divergence of southwestern European alpine plant species. *Molecular Ecology*, vol. 18, pp. 697-708

Laird, C.D., McConaughy, B.L. & McCarthy, B.J. (1969). Rate of fixation of nucleotide substitutions in evolution. *Nature*, vol. 224, pp. 149-154

Langley, C.H. & Fitch, W.M. (1974). An examination of the constancy of the rate of molecular evolution. *Journal of Molecular Evolution*, vol. 3, pp. 161-177

Li, W.-H. & Tanimura, M. (1987). The molecular clock runs more slowly in man than in apes and monkeys. *Nature*, vol. 326, pp. 93-96

Li, W.-H., Tanimura, M. & Sharp, P.M. (1987). An evaluation of the molecular clock hypothesis using mammalian DNA sequences. *Journal of Molecular Evolution*, vol. 25, pp. 330-342

Li, W.-H. (1997). *Molecular evolution*, Sinauer Associates, Sunderland, Massachussets, USA.

Linder, H.P., Hardy, C.R. & Rutschmann, F. (2005). Taxon sampling effects in molecular clock dating: An example from the African Restionaceae. *Molecular Phylogenetics and Evolution*, vol. 35, pp. 569-582

Liu, L., Yu, L., Kubatko, L., Pearl, D.K. & Edwards, S.V. (2009). Coalescent methods for estimating phylogenetic trees. *Molecular Phylogenetics and Evolution*, vol. 53, pp. 320-328

López González, G. (1993). *Sesamoides* All. In: *Flora Iberica*, Castroviejo, S., & al. (Eds.), vol. 4, pp. 475-483, Consejo Superior de Investigaciones Científicas, Madrid, Spain.

Lowe, A., Harris, S. & Ashton, P. (2004). *Ecological genetics: design, analysis and application*, Blackwell Publishing, Malden, Massachusetts, USA

Lynch, M. & Milligan, B.G. (1994). Analysis of population genetic structure with RAPD markers. *Molecular Ecology*, vol. 3, pp. 91-99

Lledó, M.D., Crespo, M.B., Fay, M.F. & Chase, M.W. (2005) Molecular phylogenetics of *Limonium* and related genera (Plumbaginaceae): Biogeographical and systematic implications. *American Journal of Botany*, vol. 92, pp. 1189-1198

Magallón, S.A. (2004). Dating lineages: Molecular and paleontological approaches to the temporal framework of clades. *International Journal of Plant Sciences*, vol. 165 (Supplement), pp. S7-S21.

Magallón, S.A. & Sanderson, M.J. (2005). Angiosperm divergence times: The effect of genes, codon positions, and time constraints. *Evolution*, vol. 59, pp. 1653-1670

Maddison, W.P. & Maddison, D.R. (2010). Mesquite: A modular system for evolutionary analysis, version 2.74. Available from http://mesquiteproject.org

Mansion, G., Selvi, F., Guggisberg, A. & Conti, E. (2009). Origin of Mediterranean insular endemics in the Boraginales: integrative evidence from molecular dating and ancestral area reconstruction. *Journal of Biogeography*, vol. 36, pp. 1282-1296

Martin, A.P. & Palumbi, S.R. (1993). Body size, metabolic rate, generation time, and the molecular clock. *Proceedings of the National Academy of Sciences of the United States of America*, vol. 90, pp. 4087-4091

Martín-Bravo, S., Meimberg, H., Luceño, M., Märkl, W., Valcárcel, V., Braüchler, C., Vargas, P. & Heubl, G. (2007). Molecular systematics and biogeography of Resedaceae based on ITS and *trnL-F* sequences. *Molecular Phylogenetics and Evolution*, vol. 44, pp. 1105-1120

Martín-Bravo, S. 2009. Sistemática, evolución y biogeografía de la familia Resedaceae. Ph.D. Dissertation, Universidad Pablo de Olavide, Sevilla, Spain

Martín-Bravo, S., Vargas, P. & Luceño, M. (2009). Is *Oligomeris* (Resedaceae) indigenous to North America? Molecular evidence for a natural colonization from the Old World. *American Journal of Botany*, vol. 96, pp. 507-518

Martín-Bravo, S., Valcárcel, V., Vargas, P. & Luceño, M. (2010). Geographical speciation related to Pleistocene range shifts in the western Mediterranean mountains (*Reseda* sect. *Glaucoreseda*, Resedaceae). *Taxon*, vol. 59, pp. 466-482

Mason-Gamer, R.J., Holsinger, K.E. & Jansen, R.K. (1995). Chloroplast DNA haplotype variation within and among populations of *Coreopsis grandiflora* (Asteraceae). *Molecular Biology and Evolution*, vol. 12, pp. 371-381

Mayr, E. (1965). Discussion, In: *Evolving Genes and Proteins*, Bryson, V. & Vogel, H.J. (Eds), pp. 293-294, Academic Press, New York, USA

McBreen, K. & Lockhart, P.J. (2006). Reconstructing reticulate evolutionary histories of plants. *Trends in Plant Sciences*, vol. 11, pp. 398-404

Médail, F. & Quézel, P. (1997). Hot-spots analysis for a conservation of plant biodiversity in the Mediterranean basin. *Annals of the Missouri Botanical Garden*, vol. 84, pp. 112-128

Médail, F. & Quézel, P. (1999). Biodiversity hotspots in the Mediterranean Basin: Setting global conservation priorities *Conservation Biology*, vol. 13, pp. 1510-1513

Médail, F. & Diadema, K. (2009). Glacial refugia influence plant diversity patterns in the Mediterranean Basin. *Journal of Biogeography*, vol. 36, pp. 1333-1345

Meudt, H.M. & Clarke, A.C. (2007) Almost forgotten or latest practice? AFLP applications, analyses and advances. *Trends in Plant Sciences*, vol. 12, pp. 106-117

Meyers, S.C. & Liston, A. (2008). The biogeography of *Plantago ovata* Forssk. (Plantaginaceae). *International Journal of Plant Sciences*, vol. 169, pp. 954-962

Milne, R.I. (2006). Northern Hemisphere Plant disjunctions: a window on Tertiary Land Bridges and Climate change? *Annals of Botany*, vol. 98, pp. 465-472

Miller, A.G. (1984). A revision of *Ochradenus*. *Notes of the Royal Botanic Garden of Edinburgh*, vol. 41, pp. 491-594

Moore, D.M. & Chater, O.A. (1971). Studies on bipolar species. I. *Carex*. *Botaniska Notiser*, vol. 124, pp. 317-334

Morrison, D.A. (2005). Networks in phylogenetic analysis: New tools for population biology. *International Journal for Parasitology*, vol. 35, pp. 567-582

Mueller, U.G. & Wolfenbarger, L.L. (1999). AFLP genotyping and fingerprinting. *Trends in Ecology and Evolution*, vol. 14, pp. 389-394

Müller Argoviensis, J. (1868). Resedaceae. In: *Prodromus Systematis Naturalis Regni Vegetabilis*, de Candolle, A.P. (Ed.), vol. 16(2), pp. 548-589, Victor Masson *et filii*, Paris, France.

Mummenhoff, K. & Franzke, A. (2007). Gone with the bird: Late Tertiary and Quaternary intercontinental long-distance dispersal and allopolyploidization in plants. *Systematics and Biodiversity*, vol. 5, pp. 255-260

Myers, N., Mittermeler, R.A., Mittermeler, C.G., da Fonseca, G.A.B. & Kent, J. (2000). Biodiversity hotspots for conservation priorities. *Nature*, vol. 403, pp. 853-858

Nei, M. (1972). Genetic distance between populations. *The American Naturalist*, vol. 106, pp. 283-292

Nei, M. (1973). Analysis of gene diversity in subdivided populations. *Proceedings of the National Academy of Sciences of the United States of America*, vol. 70, pp. 3321-3323

Nei, M. (1978). Estimation of average heterozygosity and genetic distance from a small number of individuals. *Genetics*, vol. 89, pp. 583-590

Nei, M. & Li, W.-H. (1979). Mathematical model for studying genetic variation in terms of restriction endonucleases. *Proceedings of the National Academy of Sciences of the United States of America*, vol. 76, pp. 5269-5273

Nei, M., Tajima, F. & Tateno, Y. (1983). Accuracy of estimated phylogenetic trees from molecular data. *Journal of Molecular Evolution*, vol. 19, pp. 153-170

Nei, M. (1987). *Molecular evolutionary genetics*, Columbia University Press, New York, USA.

Newton, A.C., Allnut, T.R., Gillies, A.C.M., Lowe, A.J. & Ennos, R.A. (1999). Molecular phylogeography, intraspecific variation and the conservation of tree species. *Trends in Ecology and Evolution*, vol. 14, pp. 140-145

Nieto Feliner, G. & Rosselló, J.A. (2007). Better the devil you know? Guidelines for insightful utilization of nrDNA ITS in species-level evolutionary studies in plants. *Molecular Phylogenetics and Evolution*, vol. 44, pp. 911-919

Nylander, J.A.A., Olsson, U., Alström, P. & Sanmartín, I. (2008). Accounting for phylogenetic uncertainty in biogeography: a Bayesian approach to dispersal-vicariance analysis of the thrushes (Aves: *Turdus*). *Systematic Biology*, vol. 57, pp. 257-268

Ohta, T. (1995). Synonymous and nonsynonymous substitutions in mammalian genes and the nearly neutral theory. *Journal of Molecular Evolution*, vol. 40, pp. 56-63

Olsen, K.M. (2002). Population history of *Manihot esculenta* (Euphorbiaceae) inferred from nuclear DNA sequences. *Molecular Ecology*, vol. 11, 901-911

Olsen, K.M. & Schaal, B.A. (1999). Evidence on the origin of cassava, phylogeography of *Manihot esculenta*. *Proceedings of the National Academy of Sciences of the United States of America*, vol. 96, pp. 5586-5591

Ouborg, N.J., Piquot, Y. & van Groenendael, J.M. (1999). Population genetics, molecular markers, and the study of dispersal in plants. *Journal of Ecology*, vol. 87, pp. 551-569

Page, R.D.M. & Holmes, E.C. (1998). *Molecular Evolution: a phylogenetic approach*, Blackwell Science, Oxford, UK

Palmer, J.D. (1992). Mitochondrial DNA in plant systematics: applications and limitations, In: *Molecular systematics of plants*, Soltis, P.S., Soltis, D.E. & Doyle, J.J. (Eds.), pp. 36-49, Chapman Hall, New York, USA

Pearse, I.S. & Hipp, A.L. (2009). Phylogenetic and trait similarity to a native species predict herbivory on non-native oaks. *Proceedings of the National Academy of Sciences of the United States of America*, vol. 106, pp. 18097-18102

Petit, R.J., Hampe, A. & Cheddadi, R. (2005). Climate changes and tree phylogeography in the Mediterranean. *Taxon*, vol. 54, pp. 877-885

Pleines, T., Jakob, S.S. & Blattner, F.R. (2009). Application of non-coding DNA regions in intraspecific analyses. *Plant Systematics and Evolution*, vol. 282, pp. 281-294

Popp, M., Mirré, V. & Brochmann, C. (2011). A single Mid-Pleistocene long-distance dispersal by a bird can explain the extreme bipolar disjunction in crowberries (*Empetrum*). *Proceedings of the National Academy of Sciences of the United States of America*, vol. 108, pp. 6520-6525

Posada, D. (2008). jModelTest: Phylogenetic Model Averaging. *Molecular Biology and Evolution*, vol. 25, pp. 1253-1256

Posada, D. & Crandall, K.A. (2001). Intraspecific gene genealogies: trees grafting into networks. *Trends in Ecology and Evolution*, vol. 16, pp. 37-45

Pritchard, J.K., Stephens, M. & Donnelly, P. (2000). Inference of population structure using multilocus genotype data. *Genetics*, vol. 155, pp. 945-959

Pulquério, M.J.F. & Nichols, R.A. (2007). Dates from the molecular clock: how wrong can we be? *Trends in Ecology and Evolution*, vol. 22, pp. 180-184

de Queiroz, A. (2005). The resurrection of oceanic dispersal in historical biogeography. *Trends in Ecology and Evolution*, vol. 20, pp. 68-73

Quézel, P. (1978). Analysis of the flora of Mediterranean and Saharan Africa. *Annals of the Missouri Botanical Garden*, vol. 65, pp. 479-534

Rambaut, A. (2001). RHINO: a program to estimate relative rates of substitution between specified lineages of a phylogenetic tree within a maximum likelihood framework. Available from http://evolve.zoo.ox.ac.uk

Rambaut, A. & Bromham, L. (1998). Estimating divergence dates from molecular sequences. *Molecular Biology and Evolution*, vol.15, pp. 442-448

Raven, P. H. (1972). Plant species disjunctions: a summary. *Annals of the Missouri Botanical Garden*, vol. 59, pp. 234-246

Ree, R.H., Moore, B.R., Webb, C.O. & Donoghue, M.J. (2005). A likelihood framework for inferring the evolution of geographic range on phylogenetic trees. *Evolution*, vol. 59, pp. 2299-2311

Ree, R.H. & Smith, S.A. (2008). Maximum likelihood inference of geographic range evolution by dispersal, local extinction, and cladogenesis. *Systematic Biology*, vol. 57, pp. 4-14

Ree, R.H. & Sanmartín, I. (2009). Prospects and challenges for parametric models in historical biogeographical inference. *Journal of Biogeography*, vol. 36, pp. 1211-1220

Reisz, R.R. & Müller, J. (2004). Molecular timescales and the fossil record: A paleontological perspective. *Trends in Genetics*, vol. 20, pp. 237-241

Rendell, S. & Ennos, R.A. (2003). Chloroplast DNA diversity of the dioecious European tree *Ilex aquifolium* L. (English holly). *Molecular Ecology*, vol. 12, pp. 2681-2688

Renner, S.S. (2005). Relaxed molecular clocks for dating historical plant dispersal events. *Trends in Plant Science*, vol. 10, pp. 550-558

Richardson, J.E., Fay, M.F., Cronk, Q.C.B & Chase, M.W. (2003). Species delimitation and the origin of populations in island representatives of *Phylica* (Rhamnaceae). *Evolution*, vol. 57, pp. 816-827

Ricklefs, R.E., Schwarzbach, A.E. & Renner, S.S. (2006). Rate of lineage origin explains the diversity anomaly in the world´s mangrove vegetation. *The American Naturalist*, vol. 168, pp. 805-810

du Rietz, G.E. (1940). Problems of the bipolar plant distribution. *Acta Phytogeographica Suecica*, vol. 13, pp. 215-282

Rodríguez Sánchez, F., Pérez-Barrales, R., Ojeda, F., Vargas, P. & Arroyo, J. (2008). The Strait of Gibraltar as a melting pot for plant biodiversity. *Quaternary Science Reviews*, vol. 27, pp. 2100-2117

Ronquist, F. (1997). Dispersal-vicariance analysis: A new approach to the quantification of historical biogeography. *Systematic Biology*, vol. 46, pp. 195-203

Ronquist, F. & Huelsenbeck, J.P. (2003). MrBayes 3: Bayesian phylogenetic inference under mix models. *Bioinformatics*, vol. 19, pp. 1572-1574

Rosselló, J.A., Lázaro, A., Cosín, R. & Molins, A. (2007). A phylogeographic split in *Buxus balearica* (Buxaceae) as evidenced by nuclear ribosomal markers: when ITS paralogues are welcome. *Journal of Molecular Evolution*, vol. 64, pp. 143-157

Rouchy, J.M. & Caruso, A. (2006). The Messinian Salinity Crisis in the Mediterranean Basin: a reassessment of the data and an integrated scenario. *Sedimentary Geology*, vol. 188, pp. 35-67

Rubio de Casas, R., Besnard, G., Schönswetter, P., Balaguer, L. & Vargas, P. (2006). Extensive gene flow blurs phylogeographic but not phylogenetic signal in *Olea europaea* L. *Theoretical and Applied Genetics*, vol. 113, pp. 575-583

Rutschmann, F. (2006). Molecular dating of phylogenetic trees: A brief review of current methods that estimate divergence times. *Diversity and distribution*, vol. 12, pp. 35-48

Rutschmann, F., Eriksson, T., Salim, K.A. & Conti, E. (2007). Assessing calibration uncertainty in molecular dating: The assignment of fossils to alternative calibration points. *Systematic Biology*, vol. 56, pp. 591-608

Rzhetsky, A. & Nei, M. (1993). Theoretical foundation of the minimum-evolution method of phylogenetic inference. *Molecular Biology and Evolution*, vol. 10, pp. 1073-1095

Saitou, N. & Nei, M. (1987). The neighbor-joining method: a new method for reconstructing phylogenetic trees. *Molecular Biology and Evolution*, vol. 4, pp. 406-425

Salvo, G., Ho, S.Y.W., Rosenbaum, G., Ree, R. & Conti, E. (2010). Tracing the temporal and spatial origins of island endemics in the Mediterranean region: a case study from the citrus familiy (*Ruta* L., Rutaceae). *Sytematic Biology*, vol. 59, pp. 705-722

Sanderson, M.J. (1997). A nonparametric approach to estimating divergence times in the absence of rate constancy. *Molecular Biology and Evolution*, vol. 14, pp. 1218-1231

Sanderson, M.J. & Doyle, J.A. (2001). Sources of error and confidence intervals in estimating the age of angiosperms from *rbcL* and 18S rDNA data. *American Journal of Botany*, vol. 88, pp. 1499-1516

Sanderson, M.J. (2002). Estimating absolute rates of molecular evolution and divergence times: A penalized likelihood approach. *Molecular Biology and Evolution*, vol. 19, pp. 101-109

Sanderson, M.J. (2003). r8s: Inferring absolute rates of molecular evolution and divergence times in the absence of a molecular clock. *Bioinformatics*, vol. 19, pp. 301-302

Sanderson, M.J., Thorne, J.L., Wikström, N. & Bremer, K. (2004). Molecular evidence on plant divergence times. *American Journal of Botany*, vol. 91, 1656-1665

Sanmartín, I., van der Mark, P. & Ronquist, F. (2008). Inferring dispersal: A Bayesian approach to phylogeny-based island biogeography with special reference to the Canary Islands. *Journal of Biogeography*, vol. 35, pp. 428-449

Sarich, V.M. & Wilson, A.C. (1967). Immunological time scale for hominid evolution. *Science*, vol. 158, pp. 1200-1203

Schaal, B.A., Hayworth, D.A., Olsen, K.M., Rauscher, J.T. & Smith, W.A. (1998). Phylogeographic studies in plants, problems and prospects. *Molecular Ecology*, vol. 7, pp. 465-474

Schaal, B.A. & Leverich, W.J. (2001). Plant population biology and systematics. *Taxon*, vol. 50, pp. 679-695

Schaal, B.A. & Olsen, K.M. (2000). Gene genealogies and population variation in plants, In: *Variation and Evolution in Plants and Microorganisms: Towards a New Synthesis 50 years after Stebbins*, Ayala, F.J., Fitch, W.M. & Clegg, M.T. (Eds.), pp. 235-251, National Academy of Sciences Press, Irvine, California, USA

Schierup, M.H. & Hein, J. (2000). Consequences of recombination on traditional phylogenetic analysis. *Genetics* 156, 879-891

Schluter, D., Price, T., Mooers, A.O. & Ludwig, D. (1997). Likelihood of ancestor states in adaptive radiation. *Evolution*, vol. 51, pp. 1699-1711

Shaw, A.J., Werner, O. & Ros, R.M. (2003). Intercontinental Mediterranean disjunct mosses: Morphological and molecular patterns. *American Journal of Botany*, vol. 90, pp. 540-550

Simpson, G.G. (1964). Organisms and molecules in evolution. *Science*, vol. 146, pp. 1535-1538.

Sinclair, W.T., Morman, J.D. & Ennos, R.A. (1998). Multiple origins for Scots pine (*Pinus sylvestris* L.) in Scotland: Evidence from mitochondrial DNA variation. *Heredity*, vol. 80, pp. 233-240

Sneath, P.H.A. & Sokal, R.R. (1973). *Numerical Taxonomy*, W.H. Freeman, San Francisco, California, USA

Sokal, R.R. & Michener, C.D. (1958). A statistical method for evaluating systematic relationships. *University of Kansas Science Bulletin*, vol. 38, pp. 1409-1438

Soltis, D.E., Soltis, P.S. & Milligan, B.G. (1992). Intraspecific chloroplast DNA variation: Systematic and phylogenetic implications, In: *Molecular Systematics of Plants*, Soltis, D.E., Soltis, P.S. & Doyle, J.J. (Eds.), pp. 117-150, Chapman and Hall, London

Stebbins, G.L. & Day, A. (1967). Cytogenetic evidence for long continued stability in the genus *Plantago*. *Evolution*, vol. 21, pp. 409-428

Steiper, M.E. & Young, N.M. (2008). Timing primate evolution: Lessons from the discordance between molecular and paleontological estimates. *Evolutionary Anthropology*, vol. 17, pp. 179-188

Svenning, J.C., Borchsenius, F., Bjorholm, S. & Balslev, H. (2008). High tropical net diversification drives the New World latitudinal gradient in palm (Arecaceae) species richness. *Journal of Biogeography*, vol. 35, pp. 394-406

Swofford, D.L. (2003). PAUP*. Phylogenetic Analysis Using Parsimony (*and Other Methods). Version 4. Sinauer Associates, Sunderland, Massachusetts, USA

Taberlet, P., Fumagalli, L., Wust-Saucy, A.G. & Cosson, J.F. (1998). Comparative phylogeography and postglacial colonization routes in Europe. *Molecular Ecology*, vol. 7, pp. 453-464

Tavaré, S. (1984). Line-of-descent and genealogical processes, and their application in population genetics models. *Theoretical Population Biology*, vol. 26, pp. 119-164

Taylor, P. (1958). The genus *Caylusea* St. Hil. in Tropical Africa. *Kew Bulletin*, vol. 13, pp. 283-286

Templeton, A.R., Crandall, K.A. & Sing, C.F. (1992). A cladistic analysis of phenotypic associations with haplotypes inferred from restriction endonuclease mapping and DNA sequence data. III. Cladogram estimation. *Genetics*, vol. 132, pp. 619-633

Thiv, M., van der Niet, T., Rutschmann, F., Thulin, M., Brune, T. & Linder, H.P. (2011). Old-New World and trans-African disjunctions of *Thamnosma* (Rutaceae): Intercontinental long-distance dispersal and local differentiation in the succulent biome. *American Journal of Botany*, vol. 98, pp. 76-87

Thorne, R.F. (1972). Major disjunctions in the geographic ranges of seed plants. *Quarterly Review of Biology*, vol. 47, pp. 365-411

Thorne, J.L. & Kishino, H. (2002). Divergence time and evolutionary rate estimation with multilocus data. *Systematc Biology*, vol. 51, pp. 689-702

Thorne, J.L., Kishino, H., Painter, I.S. (1998). Estimating the rate of evolution of the rate of molecular evolution. *Molecular Biology and Evolution*, vol. 15, pp. 1647-1657

Thulin, M. (1994). Aspects of disjunct distributions and endemism in the arid parts of the horn of Africa, particularly Somalia, In: *Proceedings of the XIIIth Plenary Meeting AETFAT 2*, Seyani, J.H. & Chikuni, A.C. (Eds.), pp. 1105-1119, Malawi.

Tiffney, B.H. (1985). The Eocene North Atlantic land bridge: Its importance in Tertiary and modern phytogeography of the Northern Hemisphere. *Journal of the Arnold Arboretum*, vol. 66, pp. 243-273

Tomaru, N., Takahashi, M., Tsumura, Y., Takahashi, M. & Ohba, K. (1998). Intraspecific variation and phylogeographic patterns of *Fagus crenata* (Fagaceae) mitochondrial DNA. *American Journal of Botany*, vol. 85, pp. 629-636

Tremetsberger, K., Stuessy, T.F., Kadlec, G., Urtubey, E., Baeza, C.M., Beck, S.G., Valdebenito, H.A., de Fátima Ruas, C. & Matzenbacher, N.I. (2006). AFLP phylogeny of South American species of *Hypochaeris* (Asteraceae, Lactuceae). *Systematic Botany*, vol. 31, pp. 610-626

Valcárcel, V., Fiz, O. & Vargas, P. (2003). Chloroplast and nuclear evidence for multiple origins of polyploids and diploids of *Hedera* (Araliaceae) in the Mediterranean basin. *Molecular Phylogenetics and Evolution*, vol. 27, pp. 1-20

Valdés, B. (1991). Andalucía and the Rif. Floristic links and common flora. *Botanika Chronika (Patras)*, vol. 10, pp. 117-124

Valdés Bermejo, E. (1993). *Reseda* L., In: *Flora Iherica*, Castroviejo, S. & al. (Eds.), vol. 4., pp. 440-475, Consejo Superior de Investigaciones Científicas, Madrid, Spain

Valente, L.M., Savolainen, V. & Vargas, P. (2010). Unparalleled rates of species diversification in Europe. *Proceedings of the Royal Society B*, vol. 277, pp. 1489-1496

Valente, L.M., Savolainen, V., Manning, J.C., Goldblatt, P. & Vargas, P. (2011). Explaining disparities in species richness between Mediterranean floristic regions: a case study in *Gladiolus* (Iridaceae). *Global Ecology and Biogeography*, vol. 20, pp. 881-892

Vargas, P. (2007). Are Macaronesian islands refugia of relict plant lineages?: a molecular survey. In: *Phylogeography in Southern European Refugia: evolutionary perspectives on the origins and conservation of European biodiversity*, Weiss, S. J. & Ferrand, N. (Eds.), pp. 297-314, Kluwer Academic Publishers, Dordrecht, The Netherlands

Verdcourt, B. (1969). The arid corridor between the north-east and southwest areas of Africa. *Palaeoecology of Africa*, vol. 4, pp. 140-144

Vollan, K., Heide, O.M., Lye, K.A. & Heun, M. (2006). Genetic variation, taxonomy and mountain-hopping of four bipolar *Carex* species (Cyperaceae) analysed by AFLP fingerprinting. *Australian Journal of Botany*, vol. 54, pp. 305-313

Wagner, D.B., Furnier, G.R., Saghai-Maroof, M.A., Williams, S.M., Dancik, B.P. & Allard, R.W. (1987). Chloroplast DNA polymorphisms in lodgepole and jack pines and their hybrids. *Proceedings of the National Academy of Sciences of the United States of America*, vol. 84, pp. 2097-2100

Weising, K., Nybom, H., Wolff, K. & Kahl, G. (2005). *DNA fingerprintings in plants: principles, methods and applications* (2nd edition), Taylor & Francis, Boca Raton, Florida, USA

Wen, J. & Ickert-Bond, S.M. (2009). Evolution of the Madrean-Tethyan disjunctions and the North and South American amphitropical disjunctions in plants. *Journal of Systematics and Evolution*, vol. 47, pp. 331-348

Wilkinson-Herbots, H.M. (2008). The distribution of the coalescence time and the number of pairwise nucleotide differences in the "isolation with migration" model. *Theoretical Population Biology*, vol. 73, pp. 277-288

de Winter, B. (1971). Floristic relationship between the northern and southern arid areas in Africa. *Mitteilungen der Botanischen Staatssammlung München*, vol. 10, pp. 424-437

Wolfe, K.H., Li, W.-H. & Sharp, P.M. (1987). Rates of nucleotide substitution vary greatly among plant mitochondrial, chloroplast and nuclear DNA. *Proceedings of the National Academy of Sciences of the United States of America*, vol. 84, pp. 9054–9058

Wood, C.E. (1972). Morphology and phytogeography: The classical approach to the study of disjunctions. *Annals of the Missouri Botanical Garden*, vol. 59, pp. 107-124

Wright, S. (1951). The genetical structure of populations. *Annals of Eugenics*, vol. 15, pp. 323-354

Yang, Z. (1997). PAML: A program package for phylogenetic analysis by maximum likelihood. *Computer Applications in the Biosciences*, vol. 13, pp. 555-556

Yoder, A.D. & Yang, Z. (2000). Estimation of primate speciation dates using local molecular clocks. *Molecular Biology and Evolution*, vol. 17, pp. 1081-1090

Yu, Y., Harris, A.J. & He, X. (2010). S-DIVA (Statistical Dispersal Vicariance Analysis): a tool for inferring biogeographic histories. *Molecular Phylogenetics and Evolution*, vol. 56, pp. 848-850

Zachos, F.E. (2009). Gene trees and species trees - Mutual influences and interdependences of population genetics and systematics. *Journal of Zoological Systematics and Evolutionary Research*, vol. 47, pp. 209-218

Zhang, D.X. & Hewitt, G.M. (2003). Nuclear DNA analyses in genetic studies of populations, practice, problems and prospects. *Molecular Ecology*, vol. 12, pp. 563-584.

Zhengyi, W. (1983). On the significance of Pacific intercontinental discontinuity. *Annals of the Missouri Botanical Garden*, vol. 70, pp. 577-590

van Zinderen Bakker, E.M. (1978). Quarternary vegetation changes in southern Africa, In: *Biogeography and Ecology of Southern Africa*, Werger, M.J.A. (Ed.), pp. 79–119, Junk, The Hague, The Netherlands.

Zuckerkandl, E. & Pauling, L. (1962). Molecular disease, evolution, and genetic heterogeneity, In: *Horizons in Biochemistry*, Kasha, M. & Pullman, B. (Eds.), pp. 189-225, Academic Press, New York, USA

Zuckerkandl, E. & Pauling, L. (1965). Evolutionary divergence and convergence in proteins, In: *Evolving Genes and Proteins*, Bryson, V. & Vogel, H.J. (Eds), pp. 97-166, Academic Press, New York, USA

# 15

# Aspects of the Biogeography of North American Psocoptera (Insecta)

Edward L. Mockford

*School of Biological Sciences, Illinois State University, Normal, Illinois, USA*

## 1. Introduction

The group under consideration here is the classic order Psocoptera as defined in the Torre-Bueno Glossary of Entomology (Nichols & Shuh, 1989). Although this group is unquestionably paraphyletic (see Lyal, 1985, Yoshizawa & Lienhard, 2010), these free-living, non-ectoparasitic forms are readily recognizable.

In defining North America for this chapter, I adhere closely to Shelford (1963, Fig. 1-9), but I shall use the Tropic of Cancer as the southern cut-off line, and I exclude the Antillean islands. Although the ranges of many species of Psocoptera extend across the Tropic of Cancer, the inclusion of the tropical areas would involve the comparison of relatively well-studied regions and relatively less well-studied regions.

The North America Psocoptera, as defined above, comprises a faunal list of 397 species in 90 genera and 27 families ( Table 1) . Comparisons are made here with several other relatively well-studied faunas. The psocid fauna of the Euro-Mediterranean region, summarized by Lienhard (1998) with additions by the same author (2002, 2005, 2006) and Lienhard & Baz (2004) has a fauna of 252 species in 67 genera and 25 families. As would be expected, nearly all of the families are shared between the two regions. The only two families not shared are Ptiloneuridae and Dasydemellidae, which reach North America but not the western Palearctic. Ptiloneuridae has a single species and Dasydemellidae two in North America. The rather large differences at the generic and specific levels are probably due to the much greater access that these insects have for invasion of North America from the tropics than invasion from the tropics in the Western Palearctic. In the latter region, the Sahara Desert and its eastward extension through the Middle East have severely limited northward movement from tropical sub-Saharan Africa. In North America, the copiously varied climates and topography of Mexico and the proximity of the Antillean islands to the Florida peninsula apparently have allowed numerous invasions and/or re-invasions of North America by tropical psocid taxa. Examples will be discussed below.

Suborder Trogiomorpha
  Infraorder Atropetae
    Family Lepidopsocidae
      Subfamily Echinopsocinae

                  *Cyptophania* sp. N, 2
                  *Neolepolepis caribensis* (Turner) N, 2
                  *N. occidentalis* (Mockford) N, 1
                  *N. xerica* García Aldrete N, 6
                  *Pteroxanium forcepeta* García Aldrete N, 6
                  *P. kelloggi* (Ribaga) C, 5
            Subfamily Lepidopsocinae
                  *Echmepteryx (Echmepteryx) alpha* García Aldrete N, 2
                  *E. (E.) hageni* (Packard) N, 1
                  *E. (E.) intermedia* Mockford N, 2
                  *E. (E.) youngi* Mockford N, 2
                  *E. (Thylacopsis) falco* Badonnel N, 2
                  *E. (T.) madagascariensis* (Kolbe) N, 2
                  *E. (T.)* sp. N, 2
            Subfamily Perientominae
                  *Nepticulomima* sp. N, 2
                  *Proentomum personatum* Badonnel N, 2
                  *Soa flaviterminata* Enderlein N, 2
            Subfamily Thylacellinae
                  *Thylacella cubana* (Banks) N, 2
        Family Psoquillidae
                  *Balliella ealensis* Badonnel N, 1 (?)
                  *Psoquilla marginepunctata* Hagen IC, 2
                  *Rhyopsocoides typhicola* García Aldrete N, 6
                  *Rhyopsocus bentonae* Sommerman N, 2
                  *R. disparilis* (Pearman) I
                  *R. eclipticus* Hagen N, 1
                  *R. maculosus* García Aldrete N, 6
                  *R. micropterus* Mockford N, 5
                  *R. texanus* (Banks) N, 4, 6
                  *R.* sp. nr. *bentonae* N, 2
        Family Trogiidae
                  *Cerobasis annulata* (Hagen) IC (dom)
                  *C. guestfalica* (Kolbe) N, 1, 2
                  *C.* sp. #1 N, 2
                  *C.* sp. #2 N, 1
                  *C.* sp. #3 N, 5
                  *Lepinotus inquilinus* Heyden IC (dom)
                  *L. patruelis* Pearman IC (dom)
                  *L. reticulatus* Enderlein N, 4
                  *Myrmecodipnella aptera* Enderlein N, 5
                  *Trogium braheicola* García Aldrete N, 6
                  *T. pulsatorium* (Linn.) C (dom)
    Infraorder Psyllipsocetae
        Family Psyllipsocidae
                  *Dorypteryx domestica* (Smithers) IC (dom)

        *D. pallida* Aaron IC (dom)

        *Pseudorypteryx mexicana* García Aldrete N, 4, 6

        *Psocathropos lachlani* Ribaga IC, 2

        *Psyllipsocus apache* Mockford N, 4

        *P. decoratus* Mockford N, 4

        *P. hilli* Mockford N, 4

        *P. huastecanus* Mockford N, 6

        *P. hyalinus* García Aldrete N, 6

        *P. kintpuashi* Mockford N, 5

        *P. maculatus* García Aldrete N, 4, 6

        *P. neoleonensis* García Aldrete N, 6

        *P. oculatus* Gurney I

        *P. ramburii* Selys-Longchamps C (dom & caves)

        *P. regiomontanus* Mockford N, 6

        *P. subterraneus* Mockford N, 4

Infraorder Prionoglaridetae

    Family Prionoglarididae

        *Speleketor flocki* Gurney N, 4

        *S. irwini* Mockford N, 4

        *S. pictus* Mockford N, 4

Suborder Troctomorpha

    Infraorder Nanopsocetae

    Family Pachytroctidae

        *Nanopsocus oceanicus* Pearman IC, 2

        *Pachytroctes aegyptius* Enderlein 2 (dom)

        *P. neoleonensis* Garcia Aldrete N, 6

        *Tapinella maculata* Mockford N, 4, 6

        *T. olmeca* Mockford N, 6

    Family Sphaeropsocidae

        *Badonnelia titei* Pearman I (dom)

        *Prosphaeropsocus californicus* Mockford N, 5

        *P. pallidus* Mockford N, 5

        *Sphaeropsocopsis argentina* Badonnel I

        *S. castanea* Mockford N, 6

        *Troglosphaeropsocus voylesi* Mockford N, 4

    Family Liposcelididae

        Subfamily Embidopsocinae

            *Belaphotroctes alleni* Mockford N, 4

            *B. badonneli* Mockford N, 2

            *B. ghesquierei* Badonnel N, 2

            *B. hermosus* Mockford N, 4

            *B. simberloffi* Mockford N, 2

            *Embidopsocus* Group IB

                *E. bousemani* Mockford N, 1

                *E. laticeps* Mockford N, 2

                *E. mexicanus* Mockford N, 6

     *Embidopsocus* Group IIA
      *E. citrensis* Mockford N, 2
      *E. thorntoni* Badonnel I (dom)
     *Embidopsocus* Group IIB
      *E. needhami* (Enderlein) N, 1
     *Embidopsocus* Group III
      *E. femoralis* Badonnel I (?), 2
   Subfamily Liposcelidinae
    *Liposcelis,* Section I
     Group A, Subgroup Aa
      *L. brunnea* Motschulsky C, 4
     Group A, Subgroup Ab
      *L. deltachi* Sommerman N, 4
      *L entomophila* (Enderlein) C, 2
      *L. fusciceps* Badonnel I
      *L. hirsutoides* Mockford N, 2
      *L. nasus* Sommerman N, 4
      *L. ornata* Mockford N, 2
      *L pallens* Badonnel N, 1
      *L. pallida* Mockford N, 4
      *L. villosa* Mockford N, 4
     Group B, Subgroup Bb
      *L. bicolor* (Banks) N, 1
      *L. decolor* (Pearman) C (dom)
      *L. lacinia* Sommerman N, 4
      *L. nigra* (Banks) N, 1
      *L. pearmani* Lienhard C (dom)
      *L. rufa* Broadhead C, 1, 5
      *L. silvarum* (Kolbe) N, 4
      *L. triocellata* Mockford N, 5
    Section II
     Group C
      *L. formicaria* (Hagen) N, 4
      *L. mendax* Pearman C (dom)
      *L. obscura* Broadhead C (dom)
     Group D
      *L. bostrychophila* Badonnel C, 4
      *L. corrodens* Heymons C (dom)
      *L. paeta* Pearman C (dom)
      *L. prenolepidis* (Enderlein) C, 5
      *L.* sp. N, 4
 Infraorder Amphientometae
  Family Amphientomidae
   *Lithoseopsis hellmani* (Mockford & Gurney) N, 4, 6
   *L. hystrix* (Mockford) N, 2
   *Stimulopalpus japonicus* Enderlein I, 1

*Syllisis* sp. I, 2
Family Protroctopsocidae
 *Protroctopsocus enigmaticus* Mockford N, 6
Suborder Psocomorpha
 Infraorder Archipsocetae
  Family Archipsocidae
   *Archipsocopsis frater* (Mockford) N, 2
   *A. parvula* (Mockford) N, 2
   *Archipsocus floridanus* Mockford N, 2
   *A. gurneyi* Mockford N, 2
   *A. nomas* Gurney N, 2
   *A.* sp. N, 2
   *Pararchipsocus elongatus* Badonnel et al. I
 Infraorder Epipsocetae
  Family Cladiopsocidae
   *Cladiopsocus garciai* Eertmoed I
  Family Epipsocidae
   *Bertkauia crosbyana* Chapman N, 1, 6
   *B. lepicidinaria* Chapman N, 1
   *Epipsocus petenensis* Mockford I
   *Mesepipsocus niger* (New) I
  Family Ptiloneuridae
   *Loneura* sp. N, 4
 Infraorder Caecilietae
  Superfamily Asiopsocoidea
   Family Asiopsocidae
    *Asiopsocus sonorensis* Mockford & García Aldrete N, 4
    *Notiopsocus* sp. N, 2
    *Pronotiopsocus* sp. N, 2
  Superfamily Caeciliusoidea
   Family Amphipsocidae
    *Polypsocus corruptus* Hagen N, 1, 5
   Family Caeciliusidae
    Subfamily Caeciliusinae
     Tribe Maoripsocini
      *Maoripsocus africanus* (Ribaga) C, 2
     Tribe Coryphacini
      *Valenzuela caligonus* Group
       *V. indicator* (Mockford) N, 2
      *Valenzuela confluens* Group
       *V. confluens* (Walsh) N, 1
       *V. gonostigma* (Enderlein) I
       *V. graminis* (Mockford) N, 3
      *Valenzuela fasciatus* Group
       *V. distinctus* (Mockford) I

V. mexicanus (Enderlein) N, 6
*V. nadleri* (Mockford) N, 1
*V. totonacus* (Mockford) N, 4, 6
*Valenzuela flavidus* Group
*V. atricornis* (McLachlan) I
*V. boreus* (Mockford) N, 3
*V. burmeisteri* (Brauer) N, 3
*V. croesus* (Chapman) N, 2
*V. flavidus* (Stephens) N, 1, 5
*V. hyperboreus* (Mockford) N, 3
*V. lochloosae* (Mockford) N, 3
*V. manteri* (Sommerman) N, 1
*V. maritimus* (Mockford) N, 5
*V. micanopi* (Mockford) N, 2
*V. perplexus* (Chapman) N, 4
*V. pinicola* (Banks) N, 1
*V. tamiami* (Mockford) N, 2
*Valenzuela posticus* Group
*V. posticus* (Banks) N, 1
*Valenzuela subflavus* Group
*V. incoloratus* (Mockford) N, 2
*V. juniperorum* (Mockford) N, 2
*V. subflavus* (Aaron) N, 2
*Stenocaecilius antillanus* (Banks) N, 2
*S. casarum* (Badonnel) N, 2
*S. insularum* (Mockford) N, 2
Subfamily Paracaeciliinae
*Xanthocaecilius anahuacensis* Mockford N, 6
*X. microphthalmus* Mockford N, 6
*X. quillayute* (Chapman) N, 5, (1?)
*X. sommermanae* (Mockford) N, 1
*X.* sp. N, 4
Family Dasydemellidae
*Dasydemella silvestrii* Enderlein I
*Teliapsocus conterminus* (Walsh) N, 1, 5
Family Stenopsocidae
*Graphopsocus cruciatus* (Linn.) N(?) 1, 5
*G. mexicanus* Enderlein N, 6
Infraorder Homilopsocidea
Family Lachesillidae
Subfamily Eolachesillinae
*Anomopsocus amabilis* (Walsh) N, 1
*Nanolachesilla chelata* Mockford & Sullivan N, 2
*N. hirundo* Mockford & Sullivan N, 2
*Prolachesilla mexicana* Mockford & Sullivan N, 6

*P. terricola* Mockford & Sullivan N, 4

Subfamily Lachesillinae

*Lachesilla andra* Group

  *L. andra* Sommerman N, 1

  *L. arnae* Sommerman N, 3

  *L. dona* Sommerman N, 3

  *L. kola* Sommerman N, 4

  *L. nubilis* (Aaron) N, 1

  *L. nubiloides* García Aldrete N, 4

  *L. punctata* (Banks) N, 4

  *L. texana* Mockford & García Aldrete N, 1, 4, 6

*Lachesilla centralis* Group

  *L. centralis* García Aldrete N, 5

  *L. cintalapa* García Aldrete I

  *L. perezi* García Aldrete N, 2 ,6

*Lachesilla corona* Group

  *L. albertina* García Aldrete N, 3

  *L. corona* Chapman N, 1

  *L. dispariforceps* Mockford N, 6

  *L. dividiforceps* García Aldrete N, 6

  *L. hermosa* García Aldrete N, 6

  *L. michiliensis* García Aldrete N, 4, 6

  *L. neoleonensis* García Aldrete N, 6

  *L. picticeps* Mockford N, 6

  *L. querpina* García Aldrete N, 6

  *L. regiomontana* García Aldrete N, 6

*Lachesilla forcepeta* Group

  *L. acuminiforceps* García Aldrete N, 6

  *L. alpejia* García Aldrete I

  *L. anna* Sommerman N, 1

  *L. bottimeri* Mockford & Gurney N, 2

  *L. chapmani* Sommerman N, 2

  *L. cladiumicola* García Aldrete N, 2

  *L. contraforcepeta* Chapman N, 1

  *L. cuala* García Aldrete N, 6

  *L. denticulata* García Aldrete I

  *L. floridana* García Aldrete N, 2

  *L. forcepeta* Chapman N, 1

  *L. gracilis* García Aldrete N, 2

  *L. kathrynae* Mockford & Gurney N, 4

  *L. laciniosiforceps* García Aldrete N, 6

  *L. major* Chapman N, 1

  *L. penta* Sommerman N, 2

  *L. typhicola* García Aldrete N, 1

*Lachesilla fuscipalpis* Group

                   *L. curvipila* García Aldrete N, 6
                   *L. fuscipalpis* Badonnel I
            *Lachesilla patzunensis* Group
                   *L. bifurcata* García Aldrete N, 6
            *Lachesilla pedicularia* Group
                   *L. aethiopica* (Enderlein) N, 6
                   *L. greeni* Pearman I
                   *L. huasteca* Mockford N, 4, 6
                   *L. otomi* Mockford N, 2, 6
                   *L. pacifica* Chapman N, 5
                   *L. pallida* (Chapman) N, 1
                   *L. pedicularia* (Linn.) C, 1, 5
                   *L. quercus* (Kolbe) I (dom)
                   *L. rena* Sommerman N, 4, 6
                   *L. tectorum* Badonnel N, 4, 6
                   *L. yucateca* Mockford N, 2, 6
            *Lachesilla riegeli* Group
                   *L. riegeli* Sommerman N,2
                   *L. tropica* García Aldrete N, 2
                   *L. ultima* García Aldrete N, 2
            *Lachesilla rufa* Group
                   *L. abiesicola* García Aldrete N, 6
                   *L. arida* Chapman N, 4
                   *L. aspera* García Aldrete N, 6
                   *L. braheicola* García Aldrete N, 6
                   *L. broadheadi* García Aldrete N, 6
                   *L. cupressicola* García Aldrete N, 6
                   *L. jeanae* Sommerman N, 5
                   *L. juniperana* García Aldrete N, 6
                   *L. nita* Sommerman N, 2
                   *L. pinicola* García Aldrete N, 6
                   *L. rufa* (Walsh) N, 1
                   *L. sommermanae* García Aldrete N, 6
                   *L. yakima* Mockford & García Aldrete N, 5
            *Lachesilla sclera* Group
                   *L. sulcata* García Aldrete N, 4, 6
            *Lachesilla texcocana* Group
                   *L. delta* García Aldrete N, 6
                   *L. monticola* García Aldrete N, 6
                   *L. texcocana* García Aldrete N, 4, 6
      Family Ectopsocidae
            *Ectopsocopsis cryptomeriae* (Enderlein) N, 1
            *E. decorata* Thornton & Wong I
            *Ectopsocus briggsi* McLachlan N, 5
            *E. californicus* (Banks) N, 1, 5

   *E. maindroni* Badonnel N, 2
   *E. meridionalis* Ribaga N, 1
   *E. petersi* Smithers I, 5
   *E. pumilis* (Banks) I
   *E. richardsi* (Pearman) I
   *E. salpinx* Thornton & Wong I
   *E. strauchi* Enderlein I
   *E. thibaudi* Badonnel N, 2
   *E. titschacki* Jentsch I
   *E. vachoni* Badonnel N, 4
   *E. vilhenai* Badonnel N, 2
  Family Peripsocidae
   *Kaestneriella fumosa* (Banks) N, 4
   *K. tenebrosa* Mockford & Sullivan N, 4
   *Peripsocus* Group IA
    *P. madidus* (Hagen) N, 1
    *P. milleri* (Tillyard) I, 5
    *P. pauliani* Badonnel I
    *P. subfasciatus* (Rambur) N, 1, 5
   *Peripsocus* Group IB
    *P. alachuae* Mockford N, 2
    *P. alboguttatus* (Dalman) N, 3
    *P. maculosus* Mockford N, 1
    *P. madescens* (Walsh) N, 1
    *P. minimus* Mockford N, 1
    *P. potosi* Mockford N, 4, 6
   *Peripsocus* Group IIIB
    *P. phaeopterus* (Stephens ) I
   *Peripsocus* Group IIIC
    *P. stagnivagus* Chapman N, 1
  Family Pseudocaeciliidae
   *Heterocaecilius* sp. I
   *Ophiodopelma* sp. I, 2
   *Pseudocaecilius citricola* (Ashmead) N, 2
   *P. tahitiensis* (Karny) N, 2
  Family Trichopsocidae
   *Trichopsocus clarus* (Banks) N, 5
   *T. dalii* (McLachlan) I, 2
  Family Philotarsidae
   *Aaroniella achrysa* (Banks) N, 2
   *A. badonneli* (Danks) N, 1
   *A. maculosa* (Aaron) N, 1
   *Philotarsus arizonicus* Mockford N, 4
   *P. californicus* Mockford N, 5
   *P. kwakiutl* Mockford N, 5

*P. parviceps* Roesler N, 3
*P. potosinus* Mockford N, 6
Family Elipsocidae
    *Cuneopalpus cyanops* (Rostock) I, 1, 5
    *Elipsocus abdominalis* Reuter N, 3
    *E. guentheri* Mockford N, 4
    *E. hyalinus* (Stephens) N, 3
    *E. moebiusi* Tetens I, 3
    *E. obscurus* Mockford N, 4
    *E. pumilis* (Hagen) N, 3
    *E.* sp. N, 3
    *Nepiomorpha peripsocoides* Mockford N, 2
    *Palmicola aphrodite* Mockford N, 2
    *P. solitaria* Mockford N, 2
    *P.* sp. #1 N, 2
    *P.* sp. #2 N, 2
    *Propsocus pulchripennis* (Perkins) I, 5
    *Reuterella helvimacula* (Enderlein) N, 3
Family Mesopsocidae
    *Mesopsocus immunis* (Stephens) I, 3
    *M. laticeps* (Kolbe) N, 3
    *M. unipunctatus* (Müller) N, 3
Infraorder Psocetae
Family Hemipsocidae
    *Hemipsocus chloroticus* (Hagen) I, 2
    *H. pretiosus* Banks N, 2
Family Myopsocidae
    *Lichenomima coloradensis* (Banks) N, 4
    *L. lugens* (Hagen) N, 1
    *L. sparsa* (Hagen) N, 1
    *L.* sp. #1 N, 2
    *L.* sp. #2 N, 2
    *L.* sp. #3 N, 1
    *L.* sp. #4 N, 2
    *L.* sp. #5 N, 2
    *L.* sp. #6 N, 1
    *L.* sp. #7 N, 1
    *L.* sp. #8 N, 2
    *Myopsocus antillanus* (Mockford) N, 2
    *M. eatoni* McLachlan I
    *M. minutus* (Mockford) N, 2
    *M.* sp. I, 5
Family Psocidae
    Subfamily Psocinae
        Tribe Psocini

*Atropsocus atratus* (Aaron) N, 1
*Hyalopsocus floridanus* (Banks) N, 1
*H. striatus* (Walker) N, 1
*H.* sp. #1 N, 1
*H.* sp. #2 N, 4
*Psocus crosbyi* Chapman N, 5
*P. leidyi* Aaron N, 1
*P.* sp. N, 3
Tribe Ptyctini
*Camelopsocus bactrianus* Mockford N, 5
*C. hiemalis* Mockford N, 5
*C. monticolus* Mockford N, 4
*C. similis* Mockford N, 4
*C. tucsonensis* Mockford N, 4
*C.* sp. N, 4
*Indiopsocus bisignatus* (Banks) N, 1
*I. campestris* (Aaron) N, 2
*I. ceterus* Mockford N, 2
*I. coquilletti* (Banks) N, 5
*I. infumatus* (Banks) N, 1
*I. texanus* (Aaron) N, 2
*I.* sp. #1 N, 2
*I.* sp. #2 N, 2
*I.* sp. #3 N, 2
*Loensia conspersa* (Banks) N, 4
*L. fasciata* (Fabricius) I
*L. maculosa* (Banks) N, 5
*L. moesta* (Hagen) N, 1
*L.* sp. #1 N, 2
*L.* sp. #2 N, 2, 4
*Ptycta lineata* Mockford N, 2
*P. polluta* (Walsh) N, 1
*Steleops elegans* (Banks) N, 1
*S. lichenatus* (Walsh) N, 1
*Trichadenotecnum alexanderae* Group
*T. alexanderae* Sommerman N, 1
*T. castum* Betz N, 1
*T. innuptum* Betz N, 3
*T. merum* Betz N, 1
*Trichadenotecnum circularoides* Group
*T. circularoides* Badonnel N, 2
*Trichadenotecnum desolatum* Group
*T. acutilingum* Yoshizawa et al. N, 6
*T. desolatum* (Chapman) N, 4, 6
*Trichadenotecnum quaesitum* Group

*T. cerrosillae* Yoshizawa et al. N, 6
*T. maculatum* Yoshizawa et al. N, 6
*T. neoleonense* Yoshizawa et al. N, 6
*T. quaesitum* (Chapman) N, 1
*Trichadenotecnum slossonae* Group
*T. slossonae* (Banks) N, 1
*Trichadenotecnum spiniserrulum* Group
*T. pardus* Badonnel N, 2
Tribe Metylophorini
*Metylophorus barretti* (Banks) N, 4, 6
*M. novaescotiae* (Walker) N, 1, 6
*M. purus* (Walsh) N, 1
Tribe Cerastispsocini
*Cerastipsocus trifasciatus* (Provancher) N, 1, 4, 6
*C. venosus* (Burmeister) N, 1, 2
Subfamily Amphigerontiinae
*Amphigerontia bifasciata* (Latreille) N, 3
*A. contaminata* (Stephens) I, 5
*A. infernicola* (Chapman) N, 4
*A. longicauda* Mockford & Anonby N, 4
*A. montivaga* (Chapman) N, 1, 4
*A. petiolata* (Banks) N, 4
*Blaste cockerelli* (Banks) N, 4
*B. garciorum* Mockford N, 2, 6
*B. longipennis* (Banks) N, 4
*B. opposita* (Banks) N, 1
*B. oregona* (Banks) N, 5
*B. osceola* Mockford N, 2
*B. persimilis* (Banks) N, 2
*B. posticata* (Banks) N, 2, 6
*B. quieta* (Hagen) N, 1
*B. subapterous* (Chapman) N, 4
*B. subquieta* (Chapman) N, 1
*Blastopsocus goodrichi* Mockford N, 4
*B. johnstoni* Mockford N, 4
*B. lithinus* (Chapman) N, 1
*B. semistriatus* (Walsh) N, 1
*B. variabilis* (Aaron) N, 1
*B. walshi* Mockford N, 1
*B.* sp. N, 4

The list is based on the classification used by Lienhard and Smithers (2002), modified by the work of Yoshizawa et al. (2006) and Johnson and Mockford (2003). See text (section 6) for explanation of letter/number combinations following species names.

Table 1. Synoptic List of North American Psocoptera

A very different picture is seen in comparing the North American psocid fauna with the relatively well-studied psocid fauna of China (Li, 2002, see also Lienhard, 2003). The Chinese fauna is reported to contain 1505 species in 170 genera. The North American fauna is miniscule by comparison. A minor part of this difference may be due to over splitting at the generic and specific levels in the study of the Chinese material. Most of this huge difference must be due to historico-geographic and historico-climatic differences beyond the scope of this chapter.

Another comparison of interest is that of the Mexican fauna (faunal list in Mockford & García Aldrete, 1996) and the North America fauna. Although many species of the Mexican psocid fauna have not yet been named, what must amount to the great majority have now been recognized. The Mexican list includes 642 recent species in 97 genera and 32 families. Four families of the Superfamily Electrentomoidea (Families Musapsocidae, Troctopsocidae, Manicapsocidae, and Compsocidae) and Family Dolabellopsocidae of the Infraorder Epipsocetae reach southern Mexico from the deeper tropics but fail to reach North America. These five families account for most of the generic difference, while a significant amount the species-level difference is accounted for by the huge genus *Lachesilla*, which, though well represented in North America (75 species), has more than twice as many species in Mexico (157 species).

## 2. Biological aspects of the Psocoptera affecting dispersal and distribution

Psocoptera, commonly called psocids or the bark lice, are small insects, adults from the area under consideration ranging in body length from 1 to 5 mm. They are neopterous, exopterygote, acercareous (definitions in Nichols & Schuh, 1989). Adults typically have two pairs of membranous wings, but many evolutionary lines have undergone selection for wing reduction and loss. The winged forms appear to be weak flyers (Mockford, 1962) but may be carried by wind (discussed below). Nearly all are oviparous, and although ovovivipary is known, it is restricted to only two species living in North America, both of the genus *Archipsocopsis*. Many aspects of psocid biology were treated by New (1987). Immatures (nymphs) are generally cryptic in color and sometimes in form against their substrate. They feed on epiphytic and epigaeic algae and lichens, as well, in some forms, insect eggs and remains of dead insects. Mouthparts are of the chewing type (cf. Triplehorn & Johnson, 2005) with laciniae developed as rods. The latter are thought to stabilize the heavy mandibles as they bite and chew through tough material. Some 50 species have adapted to living primarily or in part in human habitations, and some of these feed primarily on farinaceous products (cf. Mockford, 1991). Psocids of all postembryonic stages have a unique apparatus in the mouthparts allowing extraction of water from the atmosphere. This involves a complex hypopharynx with a pair of lingual sclerites that are extruded with their lower surfaces out of the mouth during periods of higher relative humidity following dry periods. Water molecules accumulate on these surfaces and pass to the foregut by capillarity and action of a cibarial pump (see discussion in Lienhard, 1998: 33 - 36). This mechanism has probably allowed some forms to survive in desert regions and others to live in heated human dwellings during winter.

## 3. Seasonality and its adaptations

Psocids living in the northern portion of North America (Canada and the northern two to three tiers of US states) pass the winter in the egg stage. Eertmoed (1978) and Glinianaya (1975) found a temperature–dependent response of females to late-summer shortening of day length as induction of winter diapausing eggs. In northern Mexico and the extreme southern United States, numerous species of psocids are active throughout the year. In southern Arizona and southern California, several species have only a single generation per year, in winter, with eggs hatching in mid November to early December and adults appearing in late December and persisting through May (Mockford, 1984b and pers. obs.) Observations by D. Young in south-central Texas suggest a similar pattern for several species in that region (personal communication). Such a pattern permits the species to pass the hot, dry summers of these areas in the relatively persistent egg stage. Although no data are available, it is likely that an environmentally induced summer egg diapause is involved. D .J. Schmidt (unpublished MS thesis, Illinois State University, 1989) determined seasonal occurrence of 27 species of Psocoptera at the Archbold Biological Station, Lake Placid, Florida (south-central Florida). Fifteen species were winter-uniseasonal (peak of abundance between December and May). Seven species were summer-uniseasonal (peak of abundance between June and November). Three species appeared to be biseasonal, with a peak in December and a peak in June. A single species appeared to be non-seasonal, and seasonality of one could not be determined. Nearly all of the larger forms (Family Psocidae) were winter-uniseasonal.

## 4. Modes of dispersal

Unfortunately, this is an area about which little is known. Earlier literature on this subject was reviewed by New (1987). A regular period of flight activity is part of the life cycle of some species of psocids, but such flights usually involve only short distances. Some species are known to become part of the aerial plankton and are carried long distances in that way. A notable case is that of the (primarily) North American species *Lachesilla pacifica* Chapman. Thornton (1964) reported that a female of this species was taken on a ship's aerial trapping device, 835 km at sea from San Francisco and thought to be a genuine air capture. The species occurs regularly along the Pacific coast from Vancouver, B.C., to southern California (Mockford, 1993). Throughout that area it is represented by sexual and thelytokous (all female) populations. Temporary populations, always of the thelytokous form, become established in central Illinois and Kentucky, far to the east of the usual range (García Aldrete, 1973; Mockford, 1993). The species was also reported from the region of Geneva, Switzerland (Lienhard, 1989), where three females were taken in two successive years. In all of these cases, long-distance wind transport is the likely means of dispersal.

Other non-human modes of psocid transport, including phoresy on birds and mammals, reviewed by New (1987) are of interest, but have not been investigated in North America.

Mockford (1991) listed and keyed 50 species of psocids known to live commonly in human habitations. Many of these are cosmopolitan in distribution and are known to be spread through human commerce. The region of origin of most of those species (those designated "C" in Table I) are not known and can only be suggested in some cases where species show close relationship to forms native or endemic to particular regions.

## 5. Native versus introduced taxa

Mockford (1993) proposed the following criteria to determine if a species has been introduced: (1) species commonly associated with human commerce are regarded as introduced unless they have an extensive out-door distribution in the study area; (2) species for which introduction into the study area has been documented and which are not, or scarcely, otherwise present are regarded as introduced; 3) species widely distributed elsewhere, and with a very limited, coastal distribution in the study area are regarded as introduced unless the distribution in the study area appears to be part of a natural distribution largely outside the study area. The species designated I and (dom) on Table I are regarded as introduced.

## 6. Distribution patterns in North American Psocoptera

Table 1, the synoptic list of taxa, includes for each species its status: native (N), intercepted at a port of entry (I, followed by no number), introduced or cosmopolitan and established in a particular area ( C or I, followed by a number), introduced or cosmopolitan but found only in human habitations [C or I, followed by (dom)]. The numbers refer to primary distribution patterns. Mockford (1993) recognized five of these for the North American psocid fauna. A sixth must be added for the extended definition of North America followed here. Following are the definitions of the patterns (see also map, Fig. 1).

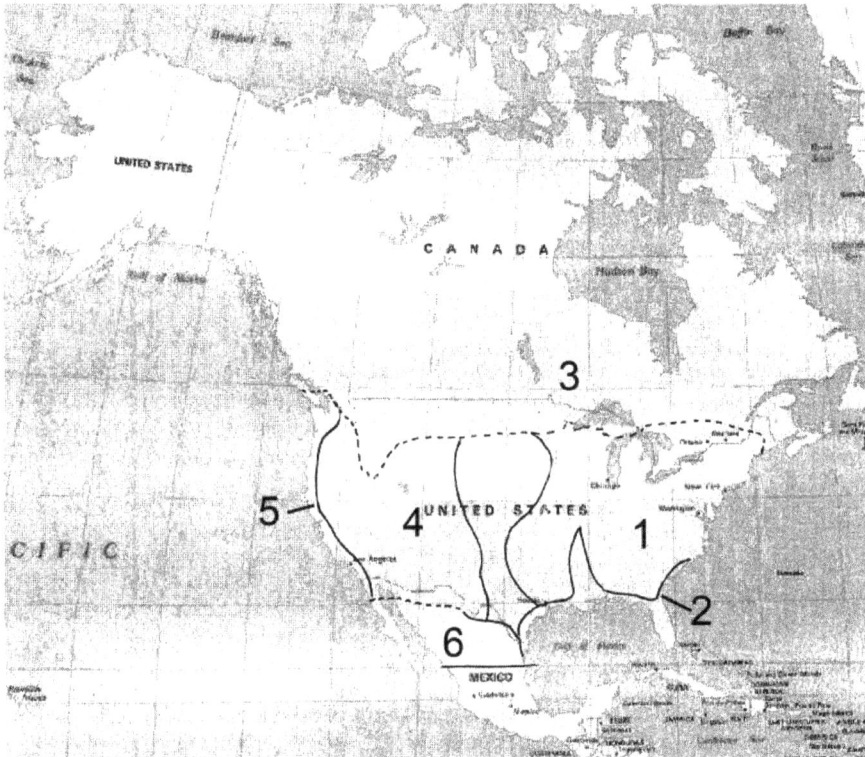

Fig. 1. Outline map of North America showing distribution patterns of Psocoptera. See text section 6 for explanation of numbers. Note an area in central United States where records are too sparse to permit assignment to a pattern.

1.  The Eastern Deciduous Forest Pattern. The area corresponds closely to the Eastern
    Deciduous Forest as delimited by Braun (1950). Based on psocid distributions, its
    northern limit remains poorly understood for lack of collecting, but it clearly extends
    north to central Ontario and northern Minnesota. Its southern limit in the east lies in
    north-peninsular Florida. From there, it extends westward just off the coastal plain to
    the forested area of northeastern Texas. This pattern is represented by 79 species (Table
    1, species designated N, 1 and I, 1) or 20.7% of the total North American fauna.
2.  The Southeastern Subtropical Pattern. This pattern is coextensive with the coastal plain
    of the Gulf states and that of the southern Atlantic states, including essentially all of
    Florida extending up the Mississippi Embayment, for some species as far as southern
    Illinois, and up the Atlantic coast, for a few species even to Long Island. This pattern
    includes 88 species (Table 1, species designated N, 2, I, 2, and C, 2) or 22.7% of the
    North American fauna. A few of these species may, in fact, be introductions, but the
    criteria for recognition of introduction, discussed above, are not met. Speciation among
    the taxa in this pattern is discussed in Section 7.
3.  The Boreal and Holarctic Pattern. This pattern is represented in North America by a
    very small array of species (Table 1, species designated N, 3 and I, 3), only 18, or 4.7% of
    the North American fauna. These are all either Holarctic in distribution or have close
    affinity with Holarctic species. Included here are all three of the North American
    species of *Mesopsocus* and five of the seven North American *Elipsocus* species.
4.  The Southwestern and Rocky Mountain Pattern. This pattern is based on taxa found in
    the southern and central Rocky Mountains, the surrounding arid lands, and the
    California Sierra Nevada. Sixty-nine species (Table 1, species designated N, 4, and C, 4)
    are included in this pattern, or 19.8% of the North American fauna.
5.  The Pacific Coast Pattern. A substantial number of species, probably including both native
    and introduced forms, have adapted to the mesic, in some areas highly humid, habitats of
    the Pacific coastal plain. Some of those taxa range eastward into the Cascade Mountains
    and a few even into the Rocky Mountains. Some of them are taxa also represented in
    Pattern 1 (see also section 7). Thirty-five species are included in this pattern (Table 1,
    species designated N, 5; C ,5, and I, 5) or 9.0% of the North American fauna.
6.  The North-Mexican Montane areas and surrounding High Plains. Unfortunately, the
    western part of this interesting pattern remains rather poorly explored, and some
    material in collections remains to be investigated. Sixty-four species are included in this
    pattern (Table 1, species designated N, 6) or 16.5% of the North American fauna.
    The percentages of the total fauna noted above only reach 91.1%. The rest of the faunal list
    consists of intercepts at ports of entry (I , and I(dom), followed by no number on Table 1),
    of which there are 34 species, and cosmopolitan domestic (i.e. in human habitations)
    species, of which there are six (C (dom) and IC (dom) on Table 1). *Psyllipsocus ramburii*
    forms something of an exception to all others, being a cosmopolitan species occurring
    commonly in human dwellings, but also occurring in caves throughout North America.

## 7. Speciation in North American Psocoptera

### 7.1 Speciation in sexual species

The overall impression of speciation in North American Psocoptera is that there has not
been a large amount of in situ differentiation. Passing through Table 1 in systematic order,
the first group suggesting speciation is the *Echmepteryx hageni* complex, with three species,
*E. hageni, E. intermedia,* and *E. youngi.* Undoubtedly, they are closely related, but they are all

part of a much larger Antillean complex, and, indeed, *E. intermedia* also occurs in Jamaica. It seems quite possible that each of the three species may have been derived separately from Antillean ancestors.

In the genus *Rhyopsocus*, the three species *R. maculosus, R. micropterus,* and *R. texanus* appear to be closely related. The speciation events that led to their separation probably occurred in northern Mexico and adjacent southern United States. These species inhabit ground litter and pack rat nests, where humidity probably remains higher than in the surrounding xeric lands. *Rhyopsocus bentonae,* a species of peninsular Florida, southern Georgia, and the Gulf coast to Texas is part of a complex, with a second species on the Florida keys and a third in southern Mexico. The two Florida species may represent range expansions at different times of more southern species with subsequent adaptive changes.

In the genus *Psyllipsocus*, there appear to be two examples of speciation within North America, but both are currently under investigation.

In the genus *Speleketor*, the speciation events leading to the establishment of the three known species may have occurred in southwestern United States, but nymphs of a presumed fourth species of *Speleketor* have been found in a cave in Nuevo León State in northern Mexico (pers. obs.).

In the Subfamily Embidopsocinae of the Family Liposcelididae, two species, *Belaphotroctes hermosus* and *B. simberloffi,* the former of woodland ground litter in southern Texas and northern Mexico, and the latter in red mangroves on the Florida Keys, appear to be closely related (Mockford, 1972). Their speciation may have resulted from breakup of a continuous range around the Gulf of Mexico. Mockford (1993) noted the close proximity of the *Embidopsocus* species of subgroup IB (see Mockford, 1993: 78). Speciation in this subgroup has resulted in two Brazilian species, a Cuban species, and three species in North America. Of the latter three, *E. bousemani* is restricted to hilltops in the extreme southwestern Appalachian Mountains and in the Ozark hills. *Embidopsocus laticeps* is a Florida and south coastal-plain species extending north into southern Georgia and west into Louisiana. The third species, *E. mexicanus,* occurs widely in southern Mexico and is known from two localities in southeastern Texas. Oddly, a single specimen was collected at Funks Grove in central Illinois. The species of *Embidopsocus* are subcorticolous, and one species in southeastern Asia was recorded as phoretic on a migratory bird (Mockford, 1967).

In the genus *Liposcelis, L. pallida* and *L. villosa* appear to be closely related and may have speciated in the mountains of southwestern United States.

In the Family Archipsocidae, *Archipsocus floridanus* and *A. nomas* are probably closely related, but both are known from areas to the south of the study area, and it is likely that their speciation did not occur within the study area.

In the Family Caeciliusidae, the large genus *Valenzuela* shows a few examples of speciation in North America. *Valenzuela nadleri* and *V. totonacus* are closely related. Both are ground litter inhabitants, the former in eastern United States, and the latter in southwestern United States and Mexico. They may have had a continuous range prior to the increasingly xeric conditions of the Mexican border and Southwest. *Valenzuela boreus* and *V. croesus* are closely related, the two showing north–south replacement in eastern United States. The form currently called *V. croesus* in the Mexican mountains may prove to be a separate species. *Valenzuela flavidus* seemingly budded off two species in North America, *V. hyperboreus* in the north and *V. maritimus* on the Pacific coast. In the genus *Xanthocaecilius*, *X. microphthalmus* appears to have budded off in an isolated mountain area in Nuevo León State from a much more widely distributed species, *X. anahuacensis*, which is found throughout much of Mexico and Central America (see Mockford, 1989).

In the Family Lachesillidae, two species of *Nanolachesilla* occur in southern Florida, both on dead leaves, but one, *N. hirundo*, is restricted to palms. Other species of this genus are known from Jamaica and southern Mexico. It is likely that these two species are not sister species and that their speciational events may have occurred elsewhere, at least in part. *Lachesilla texana* may offer an example of ongoing speciation. A population presumably isolated in the southern Appalachian Mountains shows male genitalic features somewhat different from those in the population extending from central Texas to northern Mexico (Mockford & García Aldrete, 2010). Within the larger species groups of *Lachesilla*, other North American speciational events will probably be revealed when relationships are better understood.

In the Family Philotarsidae, *Philotarsus californicus* and *P. kwakiutl* are, with little doubt, sister species. The former, in California, is replaced by the latter further north along the coast and in adjacent mountains.

In the Family Myopsocidae, investigation at the alpha–taxonomic level of the genus *Lichenomima* is still in an early stage. Little can be said about speciation in the North American representatives of this group except to note that some differentiation must have occurred in southeastern (and southwestern?) United States.

In the Family Psocidae, the species of the genus *Camelopsocus* probably speciated within the study area. Here, climate and, perhaps, elevation, probably played roles in the speciation events. Two species in southern California, *C. bactrianus* and *C. hiemalis*, and one species in southern Arizona, *C. tucsonensis*, are winter species, maturing from late December to February. The other two named species, *C. monticolus* and *C. similis* are summer species found at somewhat higher altitudes (see Mockford, 1965, 1984b).

## 7.2 Speciation involving thelytokous parthenogenesis

Mockford (1971) reviewed parthenogenesis in psocids through the literature of 1969. The overall conclusions have changed little. Parthenogenesis in psocids is always thelytoky (female-producing), and, in general, it is obligate within the strain or species in which it occurs. Betz (1983) found four morphologically distinguishable forms of *Trichadenotecnum alexanderae*. Three of these were obligatorily thelytokous forms and would not mate with males of the sexual fourth. The latter, which proved to represent the type of *T. alexanderae*, was capable of extremely limited facultative thelytoky. He named each of the thelytokous forms as a species and concluded that each was derived from a sexual ancestor. The distributions of these species overlap broadly in eastern United States. Schmidt (1992) confirmed Betz's (1983) findings concerning morphology and reproductive type, and determined the karyotype for each of the morphospecies. One of the species, *T. castum*, proved to be triploid, for which a hybrid origin was suggested.

Several other examples of speciation in North America involve production of a thelytokous species, morphologically distinct from a closely related sexual species. In the Family Philotarsidae, *Aaroniella badonneli* and *A. maculosa* form a closely related species pair in which the former is thelytokous and the latter is sexual. In this case, the range of the sexual species is more northerly than that of the thelytokous one, although the two species overlap in distribution (see also Mockford, 1979, in which *A. badonneli* is discussed under the synonym *A. eertmoedi*).

In the Family Peripsocidae, the *Peripsocus alboguttatus* species complex consists of three sexual species and one thelytokous species. The latter, *P. maculosus*, together with its close sexual congener, *P. madescens*, occurs widely throughout eastern United States and southern Canada, both species primarily inhabiting conifers. This case seems to require further investigation.

It has long been known that infection with certain bacteria, primarily of the genus *Wolbachia*, can disable sexual reproduction in various insects, so that the infected line carries on by thelytoky. Shreve et al. (2011) reviewed this area relating to Psocoptera and noted no known cases and little likelihood of such a phenomenon in this group. Investigations on *Liposcelis bostrychophila*, a common cosmopolitan stored–product pest, have shown the presence of a *Rickettsia*–type bacterium. The domestic form of this species is obligatorily thelytokous, and Perotti et al. (2006) noted that a *Rickettsia* infection in this species is obligate for egg production. Behar et al. (in press) have shown that this *Rickettsia*, identified as *R. felis*, is present in both sexual and parthenogenetic strains of *L. bostrychophila* and so does not affect the type of reproduction. A *Wolbachia* infection has been found in a thelytokous strain of *Echmepteryx hageni* (K. Johnson, pers. comm.), but it is not known to be the causative agent of the thelytoky. It is obvious that much more investigation needs to be carried out in this field.

## 8. Endemic taxa above the species level

The genus *Speleketor* of the Family Prionoglarididae appears to be endemic to North America, with three species, all occurring in southwestern United States: *S. irwini* and *S. pictus* in southern California, and *S. flocki* in Arizona and Nevada (see Mockford, 1984a). The latter species is a partial cave inhabitant. *Speleketor irwini* lives in the skirts of dead leaves on native stands of the palm, *Washingtonia filifera*, in canyons. *Speleketor pictus* is known from a single specimen collected by black light, so its habitat remains unknown. It is likely that additional species will be found, and a nymph probably representing an additional species was collected in a cave near Laguna de Sánchez, Nuevo León State, Mexico. The closest relatives of this genus are the sensitibilline psocids (genera *Sensitibilla* and *Afrotrogla*) from Namibia and South Africa (see Lienhard 2007, Lienhard et al. 2010). The entire Family Prionoglarididae appear to be ancient, and Yoshizawa et al. (2006) suggest that they may be Pangaean relics.

In the Family Sphaeropsocidae, two genera are endemic to North America: *Prosphaeropsocus* and *Troglosphaeropsocus* (Mockford, 2009). The former genus, with two known species, is found in ground litter on coastal hills in central California. The latter, represented by a single male, is from a cave in northern Arizona. Mockford (2009) presented a strong argument for sister–taxa relationship of the genera *Troglosphaeropsocus* and *Badonnelia*. This would suggest great age for the resulting taxon, as *Badonnelia* is known from a north-Holarctic, primarily domestic species, and four species from southern Chile (Badonnel, 1963, 1967, 1972). This presumably extinct stem taxon and its offspring genera may represent another Pangaean relic.

In the Family Psocidae, the genus *Camelopsocus* is nearly endemic to North America, although the ranges of two of the species extend into southern Mexico. As noted above, this genus consists currently of two summer–active species widely distributed in the mountains of western United States and Mexico, and three winter–active species in deserts of southern Arizona and coastal and near–coastal scrub in southern California. This suggests late

Miocene–Pliocene adaptation of a form derived from a mountain–inhabiting ancestor to lowland areas with summer climate becoming gradually unfavorable, followed by further geographic speciation in both lowland and upland taxa. *Camelopsocus* has no close relatives in North America and appears to be closest to the Palaearctic genus *Oreopsocus*, as represented by the Egyptian species, *O. buholzeri* Lienhard.

## 9. Human effects

Although there are no documented cases, it is possible that a low level of extinction of psocid species has been brought about by human activities, such as extensive deforestation in the East and destruction of the tall grass prairie in the Midwest. An undescribed species of *Psocus* is known from only two females collected on tamarack (*Larix laricina*) in a bog in central Wisconsin. Is this species going extinct due to human destruction of its habitat, or will further collecting effort find it in some numbers? We do not know at present. *Valenzuela graminis* is a species of tall grass prairie remnants in central Illinois. It is persistent in these prairie remnants, and it tends to find its way into restored prairies, despite the fact that females of this species are often short-winged. It is clearly not in danger of extinction.

A glance through Table I shows that numerous species have had the opportunity to expand their ranges through the agency of human commerce. This may be viewed as a form of insurance against extinction.

## 10. Status of knowledge and needed research

Distributions of most North American psocid species remain poorly known. The known distributions still tend to reflect where specialists have lived rather than the true distributions of species. Much more collecting needs to be done, especially in the northern tier of US states and southern Canada, along the Atlantic and Pacific coasts, the northern and western coasts of the Gulf of Mexico, and throughout the Southwest.

Much more needs to be done to establish phylogenetic relationships, both through morphological and molecular studies. In this respect, life history studies have lagged far behind, not being a "popular" form of research.

Although funding is difficult to find for studies on Psocoptera, it is not impossible, and careers filled with new discoveries are available.

## 11. References

Badonnel, A. 1963. Psocoptères terricoles, lapidicoles et corticicoles du Chili. Biologie de l'Amérique australe 2: 291 – 338.

Badonnel, A. 1967. Psocoptères édaphiques du Chili (2e note). Biologie de l'Amérique australe 3: 541 – 585.

Badonnel, A. 1972. Psocoptères édaphiques du Chili (3e note). (Insecta). Bulletin du Muséum national d'Histoire naturelle (3) (1) (1971), Zool. 1: 1 – 38.

Behar, A., E. Mockford, G. Opit, and S. Perlman. In press. Not necessary after all: Ecology of *Rickettsia felis* in the common household pest *Liposcelis bostrychophila* (Psocoptera: Liposcelididae).

Betz, B. W. 1983. Systematics of the *Trichadenotecnum alexanderae* species complex (Psocoptera: Psocidae) based on an investigation of modes of reproduction and morphology. Canadian Entomologist 115(10): 1329 – 1354.

Braun, E. L. 1950. Deciduous Forests of eastern North America. Hefner Co., New York.

Eertmoed, G. 1978. Embryonic diapause in the psocid, *Peripsocus quadrifasciatus*: photoperiod, temperature, ontogeny and geographic variation. Physiological Entomology 3: 197 – 207.

García Aldrete, A. N. 1973. The life history and developmental rates of *Lachesilla pacifica* Chapman (parthenogenetic form) at four levels of temperature (Psocoptera: Lachesillidae). Ciencia, México 28(2): 73 – 77.

Glinianaya, E. I. 1975. The importance of day length in regulating the seasonal cycles and diapause in some Psocoptera. Entomological Review 54: 10 –13. Original publication in Russian: Entomologicheskoe Obozrenie 54: 17 – 22.

Johnson, K. P. and E. L. Mockford. 2003. Molecular systematics of Psocomorpha (Psocoptera). Systematic Entomology 28: 409 – 416.

Li Fasheng. 2002. Psocoptera of China. Science Press, Beijing. (In Chinese with English appendix).

Lienhard, C. 1989. Zwei interessante europäische *Lachesilla*-Arten (Psocoptera: Lachesillidae). Mitteilungen der Schweizerischen Entomologischen Gesellschaft 62: 307 – 314.

Lienhard, C. 1998. Psocoptères euro-méditerranéens. Faune de France 83: xx + 517 pp., 11 plates.

Lienhard, C. 2002. Deux psoques intéressants de Corse (Psocoptera: Caeciliusidae) avec une liste des espèces oueste–paléarctiques de la famille. Revue suisse de Zoologie 109: 687 – 694.

Lienhard, C. 2003. Nomenclatural amendments concerning Chinese Psocoptera (Insecta) with remarks on species richness. Revue suisse de Zoologie 110: 695 – 721.

Lienhard, C. 2005. Description of a new beetle–like psocid (Insecta: Psocoptera: Protroctopsocidae) from Turkey showing an unusual sexual dimorphism. Revue suisse de Zoologie 112: 333 – 349.

Lienhard, C. 2006. Four interesting psocids (Psocodea: 'Psocoptera') from European parts of Russia and from the eastern Mediterranean. Revue suisse de Zoologie 113: 807 – 815.

Lienhard, C. 2007. Description of a new African genus and a new tribe of Speleketorinae (Psocodea: 'Psocoptera': Prionoglarididae). Revue suisse de Zoologie 114: 441 – 469.

Lienhard, C. and A. Baz. 2004. On some interesting psocids (Insecta: Psocoptera) from European Macaronesia. *In*: García Aldrete, A. N., C. Lienhard, and E. L. Mockford, eds., Thorntoniana. Publicaciones Especiales 20. Instituto de Biología, UNAM, México. Pp. 79 – 97.

Lienhard, C., O. Holuša, and G. Grafitti. 2010. Two new cavedwelling Prionoglarididae from Venezuela and Namibia (Psocodea: 'Psocoptera': Trogiomorpha). Revue suisse de Zoologie 117: 185 – 197.

Lienhard, C. and C. N. Smithers. 2002. Psocoptera (Insecta): World Catalogue and Bibliography. Instrumenta biodiversitatis 5: xli + 745 pp. Muséum d'histoire naturelle, Genève.

Lyal, C. H. C. 1985. Phylogeny and classification of the Psocodea, with particular reference to the lice (Psocodea: Phthiraptera). Systematic Entomology 10: 145 – 165.

Mockford, E. L. 1962. Notes on the distribution and life history of *Archipsocus frater* Mockford (Psocoptera: Archipsocidae). Florida Entomologist 45: 149 – 151.

Mockford, E. L. 1965. A new genus of hump-backed psocids from Mexico and southwestern United States (Psocoptera: Psocidae). Folia Entomologica Mexicana 11: 3 – 15.

Mockford, E. L. 1967. Some Psocoptera from plumage of birds. Proceedings of the Entomological Society of Washington 69: 307 – 309.

Mockford, E. L. 1971. Parthenogenesis in Psocids (Insecta: Psocoptera). American Zoologist 11: 327 – 339.

Mockford, E. L. 1972. New species, records, and synonymy of Florida *Belaphotroctes* (Psocoptera: Liposcelidae). Florida Entomologist 55: 153 – 163.

Mockford, E. L. 1979. Diagnoses, distribution, and comparative life history notes on *Aaroniella maculosa* (Aaron) and *A. eertmoedi* n. sp. (Psocoptera: Philotarsidae). Great Lakes Entomologist 12: 35 – 44.

Mockford, E. L. 1984a. Two new species of *Speleketor* from southern California with comments on the taxonomic position of the genus (Psocoptera: Prionoglaridae). Southwestern Naturalist 29: 169 – 179.

Mockford, E. L. 1984b. A systematic study of the genus *Camelopsocus* with descriptions of three new species (Psocoptera: Psocidae). Pan–Pacific Entomologist 60: 193 – 212.

Mockford, E. L. 1989. *Xanthocaecilius* (Psocoptera: Caeciliidae), a new genus from the Western Hemisphere: I. Description, species complexes, and species of the *quillayute* and *granulosus* complexes. Transactions of the American Entomological Society 114: 169 – 179.

Mockford, E. L. 1991. Psocids (Psocoptera). *In*: J. R. Gorham, ed. Insect and mite pests in food. An illustrated key. Vol. 2, Chapter 22, pp. 371 – 402. United States Department of Agriculture and United States Department of Health and Human Services, Agriculture Handbook No. 655.

Mockford, E. L. 1993. North American Psocoptera (Insecta). Flora and Fauna Handbook 10: xviii + 455 pp. Sandhill Crane Press, Gainesville, FL.

Mockford, E. L. 2009. Systematics of North American species of Sphaeropsocidae (Psocoptera). Proceedings of the Entomological Society of Washington 111: 666 – 685.

Mockford, E. L. and A. N. García Aldrete. 1996. Psocoptera (pp. 175 – 205). *In*: Llorente Bousquets, J. E., A. N. García Aldrete, and E. Gonzalez Soriano (eds.) Biodiversidad, taxonomía, y biogeographía de Artrópodos de México: Hacía una síntesis de su conocimiento. Universidad Nacional Autónoma de México, México, D. F., 660 pp.

Mockford, E. L. and A. N. García Aldrete. 2010. A new species of *Lachesilla* Westwood (Psocoptera: Lachesillidae) in the *andra* group with a proposed classification of the *andra* group. Zootaxa 2335: 49 – 58.

New, T. R. 1987. Biology of the Psocoptera. Oriental Insects 21: 1–109.

Nichols, S. W. and R. T. Schuh, eds. 1989. The Torre–Bueno Glossary of Entomology. The New York Entomological Society, New York City.

Perotti, M. A., H. K. Clarke, B. D. Turner, and H. R. Braig. 2006. *Rickettsia* as obligate and mycetomic Bacteria. The FASEB Journal 20: 1646–1656, pls. 1 – 4.

Schmidt, C. M. 1991. The taxonomy and phylogeny of the *Trichadenotecnum alexanderae* species complex (order Psocoptera). Doctoral thesis, Illinois State University, Department of Biological Sciences, v + 90 pp.

Shelford, V. E. 1963. The Ecology of North America. University of Illinois Press, Urbana, IL.

Shreve, S. M., E. L. Mockford, and K. M. Johnson. 2011. Elevated genetic diversity of mitochondrial genes in asexual populations of bark lice (Psocoptera: *Echmepteryx hageni*). Molecular Ecology 26: 4433-4451

Thornton, I. W. B. 1964. Airborne Psocoptera trapped on ships and aircraft. Pacific Insects 6: 285 – 291.

Triplehorn, C. A. and N. F. Johnson. 2005. Borror and Delong's Introduction to the Study of Insects, Seventh Edition. Thomson Brooks/Cole, Belmont, CA.

Yoshizawa, K., C. Lienhard, and K. P. Johnson. 2006. Molecular systematics of the Suborder Trogiomorpha (Insecta: Psocodea: 'Psocoptera'). Zoological Journal of the Linnaean Society 146: 287 – 299.

Yoshizawa, K. and C. Lienhard. 2010. In search of the sister group of the true lice: a systematic review of booklice and their relatives, with an updated checklist of Liposcelididae (Insecta: Psocodea). Arthropod Systematics and Phylogeny 68: 181 – 195.

# 16

# Composition and Distribution Patterns of Species at a Global Biogeographic Region Scale: Biogeography of Aphodiini Dung Beetles (Coleoptera, Scarabaeidae) Based on Species Geographic and Taxonomic Data

Francisco José Cabrero-Sañudo

*Departamento de Zoología & Antropología Física, Facultad de Ciencias Biológicas,*
*Universidad Complutense de Madrid,*
*Spain*

## 1. Introduction

### 1.1 Current distributions as a consequence of history

Hypotheses about ancient processes are not testable by direct observation or manipulative experiments. However, their resulting present patterns can potentially be observed, approached from an inductively point of view, and, therefore, tested. Today, many historical biogeographical hypotheses of many taxa are often drawn from phylogenetic analyses or from fossils. Although biogeographical hypotheses may be presented in those cases simply as a narrative addendum of results, they are supported by the evolutionary relationships or dating of fossils, and are generally considered valid (but see Crisp et al., 2011). Nevertheless, sometimes an evolutionary basis to explain the past biogeography of concrete species groups is not available. This could be the case of hyperdiverse taxa, for example, many groups of insects; in groups with a high diversity of species it may be difficult in the short term to have a complete phylogeny to help us answer some biogeographical questions (for example, the location of areas with a high supraspecific-taxa diversity). This could be aggravated when no significant fossils have been found. Moreover, insufficient biogeographical knowledge exacerbates this problem although such groups may have an important ecological role and interest in conservation.

Current distribution is a consequence of past historical processes, and some basic biogeographical questions can be answered by analysis of contemporary geographic distribution of a species group. Under this assumption and having only geographical and taxonomic information, we need statistically robust methods to frame testable hypotheses and provide valid, scientifically rigorous answers. The set of approaches herein presented may be especially important when dealing with a group of species for which we have little or not at all phylogenetic information, although both the alpha taxonomy (not necessarily the beta taxonomy) and the taxon distribution are well known.

Using taxonomic and geographical information available about Aphodiini dung beetles species as an example, I examine their general current distribution and variation in diversity, taking into account the six major biogeographic regions worldwide. A similar procedure already was conducted in a previous paper by Cabrero-Sañudo & Lobo (2009), although Aphodiini genera were used rather than species. The faunal similarities and structure among the regions also will be evaluated to explore biogeographic relationships. In addition to elucidating major biogeographic patterns of this group, this study also proposes hypotheses about the historical processes operating in each biogeographic region and worldwide based on the supported results.

## 1.2 Sample study group: Aphodiini dung beetles

Dung beetles are a coleopteran group of species mostly constituted by representative taxa from the Scarabaeoidea (Insecta, Coleoptera) superfamily. Together with Diptera, they are the most abundant species group at dung communities on a worldwide scale (Hanski, 1991a). While most Aphodiini species show special morphological, behavioural and ecological adaptations to the consumption of mammal excrements (mainly from ungulates), others are also known to feed on detritus, fungi, decaying plants or roots (Hanski, 1991a). These insects are of great ecological interest, as they increase the soil permeability and recycle organic matter, favouring the fertility of pastures (Bornemissza, 1976; Ridsdill-Smith & Edwards, 2011; Rougon et al., 1988). Also, they are the main controllers of hematophagous insects and disease vectors of cattle (McQueen & Beirne, 1975; Ridsdill-Smith & Edwards, 2011; Waterhouse, 1974). Moreover, dung beetles have been also used as indicator taxa in conservation studies (A.L.V. Davis et al., 2004; McGeoch et al., 2002; Nichols & Gardner, 2011).

Within Scarabaeoidea, the tribe Aphodiini (Scarabaeidae: Aphodiinae), together with Scarabaeinae and Geotrupinae (Scarabaeidae and Geotrupidae families, respectively), comprise a significant majority of the known species of dung beetles (Halffter & Edmonds, 1982; classification *sensu* Smith, 2006). Aphodiini are distributed worldwide in every biogeographical region (G. Dellacasa et al., 2001; M. Dellacasa, 1988a, 1988b, 1988c, 1991, 1995), showing a remarkably high generic and specific diversity compared to other close groups within Aphodiinae.

Since the last Aphodiini revision by G. Dellacasa et al. (2001), some genera present in previous bibliographic sources have been later reconsidered or some other new sources have contributed new genera (Ádám, 1994; Bordat, 1999, 2003, 2009; Bordat et al., 2000; M. Dellacasa & G. Dellacasa, 2000a, 2000b, 2005; M. Dellacasa et al., 2002, 2003, 2004, 2007a, 2007b, 2008, 2010, 2011; Gordon & Skelley, 2007; Hollande & Thérond, 1999; Koçak & Kemal, 2008; Masumoto & Kiuchi, 2001; Ochi & Kawahara, 2001; Skelley, 2007; Skelley et al., 2009; Stebnicka, 2000; Tarasov, 2008; Ziani, 2002), although with limited phylogenetic support. Thus, after Cabrero-Sañudo & Lobo's (2009) paper, which considered described genera through 2005, the number of genera increased by more than 18% (36 more genera, for a total of 234), so those previous results may have changed somewhat. Moreover, although genera analyses can be used to detect genealogical relationships and to answer some biogeographical questions, internal phylogenetic relationships among Aphodiini lineages are not well identified yet (Cabrero-Sañudo, 2007; Cabrero-Sañudo & Zardoya, 2004; Forshage, 2002; Smith et al., 2006). This implies that, for the Aphodiini tribe, a species-level study may be currently more reliable than a genus-level analysis in revealing biogeographical patterns, although compilation of all the taxonomic and biogeographic data at the species level is a laborious task.

## 2. Material and methods

### 2.1 Data sets

A matrix (AphoSpes) containing information about distribution and body size of every Aphodiini species was built. Data were obtained from several bibliographic sources, including original species descriptions, which are referenced in Table 1 and Appendix 1), as well as other taxonomic and biogeographic revisions (Baraud, 1985, 1992; Cabrero-Sañudo et al., 2007; G. Dellacasa & M. Dellacasa, 2006; G. Dellacasa et al., 2001; Veiga, 1998) and databases (Bisby et al., 2011; Schoolmeesters, 2011). The body size of species was calculated as the mean between minimum and maximum lengths (mm). In 6 cases it was not possible to obtain the species body length, and the mean size for the genera was used instead. Species distribution data were included according to the presence (1) and absence (0) of species for the six worldwide biogeographical regions (Palaeotropical, Australian, Nearctic, Neotropical, Oriental and Palaearctic) proposed by Cox (2001). The area of each biogeographical region was calculated using the Idrisi Kilimanjaro GIS program (Clark Labs, 2003).

### 2.2 Descriptive examination and basic analyses

Simple descriptive statistical analyses and calculations were carried out to characterize the fauna of the different biogeographical regions, using Statistica (StatSoft Inc., 2006). For each region, several data were considered, including: (1) total number of species and endemic species, (2) number of species shared with other regions, (3) mean number of regions per species, (4) mean body size per species and endemic species, (5) mean percentage of species and endemic species from genus, and (6) mean number of species and endemic species per genus. Possible correlations and differences among the considered data were analysed using nonparametric statistical tests and the relationship between the number of species and region area was also analysed, considering several potential nonlinear fits (Fattorini, 2006; Flather, 1996; Soberón & Llorente, 1993).

### 2.3 Species co-occurrence and nestedness

To confirm the existence of possible distribution patterns of regional faunas, a co-occurrence analysis was carried out to test if there was a biogeographical signal in the data set (Connor & Simberloff, 1979; Diamond, 1975). The number of species that never co-occur in the same biogeographical region (checkerboards) was estimated and the C-score was calculated as the average number of all possible checkerboard pairs (Stone & Roberts, 1990).

In order to identify the presence of nested patterns within regional faunas (Darlington, 1957), in which species-poor regions constitute a subset of those present within richer regional faunas, three different analyses were performed (Ulrich et al., 2009). The nestedness temperature of the presence-absence species matrix was calculated by means of the temperature index, which is a descriptor of the matrix disorder (0° for a completely nested matrix, 100° for a completely random matrix) (Atmar & Patterson, 1993). The BR (Brualdi) index was

Ádám, 1983, 1994
Ahrens & Stebnicka, 1997
Akhmetova & Frolov, 2008
Allibert, 1847
Aubé, 1850
Ávila, 1986
Báguena, 1930
Ballion, 1870, 1878
Balthasar, 1929, 1931, 1932a, 1932b, 1932c, 1932d, 1932e, 1933a, 1933b, 1933c, 1935a, 1935b, 1935c, 1935d, 1936, 1937a, 1937b, 1938a, 1938b, 1938c, 1939, 1941a, 1941b, 1941c, 1942a, 1942b, 1943, 1945a, 1945b, 1945c, 1946, 1952a, 1952b, 1952c, 1955, 1960a, 1960b, 1960c, 1961a, 1961b, 1963a, 1963b, 1965a, 1965b, 1965c, 1965d, 1966, 1967a, 1967b, 1970, 1971a, 1971b, 1971c, 1973
Balthasar & Hrubant, 1960
Baraud, 1971, 1973, 1975, 1976a, 1976b, 1976c, 1977, 1978, 1980, 1981a, 1981b, 1982
Barrett, 1931, 1932
Bates, 1887, 1889, 1890
Baudi di Selve, 1870
Bedel, 1904, 1907
Berlov, 1989
Berlov, Kalinina & Nikolajev, 1989
Blackburn, 1892a, 1892b, 1895, 1897, 1904
Blanco, 1986
Boheman, 1857
Bonelli, 1812
Bordat, 1983, 1984, 1985, 1986, 1988, 1989a, 1989b, 1989c, 1990a, 1990b, 1990c, 1990d, 1992a, 1992b, 1992c, 1992d, 1993, 1994a, 1994b, 1995, 1996a, 1996b, 1997a, 1997b, 1999, 2003, 2005, 2008, 2009
Bordat, Cambefort & Bruneau de Miré, 1991
Bordat, Dellacasa, G. & Dellacasa, M. 2000
Bordat, Paulian & Pittino, 1990
Boucomont, 1928, 1929, 1930, 1932, 1936
Boucomont & Gillet, 1921
Brahm, 1790
Branco & Baraud, 1984, 1988
Brisout de Barneville, 1866
Brown, 1927, 1928a, 1928b, 1928c, 1928d, 1929a, 1929b, 1929c
Brullé, 1832
Carpaneto, 1973, 1976, 1978, 1986
Carpaneto & Piatella, 1989, 1990
Cartwright, 1939, 1944a, 1944b, 1957, 1972

Castelnau, 1840
Červenka, 1994a, 1994b, 1995, 2000, 2003, 2005
Chromy, 1993
Clément, 1928, 1958a, 1958b, 1969, 1975, 1976, 1981, 1985, 1986
Clouët des Pesruches, 1896, 1898
Cooper & Gordon, 1987
Creutzer, 1799
Csiki, 1901
Daniel, J., 1902
Daniel, K., 1900
DeGeer, 1774
Deloya & Ibáñez-Bernal, 2000
Deloya & Lobo, 1995
Deloya & McCarty, 1992
Dellacasa, G., 1982, 1983a, 1983b, 1984, 1986, 1990
Dellacasa, G. & Dellacasa, M., 1997a, 1997b, 2009
Dellacasa, G. & Johnson, 1983
Dellacasa, G. & Pittino, 1985
Dellacasa, M., 1988a, 1988b, 1988c, 1991, 1995
Dellacasa, M., Dellacasa, G. & Gordon, 2007, 2008, 2009, 2011
Dellacasa, M., Dellacasa, G. & Skelley, 2010
Dellacasa, M., Gordon & Dellacasa, G., 2003, 2007
Dellacasa, M., Dellacasa, G., Gordon & Stebnicka, 2011
Dellacasa, M., Gordon, Harpootlian, Stebnicka & Dellacasa, G., 2001
D'Orbigny, 1896
Duftschmid, 1805
Emberson & Stebnicka, 2001
Endrödi, 1955, 1956a, 1956b, 1957, 1960a, 1960b, 1961, 1964, 1967a, 1967b, 1968, 1969, 1971, 1973, 1976a, 1976b, 1977a, 1977b, 1978, 1979a, 1979b, 1980, 1982, 1983a, 1983b, 1991
Erichson, 1834, 1842, 1843, 1848
Eschscholtz, 1922, 1923
Fabricius, 1775, 1781, 1787, 1792, 1798, 1801
Fairmaire, 1849, 1871, 1881, 1882, 1883, 1886, 1888, 1892, 1893a, 1893b, 1894, 1897, 1903
Fairmaire & Coquerel, 1860
Fairmaire & Germain, 1860
Faldermann, 1835a, 1835b
Fall, 1901, 1927, 1932
Fall & Cockerell, 1907
Frivaldszky, 1879
Frolov, 1997, 2001a, 2001b, 2001c, 2001d, 2001e, 2002a, 2002b, 2006

Galante, Stebnicka & Verdú, 2003
Garnett, 1920
Gebler, 1848
Germar, 1813, 1824
Germar & Kaulfuss, 1817
Gerstaecker, 1871, 1883
Gestro, 1895
Given, 1950
Gordon, 1974, 1976, 1977a, 1977b, 2006
Gordon & Howden, 1972
Gordon & Salsbury, 1999
Gordon & Skelley, 2007
Graëlls, 1847
Gridelli, 1930
Gusakov, 1997, 2004, 2006
Gyllenhal, 1808, 1827
Haldeman, 1843, 1848
Harold, 1859, 1860, 1861, 1862a, 1862b, 1863, 1866,
    1867, 1868a, 1868b, 1868c, 1869a, 1869b, 1870,
    1871a, 1871b, 1871c, 1874a, 1874b, 1875, 1876a,
    1876b, 1877a, 1877b, 1879, 1880a, 1880b, 1881
Hatch, 1971
Herbst, 1783, 1789
Heyden, 1887
Heyden & Kraatz, 1881
Hinton, 1934a, 1934b, 1934c, 1934d, 1938
Hope, 1846
Horn, 1870, 1871, 1875, 1887
Hrubant, 1961
Hubbard, 1894
Iablokov-Khnzorian, 1972
Ilcikova & Kral, 2004
Illiger, 1798, 1803
Islas, 1945, 1955a, 1955b
Jacobson, 1897, 1911
Jacquelin du Val, 1863
Johnson, 1978
Kabakov, 1996
Kabakov & Frolov, 1996
Karsch, 1881
Käufel, 1914
Kawai, 2004
Kieseritzky, 1928
Kim, 1986, 1996
Klug, 1835, 1845, 1855
Klug & Erichson, 1859
Kolbe, 1886, 1908

Kolenati, 1846
Koshantschikov, D., 1891, 1894a, 1894b, 1894c,
    1894d
Koshantschikov, W., 1910, 1911, 1912, 1913a,
    1913b, 1916
Kral, 1995, 1996, 1997a, 1997b, 1997c, 2000, 2002
Krikken & Kaas, 1984
Kugelann, 1792
Küster, 1854
Laicharting, 1781
Landin, 1949, 1956, 1959, 1967, 1974
Lansberge, 1886
Laxmann, 1770
Lea, 1923
Lebedev, 1911, 1932
LeConte, 1850, 1857, 1858, 1872, 1878
Lewis, 1895
Linell, 1896
Linnaeus, 1758, 1761, 1767
Lucas, 1846
Mannerheim, 1843, 1849, 1853
Masumoto, 1975, 1977, 1981, 1984a, 1984b, 1988,
    1991, 1992, 1996
Masumoto & Kiuchi, 1987, 2001, 2003
Maté, 2007, 2008
Medvedev, 1928, 1968a, 1968b, 1968c
Medvedev & Dzambazish, 1977
Melsheimer, 1845
Ménétriès, 1832, 1849
Miwa, 1930
Moll, 1782
Motschulsky, 1849, 1858, 1860, 1863, 1866, 1868
Müller, G., 1940, 1941, 1942
Müller, O.F., 1776
Mulsant, 1842, 1851
Mulsant & Godart, 1879
Mulsant & Rey, 1869, 1870
Nakane, 1951, 1956, 1960, 1967, 1977, 1983
Nakane & Shirahata, 1957
Nakane & Tsukamoto, 1956
Nikolajev, 1979, 1983, 1987, 1998
Nikolajev & Frolov, 1996
Nikolajev & Puntsagdulam, 1984
Nikritin, 1969, 1971, 1973, 1979
Nikritin & Kabakov, 1979
Nomura, 1973
Nomura & Nakane, 1951

Novikov, 1996
Obenberger, 1914
Ochi, 1986, 1991
Ochi & Kawahara, 2001
Ochi & Kon, 2004, 2008
Ochi, Kawahara & Kawai, 2002
Ochi, Kawahara & Kon, 2006
Olivier, 1789
Olsoufieff, 1918
Palisot de Beauvois, 1805
Panzer, 1795, 1798, 1799, 1823
Pardo-Alcaide, 1936
Paulian, 1933, 1934, 1936a, 1936b, 1938, 1939a,
    1939b, 1939c, 1942a, 1942b, 1942c, 1945, 1954,
    1980, 1984
Paulsen, 2006a, 2006b
Penecke, 1911
Péringuey, 1901, 1908
Petrovitz, 1954, 1955, 1956, 1958a, 1958b, 1959a,
    1959b, 1961a, 1961b, 1961c, 1961d, 1961e, 1961f,
    1962a, 1962b, 1963a, 1963b, 1964, 1965a, 1965b,
    1966a, 1966b, 1967a, 1967b, 1967c, 1967d, 1968a,
    1968b, 1968c, 1969a, 1969b, 1970a, 1970b, 1970c,
    1970d, 1971a, 1971b, 1971c, 1971d, 1972a, 1972b,
    1972c, 1973a, 1973b, 1974, 1975a, 1975b, 1975c,
    1976, 1980
Peyerimhoff, 1907, 1925, 1929, 1939, 1949
Pilleri, 1953
Pittino, 1978, 1984, 1988, 1995, 1997, 2001a, 2001b,
    2004
Pittino & Ballerio, 1994
Quedenfeldt, 1884
Raffray, 1877
Rakovič, 1977, 1984, 1991
Ratcliffe, 1988
Reiche, 1847
Reiche & Saulcey, 1856
Reitter, 1887a, 1887b, 1889, 1890a, 1890b, 1891,
    1892, 1894, 1895, 1897, 1898, 1899, 1900a, 1900b,
    1901, 1904, 1906a, 1906b, 1907, 1908, 1909
Robinson, 1938, 1939, 1940, 1946, 1947
Roth, 1851
Ruiz, 1998
Sahlberg, 1908
Say, 1823, 1824, 1825, 1835

Saylor, 1935, 1940
Schaeffer, 1907
Schmidt, A., 1906, 1907a, 1907b, 1908a, 1908b,
    1908c, 1908d, 1908e, 1908f, 1908g, 1909a, 1909b,
    1909c, 1909d, 1909e, 1909f, 1909g, 1909h, 1909i,
    1910, 1911a, 1911b, 1911c, 1911d, 1911e, 1912,
    1913, 1916, 1920, 1922a, 1922b
Schmidt, W., 1840
Schönherr, 1806
Schoolmeesters & Vandenheuvel, 1999
Scopoli, 1763
Seidlitz, 1891
Semenov, 1898a, 1898b, 1903a, 1903b, 1903c
Semenov & Medvedev, 1927, 1928, 1929
Sharp, 1878
Sietti, 1903
Skelley & Gordon, 1995, 2001
Skelley & Woodruff, 1991
Solsky, 1874, 1876
Stebnicka, 1973, 1975, 1978, 1981a, 1981b, 1981c,
    1981d, 1982, 1983, 1985, 1986a, 1986b, 1988a,
    1988b, 1989, 1990, 1992, 1993, 1994, 1997, 1998
Stebnicka & Galante, 1991, 1992
Stebnicka & Howden, 1994, 1995
Stebnicka & Skelley, 2005
Sturm, 1800, 1805
Tesař, 1945, 1969
Théry, 1918, 1925
Thunberg, 1818
Van Dyke, 1918, 1928, 1933
Veiga, 1984
Villiers, 1950
Všetečka, 1939
Walker, 1858, 1871
Walter & Endrödi, 1981
Waltl, 1835
Warner & Skelley, 2006
Waterhouse, 1875
Westwood, 1839
Wickham, 1913
Wiedemann, 1823
Ziani, 2002
Zinchenko, 2003

Table 1. List of references consulted about original descriptions of Aphodiini species

also used to measure the degree of nestedness, which considers the number of discrepancies (absences or presences) that must be erased to produce a perfectly nested matrix (Brualdi & Sanderson, 1999). A third calculated index, NODF (nestedness based on overlap and decreasing fill), enabled me to differentiate between portions of overall nestedness introduced by species differences (NODFr) and site differences (NODFc) (Almeida-Neto et al., 2008).

These calculations were carried out using NODF program (Ulrich, 2010). To measure these indexes, fixed row and column constraints (Gotelli, 2000) and 1000 matrices for computing confidence limits of the null model were chosen, while the rest of parameters were those recommended by the NODF program.

## 2.4 Relationships among biogeographical traits

The relationship between the number of biogeographical regions in which each species is present and its mean body size was analyzed. The independence between these variables was tested using a chi-squared test and the shape of the relationship was analyzed by a boundary test. Simulated random matrices were built by reshuffling the observed values of each pair of variables analysed with a similar number of data points, as in the original data set. Thus, the variances and distributions of the original variables were retained, while the covariance between them was eliminated. These analyses were accomplished using the *Macroecology* module of the EcoSim package (Gotelli & Entsminger, 2011) by selecting 1,000 iterations, an asymmetrical data distribution with an upper left triangle shape, constraints defined by data, and upper right boundary tests, according to the relationship studied.

## 2.5 Similarity analyses

A sequential agglomerative, hierarchical and nested clustering (SAHN; Sneath & Sokal, 1973) was carried out for a simple examination of the faunistic similarities among biogeographical regions. This analysis takes into account information on the presence-absence of each species in the six biogeographical regions. A Jaccard similarity coefficient was calculated for regional pairs and Ward's linkage rule was applied. An analysis of similarities (ANOSIM) was used to test statistically whether there was a significant difference between the groups derived from the cluster analysis. Primer v.6 software was used in these calculations (Clarke & Gorley, 2005). In this analysis, the statistic *Global R* measures the difference of mean ranks of distance between and within groups. The maximum number of possible permutations was selected ($n = 60$).

A parsimony analysis of endemicity (PAE; Rosen, 1988; Rosen & Smith, 1988) was also carried out, which allows a grouping procedure of areas as if species were synapomorphies and regions were taxa. PAE offers an opportunity to assess relationships between different faunas in the absence of more comprehensive data. A hypothetical region containing no taxa was considered as an out-group. Winclada (Nixon, 2002) and TNT programs (Goloboff et al., 2003) were used to search for the most parsimonious tree by means of a ratchet procedure and to determine confidence levels using bootstrap and Bremer support methods.

## 2.6 Mantel tests

Simple non-partial Mantel tests (Mantel, 1967) were carried out to check possible correspondences of Aphodiini species regional composition to other characteristics of the biogeographical regions. A dissimilarity matrix for the six biogeographical regions based on Aphodiini species (APH) was built from the Aphodiini species-region matrix. Similarity among regions was based on the Jaccard index, and later changed to dissimilarity (1-similarity index). According to the indicated procedure of Cabrero-Sañudo & Lobo (2009), four other different dissimilarity matrices were obtained: two additional biological traits, Scarabaeinae genera (SCA; A.L.V. Davis et al., 2002) and mammal families (MAM; Smith, 1983); one ecological trait, Bailey ecoregions (ECO; Bailey, 1998); and one historical trait, land continuity (LC; Sanmartín & Ronquist, 2004). Mantel tests were carried out to compare the five matrices using the PASSaGE program (Rosenberg & Anderson, 2011). Simple Bonferroni $P$-values adjusted for multiple statistical tests, sequential Bonferroni values and original probabilities were jointly examined in order to interpret correlation results (Moran, 2003).

## 3. Results

### 3.1 Basic data by biogeographical region

Table 2 lists the descriptive data for each biogeographical region worldwide, based on the information obtained for a total of 2,052 Aphodiini species described up to date.

According to subtribes, these species are distributed into 1,958 Aphodiina, 45 Didactyliina and 49 Proctophanina. The Palaearctic region has the highest number of Aphodiini species (almost 41% of the total), followed by the Palaeotropical (more than 36%), Nearctic (around 15%), Oriental (7%), Neotropical (6.7%) and Australian (more than 3%) regions. The Palaearctic region is the richest for Aphodiina species, while the Palaeotropical region is the richest for Didactyliina and Proctophanina species.

A relationship between the number of species and area was observed. After examining different nonlinear procedures, a simple linear fit between the number of species and area (R = 0.81, $R^2$ = 66.06%, F = 7.79, $P$ < 0.05) was the best test. The linear relationship between the number of species and area shows that around 15.94 species are added per million square kilometres. Also, the Oriental and Palaeotropical regions have a comparatively higher number of species than predicted in relation to area, while the Palaearctic, Australian, Nearctic and Neotropical regions have comparatively lower numbers (Figure 1). The variation in the number of species among biogeographical regions differs significantly from a uniform distribution (expected species [mean] = 373.17, $\chi^2$ = 1498.70, d.f. = 5, $P$ < 0.00001), but also from the number of species expected according to the previously obtained area-species relationships (species: $\chi^2$ = 545.38, d.f. = 5, $P$ < 0.0001).

The mean percentage of species from genus varies among regions (Kruskal-Wallis ANOVA by ranks test; KW = 61.06, $P$ < 0.0001), the Palaearctic and the Palaeotropical regions holding the highest values (more than 30% of species per genus). After applying Bonferroni criteria ($P$ < 0.0033), the analyses show that the percentages are significantly different among the faunas of the Palaearctic and the Neotropical, Oriental and Australian regions, the Nearctic and the Oriental regions, and among the Palaeotropical and the Oriental and Australian regions. In fact, a significant positive relationship has been observed between the mean percentage of species from genus and the number of species (R = 0.97, $R^2$ = 93.82%, F = 60.68, $P$ = 0.001), so that those regions with higher numbers of species also have a better representation of within-genus diversity.

| | Palaeotropical | Australian | Nearctic | Neotropical | Oriental | Palaearctic |
|---|---|---|---|---|---|---|
| Approximate area (x $10^6$ km²) | 22.1 | 7.7 | 22.9 | 19.0 | 7.5 | 54.1 |
| **Species** | | | | | | |
| Aphodiini | 748 | 68 | 306 | 137 | 145 | 835 |
| Aphodiina | 681 | 59 | 302 | 126 | 145 | 827 |
| Didactyliina | 26 | 0 | 3 | 10 | 0 | 8 |
| Proctophanina | 41 | 9 | 1 | 1 | 0 | 0 |
| Number of endemic species | 721 | 57 | 222 | 60 | 80 | 751 |
| Percentage of endemic species | 96.39 | 83.82 | 72.55 | 43.80 | 55.17 | 89.94 |
| Ratio of endemic/non-endemic species | 26.70 | 5.18 | 2.64 | 0.78 | 1.23 | 8.94 |
| Endemic Aphodiina | 656 | 49 | 220 | 51 | 80 | 744 |
| Endemic Didactyliina | 25 | 0 | 2 | 9 | 0 | 7 |
| Endemic Proctophanina | 40 | 8 | 0 | 0 | 0 | 0 |
| Number of regions per species (± SD) | 1.06 ± 0.38 | 1.46 ± 1.15 | 1.32 ± 0.63 | 1.67 ± 0.80 | 1.59 ± 0.86 | 1.12 ± 0.43 |
| Body size per species (± SD) (mm) | 4.81 ± 2.19 | 5.90 ± 1.99 | 5.45 ± 1.54 | 5.19 ± 1.69 | 4.92 ± 2.07 | 5.39 ± 1.90 |
| Body size per endemic species (± SD) (mm) | 4.81 ± 2.21 | 5.95 ± 2.03 | 5.44 ± 1.53 | 4.98 ± 1.96 | 4.56 ± 1.93 | 5.39 ± 1.88 |
| Number of species per genus (± SD) | 7.79 ± 13.54 | 4.00 ± 7.93 | 4.94 ± 6.50 | 2.85 ± 3.79 | 3.54 ± 3.49 | 6.90 ± 9.64 |
| Number of species per endemic genus (± SD) | 4.08 ± 4.61 | 8.50 ± 12.77 | 3.00 ± 2.83 | 1.82 ± 1.40 | 1.00 ± 0.00 | 3.79 ± 5.40 |
| Percentage of species from genus (± SD) | 30.64 ± 42.83 | 2.90 ± 15.26 | 18.73 ± 36.99 | 12.15 ± 30.74 | 5.14 ± 16.12 | 42.20 ± 46.09 |
| Percentage of endemic species from genus (± SD) | 74.57 ± 34.75 | 33.08 ± 44.69 | 57.25 ± 46.61 | 41.87 ± 47.57 | 17.83 ± 24.50 | 79.42 ± 32.52 |

Table 2. Characteristics of worldwide regional faunas of Aphodiini according to species (unless specified, values are referred to species numbers).

The mean body size of species also varies among regions (Kruskal-Wallis ANOVA by ranks test, KW = 125.01, $P < 0.0001$), and considering Bonferroni criteria ($P < 0.0033$) the size of species present in the Palaeotropical and Oriental regions is smaller and significantly differs from species size at the Palaearctic, Nearctic, Australian and Neotropical (only for the Palaeotropical) regions. Also, the mean body size of endemic species is different among regions (Kruskal-Wallis ANOVA by rank test, KW = 125.43, $P < 0.0001$): Palaeotropical and Oriental endemic species are significantly (Bonferroni corrected) smaller than those in Nearctic, Australian and Palaearctic (only those Palaeotropical) regions.

## 3.2 Endemic species

Within Aphodiini, Proctophanina is the subtribe with the highest percentage of endemic species (98%), followed by Didactyliina (96%) and Aphodiina (92%). Both Palaearctic and

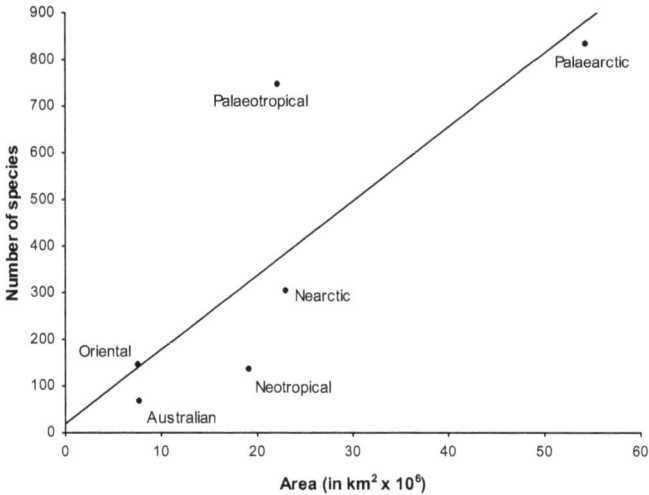

Fig. 1. Linear regression between area and richness of Aphodiini species, according to biogeographical regions ($r_s$ = 0.81; $P$ = 0.05). Number of sps. = 19.11 + 15.94•area (km$^2$ x 10$^6$)

Palaeotropical regions have the highest numbers of endemic Aphodiina species. The Palaeotropical region contains the maximum numbers of endemic species from Didactyliina and Proctophanina subtribes. Most of the Aphodiini species are only present in one biogeographical region (90%, 1,851 species), with the Palaearctic (751 species) and Palaeotropical (721 species) regions having the highest percentage (around 80%) of total endemic species (Figure 2).

The number of endemic and total species richness are related ($r_s$ = 0.996, $P$ < 0.0001), while the number of endemic and non-endemic species are not related ($r_s$ = 0.04, $P$ = 0.94). The ratio between endemic and non-endemic species is low in the Australian, Nearctic, Neotropical and Oriental regions, but is high in the Palaeartic region and very high in the Palaeotropical region (Table 2). Around 90% or more of total species in each of these two latter regions are endemic. As expected, the numbers of endemic species are significantly different from a uniform distribution among regions (expected species [mean] = 315.17, $\chi^2$ = 1746.36, d.f. = 5, $P$ < 0.0001).

Body size differs significantly between endemic and non-endemic species (Mann-Whitney U-test, U = 136.768, $n_1$ = 1,891, $n_2$ = 161, $P$ < 0.05, endemic species size = 5.15 ± 2.01 mm, non-endemic species size = 5.36 ± 1.84 mm), with non-endemic species usually larger than endemic species. Moreover, the boundary test shows that 2,048 (out of 2,052) data points have been observed within the left triangle of the relationship between species body size and number of biogeographical regions (Figure 3). However, this test confirms that the upper right-hand corner of the space is not unusually empty, and the observed number of points is not significantly lower than the number of randomly estimated points ($P$ = 0.61). There is a tendency for endemic Aphodiini species to have more variable body sizes while non-endemic species are progressively smaller as biogeographical range increases; however, this pattern is not statistically significant.

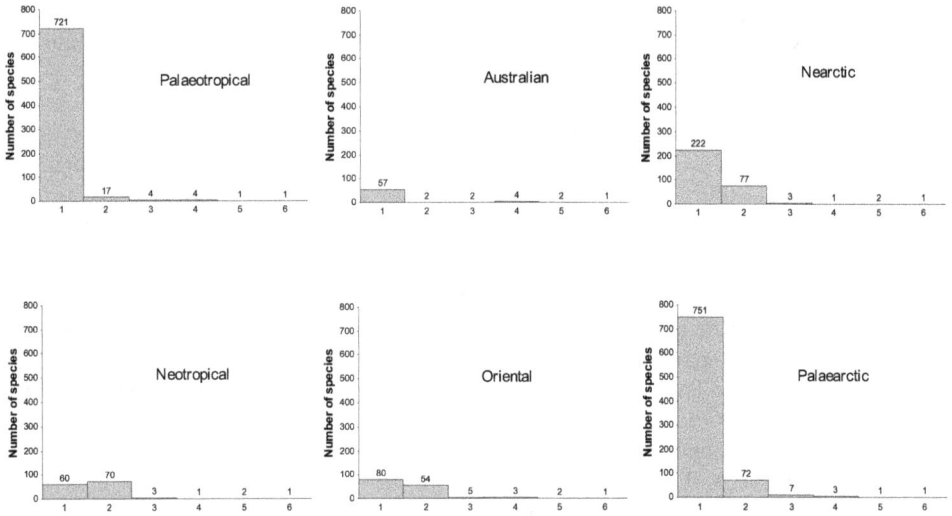

Fig. 2. Distribution of Aphodiini species in each biogeographical region, according to
categories representing the number of regions in which each species occurs

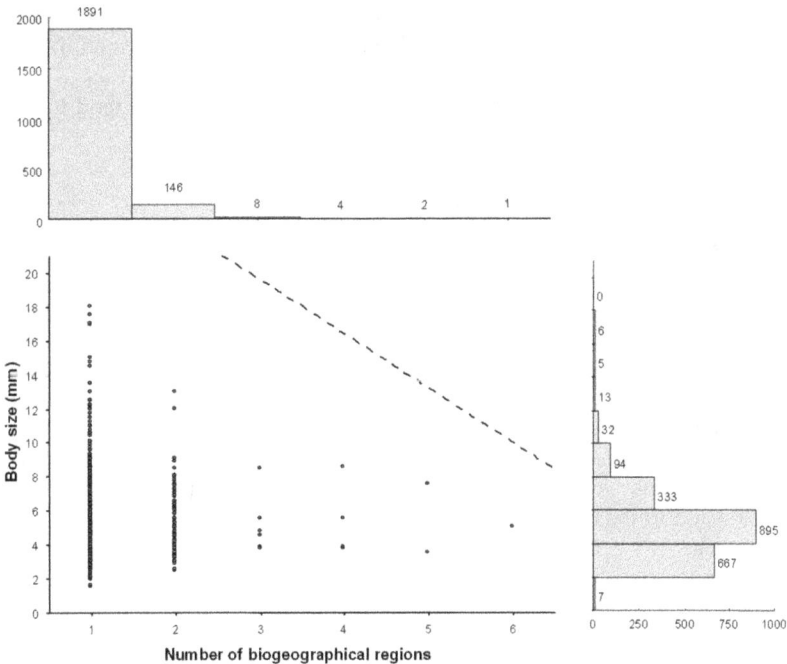

Fig. 3. Relationship between number of biogeographical regions in which Aphodiini species
are present and their body size. The broken line represents a possible constraint on this
relationship, according to a boundary test (upper-right corner of space, observed points not
significantly lower than the number of simulated points, $P = 0.61$)

The mean percentage of endemic species from genus also varies among regions (Kruskal-Wallis ANOVA by ranks test, KW = 54.66, $P < 0.0001$): again the Palaearctic and the Palaeotropical regions showed the highest numbers (more than 74% of endemic species from each genus). Taking into account Bonferroni criteria ($P < 0.0033$), Palaearctic, Palaeotropical, Nearctic and Neotropical percentages differ significantly from those of the Oriental region. A positive and significant relationship between the mean percentage of species from genus and the mean percentage of endemic species from genus also exists ($R^2$ = 0.76, $F$ = 12.51, $P < 0.05$), meaning that regions with a higher number of species from a genus will have more endemic species from that genus.

The number of species per endemic genus does not differ significantly among regions (Kruskal-Wallis ANOVA by ranks test, KW = 9.18, $P = 0.10$), but the number of species per genus differs among regions (Kruskal-Wallis ANOVA by rank test, KW = 17.64, $P < 0.005$): both Palaeotropical and Palaearctic have the highest values and differ significantly from the Neotropical region. The number of species per genus and per endemic genus is not related due to the Australian region supporting a great number of species in most of its endemic genera ($R^2$ = 0.03, $F$ = 0.12, $P > 0.05$). However, when Australian data are omitted, a significant and positive relationship is detected ($R^2$ = 0.85, $F$ = 17.57, $P < 0.05$), indicating that regions with high numbers of species per genus normally have more species in endemic genera.

The number of non-endemic Aphodiini species is not correlated to the number of species per genus ($P = 0.79$), but a marginally negative relationship exists between the number of non-endemic species and the mean number of species per endemic genus ($R^2$ = 0.56, $F$ = 5.03, $P = 0.09$). So, those regions with more widely distributed species tend to have fewer species in each of its endemic genera.

## 3.3 Non-random patterns of species distribution

The number of regions in which each species occurs (its biogeographical extent) was calculated and the distribution of the different extent categories was estimated for each biogeographical region (Fig. 2). The mean number of regions per species is reported in Table 2. The biogeographical extent of the species (Kruskal-Wallis ANOVA by ranks test, $KW$ = 391.33, $P < 0.0001$) differed significantly among biogeographical regions. Paired *post hoc* comparisons using Bonferroni criteria ($P < 0.0033$) show that biogeographical extent differs significantly among the faunas of all regions ($P < 0.001$), except those of the Palaearctic and Australian regions, the Nearctic and the Australian regions, and the Neotropical and the Oriental regions. The mean biogeographical extent is negatively and statistically significantly related to the number of species ($R^2$ = 0.83, $F$ = 18.90, $P < 0.05$) and the number of endemic species ($R^2$ = 0.86, $F$ = 24.62, $P < 0.01$) present in each region. These results suggest that regions with higher numbers of Aphodiini species usually support narrowly distributed species, and *vice versa*.

The total number of possible paired relationships among the six biogeographical regions is 63. However, only 22 different species distributions were detected (Figure 4A). For example, no species are distributed simultaneously and exclusively in the Palaeotropical and Australian regions, or in the Palaearctic and Neotropical regions. Therefore, observed frequencies do not fit the expected supposition that all possible combinations of relationships among regions are equally probable ($\chi^2$ = 165.64, d.f. = 5, $P$ < 0.0001). This result highlights the existence of concrete and non-random distribution patterns for the Aphodiini species. In fact, the co-occurrence analysis provided an observed $C$-score that was significantly higher than randomly expected ($C_{observed}$ = 0.78, $C_{expected}$ = 0.75, $P$ = 0.001). This indicates that some groups of species were repeatedly present in specific biogeographical regions, and therefore showed coincident diversity patterns (Figure 4B).

The matrix fill (number of occupied cells divided by total number of cells) given by the NODF program was only 18.18%. Nestedness temperature in the data matrix was 38.62°, which was statistically higher ($P$ < 0.001) than the estimated temperature of 34.84° (± 0.68°). Furthermore, the observed BR index was also higher than the expected value ($BR_{observed}$ = 1,360, $BR_{expected}$ = 1,280.27 ± 5.76, $P$ < 0.001). In addition, the observed NODF index was significantly lower from the estimated value ($NODF_{observed}$ = 5.37, $NODF_{expected}$ = 8.10 ± 0.15, $P$ = 0.001). Although NODFr sub-index values were similar to global NODF, observed NODFc (nestedness based exclusively on sites) was significantly higher than the expected value ($NODFc_{observed}$ = 11.24, $NODFc_{expected}$ = 8.34 ± 0.38, $P$ = 0.001). These results suggest that a genuine pattern of nestedness exists among regional faunas, although this pattern is concealed as a consequence of non-nested endemic species.

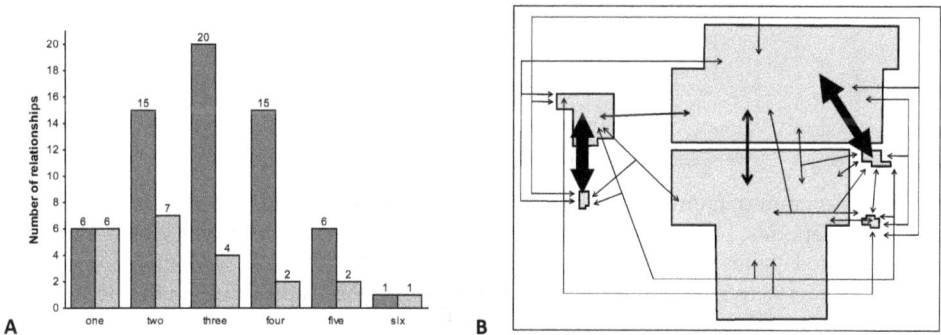

A                                                      B

Fig. 4. Patterns of distribution for Aphodiini species. (A) Expected (left) vs. observed (right) numbers of combinations of shared regions according to genera distributions; (B) Representation of the different regional relationships for Aphodiini species. Region size is proportional to the number of endemic Aphodiini species, whereas arrow and box widths are proportional to the number of species with similar patterns for each relationship

### 3.4 Similarity among biogeographical regions

The dendrogram of faunistic similarity based on Aphodiini species revealed that the Nearctic and the Neotropical regions are most similar (Figure 5A). The Palaearctic and the Oriental regions also are similar, and together are closer to the Australian region, and narrowly followed by the Palaeotropical. The ANOSIM test shows that the Neotropical-Nearctic and the Palaearctic-Oriental pairs are statistically significant. According to this analysis, the most probable similarity configuration is made up of the following three groups (Global $R$ = 0.68, $P$ = 0.08): i) Nearctic-Neotropical; ii) Palaearctic-Oriental-Australian; and, iii) Palaeotropical.

The dendrogram of faunistic similarity based on parsimony and shared species showed a unique tree (length = 2085, consistency index = 0.98, retention index = 0.80, autapomorphies included; Figure 5B). The Palaearctic-Oriental and Nearctic-Neotropical region pairs again are observed. The Palaearctic-Oriental clade is closer to the Palaeotropical region, and together are joined in a clade that is closer to the Australian region. Bootstrap validation confirmed the Palaearctic-Oriental (100%), the Palaearctic-Oriental-Palaeotropical-Australian (76%), and the Nearctic-Neotropical clades (100%). Bremer support also showed highest number of steps (>30) for the Palaearctic-Oriental and the Nearctic-Neotropical clades. Although the Palaeotropical region appears to be closer to the Palaearctic-Oriental clade than the Australian region; however, there was insufficient data for bootstrapping or Bremer supports to confirm similarity.

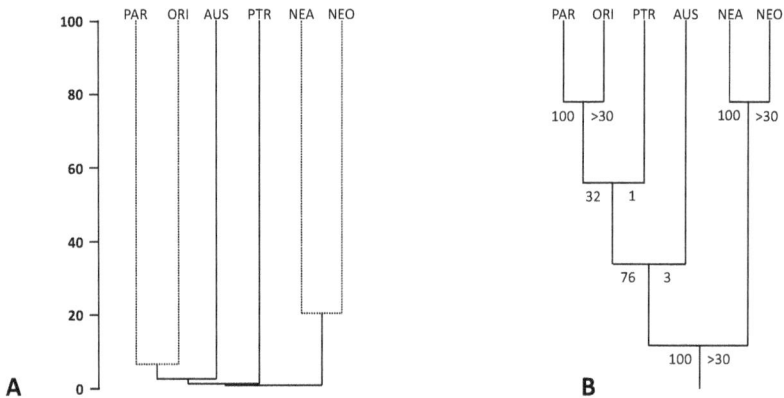

Fig. 5. Dendrogram of faunistic similarity among worldwide regions based on distributions of Aphodiini species. (A) Sequential agglomerative, hierarchical, and nested clustering analysis. Dotted clades are well supported, according to the ANOSIM test; (B) Parsimony analysis of endemicity. Left values refer to bootstrap support; right values to Bremer support. Regions: PTR, Palaeotropical; AUS, Australian; NEA, Nearctic; NEO, Neotropical; ORI, Oriental; PAR, Palaearctic

## 3.5 Independence of regional traits

The dissimilarity in the Aphodiini species composition among biogeographical regions was significantly and positively related to the genera composition of the Scarabaeinae dung beetles (Mantel test correlation coefficient ($\rho$) = 70.13%, $P < 0.005$; when both simple and sequential Bonferroni corrections are considered). A positive correlation between Aphodiini species composition and the geological time of separation among the biogeographical regions approaches significance ($\rho$ = 43.83%, $P$ = 0.059).

As showed in Cabrero-Sañudo & Lobo (2009), regional distribution of Scarabaeinae was positively related with mammal composition ($\rho$ = 68.45%, $P$ = 0.005) and the geological time of separation among biogeographical regions ($\rho$ = 76.68%, $P$ = 0.002; when both simple and sequential Bonferroni corrections are considered). Regional mammal distribution also was positively related with the geological time of separation among biogeographical regions ($\rho$ = 62.78%, $P$ = 0.006; when a sequential Bonferroni correction was used). The regional dissimilarity in ecoregions did not show any significant correlation with any biological or historical trait.

## 4. Discussion

### 4.1 Existence of non-random distribution patterns

These results imply that the faunistic composition of Aphodiini species present in different biogeographical regions follows a geographically structured pattern. Co-occurrence and nestedness analyses suggest that there is a reliable relationship among Aphodiini faunas, although each biogeographical region is singular on its own and holds more endemic species than shared ones (except for the Neotropical region, with most of its species also distributed in the Nearctic region, particularly at the Mexican Transition Zone). The most frequent distribution patterns of shared species are related to the 'Old World faunas' (Palaearctic, Oriental, Palaeotropical and Australian regions) or the 'New World faunas' (Nearctic and Neotropical regions), being these two groups of regions also supported by the similarity analyses. Mantel test revealed a nearly significant relationship between historical and biological traits, so long-term land continuity and proximity may play a unifying role in regional faunas. In fact, when higher taxa are tested (for example, genera, as in Cabrero-Sañudo & Lobo, 2009), this relationship turns significant.

The Palaearctic and Palaeotropical regions have the highest numbers of species and endemic species. These two regions have a considerable singular faunistic composition (for example, presence of more endemic lineages or genera; Cabrero-Sañudo & Lobo, 2009), or many more species than expected (as in the Palaeotropical region). This result may be partly due to the area or the environmental heterogeneity of these regions, although results show that the compositional differences do not seem to be related to ecoregional dissimilarity. This points to historical reasons that could have influenced current distribution patterns.

Taking into account all these results, two main types of regions for Aphodiini may be considered: macroevolutionary sources – those with a long history as producers or distribution centres for many lineages – and sink regions, which are those with colonization processes and recent radiations as important shapers of current faunas (Goldberg et al., 2005). Within the first type, both the Palaearctic and Palaeotropical regions may be included, although the Nearctic region to some extent and the nexus between the Nearctic and the Neotropical regions (the Mexican Transition Zone) should not be neglected. On the contrary, the Australian, Neotropical and Oriental regions, although with singular faunas, could be included within the second type, based on the lower numbers of species, endemic species and lineages (Cabrero-Sañudo & Lobo, 2009).

### 4.2 Diversification and distribution range

Regions with a lower Aphodiini diversity usually have widely distributed species. This may be due to the fact that the greatest percentage of regional Aphodiini faunas consists of endemic species, so those regions with fewer species will have a proportionately greater representation of widespread species. Widely distributed species are able to colonize more biogeographical regions because they usually have greater environmental tolerances. This could be interpreted as a variation of Rapoport's rule (the size of species distributional ranges increases with latitude; Rapoport, 1975), as one explanation for this rule is that seasonal variability fosters a greater climatic tolerance, and therefore wider latitudinal ranges (Letcher & Harvey, 1994; Stevens, 1996).

Moreover, the phylogenetic relationships among species in a biogeographic region are likely higher within the region than with species from other regions. This has been indicated by the fact that regions with large numbers of species carry a greater number of species per genus and a higher percentage of species per genus. Thus, those regions with favourable environmental conditions to accommodate a greater number of Aphodiini species also could have functioned as speciation centers (macroevolutionary source regions, as the Palaearctic and Palaeotropical).

It has been pointed that the environmental tolerance of species may be related to ecophysiological adaptations at higher taxonomic levels (Hawkins et al., 2006; Ricklefs, 2006). Although the geographic distribution of a lineage (genus) could then be related to its tolerance at a large scale, favourability and diversity of environmental conditions could promote an ecological diversification (spatial, seasonal, altitudinal, feeding, etc.) even among related species (Del Rey & Lobo, 2006; Finn & Gittings, 2003; Gittings & Giller, 1997; Hanski, 1991b). This ecological diversification would act on endemic or non-endemic lineages equally, as it has been noted that the number of species per genus within a region and the number of species per endemic genus are positively related.

### 4.3 Endemic and non-endemic species

The total species richness and the number of endemic species are related; this is probably due to a greater percentage of endemic species *versus* non-endemic ones. Actually there is a significant difference in regional species numbers between endemic and non-endemic species. A total of 1,891 out of the 2,052 Aphodiini species (92%) are endemic to a single region; less than one tenth of species are shared among regions. Widely distributed species usually correspond to those of a known wide ecological spectrum and opportunistic nature. For example, the three most widely distributed species, *Calamosternus granarius* (Linnaeus, 1767) (present in all biogeographical regions), and *Aphodius fimetarius* (Linnaeus, 1758) and *Labarrus pseudolividus* (Balthasar, 1941) (present in five biogeographical regions), show a higher climatic tolerance, are not strictly coprophagous, and can alternatively behave as saprophagous, fungivorous or cleptoparasites (Cabrero-Sañudo et al., 2010; Veiga, 1998).

The correlation between total numbers of endemic and non-endemic species is almost nonexistent, probably because world biogeographical regions have partially independent evolutionary histories. In fact, regions with the least number of non-endemic species are the Palaeotropical and the Australian (27 and 11 non-endemic species, respectively), while the other regions have from 65 to 84 non-endemic species registered. This may indicate that recent continental isolation of these two regions could have prevented introduction of many broadly distributed species. Moreover, regions with a lower number of non-endemic Aphodiini tend to have more species per endemic genus. This fact could point to competition between widespread, better adapted and more competitive species and endemic, more specialized species (although see Finn & Gittings, 2003) or certain speciation process that could occur in isolated territories (Gillespie & Roderick, 2002).

Another observed relationship is the positive relationship between the percentage of species from a genus and the percentage of endemic species from a genus, so regions with a greater number of species per genus do so at the expense of having a greater number of endemic species for that lineage. This again relates to the role as species generators of those regions that have showed favourable conditions for Aphodiini along the time.

## 4.4 Body size

Among Aphodiini genera, body size is a trait that shows no significant pattern with respect to geographical distribution, although it may be noted that some large genera are endemic to a single region (Cabrero-Sañudo & Lobo, 2009). However, species mean body size does significantly vary among different biogeographic regions. Thus, both Aphodiini species and endemic species from Palaeotropical and Oriental regions are smaller than those in other regions, especially the Australian, Nearctic or Palaearctic regions. Those species distributed in areas of temperate or cold climates (Nearctic or Palaearctic) may be larger, as a variation of Bergmann's rule for endotherm animals (Bergmann, 1847) which has also been observed for some insects (Blanckenhorn & Demont, 2004). Yet a large size also could be a consequence of the presence of ancient lineages (a possible variation of Cope's rule: taxa increase in body size over evolutionary time; Cope, 1887) or insularity (Gould & MacFadden, 2004). The first of these alternatives may occur among Palaearctic Aphodiini, as it has also been observed for Scarabaeinae in the Palaeotropical region (A.L.V. Davis et al., 2002). On the other hand, 'island' gigantism may have occurred among Australian Aphodiini, as indicated by the giant wetas of New Zealand, Madagascan cockroaches and millipedes, deep sea gastropods (McClain et al., 2006), and island vertebrates (Lomolino, 2005; but also see Meiri et al., 2008).

Nevertheless, non-endemic species are generally significantly larger than endemic ones, although endemic Aphodiini species have a greater diversity of body sizes. Aphodiini are not very good dispersers (Roslin, 2000; Roslin & Koivunen, 2001), so larger species would probably have a greater advantage to move and occupy new territories than smaller ones. In other species groups (for example, mammals or birds), larger species tend to occupy broader distributional areas (Brown, 1995).

## 4.5 Regional characteristics

The Palaearctic region presents the highest number of total and endemic Aphodiini species; however, while the Palaearctic has the highest number of species and endemics of Aphodiina, it does not support any Proctophanina. Together with the Palaeotropical region, it also shows the highest ratio of endemic/non-endemic species, indicating the great importance of endemics to the composition of its fauna. In fact, the species inhabiting the Palaearctic show the second lowest mean number of regions per species compared with those of other regions. The mean percentage of species from genus is the highest, whereas the number of species per genus is the second highest with respect to the other regions; this may be pointing to a possibly greater speciation rate for Palaearctic lineages. Also, the mean percentage of endemic species from genus is the highest in relation to the other regions. Aphodiini are usually the most common species group in dung beetle communities in the Palaearctic (Hanski, 1991b). In this region, they display the highest diversity of lineages and endemic genera (Cabrero-Sañudo & Lobo, 2009) and the greatest abundance of individuals in northern communities compared with other scarabaeid taxa (Hanski, 1991b; Gittings & Giller, 1997, 1998; Roslin & Koivunen, 2001; Finn & Gittings, 2003). The best studied area in the world has traditionally been the Western Palaearctic, in which Aphodiini taxonomy has attracted much attention, although many Aphodiini species may still remain undiscovered (Cabrero-Sañudo & Lobo, 2003). The only phylogenetic studies conducted on Aphodiini evolution were based on Iberian species within the Palaearctic (Cabrero-Sañudo & Zardoya, 2004; Cabrero-Sañudo, 2007) and showed that most of the earlier branches of Aphodiini are of Palaearctic or Holarctic in distribution. While this may support the hypothesis of a Laurasian origin of Aphodiini, the geographical and taxonomic scope of those studies was limited. Nevertheless, the oldest known fossil evidence for Aphodiini is from the Paleocene, with many other fossils from subsequent epochs (Figure 6; Krell, 2007).

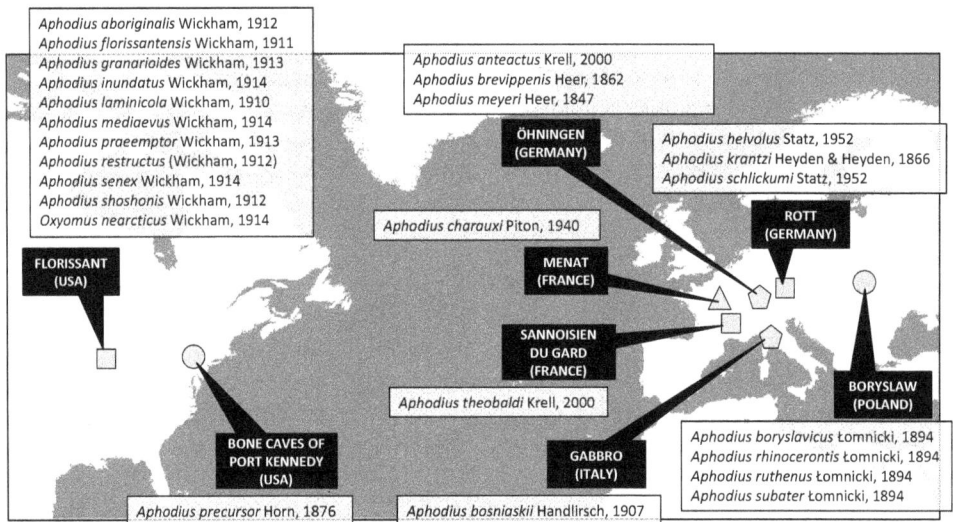

Fig. 6. Sites where extinct Aphodiini fossil species have been found, according to Krell (2007). For each site, a list of described species is provided. In most cases, the genus of the species is *Aphodius* Illiger, 1758, as they have not been yet classified into suitable genera. Geologic epoch of lagerstätten represented by geometric figures: i) triangle, Paleocene; ii) square, Eocene; iii) pentagon, Oligocene; and, iv) circle, Pleistocene.

The Palaeotropical region also shows important numbers of total and endemic species of Aphodiini. Its endemic/non-endemic species ratio is the highest and almost three times that of the next highest region, the Palaearctic. Thus, the mean number of regions per species is the lowest compared with other regions. Aphodiina is well represented, with both second highest numbers of species and endemic species. Also, both Didactyliina and Proctophanina show the highest species and endemic species richness in this region. Although Palaeotropical dung beetle communities tend to be dominated by Scarabaeinae individuals (Cambefort, 1991; A.L.V. Davis et al., 2002; Doube, 1991), some studies show that the abundance of Aphodiini in some localities can sometimes be very high, or even the highest (Bernon, 1981; Krell et al., 2003; D. Rougon & C. Rougon, 1991). However, the mean body size of Palaeotropical species is the lowest in relation to other regions, and the mean body size of endemic species is also the second lowest. The Palaeotropical region holds the second highest mean percentage of species from genus, the second highest mean percentage of endemic species from genus, and the highest mean number of species per genus. In addition, the relatively recent isolation of the Palaeotropical region with respect to other regions could explain its having the second lowest number of non-endemic species and the largest number of species per endemic genus. So, together with the Palaearctic region, the Palaeotropical region can be considered as a macroevolutionary source region, possibly acting as a refuge and/or a recent diversification centre for Aphodiini species. However, there is as yet no record for Tertiary or Quaternary Aphodiini fossils within the Palaeotropical region (Krell, 2007), possibly a consequence of limited preservation, prospecting efforts, and/or recent Aphodiini species diversification.

The Oriental region is the fourth most prominent in terms of number of species, all of which are from the subtribe Aphodiina. Although this region hosts a few more species than expected according to its area, it is the third lowest in terms of endemic species, has the second lowest endemic/non-endemic species ratio, and has one of the greatest mean species distribution ranges, sharing many species with the Palaearctic region (61 out of 65 non-endemic species). This region has the second lowest mean body size per species, the lowest mean body size of endemic species, the second lowest percentage of species from genus, the lowest mean percentage of endemic species from genus, the second lowest number of species per genus, and the lowest number of species per endemic genus. The Oriental region may have had less environmentally favourable conditions for Aphodiini species, so that widely distributed species would have been proportionately more successful. This suggests the Oriental region to be mainly a macroevolutionary sink for Aphodiini species. Studies on dung beetle communities carried out in the Oriental region frequently omit results for Aphodiini or show that they represent a small proportion of local Scarabaeidae richness (c. 4-10% of species; A.J. Davis, 2000; A.J. Davis et al., 2001; Hanski & Krikken, 1991; Shahabuddin et al., 2005). However, most of those studies were conducted on island communities, and there is a conspicuous lack of study of continental communities, particularly considering the likely species richness and abundances for Aphodiini. As for the Palaeotropical region, there is as yet no fossil evidence for Aphodiini in the Oriental region (Krell, 2007).

The Nearctic region has the highest number of Aphodiini species and endemic species in the New World, hosting representatives from the three tribes, although only Aphodiina and Didactyliina contain endemic species. In relation to the other regions, the Nearctic has the third highest number of species and endemics. Moreover, the Nearctic region supports the third highest mean percentage of species from genus, endemic species from genus, and number of species per genus. These facts define the Nearctic as a mainly macroevolutionary source region for Aphodiini. Several Nearctic species are shared with the Neotropical region (77 out of 84 non-endemic species), and many of them are endemic to the Mexican Transition Zone, although with a Nearctic or Holarctic origin (Cabrero-Sañudo et al., 2007, 2010). Aphodiini are usually the dominant species group in northern dung beetle communities in the Nearctic region (Lobo, 2000, and references therein). Also, Nearctic region total and endemic species have the second greatest mean body size per species, probably due to colder climates or presence of older lineages. This region also has displayed an ancient presence of Aphodiini, with the first Nearctic fossil records dating from the Oligocene, a little more recent than those of the Palaearctic (Krell, 2007).

The Neotropical region holds the second lowest richness for species and endemic species, and holds the least number of species on the basis of the region's land area. Although it hosts species from the three different subtribes, there are only endemic representatives from the Aphodiina and Didactyliina. It shares several Aphodiini species with the Nearctic region (all its non-endemic species), mainly endemic from the Mexican Transition Zone (Cabrero-Sañudo et al., 2007, 2010). Its ratio of endemic/non-endemic genera and the mean number of species per genus are the lowest, and the mean number of species per endemic genus is the second lowest. These facts point to the Neotropical having probably acted as a macroevolutionary sink region. Few studies on Neotropical dung beetle communities have taken Aphodiini into consideration. Although very few Aphodiini species are represented in those communities, they can be very abundant (Andresen, 2002). No Aphodiini fossil records have been found for this region.

The Australian region shows the lowest numbers of Aphodiini species and endemic species, with only two subtribes (Aphodiina and Proctophanina) represented in the region. The mean percentage of species from genus is also the lowest, whereas the mean percentage of endemic species from genus is the second lowest, after the Oriental region. However, the number of species per endemic genus is the highest, in relation to other regions. Due to its isolation, the Australian region seems to have received occasional representatives (widely distributed, generalist species) of different lineages over time. Some of these colonizing species likely led to the emergence of endemic genera, which then diversified. Consequently, the Australian region appears to have acted as a macroevolutionary sink with regard to Aphodiini lineages, but has served as a source region with regard to species. Australian species also display the largest mean body size per species and per endemic species compared with other regions, indicating a possible island gigantism or the presence of ancient lineages. With regard to abundance and species richness, Scarabaeinae dominate northern Australian dung beetle communities, while southern Australian communities are dominated by endemic species of Aphodiini (Doube et al., 1991; Steinbauer & Weir, 2007). There is as yet no fossil record for Aphodiini in this region, due to causes similar to those of the Palaeotropical, Oriental and Neotropical regions.

## 4.6 A synthesis of Aphodiini evolution and historical biogeography

The main radiation of Scarabaeoidea dates from the Mesozoic and Cenozoic ages (A.L.V. Davis et al., 2002; Krell, 2000; Scholtz & Chown, 1995), but Aphodiini probably did not separate from the Scarabaeidae main lineage and from the other Aphodiinae subfamilies until the Jurassic or Cretaceous (Krell, 2000). The radiation of coprophagous beetles presumably happened as dung from vertebrates (dinosaurs and/or small mammals) increased (Arillo & Ortuño, 2008; A.L.V. Davis, 1990, 2002; Halffter & Matthews, 1966; Jeannel, 1942; Philips, 2011). The first Aphodiini probably developed before or around the early Jurassic age (200-170 Ma), when most continents were joined in Pangea. Most Aphodiini have temperate to temperate-cold and/or subalpine preferences, so they may have arisen at the northern territories of contemporary Eurasia (Cabrero-Sañudo & Zardoya, 2004), where the climate was cool with temperate conditions, compared to the rest of the world, which was very arid and hot (Scotese, 2003). Some of the first lineages of Aphodiini could have spread later to other Pangean southern territories before their break-up. After the fragmentation of Pangea (middle Jurassic to early Cretaceous; 160-130 Ma), most

Aphodiini would have remained within Laurasia, although it is possible that a few Aphodiini lineages survived on Gondwanan continents, as shown by some genera and species distributions.

Due to the extinction of dinosaurs (K/T boundary, late Cretaceous, c. 66 Ma) dung from mammals became increasingly more common, providing a resource that could be consumed gradually by new Aphodiini taxa. Eurasia and North America approached each other and were intermittently connected (from late Cretaceous to Eocene periods, 66-38 Ma; Scotese, 2003), helping explain why the first Aphodiini fossils registered in North America date from these ages. The faunas from the rest of the regions probably had little contact in these periods.

During the Miocene (26-12 Ma), Eurasia and Africa collided and a secondary radiation of Aphodiini (similar to that of Scarabaeinae; A.L.V. Davis & Scholtz, 2001; A.L.V. Davis et al., 2002) may have resulted as a consequence of mixing faunas and the establishment of new dispersal routes between the two continents (Potts & Behrensmeyer, 1993). This interchange culminated during the late Miocene period (12 Ma), when the Indian peninsula collided against Eurasia and several Aphodiini taxa probably colonized that territory. Also, prairies and savannas became more common as aridity and climate cooling increased (Cambefort, 1991b; Scotese, 2003), and the radiation of Artiodactyla (Bovini) (30+ Ma; Cumming, 1982; Silva & Downing, 1995) brought new high-quality soft-fibrous droppings. Mantel tests results highlight the relationships between Aphodiini and Scarabaeinae, mammal faunas, and land connectivity, all of which are probably related to these events.

An increase in prairie lands also occurred in North America, and the Beringian land bridges permitted the passage of Bovini and other mammals from Eurasia to the Nearctic region during the Miocene and Pliocene (Potts & Behrensmeyer, 1993). However, the most important American event was the closure of the Isthmus of Panama during the late Miocene and Pliocene (13-7 Ma; Coates et al., 2004), which caused the Great American Interchange of species (Webb, 1985), and also range expansions of many dung beetle taxa, such as the Scarabaeinae (A.L.V. Davis & Scholtz, 2001). A certain number of current Neotropical Aphodiini species are the likely survivors of a specialized and hardly diversified fauna previously present at this region.

The principal characteristic of the Australian continent has been its prolonged isolation from the rest of the regions, and the lack of placental mammals until the Pleistocene (2.5 Ma-10 ka; Cox, 2000). Hence, older Australian Aphodiini lineages would have been adapted to exploiting excrements from marsupials. Land connections during the Pleistocene between Eurasia and Australia probably allowed the dispersal of mammals towards the latter region, together with a number of Aphodiini species associated with them. The mass extinction of monotremes and marsupials (Murray, 1984) may have caused the extinction of some endemic Aphodiini, but newcomer Aphodiini species may have proliferated with the advent of soft-fibrous droppings.

Middle-late Pleistocene (420-18 ka) glacial-interglacial cycles (Imbrie et al., 1993), could have played an important role as modifying factors of Aphodiini distributions and diversity, especially in the Holarctic regions (Hanski, 1991b). Fossil evidence (Coope, 1978, 1990; Coope & Angus, 1975; Lindroth, 1948) confirms that insects underwent range shifts during the Pleistocene in relation to changes in climate and vegetation. Thus, in the Northern

Hemisphere, Aphodiini should have shifted their distribution ranges southwards during glacial periods, and northwards during interglacials (Cabrero-Sañudo & Lobo, 2006). The mixing of fauna should then have been more frequent among Old World regions and between New World regions, as the ice sheets interrupted the interchanges across Beringia. The Oriental region probably was isolated by the glaciated ice sheets of mountain ranges, and therefore may not have served as a refuge for northern Aphodiini lineages. This possibly favoured the role of the Palaeotropical region as a refuge and recent diversification centre for a great number of Old World Aphodiini lineages, helping explain its high levels of endemism.

During the Holocene (10 ka to the present), cattle, horses and other domesticated animals, as well as human movements, have contributed particularly to the dispersion of Aphodiini. Human-induced changes have been especially important in the Western Palaearctic during the late Quaternary (Birks, 1986; Hanski, 1991b) and over the past few centuries for the other regions (P.A. Delcourt & H.R. Delcourt, 1987; Doube et al., 1991; Kohlmann, 1991; Mirol et al., 2003). Recent changes in soil uses, modifications through livestock and agricultural practices, chemical contamination and urban development have negatively influenced Aphodiini diversity, distribution and populations (Barbero et al., 1999; Gittins & Giller, 1999; Gittings et al., 1994; Hutton & Giller, 2003; Lobo et al., 1997, 2001, 2006; Lumaret, 1986, 1990; Lumaret & Kirk, 1911; Lumaret & Martínez, 2005; Lumaret et al., 1993; Romero-Samper & Lobo, 2006; Roslin & Koivunen, 2001).

## 5. Conclusions

Even when phylogenetic information is available, the lack of comprehensive fossil information and shifts in the distribution of species makes it extremely difficult to disentangle past dispersal patterns, complicating the formulation of reliable hypotheses that allow explanation of current distribution by means of past events (Gaston & Blackburn, 1996; Losos & Glor, 2003; Pulquério & Nichols, 2007; Thomas et al., 2006). One of the main challenges for biogeographers continues to be the formulation of reliable hypotheses about the underlying historical processes based on present-day biogeographical data. In this chapter, through statistical tests and a simple methodology, I have attempted to show that it is possible to identify signatures of the processes from which current distributional patterns originated, and to elucidate a likely past biogeography of the Aphodiini.

In relation to Aphodiini, it has been suggested that Palaeotropical and Palaeartic regions, together with the Nearctic, have been primary diversification centres after the break-up of Pangea. Consequently, these three regions may have acted jointly as macroevolutionary source regions in different times, also sustaining migration processes and extinctions that obscure linkage between past events and present-day distributions. Future phylogenetic data are needed to more completely resolve taxonomic issues and confirm internal relationships among the Aphodiini lineages. Such results also will help to confirm or to reject the hypotheses herein presented.

## 6. Acknowledgment

This chapter was supported by a MCI project (CGL 2010-16944) and an UCM-BSCH project (GR35/10-a). I thank Fernando Zagury Vaz de Mello and Paul Schoolmeesters, who indirectly provided most of the Aphodiini description references by means of the bibliography databases they have compiled. I wish to dedicate this chapter to Julieta, my newborn daughter and future 'bug lover' (I hope so…).

# 7. References

Ádám, L. (1994). A check-list of the Hungarian Scarabaeoidea with the description of ten new taxa. *Folia Entomologica Hungarica*, Vol. 55, pp. 5-17, ISSN 0373-9465

Almeida-Neto, M., Guimarães, P., Guimarães, Jr.P.R., Loyola, R.D. & Ulrich, W. (2008). A consistent metric for nestedness analysis in ecological systems: reconciling concept and quantification. *Oikos*, Vol. 117, pp. 1227-1239, ISSN 0030-1299

Andresen, E. (2002). Dung beetles in a Central Amazonian rainforest and their ecological role as secondary seed dispersers. *Ecological Entomology*, Vol. 27, pp. 257-270, ISSN 0307-6946

Arillo, A. & Ortuño, V.M. (2008). Did dinosaurs have any relation with dung-beetles (the origin of coprophagy). *Journal of Natural History*, Vol. 42, pp. 1405-1408, ISSN 0022-2933

Atmar, W. & Patterson, B.D. (1993). The measure of order and disorder in the distribution of species in fragmented habitat. *Oecologia*, Vol. 96, pp. 373-382, ISSN 0029-8549

Bailey, R.G. (1998). *Ecoregions: the ecosystem geography of the oceans and continents*. Springer Verlag, ISBN 0387983112, New York, NY

Baraud, J. (1985). *Coleópteres Scarabaeoidea. Faune du Nord de l'Afrique, du Maroc au Sinaï*. Encyclopédie Entomologique 46, Lechevalier, ISBN 272050517X, Paris, France

Baraud, J., (1992). *Coléoptères Scarabaeoidea d'Europe. Faune de France 78*. Féderation Française des Sociétés de Sciences Naturelles, ISBN 2903052123, Paris, France

Barbero, E.; Palestrini, C. & Rolando, A. (1999). Dung beetle conservation: effects of habitat and resource selection (Coleoptera: Scarabaeoidea). *Journal of Insect Conservation*, Vol. 3, pp. 75-84, ISSN 1366-638X

Bergmann, C. (1847). Über die Verhältnisse der Wärmeökonomie der Thiere zu ihrer Grösse. Göttinger Studien, Vol. 3, pp. 595-708

Bernon, G. (1981). *Species abundance and diversity of the Coleoptera component of a South African cow dung community, and associated insect predators*. PhD Thesis, Bowling Green State University, Bowling Green, OH

Birks, H.J.B. (1986). Late-Quaternary biotic changes in terrestrial and lacustrine environments, with particular reference to north-west Europe. In: *Handbook of Holocene palaeoecology and palaeohydrology*, B.E. Berglund (Ed.), 39-56, Wiley, ISBN 0471906913, New York, NY

Bisby, F.A.; Roskov, Y.R.; Orrell, T.M.; Nicolson, D.; Paglinawan, L.E.; Bailly, N.; Kirk, P.M.; Bourgoin, T.; Baillargeon, G. & Ouvrard, D. (2011). Species 2000 & ITIS Catalogue of Life, 26th July 2011. In: *Species 2000*, 15.08.2011, Available from: www.catalogueoflife.org/col/

Blanckenhorn, W.U. & Demont, M. (2004). Bergmann and converse Bergmann latitudinal clines in arthropods: two ends of a continuum? Integrative and Comparative Biology, Vol. 44, pp. 413-424, ISSN 1540-7063

Bordat, P. (1999) *Ammoecioides*, nouveau genre et ses espèces (Coleoptera, Scarabaeoidea, Aphodiidae). *Nouvelle Revue d'Entomologie*, Vol. 16, pp. 161-182, ISSN 0374-9797

Bordat, P. (2003). *Haroldaphodius* et *Euhemicyclium* nouveaux genres d'Aphodiinae et leurs espèces. *Nouvelle Revue d'Entomologie*, Vol. 19, pp. 235-248, ISSN 0374-9797

Bordat, P. (2009). Nouveaux taxons afrotropicaux dans la famille Aphodiidae. *Nouvelle Revue d'Entomologie*, Vol. 25, pp. 123-144, ISSN 0374-9797

Bordat, P.; Dellacasa, G. & Dellacasa, M. (2000). Revision du genre *Macroretrus* et description d'une nouvelle espece. *Macroretroides lumareti n.g.,n.sp. Nouvelle Revue d'Entomologie*, Vol. 17, pp. 107-122, ISSN 0374-9797

Bornemissza, G.F. (1976). The Australian Dung Beetle Project 1965-1975. *Australian Meat Research Committee Review*, Vol. 30, pp. 1-30, ISSN 0311-0842

Brown, J.H. (1995). *Macroecology*. University of Chicago Press, ISBN 0-226-07614-8, Chicago, IL

Brualdi, R.A. & Sanderson, J.G. (1999). Nested species subsets, gaps, and discrepancy. *Oecologia*, Vol. 119, pp. 256-264, ISSN 0029-8549

Cabrero-Sañudo, F.J. (2007). The phylogeny of Iberian Aphodiini species (Coleoptera, Scarabaeoidea, Scarabaeidae, Aphodiinae) based on morphology. *Systematic Entomology*, Vol. 32, pp. 156-175, ISSN 0307-6970

Cabrero-Sañudo, F.J. & Lobo, J.M. (2003). Estimating the number of species not yet described and their characteristics: the case of Western Palaearctic dung beetle species (Coleoptera, Scarabaeoidea). *Biodiversity and Conservation*, Vol. 12, pp. 147-166, ISSN 0960-3115

Cabrero-Sañudo, F.J. & Lobo, J.M. (2006). Determinant variables of Iberian Peninsula Aphodiinae diversity (Coleoptera, Scarabaeoidea, Aphodiidae). *Journal of Biogeography*, Vol. 33, pp. 1021-1043, ISSN 0305-0270

Cabrero-Sañudo, F.J. & Lobo, J.M. (2009). Biogeography of Aphodiinae dung beetles based on the regional composition and distribution patterns of genera. *Journal of Biogeography*, Vol. 36, pp. 1474-1492, ISSN 0305-0270

Cabrero-Sañudo, F.J. & Zardoya, R. (2004). Phylogenetic relationships of Iberian Aphodiini (Coleoptera: Scarabaeidae) based on morphological and molecular data. *Molecular Phylogenetics and Evolution*, Vol. 31, pp. 1084-1100, ISSN 1055-7903

Cabrero-Sañudo, F.J.; Dellacasa, M.; Martínez, I., & Dellacasa, G. (2007). Estado actual de conocimiento de los Aphodiinae mexicanos (Coleoptera, Scarabaeoidea, Aphodiidae), In: *Escarabajos, diversidad y conservación biológica. Ensayos en homenaje a Gonzalo Halffter. Monografías 3er Milenio, vol. 7*, M. Zunino & A. Melic (Eds.), 69-91, Sociedad Entomológica Aragonesa, ISSN 978-84-935872-1-5, Zaragoza, Spain

Cabrero-Sañudo, F.J.; Dellacasa, M.; Martínez, I.; Lobo, J.M. & Dellacasa, G. (2010). Distribución de las especies de Aphodiinae (Coleoptera, Scarabaeidae, Aphodiidae) en México. *Acta Zoológica Mexicana*, Vol. 26, pp. 323-399, ISSN 0065-1737

Cambefort, Y. (1991). Dung beetles in tropical savannas. In: *Dung beetle ecology*, I. Hanski & Y. Cambefort (Eds.), 156-178. Princeton University Press, ISBN 0691087393, Princeton, NJ

Carpaneto, G.M. & Mazziotta, A. (2007). Inferring species decline from collection records: roller dung beetles in Italy (Coleoptera Scarabaeidae). *Diversity and Distributions*, Vol. 13, pp. 903-919, ISSN 1366-9516

Clark Labs (2003). *Idrisi Kilimanjaro*. Clark University, Worcester, MA

Clarke, K.R. & Gorley, R.N. (2005). PRIMER version 6. In: *PRIMER-Emultivariate Statistics for Ecologists*, 15.08.2011, Available from: http://www.primer-e.com/

Coates, A.G.; Collins, L.S.; Aubry, M.P. & Berggren, W.A. (2004). The geology of the Darien, Panama, and the late Miocene-Pliocene collision of the Panama arc with northwestern South America. *Geological Society of America Bulletin*, Vol. 116, pp. 1327-1344, ISSN 1050-9747

Connor, E.F. & Simberloff, D. (1979). The assembly of species communities: chance or competition? *Ecology*, Vol. 60, pp. 1132-1140, ISSN 0012-9658

Coope, G.R. (1978). Constancy of insect species versus inconstancy of Quaternary environments. In: *Diversity of insect faunas*, L.A. Mound & N. Waloff (Eds.), 176-187, Blackwell, ISBN 0632003529, Oxford, UK

Coope, G.R. (1990). The invasion of Northern Europe during the Pleistocene by Mediterranean species of Coleoptera. In: *Biological invasions in Europe and the Mediterranean basin*, F. Di Castri, A.J. Hansen & M. Debussche (Eds.), 203-215, Kluwer, ISBN 079230411X, Dordrecht, Netherlands

Coope, G.R. & Angus, R.B. (1975). An ecological study of a temperate interlude in the middle of the last glaciation, based on fossil Coleoptera from Isleworth, Middlesex. *Journal of Animal Ecology*, Vol. 44, pp. 365-391, ISSN 0021-8790

Cope, E.D. (1887). *The origin of the fittest: essays on evolution*. Appleton, New York, NY

Cox, C.B. (2000). Plate tectonics, seaways and climate in the historical biogeography of mammals. *Memórias do Instituto Oswaldo Cruz*, Vol. 95, pp. 509-516 , ISSN 0074-0276

Cox, C.B. (2001). The biogeographic regions reconsidered. *Journal of Biogeography*, Vol. 28, pp. 511-523, ISSN 0305-0270

Crisp, M.D.; Trewick, S.A. & Cook, L.G. (2011). Hypothesis testing in biogeography. *Trends in Ecology and Evolution*, Vol. 26, pp. 66-72, ISSN 0169-5347

Cumming, D.H.M. (1982). The influence of large herbivores on savanna structure in Africa. Ecology of tropical savannas. In: *Ecology of tropical savannas*, B.J. Huntley & B.H. Walker (Eds.), 429-453. Springer-Verlag, ISBN 0387118853, Berlin, Germany

Darlington, P.J. (1957). *Zoogeography: the geographical distribution of animals*. Wiley, ISBN 0898741092, New York, NY

Davis, A.J. (2000). Species richness of dung-feeding beetles (Coleoptera: Aphodiidae, Scarabaeidae, Hybosoridae) in tropical rainforest at Danum Valley, Sabah, Malaysia. *The Coleopterists Bulletin*, Vol. 54, pp. 221-231, ISSN 010-065X

Davis, A.J.; Holloway, J.D.; Huijbregts, H.; Krikken, J.; Kirk-Spriggs, A.H. & Sutton, S.L. (2001). Dung beetles as indicators of change in the forests of northern Borneo. *Journal of Applied Ecology*, Vol. 38, pp. 593-616, ISSN 0021-8901

Davis, A.L.V. (1990). *Climatic change, habitat modification and relative age of dung beetle taxa (Coleoptera: Scarabaeidae, Hydrophilidae, Histeridae, Staphylinidae) in the southwestern Cape*. PhD Thesis, University of Cape Town, South Africa

Davis, A.L.V. & Scholtz, C.H. (2001). Historical vs. Ecological factors influencing global patterns of scarabaeine dung beetle diversity. *Diversity and Distributions*, Vol. 7, pp. 161-174, ISSN 1366-9516

Davis, A.L.V.; Scholtz, C.H. & Philips, T.K. (2002). Historical biogeography of scarabaeine dung beetles. *Journal of Biogeography*, Vol. 29, pp. 1217-1256, ISSN 0305-0270

Davis, A.L.V.; Scholtz, C.H.; Dooley, P.W.; Bharm, M. & Kryger, U. (2004). Scarabaeinae dung beetles as indicators of biodiversity, habitat transformation and pest control chemicals in agro-ecosystems. *South African Journal of Animal Science*, Vol. 100, pp. 415-424, ISSN 0375-1589

Del Rey, L. & Lobo, J.M. (2006). Distribución observada y potencial del género *Acrossus* Mulsant (Coleoptera, Scarabaeoidea, Aphodiidae) en la Península Ibérica. *Boletín de la Sociedad Entomológica Aragonesa*, Vol. 39, pp. 285-291, ISSN 1134-6094

Delcourt, P.A. & Delcourt, H.R. (1987). *Long-term forest dynamics of the temperate zone. Ecological Studies 63.* Springer-Verlag, ISBN 0387964959, New York, NY

Dellacasa, G. & Dellacasa, M. (2006). *Coleoptera Aphodiidae Aphodiinae. Fauna d'Italia XLI.* Calderini, ISBN 8850652038, Bologna, Italy

Dellacasa, G.; Bordat, P. & Dellacasa, M. (2001). A revisional essay of world genus-group taxa of Aphodiinae (Coleoptera Aphodiidae). *Memorie della Società Entomologica Italiana*, Vol. 79, pp. 1-482, ISSN 0037-8747

Dellacasa, M. (1988a [1987]). Contribution to a world-wide catalogue of Aegialiidae Aphodiidae Aulonocnemidae Termitotrogidae. *Memorie della Società Entomologica Italiana*, Vol. 66, pp. 1-455, ISSN 0037-8747

Dellacasa, M. (1988b). Contribution to a world-wide catalogue of Aegialiidae, Aphodiidae, Aulonocnemidae, Termitotrogidae (Part II). *Memorie della Società Entomologica Italiana*, Vol. 67, pp. 1-231, ISSN 0037-8747

Dellacasa, M. (1988c). Contribution to a world-wide catalogue of Aegialiidae, Aphodiidae, Aulonocnemidae, Termitotrogidae. Addenda et corrigenda. First note. *Memorie della Società Entomologica Italiana*, Vol. 67, pp. 291-316, ISSN 0037-8747

Dellacasa, M. (1991). Contribution to a world-wide catalogue of Aegialiidae, Aphodiidae, Aulonocnemidae, Termitotrogidae (Coleoptera, Scarabaeoidea). Addenda et corrigenda (second note). *Memorie della Società Entomologica Italiana*, Vol. 70, pp. 3-57, ISSN 0037-8747

Dellacasa, M. (1995). Contribution to a world-wide catalogue of Aegialiidae, Aphodiidae, Aulonocnemidae, Termitotrogidae (Coleoptera Scarabaeoidea). Addenda et Corrigenda (third note). *Memorie della Società Entomologica Italiana*, Vol. 74, pp. 159-232, ISSN 0037-8747

Dellacasa, M. & Dellacasa, G. (2000a). Systematic revision of the genus *Erytus* Mulsant & Rey, 1870, and description of the new genus *Sahlbergianus* (Coleoptera: Aphodiidae). *Frustula Entomologica*, Vol. 23, pp. 109-130, ISSN 0532-7679

Dellacasa, G.& Dellacasa, M. (2000b). Systematic revision of the genera *Euheptaulacus* and *Heptaulacus* with description of the new genus *Pseudoheptaulacus. Elytron*, Vol. 14, pp. 3-37, ISSN 0214-1353

Dellacasa, M. & Dellacasa, G. (2005). Comments on some systematic and nomenclatural questions in Aphodiinae with descriptions of new genera and on Italian taxa. *Memorie della Società Entomologica Italiana*, Vol. 84, pp. 45-99, ISSN 0037-8747

Dellacasa, M.; Dellacasa, G. & Bordat, P. (2002). Systematic redefinition of taxa belonging to the genera *Ahermodontus* Báguena, 1930 and *Ammoecius* Mulsant, 1842, with description of the new genus *Vladimirellus* (Coleoptera: Aphodiidae). *Acta Zoologica Academiae Scientiarum Hungaricae*, Vol. 48, pp. 269-316, ISSN 0001-7264

Dellacasa, M.; Dellacasa, G. & Gordon, R.D. (2007a). *Ferrerianus*, new genus for *Aphodius biimpressus* Schmidt, 1909 (Scarabaeoidea: Aphodiidae). *Insecta Mundi*, Vol. 11, pp. 1-4, ISSN 0749-6737

Dellacasa, M.; Dellacasa, G. & Gordon, R.D. (2008). *Agrilinellus*, new genus and four new species of Mexican Aphodiini. *Insecta Mundi*, Vol. 53, pp. 1-16, ISSN 0749-6737

Dellacasa, M.; Dellacasa, G. & Skelley, P.E. (2010). *Neotrichaphodioides*, new genus of Neotropical Aphodiini, with description of a new species from Peru (Scarabaeoidea: Scarabaeidae: Aphodiinae). *Insecta Mundi*, Vol. 133, pp. 1-12, ISSN 0749-6737

Dellacasa, M.; Gordon, R.D. & Dellacasa, G. (2003). *Jalisco plumipes*, new genus and new species of Mexican Aphodiini. *Insecta Mundi*, Vol. 17, pp. 69-71, ISSN 0749-6737

Dellacasa, M.; Gordon, R.D. & Dellacasa, G. (2004). *Neotrichonotulus*, a new genus for three Mexican Aphodiini. *Acta Zoológica Mexicana*, Vol. 20, pp. 1-7, ISSN 0065-1737

Dellacasa, M.; Gordon, R.D. & Dellacasa, G. (2007b). *Pseudogonaphodiellus zdzislawae*, new genus and new species of Mexican Aphodiini. *Acta Zoologica Cracoviensia*, Vol. 50B, pp. 139-142, ISSN 1895-3131

Dellacasa, M.; Dellacasa, G.; Gordon, R. & Stebnicka, Z.T. (2011). *Skelleyanus eremita* new genus and new species of Mexican Aphodiini (Coleoptera: Aphodiidae). *Acta Zoologica Cracoviensia*, Vol. 54B, pp. 1-4, ISSN 1895-3131

Diamond, J.M. (1975). Assembly of species communities, In: *Ecology and evolution of communities*, J.M. Diamond (Ed.), 342-444, Harvard University Press, ISBN 0674224442, Cambridge, MA

Doube, B.M.A. (1991). Dung beetles of southern Africa. In: *Dung Beetle Ecology*, I. Hanski & Y. Cambefort (Eds.), 133-155, Princeton University Press, ISBN 0691087393, Princeton, NJ

Doube, B.M.; Macqueen, A.; Ridsdill-Smith, T.J. & Weir, T.A. (1991). Native and introduced dung beetles in Australia. In: *Dung Beetle Ecology*, I. Hanski & Y. Cambefort (Eds.), 255-278, Princeton University Press, ISBN 0691087393, Princeton, NJ

Fattorini, S. (2006). Detecting biodiversity hotspots by species-area relationships: a case study of Mediterranean beetles. *Conservation Biology*, Vol. 20, pp. 1169-1180, ISSN 0888-8892

Finn, J.A. & Gittings, T. (2003). A review of competition in north temperate dung beetle communities. *Ecological Entomology*, Vol. 28, pp. 1-13, ISSN 0307-6946

Flather, C.H. (1996). Fitting species-accumulation functions and assessing regional land use impacts on avian diversity. *Journal of Biogeography*, Vol. 23, pp. 155-168, ISSN 0305-0270

Forshage, M. (2002). *State of knowledge of dung beetle phylogeny - a review of phylogenetic hypotheses regarding Aphodiinae (Coleoptera; Scarabaeidae)*. MSc Thesis, Department of Systematic Zoology, Evolutionary Biology Center, Uppsala University, Uppsala, Sweden

Gaston, K.J. & Blackburn, T.M. (1996). The tropics as a museum of biological diversity: an analysis of the New World avifauna. *Proceedings of the Royal Society B: Biological Sciences*, Vol. 263, pp. 63-68, ISSN 0962-8452

Gillespie, R.G. & Roderick, G.K. (2002). Arthropods on islands: colonization, speciation, and conservation. *Annual Review of Entomology*, Vol. 47, pp. 595-632, ISSN 0066-4170

Gittings, T. & Giller, P.S. (1997). Life history traits and resource utilisation in an assemblage of north temperate *Aphodius* dung beetles (Coleoptera: Scarabaeinae). *Ecography*, Vol. 20, pp. 55-66, ISSN 0906-7590

Gittings, T. & Giller, P.S. (1998). Resource quality and the colonisation and succession of coprophagous dung beetles. *Ecography*, Vol. 21, pp. 581-592, ISSN 0906-7590

Gittings, T. & Giller, P.S. (1999). Larval dynamics in an assemblage of *Aphodius* dung beetles. *Pedobiology*, Vol. 43, pp. 439-452

Gittings, T.; Giller, P.S. & Stakelum, G. (1994). Dung decomposition in contrasting temperature pastures in relation to dung beetle and earthworm activity. *Pedobiology*, Vol. 38, pp. 455-474

Goldberg, E.E.; Roy, K.; Lande, R. & Jablonski, D. (2005). Diversity, endemism, and age distributions in macroevolutionary sources and sinks. *The American Naturalist*, Vol. 165, pp. 623-633, ISSN 0003-0147

Goloboff, P.; Farris, J. & Nixon, K. (2003). T.N.T.: Tree Analysis Using New Technology. Program and documentation. In: *Zoologisk Museum, Statens Naturhistoriske Museum, Københavns Universitet*, 15.08.2011, Available from: www.zmuc.dk/public/phylogeny

Gordon, R.D. & Skelley, P.E. (2007). A monograph of the Aphodiini inhabiting the United States and Canada. *Memoirs of the American Entomological Institute*, Vol. 79, pp. 1-580, ISBN 978-1-887988-23-0

Gotelli, N.J. (2000). Null model analysis of species co-occurrence patterns. *Ecology*, Vol. 81, pp. 2606-2621, ISSN 0012-9658

Gotelli, N.J. & Entsminger, G.L. (2011). EcoSim: Null models software for ecology, version 7.72. In: *Acquired Intelligence Inc. and Kesey-Bear*, 15.08.2011, Available from: http://garyentsminger.com/ecosim/index.htm.

Gould, G.C. & MacFadden, B.J. (2004). Gigantism, dwarfism, and Cope's rule: "Nothing in Evolution makes sense without a phylogeny". *Bulletin of the American Museum of Natural History*, Vol. 285, pp. 219-237, ISSN 0003-0090

Halffter, G. & Edmonds, D. (1982). *The Nesting Behaviour of Dung Beetles (Scarabaeinae). An Ecological and Evolutive Approach*. Instituto de Ecología, A.C., Xalapa, Veracruz

Halffter, G. & Matthews, E.G. (1966). The natural history of dung beetles of the subfamily Scarabaeinae (Coleoptera: Scarabaeinae). *Folia Entomológica Mexicana*, Vol. 12-14, pp. 1-312, ISSN 0430-8603

Hanski, I. (1991a). The dung insect community. In: *Dung Beetle Ecology*, I. Hanski & Y. Cambefort (Eds.), 5-21, Princeton University Press, ISBN 0691087393, Princeton, NJ

Hanski, I. (1991b). North temperate dung beetles. In: *Dung Beetle Ecology*, I. Hanski & Y. Cambefort (Eds.), 75-96, Princeton University Press, ISBN 0691087393, Princeton, NJ

Hanski, I. & Krikken, J. (1991). Dung beetles in tropical forests in South-East Asia. In: *Dung Beetle Ecology*, I. Hanski & Y. Cambefort (Eds.), 179-197, Princeton University Press, ISBN 0691087393, Princeton, NJ

Hawkins, B.A.; Diniz-Filho, J.A.F.; Jaramillo, C.A. & Soeller, S.A. (2006). Post-eocene climate change, niche conservatism, and the latitudinal diversity gradient of New World birds. *Journal of Biogeography*, Vol. 33, pp. 770-780, ISSN 0305-0270

Hollande, A. & Thérond, J. (1998) *Aphodiidae du Nord de l'Afrique (Coleoptera, Scarabaeoidea). Monographie 31*. Museo Regionale di Scienze Naturali di Torino, ISBN 8886041187, Torino, Italy

Hutton, S. & Giller, P.S. (2003). The effects of the intensification of agriculture on northern temperate dung beetle communities. *Journal of Applied Ecology*, Vol. 40, pp. 994-1007, ISSN 0021-8901

Imbrie, J.; Berger, A.; Boyle, E.A.; Clemens, S.C.; Duffy, A.; Howard, W.R.; Kukla, G.; Kutzbach, J.; Martinson, D.G.; McIntyre, A.; Mix, A.C.; Molfino, B.; Morley, J.J.; Peterson, L.C.; Pisias, N.G.; Prell, W.L.; Raymo, M.E.; Shackleton, N.J. & Toggweiler, J.R. (1993). On the structure and origin of major glaciation cycles, 2, The 100,000-year cycle. *Paleoceanography*, Vol. 8, pp. 699-735, ISSN 0883-8305

Jeannel, R. (1942). *La genesè des faunes terrestres*. Presses Université de France, Paris

Koçak, A.O. & Kemal, M. (2008). Nomenclatural notes on some genus-group names in the order Coleoptera. *Miscellaneous Papers*, Vol. 144, pp. 5, ISSN 1015-8235

Kohlmann, B. (1991) Dung beetles in subtropical North America. In: *Dung Beetle Ecology*, I. Hanski & Y. Cambefort (Eds.), 116-132, Princeton University Press, ISBN 0691087393, Princeton, NJ

Krell, F.T. (2000). The fossil record of Mesozoic and Tertiary Scarabaeoidea (Coleoptera: Polyphaga). *Invertebrate Taxonomy*, Vol. 14, pp. 871–905, ISSN 0818-0164

Krell, F.T. (2007). Catalogue of fossil Scarabaeoidea (Coleoptera: Polyphaga) of the Mesozoic and Tertiary, Version 2007. Denver Museum of Nature and Science Technical Report, Vol. 2007-2008, pp. 1-79

Krell, F.T.; Krell-Westerwalbesloh, S.; Weiß, I.; Eggleton, P. & Linsenmair, K.E. (2003). Spatial separation of Afrotropical dung beetle guilds: a trade-off between competitive superiority and energetic constraints (Coleoptera: Scarabaeidae). *Ecography*, Vol. 26, pp. 210-222, ISSN 0906-7590

Legendre, P. & Legendre, L. (1998). *Numerical ecology*, 2nd ed. Elsevier Science, ISBN 978-0-444-89250-8, Amsterdam, Netherlands

Letcher, A.J. & Harvey, P.H. (1994) Variation in geographical range size among mammals of the Palearctic. The American Naturalist, Vol. 144, pp. 30-42, ISSN 0003-0147

Lindroth, C.H. (1948). Interglacial insect remains from Sweden. *Sveriges Geologiska Undersökning*, Vol. 42, pp. 1-29, ISSN 0082-0024

Lobo, J.M. (2000). Species diversity and composition of dung beetle (Coleoptera: Scarabaeoidea) assemblages in North America. *The Canadian Entomologist*, Vol. 132, pp. 307-321, ISSN 0008-347X

Lobo, J.M.; Hortal, J. & Cabrero-Sañudo, F.J. (2006). Regional and local influence of grazing activity on the diversity of a semi-arid dung beetle community. *Diversity and Distributions*, Vol. 12, pp. 111-123, ISSN 1366-9516

Lobo, J.M.; Lumaret, J.P. & Jay-Robert, P. (2001). Diversity, distinctiveness and conservation status of the Mediterranean coastal dung beetle assemblage in the Regional Natural Park of the Camargue (France). *Diversity and Distributions*, Vol. 7, pp. 257-270, ISSN 1366-9516

Lobo, J.M.; Sanmartín, I. & Martín-Piera, F. (1997). Diversity and spatial turnover of dung beetles (Col., Scarabaeoidea) communities in a protected area of south Europe (Doñana National Park, Huelva, Spain). *Elytron*, Vol. 11, pp. 71-88, ISSN 0214-1353

Lomolino, M.V. (2005). Body size evolution in insular vertebrates: generality of the island rule. *Journal of Biogeography*, Vol. 32, pp. 1683-169, ISSN 0305-0270

Losos, J.B. & Glor, R.E. (2003). Phylogenetic comparative methods and the geography of speciation. *Trends in Ecology and Evolution*, Vol. 18, pp. 220-227, ISSN 0169-5347

Lumaret, J.P. (1986). Toxicité de certains helminthicides vis-avis des insectes coprophages et consequences sur la disparition des excre'ments de la surface du sol. *Acta Oecologica*, Vol. 7, pp. 313-324, ISSN 1146-609X

Lumaret, J.P. (1990). *Atlas des Coléoptères Scarabeides Laparosticti de France. Múseum National d'Histoire Naturelle, Inventaires de Faune et de Flore, fascicule 1*. Secretariat de la Faune et de la Flore, ISBN 2865150577, Paris, France

Lumaret, J.P. & Kirk, A.A. (1991). South temperate dung beetles. In: *Dung Beetle Ecology*, I. Hanski & Y. Cambefort (Eds.), 97-115, Princeton University Press, ISBN 0691087393, Princeton, NJ

Lumaret, J.P. & Martínez, I. (2005). El impacto de productos veterinarios sobre insectos coprófagos: consecuencias sobre la degradación del estiércol en pastizales. *Acta Zoológica Mexicana*, Vol. 21, pp. 137-148, ISSN 0065-1737

Lumaret, J.P.; Galante, E.; Lumbreras, C.; Mena, J.; Bertrand, M.; Bernal, J.L.; Cooper, J.L.; Kadiri, N. & Crowe, D. (1993). Field effects of ivermectin residues on dung beetles. *Journal of Applied Ecology*, Vol. 30, pp. 428-436, ISSN 0021-8901

Masumoto, K. & Kiuchi, M. (2001). A new apterous Aphodiine (Coleoptera, Scarabaeidae) from Southwest China, with proposal of a new subgenus. *Elytra*, Vol. 29, pp. 119-123, ISSN 0387-5733

McClain, C.R.; Boyer, A.G. & Rosenberg, G. (2006). The island rule and the evolution of body size in the deep sea. *Journal of Biogeography*, Vol. 33, pp. 1578-1584, ISSN 0305-0270

McGeoch, M.A.; Van Rensburg, B.J. & Botes, A. (2002). The verification and application of bioindicators: a case study of dung beetles in a savanna ecosystem. *Journal of Applied Ecology*, Vol. 39, pp. 661-672, ISSN 0021-8901

McQueen, A. & Beirne, B.P. (1975). Influence of other insects on production of horn fly, *Haematobia irritans* (Diptera: Muscidae) from cattle dung in south-central British Columbia. *Canadian Journal of Plant Science*, Vol. 107, pp. 1255-1264, ISSN 0008-4220

Meiri, S.; Cooper, N. & Purvis, A. (2008). The island rule: made to be broken? *Proceedings of the Royal Society, Series B*, Vol. 275, pp. 141-148, ISSN 0962-8452

Mirol, P.M.; Giovambattista, G.; Lirón, J.P. & Dulout, F.N. (2003). African and European mitochondrial haplotypes in South American Creole cattle. *Heredity*, Vol. 91, pp. 248-254, ISSN 0018-067X

Moran, M.D. (2003). Arguments for rejecting the sequential Bonferroni in ecological studies. *Oikos*, Vol. 100, pp. 403-405, ISSN 0030-1299

Murray, P. (1984). Extinctions down under: a bestiary of extinct Australian late Pleistocene monotremes and marsupials. In: *Quaternary extinctions: a prehistoric revolution*, P.S. Martin & R.G. Klein (Eds.), 600-628, University of Arizona Press, ISBN 0816511004, Tucson, AZ

Nichols, E.S. & Gardner, T.A. (2011). Dung beetles as a candidate study taxon in applied biodiversity conservation research. In: *Ecology and evolution of dung beetles*, L.W. Simmons & T.J. Ridsdill-Smith (Eds.), 267-291, Wiley-Blackwell, ISBN 978-1-4443-3315-2, West Sussex, UK

Nixon, K.C. (2002). WinClada, version 1. In: *Cladistics.com*, 15.08.2011, Available from: http://www.cladistics.com

Ochi, T. & Kawahara, M. (2001). A new subgenus and species of the genus *Aphodius* Illiger from Central Japan (Coleoptera, Aphodiidae). *Kogane*, Vol. 2, pp. 59-63, ISSN 1346-0943

Philips, T.K. (2011). The evolutionary history and diversification of dung beetles. In: *Ecology and evolution of dung beetles*, L.W. Simmons & T.J. Ridsdill-Smith (Eds.), 21-46, Wiley-Blackwell, ISBN 978-1-4443-3315-2, West Sussex, UK

Potts, R. & Behrensmeyer, A.K. (1993). Late Cenozoic terrestrial ecosystems. In: Terrestrial ecosystems through time, A.K. Behrensmeyer, J.D. Damuth, W.A. DiMichele, R. Potts, H.D. Sues & S.L. Wing (Eds.), 419-541, University of Chicago Press, ISBN 0226041557, Chicago, IL

Pulquério, M.J.F. & Nichols, R.A. (2007). Dates from the molecular clock: how wrong can we be? *Trends in Ecology and Evolution*, Vol. 22, pp. 180-184, ISSN 0169-5347

Rapoport, E.H. (1975). *Areografía. Estrategias Geográficas de las Especies*. Fondo de Cultura Económica, Mexico

Ricklefs, R.E. (2006). Evolutionary diversification and the origin of the diversity-environment relationship. *Ecology*, Vol. 87, pp. 3-13, ISSN 0012-9658

Ridsdill-Smith, T.J. & Edwards, P.B. (2011). Biological control: ecosystem functions provided by dung beetles. In: *Ecology and evolution of dung beetles*, L.W. Simmons & T.J. Ridsdill-Smith (Eds.), 245-266, Wiley-Blackwell, ISBN 978-1-4443-3315-2, West Sussex, UK

Romero-Samper, J. & Lobo, J.M. (2006). Los coleópteros escarabeidos telecópridos del Atlas Medio (Marruecos): influencia del tipo de hábitat, altitud y estacionalidad y relevancia en las comunidades coprófagas (Coleoptera, Scarabaeidae). *Boletín de la Sociedad Entomológica Aragonesa*, Vol. 39, pp. 235-244, ISSN 1134-6094

Rosen, B.R. (1988). Biogeographical patterns: a perceptual overview, In: *Analytical biogeography: an integrated approach to the study of animal and plant distributions*, A.A. Myers & P.S. Giller (Eds.), 437-481, Chapman & Hall, ISBN 0412400502, London, UK

Rosen, B.R. & Smith, A.B. (1988). Tectonics from fossils? Analysis of reef-coral and sea-urchin distributions from the late Cretaceous to recent, using a new method, In: *Gondwana and Tethys. Special Publication 37*, M.G. Audley-Charles & A. Hallam (Eds.), 275-306, Geological Society of London, ISBN 0-19-854448-0, London, UK

Rosenberg, M.S. & Anderson, C.D. (2011). PASSaGE: Pattern Analysis, Spatial Statistics and Geographic Exegesis. Version 2. *Methods in Ecology & Evolution*, Vol. 2: pp. 229-232, ISSN 2041-210X

Roslin, T. (2000). Dung beetle movement at two spatial scales. *Oikos*, Vol. 91, pp. 323-335, ISSN 0030-1299

Roslin, T. & Koivunen, A. (2001). Distribution and abundance of dung beetles in fragmented landscapes. *Oecologia*, Vol. 127, pp. 69-77, ISSN 0029-8549

Rougon, D. & Rougon, C. (1991). Dung beetles of the Sahel region. In: *Dung Beetle Ecology*, I. Hanski & Y. Cambefort (Eds.), 230-241, Princeton University Press, ISBN 0691087393, Princeton, NJ

Rougon, D.; Rougon, C.; Trichet, J. & Levieux, J. (1988). Enrichissement en materie organique d'un sol sahélien au Niger par les insects coprophages. *Revue d'Ecologie et de Biologie du Sol*, Vol. 25, pp. 413-434, ISSN 0035-1822

Sanmartín, I. & Ronquist, F. (2004). Southern hemisphere biogeography inferred by event-based models: plant versus animal patterns. *Systematic Biology*, Vol. 53, pp. 216-243, ISSN 1063-5157

Scholtz, C.H. & Chown, S.L. (1995). The evolution of habitat use and diet in the Scarabaeoidea: a phylogenetic approach. In: *Biology, phylogeny, and classification of Coleoptera: papers celebrating the 80th birthday of Roy A. Crowson*, J. Pakaluk & S.A. Ślipinski (Eds.), 355-374, Museum i Instytut Zoologii PAN, ISBN 83-85192-34-4, Warszawa

Schoolmeesters, P. (2011). Scarabs: World Scarabaeidae Database (version May 2011). In: *Species 2000*, 15.08.2011, Available from: www.catalogueoflife.org/col/

Scotese, C.R. (2003). Palcomap project home page. In: *Paleomap Project*, 15.08.2011, Available from: http://www.scotese.com

Shahabuddin; Schulze, C.H. & Tscharntke, T. (2005). Changes of dung beetle communities from rainforests towards agroforestry systems and annual cultures in Sulawesi (Indonesia). *Biodiversity and Conservation*, Vol. 14, pp. 863-877, ISSN 0960-3115

Silva, M. & Downing, J.A. (1995). *CRC handbook of mammalian body masses*. CRC Press, ISBN 0849327903, Boca Raton, FL

Skelley, P.E. (2007). *Ozodius, n. gen.*, for the Australian members of the genus *Drepanocanthoides* Schmidt (Coleoptera: Scarabaeidae: Aphodiinae). *Papers in Entomology*, Vol. 15, pp. 1-2, ISSN 0749-6737

Skelley, P.E.; Dellacasa, M. & Dellacasa, G. (2009). *Gordonius rhinocerillus*, a new genus and species of Colombian Aphodiini. *Insecta Mundi*, Vol. 67, pp. 1-6, ISSN 0749-6737

Smith, C.H. (1983). A system of world mammal faunal regions. I. Logical and statistical derivation of the regions. *Journal of Biogeography*, Vol. 10, pp. 455-466, ISSN 0305-0270

Smith, A.B.T. (2006). A review of the family group names for the superfamily Scarabaeoidea (Coleoptera) with corrections to nomenclature and a current classification. *Coleopterists Society Monograph*, Vol. 5, pp. 144-204, ISSN 1934-045

Smith, A.B.T.; Hawks, D.C. & Heraty, J.M. (2006). An overview of the classification and evolution of the major scarab beetle clades (Coleoptera: Scarabaeoidea) based on preliminary molecular analyses. *Coleopterists Society Monographs*, Vol. 5, pp. 35-46, ISSN 1934-045

Sneath, P.H.A. & Sokal, R.R. (1973). *Numerical taxonomy*. W.H. Freeman & Co., ISBN 0716706970, San Francisco, CA

Soberón, J. & Llorente, J. (1993). The use of species accumulation functions for the prediction of species richness. *Conservation Biology*, Vol. 7, pp. 480-488, ISSN 0888-8892

StatSoft, Inc. (2006). *Statistica electronic manual*. StatSoft Inc., Tulsa, OK

Stebnicka, Z. (2000). A new genus for Nearctic *Pleurophorus ventralis* with phylogenetic inferences. *Acta Zoologica Cracoviensia*, Vol. 43, pp. 287-291, ISSN 1895-3131

Steinbauer, M.J. & Weir, T.A. (2007). Summer activity patterns of nocturnal Scarabaeoidea (Coleoptera) of the southern tablelands of New South Wales. *Australian Journal of Entomology*, Vol. 46, pp. 716, ISSN 1326-6756

Stevens, G.C. (1996). Extending Rapoport's rule to Pacific marine fishes. *Journal of Biogeography*, Vol. 23, pp. 149-154, ISSN 0305-0270

Stone, L. & Roberts, A. (1990). The checkerboard score and species distributions. *Oecologia*, Vol. 85, pp. 74-79, ISSN 0029-8549

Tarasov, S.I. (2008). A revision of *Aphodius* Illiger, 1798 subgenus *Amidorus* Mulsant et Rey, 1870 with description of the new subgenus *Chittius* (Coleoptera: Scarabaeidae). *Russian Entomological Journal*, Vol. 17, pp. 177-192, ISSN 0132-8069

Thomas, J.A.; Welch, J.J.; Woolfit, M. & Bromham, L. (2006). There is no universal molecular clock for invertebrates, but rate variation does not scale with body size. *Proceedings of the National Academy of Sciences USA*, Vol. 103, pp. 7366-7371, ISSN 0027-8424

Ulrich, W. (2010). NODF. In: *Macroecology Research Group, Department of Animal Ecology, Nicolaus Copernicus University*, 15.08.2011, Available from: http://www.home.umk.pl/~ulrichw/?Research:Software:NODF

Ulrich, W.; Almeida-Neto, M. & Gotelli, N.G. (2009). A consumer's guide to nestedness analysis. *Oikos*, Vol. 118, pp. 3-17, ISSN 0030-1299

Veiga, C.M. (1998). *Los Aphodiinae (Coleoptera, Aphodiidae) ibéricos*. PhD Thesis. Departamento de Biología Animal I, Facultad de Ciencias Biológicas, Universidad Complutense de Madrid, Madrid, Spain

Waterhouse, D.F. (1974). The biological control of dung. *Scientific American*, Vol. 230, pp. 101-109, ISSN 0036-8733

Webb, S.D. (1985). Late Cretaceous mammal dispersals between the Americas. In: *The great American biotic interchange*, F.G. Stehli & S.D. Webb (Eds.), 357-386, Plenum Press, ISBN 030642021X, New York, NY

Ziani, S. (2002). A new genus and species of Aphodiini inhabiting burrows of small mammals in Lebanon's mountains. *Zoology in the Middle East*, Vol. 27, pp. 101-106, ISSN 0939-7140

# Permissions

The contributors of this book come from diverse backgrounds, making this book a truly international effort. This book will bring forth new frontiers with its revolutionizing research information and detailed analysis of the nascent developments around the world.

We would like to thank Lawrence E. Stevens, PhD, for lending his expertise to make the book truly unique. He has played a crucial role in the development of this book. Without his invaluable contribution this book wouldn't have been possible. He has made vital efforts to compile up to date information on the varied aspects of this subject to make this book a valuable addition to the collection of many professionals and students.

This book was conceptualized with the vision of imparting up-to-date information and advanced data in this field. To ensure the same, a matchless editorial board was set up. Every individual on the board went through rigorous rounds of assessment to prove their worth. After which they invested a large part of their time researching and compiling the most relevant data for our readers. Conferences and sessions were held from time to time between the editorial board and the contributing authors to present the data in the most comprehensible form. The editorial team has worked tirelessly to provide valuable and valid information to help people across the globe.

Every chapter published in this book has been scrutinized by our experts. Their significance has been extensively debated. The topics covered herein carry significant findings which will fuel the growth of the discipline. They may even be implemented as practical applications or may be referred to as a beginning point for another development. Chapters in this book were first published by InTech; hereby published with permission under the Creative Commons Attribution License or equivalent.

The editorial board has been involved in producing this book since its inception. They have spent rigorous hours researching and exploring the diverse topics which have resulted in the successful publishing of this book. They have passed on their knowledge of decades through this book. To expedite this challenging task, the publisher supported the team at every step. A small team of assistant editors was also appointed to further simplify the editing procedure and attain best results for the readers.

Our editorial team has been hand-picked from every corner of the world. Their multi-ethnicity adds dynamic inputs to the discussions which result in innovative outcomes. These outcomes are then further discussed with the researchers and contributors who give their valuable feedback and opinion regarding the same. The feedback is then collaborated with the researches and they are edited in a comprehensive manner to aid the understanding of the subject.

Apart from the editorial board, the designing team has also invested a significant amount of their time in understanding the subject and creating the most relevant covers. They scrutinized every image to scout for the most suitable representation of the subject and create an appropriate cover for the book.

The publishing team has been involved in this book since its early stages. They were actively engaged in every process, be it collecting the data, connecting with the contributors or procuring relevant information. The team has been an ardent support to the editorial, designing and production team. Their endless efforts to recruit the best for this project, has resulted in the accomplishment of this book. They are a veteran in the field of academics and their pool of knowledge is as vast as their experience in printing. Their expertise and guidance has proved useful at every step. Their uncompromising quality standards have made this book an exceptional effort. Their encouragement from time to time has been an inspiration for everyone.

The publisher and the editorial board hope that this book will prove to be a valuable piece of knowledge for researchers, students, practitioners and scholars across the globe.

# List of Contributors

**Hugo H. Mejía-Madrid**
Universidad Nacional Autónoma de México/Facultad de Ciencias, México

**Khalid Al Mutairi, Mashhor Mansor and Asyraf Mansor**
School of Biological Sciences, Universiti Sains Malaysia, Penang, Malaysia

**Saud L. Al-Rowaily**
Department of Plant Production, College of Agricultural & Food Sciences, King Saud University, Riyadh, Saudi Arabia

**Magdy El-Bana**
Department of Plant Production, College of Agricultural & Food Sciences, King Saud University, Riyadh, Saudi Arabia
Department of Biological Sciences, Faculty of Education at El-Arish, Suez Canal University, El-Arish, Egypt

**Toshiharu Mita**
Laboratory of Entomology, Faculty of Agriculture, Tokyo University of Agriculture, Atsugi, Japan

**Yukiko Matsumoto**
National Institute of Agrobiological Sciences, Owashi, Tsukuba, Japan

**Sachiyo Sanada-Morimura and Masaya Matsumura**
National Agricultural Research Center for Kyushu Okinawa Region, Koshi, Japan

**Raúl Contreras-Medina**
Escuela de Ciencias, Universidad Autónoma "Benito Juárez" de Oaxaca (UABJO), México

**Isolda Luna-Vega**
Laboratorio de Biogeografía y Sistemática, Departamento de Biología Evolutiva, Facultad de Ciencias, Universidad Nacional Autónoma de México (UNAM), México

**Eusebio Cano Carmona**
Dpto. Biología Animal, Biología Vegetal y Ecología, Área de Botánica, Universidad de Jaén, Spain

**Ana Cano Ortiz**
Dpto. Sostenibilidad, INTERRA, Ingeniería y Recursos S.L., Spain

**Alberto S. Taylor B and Jorge Mendieta**
Universidad de Panamá, Departamento de Botánica, Panamá

**Jody L. Haynes**
IUCN/SSC Cycad Specialist Group, USA

**Dennis W. Stevenson,**
The New York Botanical Garden, USA

**Gregory Holzman**
Pacific Cycad Nursery, USA

**Benny K.K. Chan**
Biodiversity Research Center, Academia Sinica, Taipei, Taiwan
Institute of Ecology and Evolutionary Biology, National Taiwan University, Taipei, Taiwan

**Pei-Fen Lee**
Institute of Ecology and Evolutionary Biology, National Taiwan University, Taipei, Taiwan

**Marcela A. Vidal and Helen Díaz-Páez**
Departamento de Ciencias Básicas, Facultad de Ciencias, Universidad del Bío-Bío, Chile
Departamento de Ciencias Básicas, Universidad de Concepción, Campus Los Ángeles, Chile

**Lawrence E. Stevens**
Biology Department, Museum of Northern Arizona, Flagstaff, Arizona, USA

**Isolda Luna-Vega**
Laboratorio de Biogeografía y Sistemática, Departamento de Biología Evolutiva, Facultad de Ciencias, Universidad Nacional Autónoma de México (UNAM), México

**Raúl Contreras-Medina**
Escuela de Ciencias, Universidad Autónoma "Benito Juárez" de Oaxaca (UABJO), México

**Monica Axini**
"Monachus" Group of Scientific Research and Ecological Education, Romania

**Patricio De los Ríos-Escalante**
Universidad Católica de Temuco, Facultad de Recursos Naturales, Escuela de Ciencias Ambientales, Chile

**Alfonso Mardones Lazcano**
Universidad Católica de Temuco, Facultad de Recursos Naturales, Escuela de Acuicultura, Chile

**Melissa Sánchez-Herrera and Jessica L. Ware**
Department of Biology, Rutgers The State University of New Jersey, Newark Campus, USA

**Santiago Martín-Bravo**
Department of Molecular Biology and Biochemical Engineering, Pablo de Olavide University, Spain

**Marcial Escudero**
Department of Molecular Biology and Biochemical Engineering, Pablo de Olavide University, Spain
The Morton Arboretum, Lisle, USA

**Edward L. Mockford**
School of Biological Sciences, Illinois State University, Normal, Illinois, USA

**Francisco José Cabrero-Sañudo**
Departamento de Zoología & Antropología Física, Facultad de Ciencias Biológicas, Universidad Complutense de Madrid, Spain